Metallo-organic Chemistry

DATE DUE

MAR 17 1988		5/30	5, 23
MAR 11 1989		4/4	6 17
		4/18	5/59
1/25	12am	3·12	11:46a
1/30	11:05p		
3/1	8:30p	RETURNED	
4/8	7:55p	AUG 10 1997	
		SEP 16 1998	
JUN 14 1989			
RETURNED			
DEC 12 1989			
1/23	7.23p		
1/24	12:14p		
	9:24		
2/5	12am		
RETURNED			
JUL 14 1990			
3/30	3:23pa		

DEMCO NO. 38-298

Metallo-organic Chemistry

Anthony J. Pearson
Department of Chemistry
Case Western Reserve University

A Wiley–Interscience Publication

JOHN WILEY & SONS

Chichester · New York · Brisbane · Toronto · Singapore

Library of Congress Cataloging in Publication Data:

Pearson, Anthony J.
  Metallo-organic chemistry.

  Includes bibliographical references and index.
  1. Organometallic compounds. I. Title.
QD411.P43   1985   547'.05 84–3702

ISBN 0 471 90440 6 (cloth)
ISBN 0 471 90446 5 (paper)

British Library Cataloguing in Publication Data:

Pearson, Anthony J.
  Metallo-organic chemistry.
  1. Organometallic compounds   2. Chemistry,
  Organic—Synthesis
  I. Title
  574'.050459   QD411

ISBN 0 471 90440 6 (cloth)
ISBN 0 471 90446 5 (paper)

Photosetting by Thomson Press (India) Ltd., New Delhi and
printed by Page Bros. (Norwich) Ltd.

Contents

Preface   ix

Introduction   xi

1. Bonding in Transition Metal Organometallic Complexes   1
   - The M—CO bond   2
   - Metal–carbene (and carbyne) complexes   5
   - η^2-Alkene complexes   8
   - η^3-Allyl complexes   9
   - η^4-Diene complexes   11
   - η^5-Dienyl Complexes (excluding metallocenes)   15
   - Metallocenes and arene sandwich compounds   17
   - The 18-electron rule   17
   - Reference   20

2. Some Common Properties and Reactions of Transition Metal Organometallic Complexes   21
   - Fluxionality and dynamic equilibria   21
   - Fluxionality in tricarbonyl (diene) iron complexes   21
   - Fluxionality in η^2-olefin complexes   23
   - Fluxionality and dynamic equilibria in π-allyl complexes   24
   - Tricarbonyl cyclooctatetraene metal complexes   28
   - Oxidative addition reactions   29
     - Oxidative addition reactions of d^{10} complexes   30
     - Oxidative addition reactions of d^8 complexes   30
     - Oxidative addition reactions of d^7 complexes   33
   - Reductive elimination from transition metal σ-complexes   34
   - α- and β-elimination reactions   37
   - Olefin metathesis   40
     - Single-cross experiments   42
     - Double-cross experiments   43
     - Typical metathesis catalysts   44
     - Stereochemical aspects of the olefin metathesis reaction   45
     - Metathesis of functionalized alkenes   46
   - Insertion reactions   47

Carbon monoxide insertions 48
Sulphur dioxide insertion reactions 53
Important catalytic process based on carbonyl insertion 56
Hydroformylation (the 'oxo' reaction) 56
Acetic acid synthesis via carbonyl insertion 58
Conversion of alkenes to carboxylic acids and esters 59
Stabilization of reactive or unstable molecules 61
Cyclobutadiene 61
Trimethylenemethane 64
5,6-dimethylenecyclohexa-1,3-diene 66
Cyclopentadienone 68
Fulvene complexes 69
Cyclohexa-2,4-dien-1-one 72
Polymerization and oligomerization 74
Ziegler–Natta polymerization of alkenes 74
1,3-diene polymerization 78
Oligomerization 80
The Wacker process 87
Olefin epoxidation 88
Catalytic hydrogenation 89
Activation by oxidative addition—Wilkinson's catalyst 89
Heterolytic addition of hydrogen 91
Homolytic addition of H_2 to metal catalyst 93
Asymmetric hydrogenation 94
Effect of transition metal complexation on the stereochemical
course of electrocyclic reactions 96
References 97
3. σ-Alkyl Complexes 100
Homoleptic complexes—preparation and reactions 100
Neutral complexes 100
Anionic complexes 103
Non-homoleptic complexes 105
Preparation by the Grignard method 106
Preparation using insertion reactions 109
Preparation using elimination or de-insertion reactions 113
Preparation by reaction of metalanions with RX 114
Preparation using oxidative addition reactions 115
Physical and spectral properties of transition metal alkyls 116
Reactions of σ-alkyl complexes 117
Mechanistic aspects of reactions with electrophiles 118
Some specific transition metal alkyl σ-complexes 121
Organocopper reagents 121
Allyl–F_p σ-complexes 124
Aryl and vinyl palladium σ-complexes 127
Metallacycles 130

References 134
4. Carbene and Carbyne Complexes 136
   Carbene complexes 136
      Synthesis of stable carbene complexes 136
      Carbene complexes as reactive intermediates 144
      Reactions of carbene complexes and applicability to organic
         synthesis 149
      Reactions of alkylidene complexes 154
   Carbyne complexes 157
      Reactions of carbyne complexes 159
   References 162
5. η^2-Alkene and η^2-Alkyne Complexes 163
   η^2-Alkene complexes 163
      Iron, ruthenium and osmium group 163
      Manganese, technetium and rhenium 174
      Cobalt, rhodium and iridium 176
      Nickel, palladium and platinum 182
   η^2-Alkyne complexes 192
   References 199
6. η^3-Allyl Complexes 201
   η^3-Allyl complexes of nickel and palladium 201
      Allylnickel complexes 201
      Allylpalladium complexes 209
   Titanium, zirconium and hafnium 223
   Vanadium, niobium and tantalum 224
   Chromium, molybdenum and tungsten 225
      LUMO coefficients for *exo/endo* conformers 231
   Manganese, technetium and rhenium 233
   Iron, ruthenium and osmium 235
   Cobalt, rhodium and iridium 237
   General comments 244
   References 245
7. η^4-Diene Complexes 247
   Introductory comments 247
   Transition metal complexes of chelating, non-conjugated dienes 247
      Nickel, platinum and palladium 247
      Cobalt, rodium and iridium 252
      Iron and ruthenium 254
      Other metals 256
   Complexes of conjugated dienes 257
      Iron 257
      Other metals 275
   References 276
8. η^5-Dienyl Complexes 278
   Non-cyclically conjugated dienyl complexes 278

Iron, ruthenium and osmium 278
Synthetic applications of tricarbonyl (cyclohexadienyl) iron
complexes 293
Cyclohexenone γ-cation equivalents 293
Tricarbonyl (3-methoxycyclohexadienyl) iron hexafluoro-
phosphate and related compounds 304
Cobalt, rhodium and iridium 308
η^5-Cyclopentadienyl metal complexes 310
Ferrocene, ruthenocene and osmocene 310
Other metallocenes 325
References 345
9. η^6-Arene and η^6-Triene Complexes 348
Arene complexes 348
Chromium, molybdenum and tungsten 348
Titanium and zirconium 363
Vanadium, niobium and tantalum 364
Manganese and rhenium 366
Iron, ruthenium and osmium 370
Cobalt, rhodium and iridium 374
Nickel, platinum and palladium 377
η^6-Triene and derived complexes 377
Seven-membered rings 377
Eight-membered rings 385
Cyclononatriene complexes 387
Cyclododecatriene complexes 387
References 388

Index 391

Preface

The purpose of this book is to provide a basic coverage of the chemistry of organo-transition metal complexes and to set this in the context of their application in synthetic organic chemistry. I have tried to cover the basic chemical aspects in such a way that patterns of reactivity can be seen and I have tried to explain certain of these phenomena using a frontier molecular orbital approach, which at present is the most appropriate method. Hopefully, the organization of material will be useful to the advanced undergraduate, to graduate students of organic chemistry and to all synthetic organic chemists whose education has so far neglected organometallic chemistry.

With the explosive growth of organometallic chemistry still continuing, it has been impossible to avoid omissions. To those experts who might see the subject in a different light I can offer no apologies for presenting what I perceive as the salient features. All omissions are entirely my own responsibility, and are due to an effort to keep the size of the book reasonable and yet give a clear and adequate coverage of the material.

The first chapter gives a brief introduction to concepts of bonding in metal–olefin and related complexes designed to give the reader the necessary background to understand the later frontier molecular orbital descriptions. The second chapter gives a broad coverage of reactions and physical phenomena, e.g. oxidative addition, fluxionality and dynamic equilibrium, which crop up time and time again throughout organometallic chemistry, and are especially important for a general understanding of the subject. This chapter is rounded off by consideration of some catalytic processes which illustrate the importance of the chemical properties discussed. The remaining chapters are meant to give a fairly detailed account of the chemistry of coordinated ligands, arranged in chapters according to ligand type. Within each chapter I have tried to point out patterns of reactivity and the factors which control reactivity, where known. The potential usefulness of certain complexes to the synthetic organic chemist is illustrated at appropriate points by examples of, e.g., natural product syntheses chosen from the recent literature.

I started writing this book whilst at Cambridge, during 1981–82, and the typescript was completed at Case Western Reserve University in 1982. Coverage of the literature is up to mid-1982, but much of the source material included review articles, etc. Hopefully, the audience for whom I have tried to cater will be happy with the presentation given.

23 November 1983

A. J. Pearson

Introduction

The explosive development of organo-metallic chemistry over the past 30 years, and particularly within the last 20 years, has made it extremely difficult to keep abreast of all the literature. This is particularly so for the newcomer to this area, and it is now a major effort to gain a good working knowledge of basic organometallic chemistry. The result is that a large number of organic chemists are turned away from pursuing organometallic studies, since there appears to be such a diversity of reactivity and a multitude of odd, if interesting, structures. Consequently, only a few major developments have been made in the application of transition metal organometallics to complex organic synthesis. It is my hope that this book will provide a starting point, particularly for the organic chemist who is seeking new areas for research. With this in mind, I have attempted a broad coverage of organo-metallic chemistry, classified according to ligand type. In many cases I have tried to illustrate the chemistry of certain complexes by drawing attention to established, or potential, application to organic synthesis. Wherever possible, I have attempted to rationalize reactivity of organometallic complexes using frontier molecular orbital (FMO) arguments, which appears to be the most reasonable approach to mechanism at the molecular level.

In Chapter 1 a brief account is given of bonding between unsaturated ligands and transition metal groups. There are a number of possible approaches to this, but I have tried to adopt a simple pictorial approach which is useful for the FMO descriptions later. Chapter 2 covers a wide range of basic reactivity patterns found in organometallic synthesis which are particularly important for understanding homogeneous catalysis. Illustrations of a number of important catalytic processes are presented. The remaining chapters give a systematic description of reactions of organometallic complexes classified according to ligand type. Within each chapter I have first discussed in some detail the metal(s) most commonly encountered in association with the particular ligand, and this is followed by a less detailed account of other metal complexes. Therefore, the arrangement of material according to metal is not necessarily the same for each chapter.

It is my sincere hope that this book will provide useful source material for advanced undergraduate and graduate students, as well as for experienced workers who are newcomers to the area.

CHAPTER 1

Bonding in Transition Metal Organometallic Complexes

The purpose of this chapter is to introduce the student to fundamental ideas of the ways in which orbitals of unsaturated organic molecules interact with those of a transition metal in the formation of a stable transition metal complex. Many descriptions have appeared in text books and the research literature, all evolving from the basic Dewar–Chatt–Duncanson approach and at various levels of sophistication. Currently much effort is being directed towards the development of molecular orbital techniques, and in this respect we may note particularly the work of Hoffmann. It is this approach which seems the most useful in trying to develop some kind of theoretical rationalization for reactivity of transition metal π-complexes. It is conceptually simple and sets out to determine the shapes of orbitals available for bonding in various metal complex fragments, ML_n. As a simple example consider the reaction shown in equation (i):

$$Fe(CO)_5 \ + \ CH_2 = CH_2 \ \longrightarrow \ \begin{matrix} CH_2 \\ \parallel \\ CH_2 \end{matrix} - Fe(CO)_4 \ + \ CO \quad (1)$$

A carbonyl ligand in pentacarbonyliron has been displaced by ethene, formally a two-electron donor, to give tetracarbonylethene iron. This probably occurs by prior dissociation of $Fe(CO)_5$, an 18-electron species,* to give $Fe(CO)_4$, a reactive 16-electron species,* followed by reaction of this with the ethene ligand. It seems logical, therefore, to do the same thing in conceptual terms. Thus, we can imagine that bonding of ethene to the $Fe(CO)_4$ fragment can be effectively described by determining the nature of the orbitals available on the $Fe(CO)_4$ fragment which interact in an energetically favorable way with the orbitals of the alkene. As we shall see, it is the highest occupied molecular orbital (HOMO) and lowest unoccupied molecular orbital (LUMO) of each entity which are important. These are the frontier molecular orbitals (FMOs), and interaction between them can be expected to lead to FMOs of the metal complex. First we shall consider the simplest system, the metal—carbonyl bond, and then go on to develop bonding models for more complicated systems.

---

* See later for definition.

THE M—CO BOND

For this discussion it is convenient to consider a simple metal hexacarbonyl, such as $W(CO)_6$, which has octahedral symmetry. The carbon monoxide molecule is taken to have a lone pair of electrons in an sp hybrid on carbon, and the six ligands lie along the Cartesian axes with these 'lone pairs' pointing directly at the metal. The initial result, in crystal field terms, is that the energy levels of the metal nd orbitals are split, the d_{z^2} and $d_{z^2-y^2}$ being raised in energy rather more than the d_{xy}, d_{xz} and d_{yz} orbitals, giving two high-energy e_g orbitals and three low-lying t_{2g} orbitals. Since tungsten(0) has six valence shell electrons (d^6), these can all be accommodated in the t_{2g} set, whilst the e_g orbitals together with the $(n+1)$s and three $(n+1)$p orbitals are formally empty. Hence there are six empty metal orbitals, of roughly equal energy, which are available to interact with the filled orbitals from six CO ligands. We could either combine these in the 'raw' state, thereby obtaining twelve MOs, six bonding and six antibonding, which describe the M—CO σ-bonding, or else we could first of all make a linear combination of metal orbitals, to give six equivalent d^2sp^3 hybrids which then interact with ligand orbitals. We shall adopt the latter approach. These effects on the metal d, s and p orbitals are shown as energy diagrams in Figure 1.1. The energy level diagram depicting the interactions between ligand and metal σ-bonding orbitals is given in Figure 1.2

Thus, interaction between each CO orbital and a metal hybrid results in a net lowering in energy of the system, since the electrons enter the M—CO σ-bonding orbitals. This is not the complete picture, however, since the CO ligands also have some empty low-lying π^* orbitals, which are of correct symmetry to interact with the metal t_{2g} orbitals. This interaction, together with the energy level diagram, is shown in Figure 1.3. Of course, each t_{2g} orbital can interact with two π^* orbitals on CO ligands which are *trans* to each other, but this is not shown.

In this way, even more stabilization is gained, since these electrons also are now in a lower energy level. A ligand which interacts in this way is called a σ-donor π-acceptor or π-acid ligand. The overall effect is called a synergic interaction. Is there any way in which this type of interaction can be detected? Inspection of the

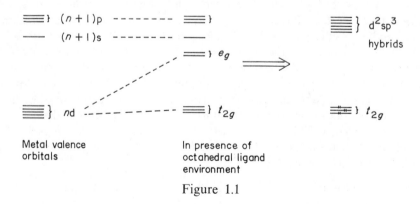

Figure 1.1

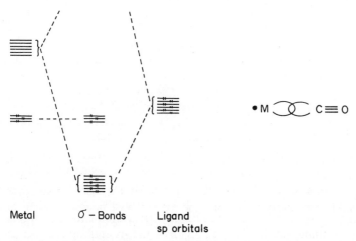

Metal σ – Bonds   Ligand
                                sp orbitals

Figure 1.2   Energy level diagram depicting interaction between ligand and metal σ-bonding orbitals

Figure 1.3   Metal—CO π-bonding

interaction depicted in Figure 1.3 reveals that the resulting molecular orbital has some character of the carbonyl π^* orbital. Therefore, the electronic population of this MO is effectively a population of the π^* orbital. This is expected to lead to a reduction in the bond order of the CO ligand compared with free carbon monoxide, and this is the kind of change which might be detected by infrared spectroscopy. It is in fact found that the CO stretching frequency of a variety of metal carbonyl complexes is appreciably lower than that of free carbon monoxide. Some typical examples are given in Table 1.1, supporting the above bonding scheme.

Let us consider what happens when a CO ligand is replaced by, for example, a phosphine or arsine ligand, typical examples in organometallic chemistry being triphenylphosphine or triphenylarsine. Again, we have an atom (P or As) which can donate its lone pair into empty metal d^2sp^3 hybrid orbitals, and at the same time receive electrons by back-donations from the t_{2g} levels, this time into empty phosphorus (or arsenic) d orbitals. However, the energy levels associated with these ligand orbitals will be different from those of the carbon monoxide. Since a higher quantum shell is involved we should expect that both the lone pair orbital

4

Table 1.1 ν_{CO} Values.

Compound	ν_{CO} (cm^{-1})
CO	2143
$Cr(CO)_6$	2100, 2000, 1985
$Mn_2(CO)_{10}$	2046, 2015, 1984
$Fe(CO)_5$	2022, 2000
$Fe(CO)_3PPh_3$	1887

and empty π-acceptor orbital will be of higher energy. Hence the lone pair orbital is closer in energy to the metal e_g set, and perturbation theory tells us that this will produce a better interaction. On the other hand, the ligand π-acceptor level is now much further away from the t_{2g} levels and so a poorer interaction results, these being shown in Figure 1.4. The resultant MOs in both cases therefore are more metal-like. In other words, the metal has effectively received more charge from the ligand σ-donor levels, and has lost less charge by back-donation into the ligand π-acceptor levels. We call the phosphine ligand a better σ-donor but poorer π-acceptor.

How will this effect manifest itself? The easiest thing to examine is the effect on CO infrared frequencies. Imagine a CO and triphenylphosphine both interacting with the same metal t_{2g} orbital and compared this with two CO ligands interacting. We can illustrate this best by examining the bonding in a stepwise fashion. Since the MO resulting from interaction of the CO with metal is slightly lower in energy than the corresponding PPh_3-based orbital, it is probable that a second CO ligand will have a stronger bonding interaction with the *latter* (see Figure 1.5). The result will be an effectively greater population of the carbonyl π^* orbital when the other ligand is a phosphine, which should result in a lowering of the CO infrared stretching frequency. Inspection of Table 1.1 reveals that this is indeed the case. An alternative explanation is that the presence of a poor π-acceptor leads to an increased electron density on the metal with consequent increased donation into the remaining CO π^* orbitals.

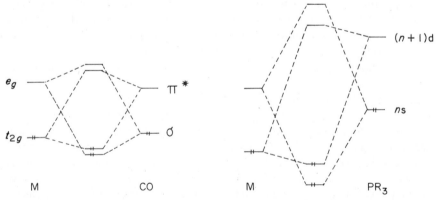

Figure 1.4   Comparison between CO and PR_3 ligands (schematic only)

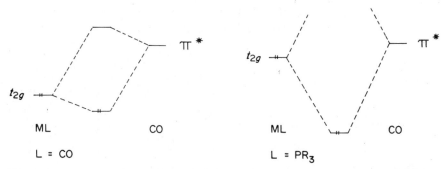

Figure 1.5   Comparison of ML fragment π-bonding interaction with CO ligand

The effect of changing the σ-donor strength of a ligand is extremely difficult to determine as an independent phenomenon since most changes in reactivity appear to be better explained using the metal t_{2g} levels, e.g. metal basicity, reaction with electrophiles (see later chapters). We shall not discuss these here.

The above synergic interaction is the basis for bonding in most low oxidation state transition metal complexes, and indeed can lead to the formation of negative oxidation states as in, for example, $Na_2Fe(CO)_4$. The stabilization is mainly due to the transfer of charge from metal to the π-acid ligand.

METAL–CARBENE (AND CARBYNE) COMPLEXES

These will be discussed in more detail in Chapter 4. The bonding is similar to that encountered above for metal carbonyls. Hence we can picture a carbene ligand as having a pair of electrons in a carbon sp^2 hybrid orbital, together with an empty p orbital, rather similar to a carbonium ion. Synergic interaction can then occur with metal orbitals as shown in Figure 1.6.

There is an appreciable body of information consistent with this bonding formulation. The carbenoid CR_2 unit is known from X-ray structure determination to be planar with only slight deviation from sp^2 geometry. In the tantalum carbene complex **1** the metal—carbene bond distance is considerably shorter than the Ta—methyl distance. Also, rotation about the Ta=CH_2 bond is

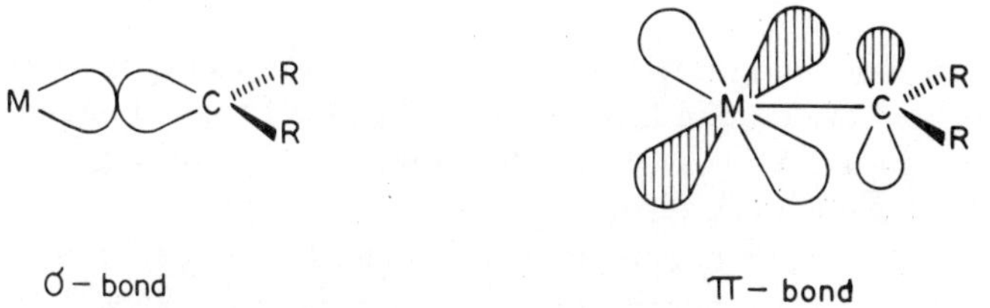

σ – bond   π – bond

Figure 1.6   Metal–Carbene bonding

found to be slow on the ^{1}H NMR time-scale, both of these observations suggesting considerable double bond character.

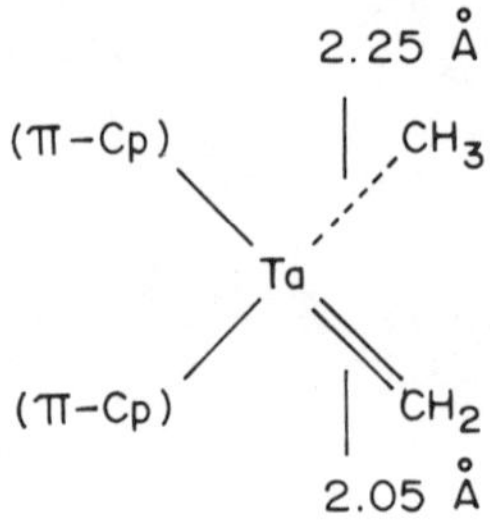

2.25 Å

(π–Cp)　　　CH$_3$

Ta

(π–Cp)　　　CH$_2$

2.05 Å

(1)

A large number of carbene complexes are known which contain heteroatom substituents on the carbene carbon atom, most commonly oxygen and nitrogen. The lone pair of electrons on these substituents will overlap with the formally vacant carbene p orbital, giving some ligand stabilization. However, a secondary effect is to weaken the metal–carbene back-donation and thus reduce the metal—carbon double bond character. In valence bond terms this is readily accounted for by the canonical forms shown below.

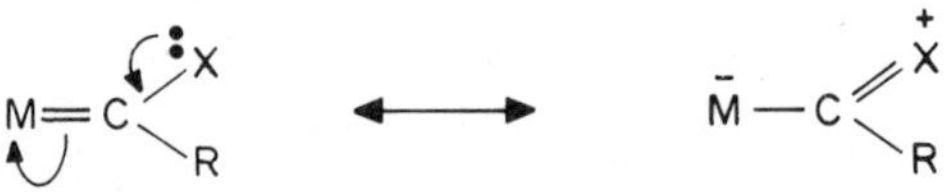

$$M=C\overset{X}{\underset{R}{<}} \quad \longleftrightarrow \quad \overset{-}{M}-C\overset{+X}{\underset{R}{<}}$$

In molecular orbital terms the interaction between heteroatom and carbon p orbital produces two new orbitals and net stabilization, but the vacant (π*-antibonding) combination is now higher in energy and shifted further away from the metal t_{2g} orbitals, thereby causing a reduction in bonding interaction (Figure 1.7) and concomitant reduction in M—C double bond character. The evidence for this interaction comes from observation of restricted rotation about the C—N bond in alkylamino-substituted complexes such as **2**, similar to that found in unsymmetrical amides.

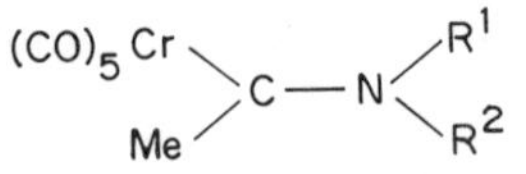

$$(CO)_5 Cr \overset{}{\underset{Me}{>}} C - N \overset{R^1}{\underset{R^2}{<}}$$

(2)

The electron-donating power of the heteroatom is, as expected, N > O, and there is infrared evidence in favor of this. Thus the carbonyl ligand stretching frequency of a $(CO)_x M$—$C(X)R$ usually shows a slight decrease on going from X = OR′ to X = NR′$_2$ (see also Chapter 4).

Transition metal carbyne complexes, exemplified by the chromium complex **3**, have a similar bonding arrangement to carbene derivatives but with an extra dπ–pπ interaction. Formally, they are similar to acetylenes in which one of the

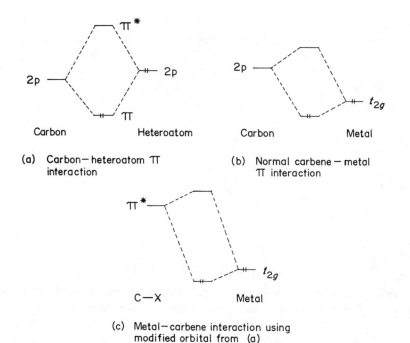

(a) Carbon–heteroatom π interaction

(b) Normal carbene–metal π interaction

(c) Metal–carbene interaction using modified orbital from (a)

Figure 1.7 Interaction of heteroatom with metal–carbon system

σ–bond π_z bond π_y bond

Figure 1.8 Bonding in metal–carbyne complexes

acetylene carbon atoms is replaced by a metal. The bonding scheme is shown in Figure 1.8.

(3)

(4)

We might be led by this to expect an M—C—R angle of 180°, and this is observed in some cases. However, there are a number of examples of complexes

8

having slightly bent structures, e.g. **4**. This indicates a small re-hybridization of the carbon from sp to sp^2, but there is no clear rationalization of this phenomenon at present. There is ample X-ray structural evidence in support of the formal M—C triple bond character. For example, the tantalum complex **5** has a metal—carbon bond length (1.849 Å) approximately 0.2 Å shorter than the carbene complex bond in **1**.

$$Cl\text{''''''}Ta \equiv C - R$$
$$PMe_3$$

(5)

η^2-ALKENE COMPLEXES

We shall use a similar approach to that of Hoffmann to describe the bonding in these complexes. A simple example is tetracarbonyletheneiron [equation (1)]. We imagine that the metal—carbonyl σ-bonds are formed by overlap between filled carbon sp hybrids and empty metal d^2sp^3 hybrids. This leaves two unused d^2sp^3 orbitals. Now, iron(0) is d^8, so that six electrons enter the three t_{2g} orbitals, the remaining two being accommodated in the spare d^2sp^3 orbitals. We can now make the combination between ethene p orbitals and iron orbitals depicted in Figure 1.9.

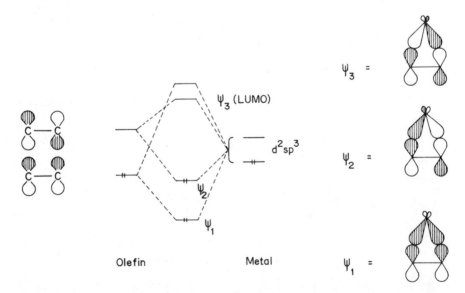

Figure 1.9

This gives two bonding orbitals and two antibonding orbitals, so that the four electrons, two from ethene and two from iron, can be readily accommodated in the bonding set. Also included in Figure 1.9 is a diagram of the LUMO (ψ_3) which is, of course, the antibonding combination between ethene LUMO and metal hybrids. We shall find this very useful when discussing the chemistry of such complexes (later). The net result corresponds to the notion of synergic interaction, i.e. donation from olefin into empty metal orbitals (to give ψ_1), together with back-donation from filled metal orbitals into empty olefin orbital (to give ψ_2). Olefins are therefore fairly good π-acid ligands, and tend to form complexes with low oxidation state metals. In a number of cases the interaction to give ψ_2 is so great that the π character of the complexed olefin is completely lost, the resultant complex resembling a metallocyclopropane. Substituents on the olefin which raise its HOMO and lower its LUMO energy levels will lead to stronger interaction with the metal orbitals. In this way, electron-withdrawing substituents on the olefin give complexes which have a metallocyclopropane structure (compare complexes **6**, **7** and **8**). These may be compared with bond lengths for ethene (1.34 Å).

(6)

(7)

(8)

Variation of the metal produces similar effects, possibly owing to the size and energy levels of the d orbitals.

Consistent with the expected loss of C=C double bond character on complexation, the infrared spectra of olefin π-complexes show no absorption in the $1600\,cm^{-1}$ region expected for a free olefin. Most complexes have a weak absorption at *ca.* $1500\,cm^{-1}$ assignable to the coordinated olefin C=C stretching mode.

η^3-ALLYL COMPLEXES

A very similar bonding picture to that for olefin complexes can be built up for π-allyl derivatives, a simple example being the tetracarbonyliron cationic derivative **9**. Placement of charge in these systems in fairly arbitrary, but the

$$H_2C \diagdown \overset{Fe(CO)_4}{\underset{+}{\diagup}} CH \qquad BF_4{}^-$$
$$H_2C \diagup$$

(9)

example quoted shows chemical properties more consistent with an allyl cation, so **9** is a convenient representation (see below). We can use the same metal bonding fragment as above, and make appropriate combinations as shown in Figure 1.10. Combination of the three allyl MOs with two metal AOs gives five complex MOs. With the arrangement used here the allyl ψ_3 is non-bonding. There are four electrons [two from Fe(CO)$_4$ and two from allyl cation] which enter the two lowest MOs. This corresponds effectively to a synergic interaction whereby electrons are donated from the allyl cations HOMO (ψ_1) to empty metal orbital, with back-donation from metal to allyl LUMO (ψ_2). For cationic complexes, the back-donation will be much more important than in neutral complexes, since with the charged ligand both ψ_1 and ψ_2 are of lower energy, leading to a strong interaction with ψ_2, now closer to the metal orbitals in energy, and a weaker interaction with ψ_1, now further away. The effect of this, of course, is

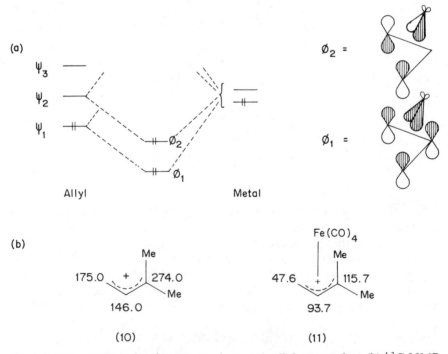

Figure 1.10 (a) Bonding interaction in metal–allyl π-complex; (b) ^{13}C NMR data—representative example (δ p.p.m. down field Me$_4$Si)

to reduce the overall positive charge on the ligand. Also, the population of allyl LUMO leads to a selective decrease of charge at the terminal carbon atom, and both of these effects are shown by the ^{13}C NMR data for the free allyl cation **10** compared with the complex **11**. (^{13}C shieldings in conjugated systems are a useful guide to π-electron densities, in the absence of inductive effects.)

η^4-DIENE COMPLEXES

The $Fe(CO)_4$ fragment forms 18-electron complexes with monoolefin (two-electron) ligands, so we should expect that further loss of CO would give an $Fe(CO)_3$ group capable of forming stable *diene* complexes. Both conjugated and unconjugated diene complexes are in fact known, but since the latter can be treated as diolefin ligands, we shall concern ourselves here with only conjugated diene complexes. Examples of both types will be found in Chapter 7. Conjugated diene complexes are particularly interesting since, as we shall see, the synergic interaction introduced above leads to certain effects which can be subjected to fairly rigorous experimental testing. Let us first examine the bonding properties of the $Fe(CO)_3$ fragment. Assuming a d^2sp^3 hybridization, coupled with pseudo-octahedral symmetry of the diene–$Fe(CO)_3$ complex, we can see that there will be three hybrids available for bonding, and these will contain two electrons. If the diene is arranged as shown in Figure 1.11 then we can imagine metal hybrid combinations which will interact with the diene MOs. For simplicity, we shall draw the interactions as though we were viewing the complex from above the metal, the tops of the diene p orbitals being seen. The shading represents the phase of the wave function (+ or −) on the metal side of the ligand. This then leads to the combinations shown in Figure 1.12, for butadieneirontricarbonyl **12**.

Again, this gives the (synergic) result of donation from occupied diene orbitals into empty metal orbitals, together with back-donation from metal into the empty diene LUMO. What is the result of this? We can see that this process corresponds effectively to partial removal of electrons from the diene HOMO, and partial population of the diene LUMO, similar to what happens when we form an excited state. Thus, we might expect that the π-bond order distribution in the complexed diene will resemble more that of the excited molecule than the ground state. The two are shown below. Since there is a correlation between π-bond orders and bond lengths in many olefinic systems, we should expect that the

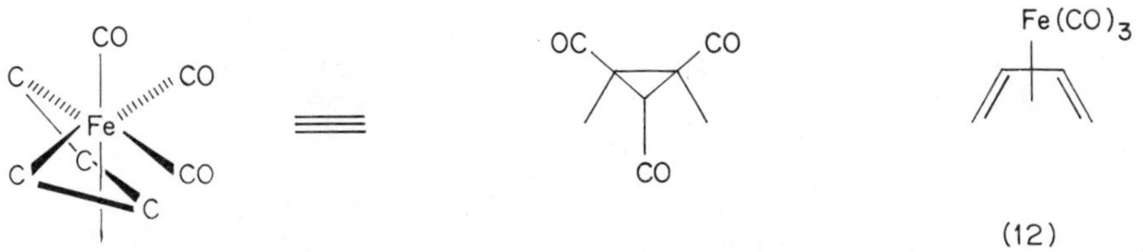

(12)

Figure 1.11   Pseudo-octahedral symmetry of butadiene–$Fe(CO)_3$ **(12)**

12

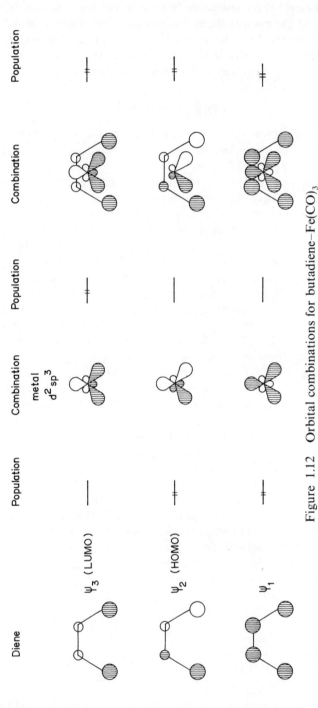

Figure 1.12 Orbital combinations for butadiene–Fe(CO)$_3$

terminal bonds of free butadiene will be shorter than the central bond, whilst the central bond in the complexed molecule should be shorter than, or of similar length to, the terminal C—C bonds. This is exactly as observed (see diagram below).

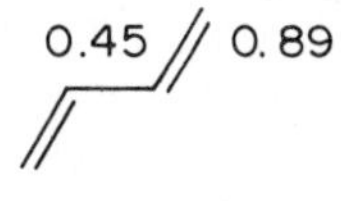

0.45 // 0.89   0.67 // 0.45

Ground state :   Excited state :
π–bond orders   π–bond orders

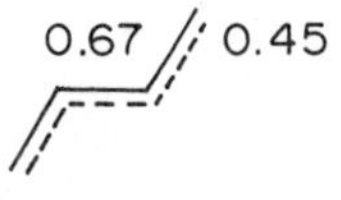

1.48 Å // 1.34 Å   Fe(CO)₃
                          1.45 Å
                          1.41 Å

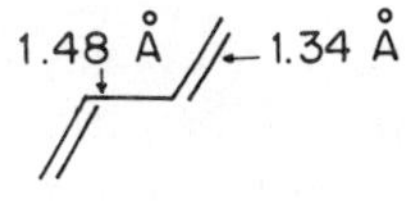

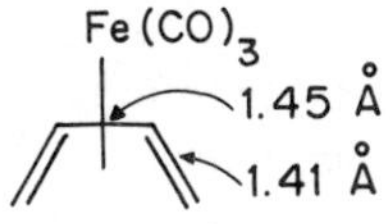

Bond lengths

There is also evidence for population of the diene LUMO from carbon-13 NMR substituent effects. It is fairly well established that for organic molecules the transmission of ^{13}C NMR substituent effects through unsaturated systems is related to the π-bond order of intervening bonds. If we examine the ^{13}C NMR spectrum of methyl vinyl ether **13** compared with ethene, we find that the β-carbon atom experiences a large upfield shift, owing to increased π-electron density by the resonance interaction shown below. The upfield shift, obtained by taking the difference, is termed the β-substituent effect. In the case of buta-1,3-diene, there is a distribution of π-bond orders shown above, so that if we compare

(13)

the ^{13}C NMR spectrum of 2-methoxybutadiene **14** with that of butadiene itself, we should expect an appreciable upfield shift of C-1 and a smaller upfield shift of C-3, and this is indeed observed (Table 1.2). When we examine butadiene—

Table 1.2. ^{13}C NMR β- and γ-methoxy substituent effects for 2-methoxybutadiene and its $Fe(CO)_3$ complex (p.p.m.).

Compound	β(C-1)*	β(C-3)*	γ(C-4)*
2-Methoxybutadiene (14)	− 31.1	− 4.3	− 3.4
2-Methoxybutadiene−Fe(CO)₃ (15)	− 6.4	− 19.2	− 9.2

* Negative values indicate upfield shift of the denoted carbon atom relative to butadiene or butadiene−Fe(CO)₃.)

Fe(CO)$_3$ derivatives, we now discover that the β-effects produced by a 2-methoxy substituent on the diene are very different. Thus, comparison between tricarbonyl(2-methoxybutadiene)iron **15** and the parent complex **12** reveals a large upfield β-effect at C-3 and only a small upfield effect at C-1. Therefore, it appears that the above notion of partial population of the diene excited state is correct.

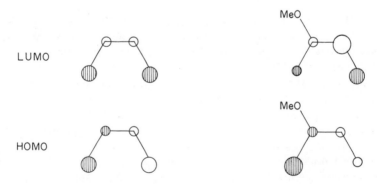

Another useful way of rationalizing the β-effect is to examine the effect of the 2-methoxy substituent on the HOMO of each system. Comparing the HOMO of **14** with free butadiene we see a large increase in relative coefficient at C-1 and very little change at C-3 (Figure 1.13). Now, from the above discussion we expect that the diene component in the HOMO of the complex corresponds roughly to the diene LUMO. Comparison of the coefficients for this orbital reveals a large coefficient at C-3 for 2-methoxybutadiene, but only a small coefficient for C-1. Also note that we expect a large coefficient at C-4 in the diene LUMO, but only a small one for diene HOMO. Compare this latter result with the ^{13}C NMR effects at this atom (γ-effect) and it will be seen that the Fe(CO)$_3$ complex shows an appreciable upfield γ-effect whereas that for free diene is small. In fact, these effects are important in considering the reactivity of substituted diene complexes, as we shall see later.

Before we leave this discussion, it is interesting to consider an alternative description which has been discussed in the literature. At one time it was thought that the bonding was a combination of σ- and π-bonding as shown in structure **16**, which is a plausible explanation of the above data. However, it is found that the geminal ^{13}C–H coupling constants for the terminal carbon atom of the

Figure 1.13 HMO coefficients for butadiene and 2-methoxybutadiene (Houk diagram, not to scale)

complexed diene are almost identical (158–161 Hz) with those obtained for sp^2 hybridized carbon atoms in olefinic systems (*ca.* 160 Hz), and very different to the C–H couplings obtained for saturated (sp^3) molecules (*ca.* 130 Hz). This appears to rule out a structure such as **16**.

(16)

η^5-DIENYL COMPLEXES (EXCLUDING METALLOCENES)

These complexes are known as either cationic species, e.g. the tricarbonyl(cyclohexadienyl)iron complex **17**, which are stable, neutral species, e.g. **18**, or anionic species, e.g. **19**, the latter usually being formed as intermediates during the addition of nucleophiles to an arene complex (see Chapter 10).

(17)   (18)

(19)   (20)

Again, the placing of formal charge is fairly arbitrary, since it is delocalized over the metal–ligand system, but usually the reactivity associated with that charge is found on the dienyl system. For example, nucleophile addition to **17** occurs specifically at the dienyl ligand (see Chapter 8). In this section we shall consider the bonding in the tricarbonylpentadienyliron cation **20**, since we are by now familiar with the bonding properties of the $Fe(CO)_3$ fragment. Whilst there are now several MO treatments in the literature, there are still a number of unanswered questions, which of course makes the discussion all the more interesting. We shall meet these as we proceed, and further in Chapter 8. The dienyl cation has four electrons accommodated in ψ_1 and ψ_2, and the $Fe(CO)_3$ fragment possesses three hybrids containing two electrons. The major interactions expected are shown in Figure 1.14, and involve the now well known synergic interaction, *viz.* donation from ψ_1 and ψ_2 (the latter being the stronger owing to its closeness in energy to empty metal d orbital) to the metal, together with back-donations from metal into dienyl LUMO (ψ_3). It seems likely on energy level considerations that the latter will be the most important interaction, leading to a delocalization of positive charge on to the iron.

We have only shown the bonding orbitals. In fact, probably the most important orbital from the point of view of reactivity of these particular complexes (nucleophile addition) is the LUMO of the complex. Unfortunately, calculations have not revealed the exact nature of this orbital, and indeed it is found that there are a number of contenders, all very close in energy. However, most have a nodal surface between the metal and the dienyl system, and an example (ϕ_4) is included in Figure 14. We shall consider the significance of this later.

The evidence for back-donation into the dienyl LUMO again comes from infrared and ^{13}C NMR data. Thus it is found that the $Fe(CO)_3$ group absorbs at *ca.* 2150 and 2050 cm^{-1} compared with the values of 2080 and 1970 cm^{-1} found in neutral diene–$Fe(CO)_3$ complexes. This is because the gain of positive charge by iron lowers the energy of the t_{2g} orbitals which now, according to perturbation theory, will have weaker interaction with the high-lying carbonyl π^* orbitals. The decreased population of the π^* orbital leads to an increased π-bond order and therefore a higher stretching frequency. The population of dienyl LUMO is reflected in the ^{13}C NMR, shown below for uncomplexed and complexed systems. There is a general upfield shift of all carbons in going from free to complexed systems, presumably owing to paramagnetic screening by the metal, and also reflecting overall back-donations, but also the distribution of ^{13}C shifts

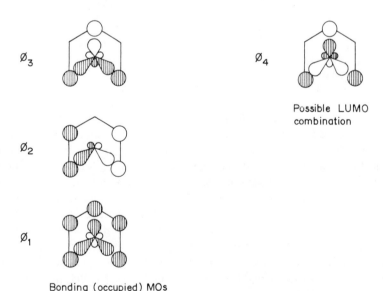

Possible LUMO combination

Bonding (occupied) MOs

Figure 1.14   Dienyl bonding to $Fe(CO)_3$ fragment

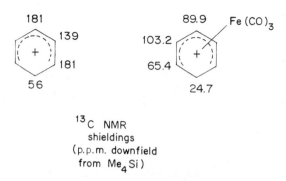

^{13}C NMR
shieldings
(p.p.m. downfield
from Me$_4$Si)

resembles the coefficients of the dienyl LUMO. The chemical shifts in the uncomplexed system relate to total electron (or charge) densities.

METALLOCENES AND ARENE SANDWICH COMPOUNDS

There is in fact a choice of bonding a description adopted for these complexes. Since there is no reason *a priori* to use d^2sp^3 metal hybrids, or simple s, p, d orbitals, we could in principle use both the t_{2g} and the e_g orbitals for overlap with cyclopentadienyl ligands. In fact, the currently used textbook description does exactly this. Thus, we envisage overlap between the cyclopentadienyl ψ_1 orbital(s) and d_{z^2}, p_z and s orbitals of the metal, as shown in Figure 1.15. Then the remaining orbitals interact with metal orbitals of appropriate symmetry, d_{xz}, d_{yz}, etc. Using this model there would appear to be very small overlap between ligand orbitals and the metal orbitals in the xy plane (d_{xy}, $d_{x^2-y^2}$, p_x, p_y), even though we have included these in Figure 1.15.

The alternative description, using d^2sp^3 hybrid combinations, is shown using the Houk-type representation in Figure 1.16. We should keep in mind that, owing to the arrangement of these orbitals, the t_{2g} set is essentially orthogonal, the cyclopentadienyl ligand no longer being parallel to the xy plane. For ferrocene, we can count iron as Fe^{II} and the cyclopentadienyl anion is a six-electron donor. The iron hybrids are therefore nominally empty, so the twelve electrons go into duplicate pairs of the bonding MOs shown.

In fact, the basic picture is the same, since the same orbitals are filled in the process, and there is no technique at present which will allow differentiation between these two alternative descriptions [the reactivities of these complexes do not permit differentiation either (see later)].

The same sort of bonding pictures can be built up for bis(arene)–metal complexes. This is left as an exercise for the student.

THE 18-ELECTRON RULE

The observant reader will have noticed in the preceding discussion that the arrangement of ligands around a particular metal remains fairly constant throughout a series of complexes. For example, conversion of $Fe(CO)_5$ into (η^2-

18

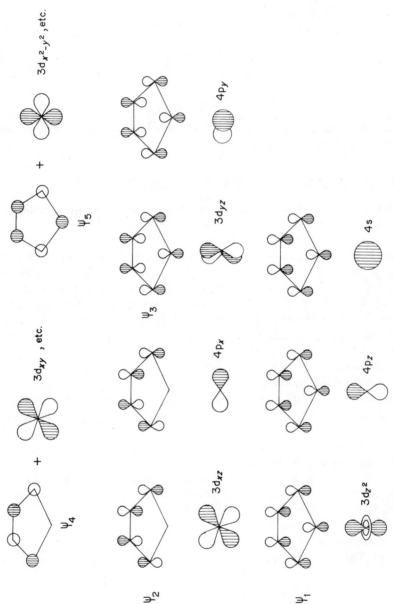

Figure 1.15 Bonding between cyclopentadienyl ligand and metal

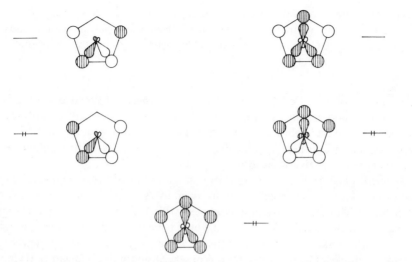

Figure 1.16 Cyclopentadienyl—metal bonding using hybrid orbital description

ethene)$Fe(CO)_4$ involves displacement of one CO ligand, which may be regarded as donating two electrons into the coordination sphere of the metal atom, with an olefin ligand, also a two-electron donor. Similarly, conversion of $Fe(CO)_5$ into butadiene–$Fe(CO)_3$ involves replacement of two CO ligands with a four-electron ligand, butadiene. Thus, the electronic environment of the metal atom remains constant. We can sum up the total number of valence shell electrons surrounding the metal as follows:

$Fe(CO)_5$:	Fe(0)	8e
	$5 \times CO$	10e
		18e
ethene–$Fe(CO)_4$:	Fe(0)	8e
	$4 \times CO$	8e
	C_2H_4	2e
		18e
butadiene–$Fe(CO)_3$:	Fe(0)	8e
	$3 \times CO$	6e
	butadiene	4e
		18e

Thus, for a stable arrangement we note that the combination of metal and ligand electrons results in a valence shell configuration of 18 electrons, exactly that for the next rare gas, in this case krypton. During the reaction which leads to displacement of CO ligand by olefin, the first step is loss of carbon monoxide:

$$Fe(CO)_5 \rightleftharpoons Fe(CO)_4 + CO$$

The initial product, $Fe(CO)_4$, is a 16-electron species, termed coordinatively unsaturated, and has only fleeting existence at normal temperatures, although it may be trapped and characterized spectroscopically at low temperature. It adds on an olefin, or another ligand, to regain the stable 18-electron configuration, becoming coordinatively saturated.

Addition of a further two-electron donor ligand to $Fe(CO)_5$ would result in the formation of a complex which contains formally 20 electrons in the iron valence shell. Since there are only five non-occupied orbitals in the available metal set for interaction with ligands, a sixth ligand would have to interact with orbitals in the next quantum shell of iron, obviously of higher energy. Consequently, 20-electron complexes of this type are unknown. Suppose that instead of trying to attach a further ligand we simply try to add more electrons to the complex, e.g. by electrochemical reduction. Since all of the bonding and non-bonding orbitals of the complex are fully occupied, any extra electrons *must* populate an antibonding orbital, leading to considerable destabilization of the complex. As a consequence, such complexes are extremely rare.

In summary, it can be seen that for first-row transition elements, a rule of thumb is that 18-electron complexes are favored, but that complexes containing fewer electrons can be formed as reactive intermediates. Complexes with more than 18 electrons in the valence shell are unstable and therefore not formed under normal circumstances. Second-row elements tend to form more stable 16-electron complexes than do first-row elements, some of these being isolable.

This rule has important consequences in organometallic chemistry. It is useful to remember that a vacant coordination site is an exceedingly important property of a homogeneous catalyst, and that coordinatively unsaturated complexes are especially reactive in migration and oxidative addition reactions. These processes are very common steps in a catalytic cycle, as will be seen later.

It should be pointed out that as d orbital energies are reduced, leading to increased $nd \rightarrow (n+1)p$ promotion energies, there is a tendency for 18-electron molecules to be less favorable. This is most noticeable towards the end of the transition series, and in particular 16-electron compounds are common for nickel, palladium and platinum. This again has important consequences for catalytic reactions.

REFERENCE

For a detailed account of bonding of unsaturated organic molecules to transition metals, see D. M. P. Mingos, in *Comprehensive Organometallic Chemistry* (Ed. G. Wilkinson, F. G. A. Stone and E. W. Abel), Pergamon Press, Oxford, 1982, Vol. 3, Ch. 1.

CHAPTER 2

Some Common Properties and Reactions of Transition Metal Organometallic Complexes

In this chapter we discuss the common processes which occur in transition metal complexes and which account for many of the fundamental steps in catalytic sequences.

FLUXIONALITY AND DYNAMIC EQUILIBRIA

Fluxionality can be loosely defined as the continual reorganization of ligands around a transition metal which does not involve discrete ligand dissociation. By a dynamic equilibrium we mean a ligand reorganization which involves either partial or total dissociation of the ligand from the metal, followed by a recomplexation in a different manner. We can draw an analogy between these and conformational equilibrium (fluxionality) or configurational change (dynamic equilibrium) in organic compounds. This will become clearer as the discussion proceeds. Both of these processes manifest themselves during NMR spectroscopic investigation, and both are important as a basis for understanding catalytic reactions. They both provide the means by which ligands which are initially non-adjacent, and therefore not suitably disposed for reaction on the metal template, become suitably juxtaposed for a reaction to occur. The observation of dynamic processes provides evidence for partial dissociation of certain ligands which is usually necessary before coordination can occur with another molecule involved in a reaction. We shall now consider examples of each phenomenon.

FLUXIONALITY IN TRICARBONYL (DIENE) IRON COMPLEXES*

The X-ray analysis of butadiene–$Fe(CO)_3$ shows a rigid structure in which two of the CO ligands are in equivalent environments, while the third CO is unique (see Chapter 1). Consequently, if this conformation were adhered to in solution,

---

* See Chapter 1 for structures.

we should expect the ^{13}C NMR spectrum to show two peaks with an integrated intensity ratio of 2:1, corresponding to these ligands. In fact, when the ^{13}C NMR spectrum is measured at room temperature, only one peak is observed corresponding to the CO ligands, at 212.8 p.p.m. (downfield from Me_4Si). However, on cooling to 195 K, the single peak splits into the pattern expected for the idealized structure. It is found that whatever process is leading to equivalence has an activation energy of 9.5 kcal mol^{-1}. The line spectra are shown in Figure 2.1.

There are two possible explanations for this phenomenon: either dissociation, followed by reassociation of one CO ligand at normal temperatures, or a process of rotation of the $Fe(CO)_3$ which results in averaging of the magnetic environments experienced by each carbonyl. The former seems unlikely. If it were the case then we might expect that incorporation of, say, triphenylphosphine into a solution of the complex at room temperature would result in rapid replacement of a CO ligand by the phosphine, for such a low activation energy dissociation. This is not the case. While the conversion of butadiene–$Fe(CO)_3$ to butadiene–$Fe(CO)_2PPh_3$ can be achieved, it will only occur at elevated temperature (*ca.* 120–150 °C) and very slowly, or under irradiation with ultraviolet light.

The mechanism of this process of fluxionality, i.e. apparent rotation without bond rupture, is not well understood. It does not appear to be a case of pseudorotation, since this is normally prohibited by chelating ligands which form a small 'metallocycle' ring (the diene in this case). Pseudorotation is in fact an appropriate mechanism for the equivalence shown by $Fe(CO)_5$. This compound has a trigonal bipyramidal structure, so we should expect a ^{13}C NMR spectrum to distinguish the two axial from the three equatorial ligands. In fact a singlet

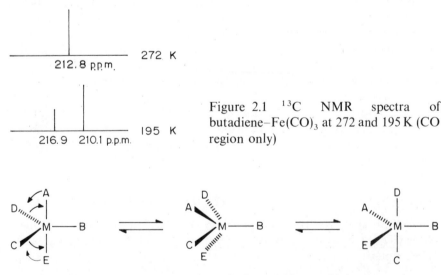

212.8 p.p.m.   272 K

216.9   210.1 p.p.m.   195 K

Figure 2.1 ^{13}C NMR spectra of butadiene–$Fe(CO)_3$ at 272 and 195 K (CO region only)

Figure 2.2   Pseudorotation of trigonal bipyramidal system

is observed. The pseudorotation mechanism proposed for this process, shown in Figure 2.2, involves vibrational deformations of the axial and equatorial ligands, so that they become exchanged.

FLUXIONALITY IN η^2-OLEFIN COMPLEXES

At ambient temperature, the proton NMR spectrum of $CpRh(C_2H_4)_2$ in chloroform solution shows three peaks: δ 5.15, doublet (coupled to rhodium) (5H) corresponding to the cyclopentadienyl protons, and two broad absorptions at 2.77 (4H) and 1.12 (4H) corresponding to the ethylene protons. Cooling the solution to $-20\,^\circ$C causes the absorptions at 2.77 and 1.12 to be each split into two pairs of doublets, while on heating to $57\,^\circ$C, the ethene bands collapse to a single peak at δ1.93. Thus, at higher temperature, there is rapid proton equilibration (NMR spectra shown schematically in Figure 2.3). The activation energy for the process is 6.2 kcal mol^{-1}.

In trying to establish the mechanism for this equivalence, several alternatives were considered:[1]

(a) simple proton exchange involving C—H bond breakage but not ethene—Rh bond rupture;

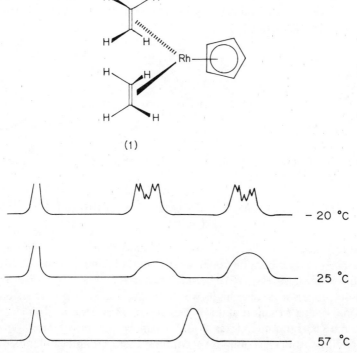

(1)

Figure 2.3 Variable temperature ^{1}H NMR spectra for $C_pRh(C_2H_4)_2$ (schematic only)

(b) proton exchange by non-classical tunnelling mechanism;

(c) rapid dissociation of ethene molecules followed by re-association either on a different complex or in different orientation on the same complex;

(d) exchange of ethylene between two complexes without dissociation;

(e) rapid rotation of each ethene about its C—C bond axis;

(f) rapid rotation of each ethene about the ethene—rhodium bond axis.

The first four mechanisms were excluded experimentally. Thus, it was found that heating solutions of **1** in CH_3OD gave no H–D exchange under conditions where rapid equilibration in the complex occurred, ruling out mechanism (a). Also, mechanism (a) is ruled out by the observation that no equivalence of the olefinic protons of $CpRh(CH_2\!=\!CHCH_2CH\!=\!CH_2)$ is observed even at 100 °C. The tunnelling mechanism (b) was excluded by deuterium NMR studies on $CpRh(C_2D_4)_2$, when it was found that equilibration of the deuterium was as rapid as the proton equilibration in **1**. Since no exchange between coordinated C_2H_4 and free C_2D_4 occurred at 100 °C, this rules out (c), while (d) is ruled out by the lack of dependence of equilibration rate on complex concentration.

It is expected that the rupture of both σ- and π-coordination between olefin and metal, which must accompany mechanism (e), would have an activation energy considerably higher than that observed (compare with the bond energy of silver–olefin complexes of 20–30 kcal[2]) and it was therefore concluded that rotation of ethene about the coordination bond, i.e. mechanism (f), with almost continued overlap between rhodium d orbitals and ethylene MOs, is the actual mechanism of equilibration (Scheme 2.1).

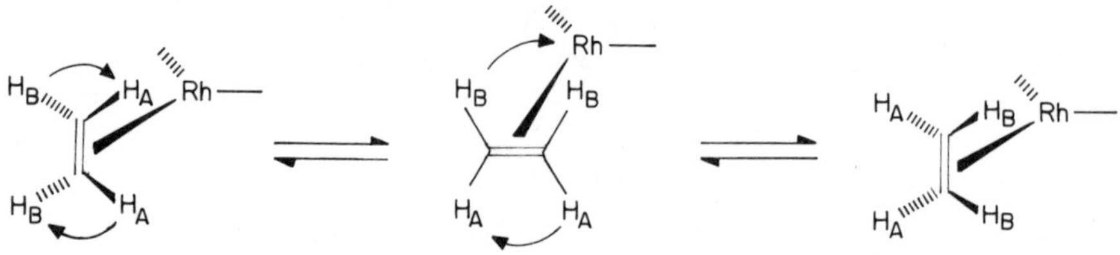

Scheme 2.1   Rotation of coordinated ethene about $Rh\!-\!C_2H_4$ bond

FLUXIONALITY AND DYNAMIC EQUILIBRIA IN π-ALLYL COMPLEXES

A large number of π-allyl complexes show equivalence by ligand protons in the NMR spectrum and both fluxional (ligand rotation) and dynamic (π-allyl/σ-allyl interconversion) processes have been characterized. When the activation energy for these processes is very high, we often observe simple conversion of one rotational isomer to a more stable one at elevated temperature. For example, the π-allyl molybdenum complexes **2** and **4** can be obtained almost pure (85:15 mixture of *endo* and *exo* complexes), but on standing at ambient temperature the NMR spectrum shows a gradual change due to slow conversion of the *endo* to the *exo* complex. At equilibrium the latter, more stable, compound predominates.[3]

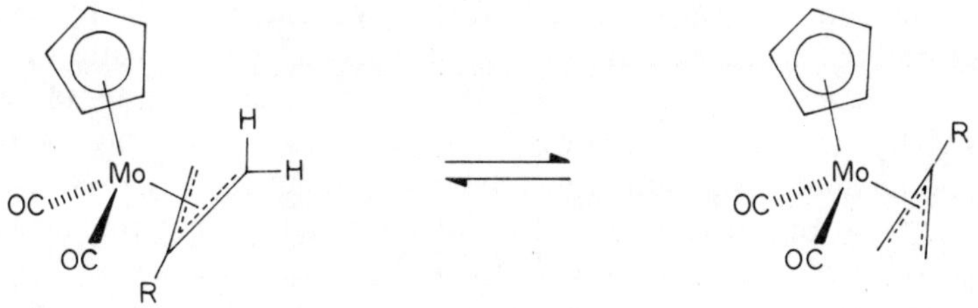

(2) *endo*   (3) *exo*

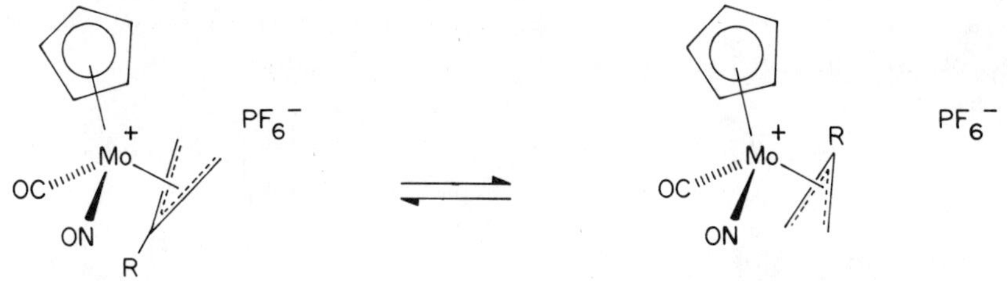

(4) *endo*   (5) *exo*

The homoleptic complex $Zr(allyl)_4$ may be prepared from anhydrous $ZrCl_4$ and C_3H_7MgCl under nitrogen at $-80\,°C$. The proton NMR spectrum has been studied over the temperature range -74 to $0\,°C$.[4] At low temperature the spectrum is of the AM_2X_2 type, typical of a fixed π-allyl complex having different signals for H-la, H-lb and H-2. As the temperature is raised the spectrum changes to the AX_4 type, i.e. a process is now occurring to effect rapid exchange of the terminal allyl methylene protons. The spectra are shown schematically in Figure 2.4.

The mechanism of equivalence of H-la and H-lb cannot be simple rotation of the allyl ligand, since in such a process these protons maintain their stereochemical integrity. The most plausible mechanism for the process is a partial dissociation of the π-allyl ligand to give a σ-allyl complex, followed by rotation of the allyl ligand about the carbon—carbon σ-bond and reformation of the π-allyl complex, as shown below. This effects the exchange of the terminal methylene protons a and b. The absence of a concentration dependence for the equilibration process rules out any intermolecular exchange of allylic ligands.

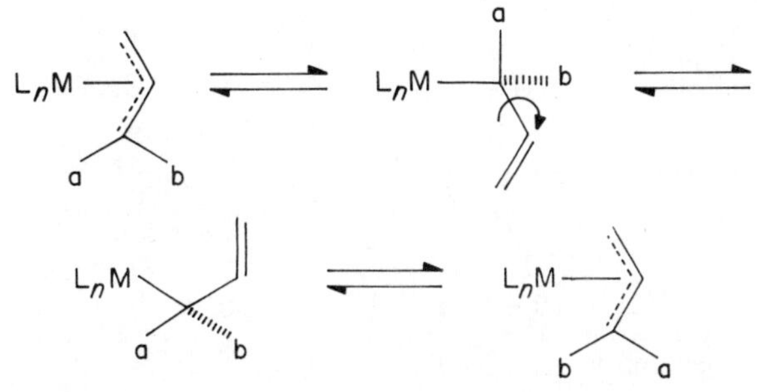

The NMR spectrum of trisallylrhodium, $Rh(\pi\text{-allyl})_3$, shows a slightly more complex temperature dependence. At $-74\,°C$ three distinct AM_2X_2 patterns are observed, but on warming to $10\,°C$ the spectrum obtained consists of two equivalent AM_2X_2 patterns and one unique AM_2X_2 pattern. This is interpreted as being due to fluxional rotation of one allyl group about the allyl—Rh axis, making the other two allyl ligands equivalent. At still higher temperature further averaging occurs which is consistent with $\sigma \rightleftharpoons \pi\text{-allyl}$ interconversions.[5]

Diallylnickel shows a slightly different behavior, in that it does not undergo a change to the AX_4 spectrum. At low temperature there are two AM_2X_2 species visible in a ratio of 3:1, and as the temperature is raised this ratio simply changes to ca. 2.2:1. Thus, there is no σ- to π-allyl interconversion and it is suggested that the following (conformational, or fluxional) equilibrium is involved:

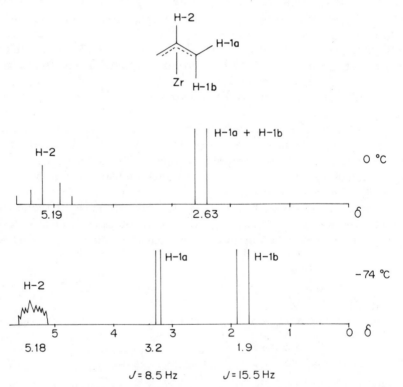

It is possible that the position of equilibrium is temperature dependent. The

Figure 2.4 Schematic variable temperature ^{1}HNMR spectra of $(allyl)_4Zr$

presence of ligands or solvent which may act as a ligand can affect the equilibration process involving formation of a σ-allyl complex, since this electron-deficient species is stabilized by coordination. Obviously, the only way in which added ligand can affect the fluxional process of 18-electron complexes is by replacement of an existing ligand, thereby leading to a different complex which has a low energy barrier to π-allyl rotation. There do not appear to be any well characterized examples of such a phenomenon.

Another consequence of the dynamic equilibrium of σ- and π-allyl complexes is loss of stereochemical integrity in the allyl group. This is best shown by the following example, where a single diastereoisomeric complex is obtained by crystallization from carbon tetrachloride at $-72\,^\circ$C. The optical rotation remains fairly stable at $-72\,^\circ$C, but on warming the rotation changes. Since the optically active amine ligand is unchanged, this indicates epimerization of the π-allyl–metal system.[6]

Asymmetric π-allyl complexes of type **6** have also been found to exhibit dynamic ^{1}H, ^{13}C and ^{31}P NMR spectra indicative of intramolecular rearrangement barriers of the order 12 kcal mol^{-1}, although an added complexity is observed. At low temperature non-equivalence of the two ends of the allyl group is indicated by an ABCDX pattern for the protons in the ^{31}P-decoupled spectrum. When the temperature is raised this changes to an A_2B_2X pattern of a symmetrical $^3\eta$-allyl complex, owing to rotation of the allyl ligand, which places the A, B and X protons in their averaged environments. In the diphos complexes the diphos methylene protons appear as an ABCD pattern at $-100\,^\circ$C, but on warming to 30 $^\circ$C they become an A_2B_2 set. This does not occur for all chelating phosphorus ligands, and is probably due to an interconversion of structures **6a** and **6b**.[7]

(6a)   (6b)

Thus, a range of effects of varying complexity may occur in the NMR spectra of π-allyl complexes, but these are usually explicable in terms of relatively simple motions of the π-allyl and/or other ligands.

TRICARBONYL CYCLOOCTATETRAENE (COT) METAL COMPLEXES
(METAL = IRON, RUTHENIUM, OSMIUM)

The first of these to be prepared was $(cot)Fe(CO)_3$. In the crystal the $Fe(CO)_3$ unit is bonded in a 1,2,3,4-tetrahapto fashion, but the NMR spectrum shows only a single line, suggesting rapid movement of $Fe(CO)_3$ around the ring somehow. A single line remains down to very low temperatures owing to the small activation energy. Activation energies for M = Ru and Os are higher, so that as the temperature is lowered a splitting becomes apparent. For M = Ru a spectrum consisting of the four multiplets expected for a tetrahapto structure is observed at *ca.* $-145\,°C$. For M = Os the limiting spectrum is obtained at $-100\,°C$. It is assumed that the equivalence mechanism is the same for all three. From analysis of the line shape changes with temperature, it is concluded that the equivalence occurs by a series of 1,2 shifts of the $M(CO)_3$ group around the ring.

For $(cot)M(CO)_3$ complexes (M = Cr, Mo, W), sharp complex spectra are observed for all three compounds at $-40\,°C$, which broaden and collapse as temperature is raised, giving a single line at 50–100 °C. Because of complexity due to spin–spin coupling dynamical analysis is not possible. However, line shape changes for 1,3,5,7-tetramethylcyclooctatetraene–$Cr(CO)_3$ as the temperature is raised are very selective. Between -23 and $+50\,°C$, two of the ring-proton resonances collapse and coalesce, while the other two remain unchanged. At the same time, all of the methyl resonances broaden and then coalesce pairwise. Either a 1,2 or a 1,4 shift mechanism can be invoked to explain the data, but the 1,2 shift is considered more plausible.

OXIDATIVE ADDITION REACTIONS

Oxidative additions of transition metal organometallic complexes are exceedingly important in that they provide methods for the synthesis of certain types of complex, e.g. σ-alkyl derivatives, and they frequently occur as key steps in catalytic cycles. Discussions of the reaction have appeared in several reviews.[8]

The reaction may be defined as follows: reaction of a metal complex in which the formal oxidation state of the metal is n, with an addendum X–Y, to give a new complex, X–M–Y or similar, in which the formal oxidation state of the metal is increased to $n + 2$. There are some slight variations on this, as will become apparent during the discussion. The reactant X–Y is commonly a halide derivative, e.g. HX (X = Cl, Br, I), RX (R = alkyl, vinyl, aryl; X = Cl, Br, I), and in some cases molecules such as H–H and Ar–H are involved. Examples of each will be found throughout the text.

Since the addition of the species X–Y to the metal results in the formation of two new bonds, M–X and M–Y, this leads to an increase in coordination number, and so the reaction invariably does not occur for coordinatively saturated complexes (see Chapter 1).

Some examples of typical reactions are given below which suffice to illustrate this. The reactions are given in general schematic terms first, followed by specific examples.

(a) $ML_2 \longrightarrow ML_4 \longrightarrow ML_6$

   d^{10} (14e) d^8 (16e) d^6 (18e)

e.g. $Pt^0(PPh_3)_2 \xrightarrow{X-Y} Pt^{II}(PPh_3)_2(X)(Y) \xrightarrow{X-Y} Pt^{IV}(PPh_3)_2(X)_2(Y)_2$

   X–Y = CH_3I, $PhCH_2Br$, Ph_3SnCl, etc.

(b) $ML_4 \longrightarrow ML_6$

   d^8 (16e) d^6 (18e)

e.g. $Ir^I(CO)(PPh_3)_2Cl \xrightarrow{X-Y} Ir^{III}(X)(Y)(CO)(PPh_3)_2Cl$

   X–Y = H_2, Cl_2, HCl, CH_3I, RSO_2Cl, RHgCl, R_3SiH, etc.

(c) $ML_5 \xrightarrow[+L]{-e} ML_6$

   d^7 (17e) d^6 (18e)

e.g. $2[Co^{II}(CN)_5]^{3-} \xrightarrow{X-Y} [Co^{III}(CN)_5X]^{3-} + [Co^{III}(CN)_5Y]^{3-}$

   X–Y = H_2, Br_2, HOH, HOOH, CH_3I, etc.

[The last example is peculiar in that a one-electron oxidation of the metal is involved, as opposed to the usual two-electron process. There is evidence to suggest that this occurs by a free radical mechanism (see later)].

Thus, it can be seen that the final result of the oxidative addition is usually the formation of an 18-electron complex, or a 16-electron complex for metals preferring this (e.g. Pt, Pd). Consequently, if a coordinatively saturated complex (18-electron) is subjected to oxidative addition conditions, it will not react in a *concerted* fashion with the addendum X–Y unless there is prior loss of a ligand to give a 16-electron complex. Alternatively, the 18-electron complex may react in a non-concerted manner to give 18-electron intermediates which subsequently undergo loss of an existing ligand and gain of the new ligand. In schematic terms, these alternatives may be represented as shown below:

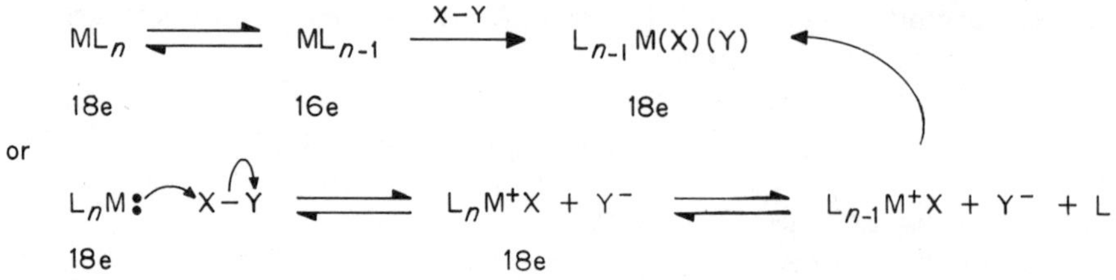

$$ML_n \rightleftharpoons ML_{n-1} \xrightarrow{\text{X–Y}} L_{n-1}M(X)(Y) \longleftarrow$$

$$\text{18e} \qquad\qquad \text{16e} \qquad\qquad\qquad \text{18e}$$

or

$$L_nM: \overset{\frown}{}X–Y \rightleftharpoons L_nM^+X + Y^- \rightleftharpoons L_{n-1}M^+X + Y^- + L$$

$$\text{18e} \qquad\qquad\qquad\qquad \text{18e}$$

We now consider the reactions of d^{10}, d^8 and d^7 complexes, in that order.

Oxidative Addition Reactions of d^{10} Complexes

Compounds having a formal d^{10} configuration have been found to undergo unusually facile oxidative addition reactions. The most extensively studied are complexes of Pt(0), notably $Pt(PPh_3)_3$ and $Pt(PPh_3)_2C_2H_4$. These react with a variety of molecules (X–Y = CH_3I, $PhCH_2Br$, Ph_3SnCl) to give adducts of type $Pt(PPh_3)_2(X)(Y)$.

The actual addition occurs to $Pt(PPh_3)_2$, a 14-electron species formed by ligand dissociation, except for hydrogen halides which appear to first protonate the PtL_3 to give $HPtL_3^+Cl^-$. Palladium complexes undergo similar reactions which turn out to be very useful for organic syntheses involving carbon—carbon bond formation (see Chapter 3), e.g.

Note that 16-electron complexes of Pt and Pd are stable.

Oxidative Addition Reactions of d^8 Complexes

These are by far the most commonly encountered examples of oxidative addition reactions, resulting in the formation of octahedral d^6 complexes by addition of a covalent molecule to the coordination sphere. Coordinatively unsaturated d^8 complexes, usually square planar, reversibly add RX, RCOX, MX, X_2, O_2, $R-C\equiv C-R$, and $RR'C=CR^2R^3$ (X = halogen). Coordinatively saturated d^8 complexes, which are usually trigonal bipyramidal, react with highly polarizable addenda, e.g. HX, HgX_2, X_2, in a two-step mechanism (see above).

Most extensive studies have been made on square-planar iridium(I) complexes of the type $Ir(CO)L_2Y$ [Y = Cl, Br, I; L = PPh_3, PPh_2CH_3, $PPh(CH_3)_2$], commonly known as Vaska's complexes. The reactions are often stereospecific; both *cis* and *trans* modes of addition have been observed, depending on the reagent and the medium. Some examples are given below.

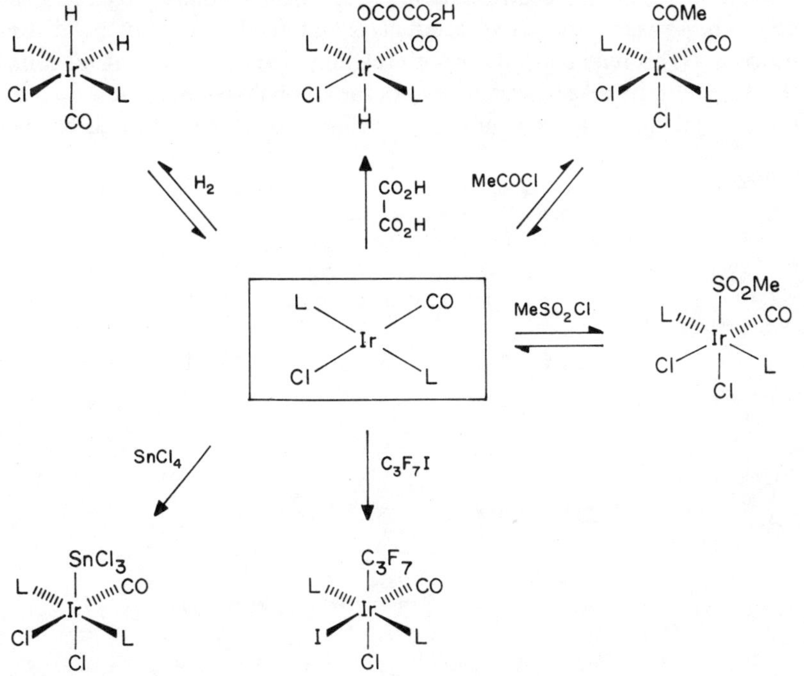

The addition of H_2 appears to be a concerted process, as shown by the negative entropy of activation ($\Delta S^{\ddagger} = -23$ e.u.; $\Delta H^{\ddagger} = 10.8$ kcal mol^{-1}), and results in the formation of the *cis* adduct. Such a concerted addition probably involves frontier orbital interactions of the type shown below, i.e. interaction of the H_2 bonding MO with empty metal d^2sp^3 hybrid orbital, together with interaction between H_2 antibonding MO and a filled metal d orbital.

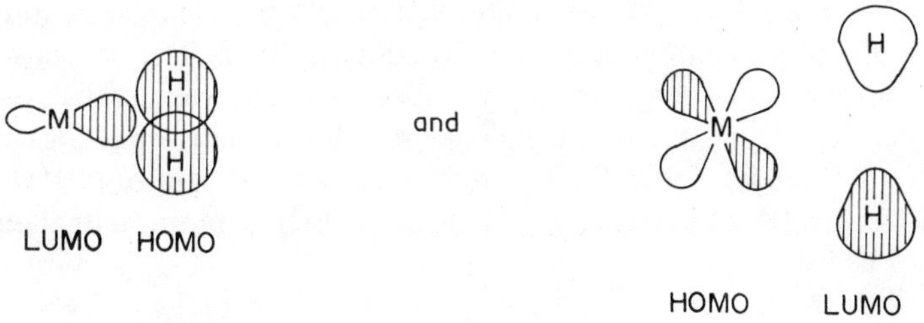

This type of HOMO + LUMO interaction is very reminiscent of concerted organic reactions, e.g. cycloaddition, and is consistent with the requirement of a low-lying unfilled metal orbital of the type found in coordinatively unsaturated complexes.

There still appears to be some confusion regarding the exact mechanism of oxidative addition of alkyl halides. In fact, there is no reason to suppose that all alkyl halides react by the same mechanism, since it is well known from organic chemistry that concerted (S_N2) or non-concerted, dissociative (S_N1) mechanisms can be followed depending on the nature of the alkyl halide. Indeed, there are cases where oxidative addition occurs in a *cis*, and probably concerted fashion, and others where the *trans* product is obtained, probably by dissociation processes. This may depend on both the alkyl halide and the metal complex. A third mechanism involves homolysis of the C—halogen bond, followed by a free radical reaction. The three possible mechanisms are summarized below.

Concerted:

$$L_nM\overset{R}{\underset{I}{\diagup}} \longrightarrow L_nM\overset{R}{\underset{I}{\diagdown}}$$

cis

Non-concerted:

$$L_nM\!:\!\overset{\frown}{R-I} \longrightarrow L_nM^+R + I^- \longrightarrow L_nM\overset{R}{\underset{I}{\diagdown}}$$

cis or *trans*

Radical:

$$L_nM + RX \underset{\text{slow}}{\rightleftharpoons} L_n\overset{\bullet}{M}X + R\bullet$$

$$L_n\overset{\bullet}{M}X + R\bullet \overset{\text{fast}}{\longrightarrow} L_nM(R)(X)$$

cis or *trans*

In recent years, increasing evidence has accumulated for the occurrence of the free-radical reaction for a number of metal complexes. Such evidence includes (a) initiation by O_2, AIBN, benzoyl peroxide, etc.; (b) retardation by radical scavengers, e.g. duroquinone, hydroquinone, galvinoxyl; (c) racemization of the alkyl halide; (d) increase in rate for halide having electronegative substituents; (e) spin trapping experiments with BuNO.

The rationalization of all the possible mechanisms is still incomplete, and clearly different mechanisms may be adopted by different complexes, and even by the same complex reacting with different addenda X–Y. The propensity of d^8 metal complexes to undergo oxidative addition follows the familiar pattern of transition metal chemistry, increasing on descending a triad or passing from right to left in the Periodic Table. For example, it is easier to oxidize Pt(II) to Pt(IV) than Ni(II) to Ni(IV), but oxidation of Os(0) to Os(II) is more facile than that of Pt(II) to Pt(IV).

The other ligands in a metal complex may also have a dramatic effect, better σ-donor or poorer π-acceptor ligands leading to enhanced oxidative addition owing to increased availability of metal d electrons. Compare the following, which illustrate different metal and different ligand effects.

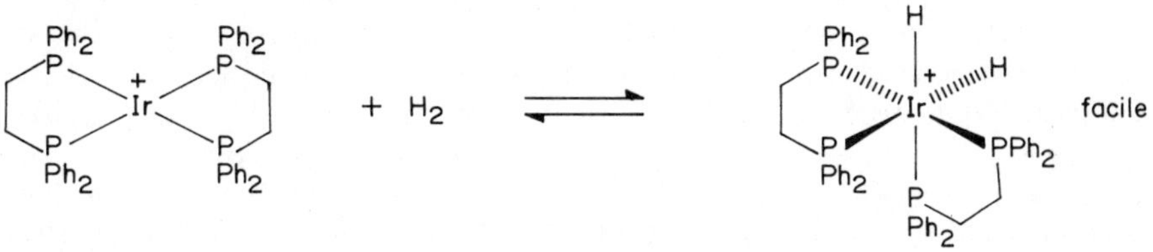

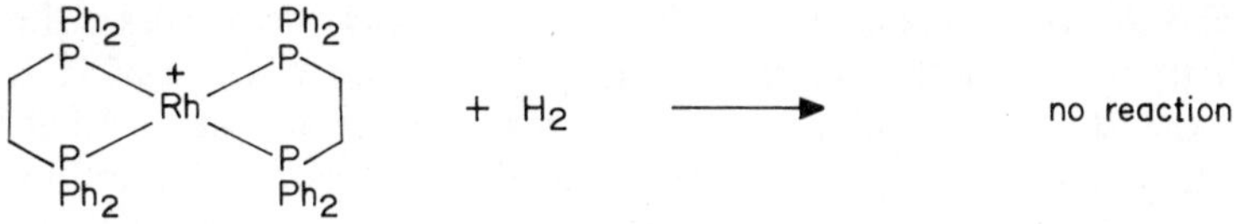

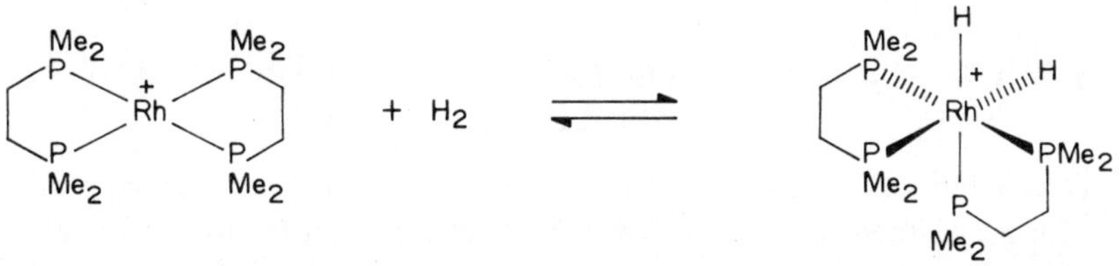

As might be expected, the ease with which ligand dissociation can occur from coordinatively *saturated* d^8 complexes also affects the ease of oxidative addition reactions. On ascending a triad or passing from right to left within Group VIII, the trend is for a d^8 complex to become coordinatively saturated. For example, the stabilities of $M(CO)_5$ complexes towards dissociation decrease in the order $Fe > Ru > Os$, as shown by the conditions necessary for oxidative addition of H_2:

$$Fe(CO)_5 \; + \; H_2 \; \xrightarrow{\; >160 \;^\circ C \;} \; H_2Fe(CO)_4$$

$$Os(CO)_5 \; + \; H_2 \; \xrightarrow{\; 80 \;^\circ C \;} \; H_2Os(CO)_4$$

On these bases we expect the most reactive d^8 compounds to be found in the lower left-hand corner of Group VIII among four-coordinate complexes containing activating ligands.

Oxidative Addition Reactions of d^7 Complexes

With the exception of reactions with H_2 and H_2O, most of these proceed through a stepwise free-radical mechanism (see above). For the coordinatively saturated derivatives, e.g. $Co(CN)_5{}^{3-}$, this clearly avoids the formation of 19-electron complexes:

$$L_n M \quad + \quad ClR \quad \longrightarrow \quad L_n MCl \quad + \quad R\bullet$$
$$\text{17e} \qquad\qquad\qquad\qquad \text{18e}$$

$$L_n M \quad + \quad R\bullet \quad \longrightarrow \quad L_n MR$$
$$\text{17e} \qquad\qquad\qquad \text{18e}$$

Evidence for the radical mechanism comes mainly from the observation that the reactivity patterns are similar to those for halogen radical abstraction by other atoms and free radical reagents (e.g. Na atoms and CH_3 radicals), i.e. the order of reactivity is $RCl < RBr < RI$ and $CH_3I < C_2H_5I < Me_2CHI < PhCH_2I$.

Interest in the study of oxidative addition reactions of low-spin cobalt complexes, notably reactions leading to organocobalt products, is enhanced by certain parallels with corresponding reactions of vitamin B_{12} derivatives and consequently by their possible relevance as vitamin B_{12} model systems.

REDUCTIVE ELIMINATION FROM TRANSITION METAL σ-COMPLEXES

Reductive elimination has been discussed in the review by Braterman and Cross.[9] For a considerable time it was widely believed that transition metal—carbon σ-bonds were inherently unstable in the absence of π-acceptor ligands. The M—C bond was considered to be stabilized by increasing the energy gap between filled and empty d orbitals on the metal, a function of ligand π-acceptor strength. However, it is now recognized that transition metal—carbon σ-bonds are thermodynamically of similar strength to main group metal—carbon bonds, and that the lability of the transition metal derivatives is entirely due to the existence of low activation energy pathways for their decomposition owing to availability of variable oxidation states denied the main group metals. Thus, the instability is a kinetic phenomenon, and if steps are taken to prepare complexes in which the decomposition pathways are eliminated or restricted, stable compounds result.

In general, reductive elimination is the opposite of oxidative addition, being the elimination of two ligands with concomitant reduction by two units of both the formal oxidation number and the coordination number of the metal. Illustrative examples are as follows:

$$(Ph_3P)AuMe_3 \quad \longrightarrow \quad (Ph_3P)AuMe \quad + \quad C_2H_6$$

$$(PhMe_2P)_2 PtMe_3 I \quad \longrightarrow \quad (PhMe_2P)_2 PtMe (I) \quad + \quad C_2H_6$$

In both of these reactions two methyl groups are split off the complex in a carbon—carbon bond-forming process to give ethane. The concerted nature of these reactions has been confirmed by various studies, including labelling experiments, which show the reaction to be intramolecular, and kinetic studies, which show a first-order reaction with an activation energy of about $31 \, kcal \, mol^{-1}$. Thus, bond making accompanies bond breaking, and no free-living high-energy intermediates such as free radicals, carbanions or carbonium ions are involved. It is a well established process of (1, 1) elimination of two groups at the same metal center.

Reductive elimination frequently follows oxidative addition, and many well known catalytic cycles depend on the interplay of both phenomena (see later). Some examples of non-catalytic processes are as follows:

$$(Ph_3P)AuCH_2SiMe_3 \xrightarrow{MeI} [(Ph_3P)AuIMe(CH_2SiMe_3)] \longrightarrow$$

$$(Ph_3P)AuI \quad + \quad EtSiMe_3$$

$$(Ph_3P)AuOSiMe_3 \xrightarrow{MeI} [(Ph_3P)AuIMe(OSiMe_3)] \longrightarrow$$

$$(Ph_3P)AuI \quad + \quad MeOSiMe_3$$

$$(Et_3P)_2PtCl_2 \xrightarrow{H_2} (Et_3P)_2Pt(H)_2Cl_2 \longrightarrow$$

$$(Et_3P)_2Pt(H)Cl \quad + \quad HCl$$

The spontaneous loss of alkene from a metal coordination sphere, dimerizations, polymerization and ring closures may be formally regarded as equivalent to reductive elimination processes:

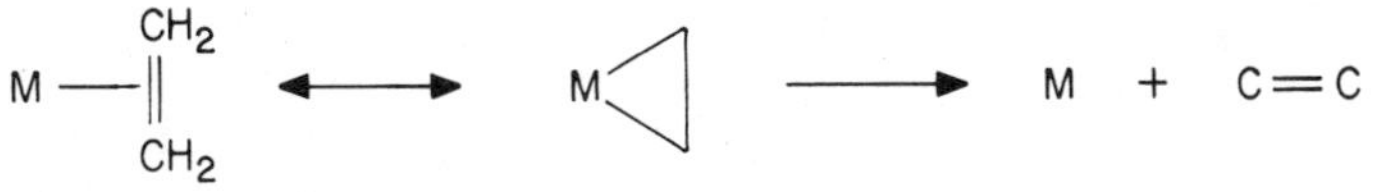

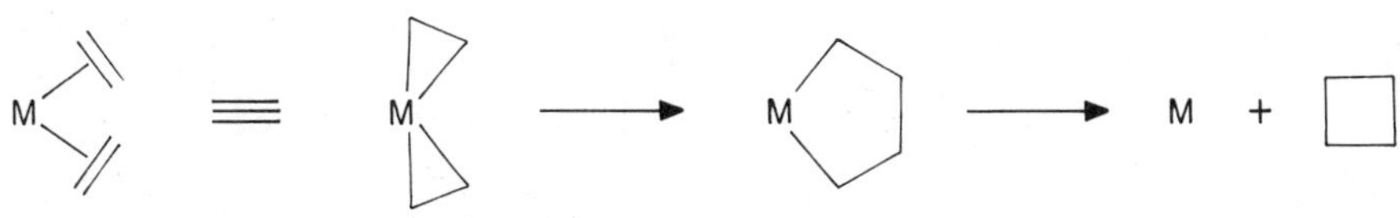

It is readily appreciated that the number of electrons formally donated to the metal is reduced by two in these elimination reactions, e.g. 18-electron complexes are converted to 16-electron complexes. Consequently, the reaction most easily occurs for coordinatively saturated complexes of metals which will form 16-electron species. The less favorable the latter become, the more difficult will be reductive elimination. Alternatively, compounds in which the orbital occupation is higher than usual for the metal concerned will be more prone to undergo elimination reactions, and any reaction or change in conditions which increases orbital occupation is likely to be followed by a counter balancing elimination. Such effects are most noticeable towards the end of the transition series (see also Chapter 1). The formal oxidation number is less important than electron availability at the metal. The following example serves to illustrate these remarks:

$$(\text{bipy})\text{NiEt}_2 \longrightarrow \text{bipy} + \text{Ni} + \text{C}_4\text{H}_{10}$$

16e

$$E_a = 65.5 \text{ kcal mol}^{-1}$$

$$\Big\downarrow \quad \text{H}_2\text{C}=\text{CHX}$$

$$(\text{bipy})\text{NiEt}_2(\text{CH}_2=\text{CHX}) \longrightarrow (\text{bipy})\text{Ni}(\text{CH}_2=\text{CHX}) + \text{C}_4\text{H}_{10}$$

18e

$$E_a = 15.5 \text{ kcal mol}^{-1}$$

Ready explanation of a wide range of useful chemical transformations employing transition metal catalysts is possible using a reductive elimination step. An appreciation is also gained as to why some metals catalyse reactions involving this process whilst others which are less prone to reductive elimination do not. The selective cross-coupling of alkyl halides and Grignard reagents catalyzed by Ni(II) phosphine complexes involves reductive elimination from a transient 18-electron nickel complex:

$$(\text{diphos})\text{NiCl}_2 \xrightarrow{\text{R}'\text{MgX}} (\text{diphos})\text{NiR}'_2 \xrightarrow{\text{R}^2\text{X}} (\text{diphos})\text{NiR}^2\text{X} + \text{R}'\text{R}'$$

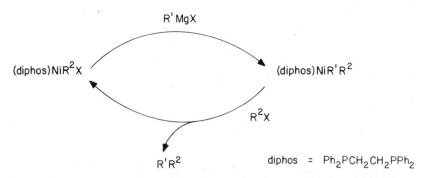

diphos $= \text{Ph}_2\text{PCH}_2\text{CH}_2\text{PPh}_2$

In the coinage group, stable 14-electron complexes are common, and it is found that even 16-electron species undergo facile eliminations. Also, because of the lanthanide contraction, elements of the second row, such as Ru, Rh and Pd, show a greater tendency to form 16-electron complexes, and third-row elements show an even more pronounced ease of formation of such compounds. Consequently, we might expect more facile reductive elimination from 18-electron complexes of these elements than their first-row counterparts. Fairly often, however, identification of a reductive elimination from a stable complex is difficult, since invariably other modes of decomposition can, and do operate. This is particularly apparent for osmium complexes.[10] For example, cis-$Os(CO)_4Me_2$ is extraordinarily stable and does not undergo simple reductive elimination. Similarly, $Os(CO)_4H_2$ and $Os(CO)_4(H)Me$ are both found to undergo elimination (of H_2 and CH_4, respectively) which is complex and, at least in the latter case, bimolecular. In considering the possibilities for the occurrence of simple intramolecular reductive elimination, it must also be borne in mind that the reaction becomes disfavored as the energy of the remaining fragment increases, and that in such cases other elimination processes become predominant. In support of this, it may be noted that the $Os(CO)_4$ fragment has been found to be of high energy, and consequently does not form complexes which undergo reductive elimination as readily as might have been expected.

α- AND β-ELIMINATION REACTIONS

Of these two modes of decomposition of transition metal σ-alkyl complexes, both involving initial hydrogen transfer from ligand to metal with the concomitant creation of an unsaturated derivative of some type, the better characterized is the β-hydride elimination. α-Elimination has been recognized more recently.

Discussions of β-elimination have appeared in reviews.[9,11] The reaction follows the course outlined in the following scheme:

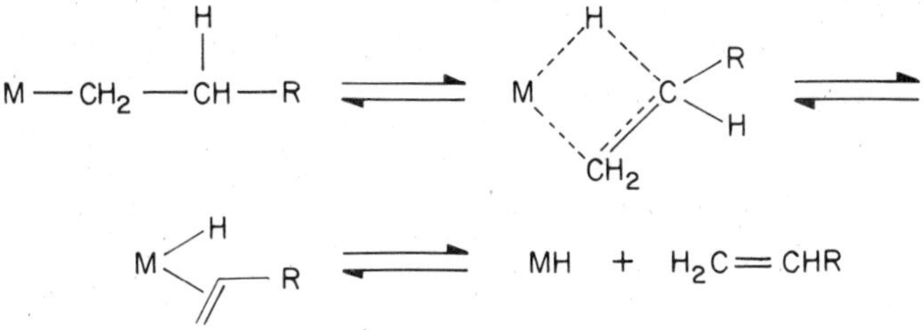

The overall result is analogous to E_2 or E_1 elimination, found in organic chemistry, the metal taking the place of a leaving group. The major differences are that the olefin formed in the reaction often forms an initial η^2-alkene complex with the metal which then dissociates, and that the sequence of events is often reversible.

Deuterium labelling studies have confirmed the above mechanistic concept, i.e. transfer of the β-hydrogen to metal, e.g.

$$(Bu_3P)CuCH_2CD_2Et \longrightarrow (Bu_3P)CuD + H_2C{=}CDEt$$

However, when a rapid reverse reaction operates there is effective mixing of H and D. We shall consider in more detail the thermal decomposition of di-n-butylbis(triphenylphosphine)platinum(II).[12]

During this reaction, it was found that no butenes other than but-1-ene were formed, and also no octane was observed. The latter observation excludes the possibility that butane and but-1-ene are formed by disproportionation of free *n*-butyl radicals produced by homolysis of the Pt—C bonds. The rate of reaction was found to be depressed by the addition of triphenylphosphine, suggesting that dissociation of the phosphine ligand is a necessary occurrence prior to the rate-determining step. This was surprising because at the time three-coordinate platinum intermediates were not commonly implicated in platinum chemistry, although both two- and three-coordinate platinum compounds are well known. It was also found that dibutylplatinum derivatives containing chelating diphosphine ligands were stable to decomposition, in support of this conclusion.

In fact it is found that in general β-elimination is suppressed by the presence of strong π-acceptor ligands in a complex because these are less prone to dissociate and give a vacant coordination site.

Mechanistically, the study of deuterium-labelled butyl–Pt derivatives does not give clear evidence for the β-elimination mechanism since scrambling of the label occurs, giving butene and butane with deuterium in all positions. This was taken as indicative that the elimination of platinum hydride from butyl complexes and its readdition to coordinated but-1-ene are both more rapid reactions than the reductive elimination of *n*-butane from intermediates containing hydride and *n*-butyl groups σ-bonded to platinum. The proposed mechanism is as follows:

This type of process is featured in many reactions of great synthetic value, such as the formation of transition metal hydrides from alkoxides, etc. It also features prominently in such catalytic reactions as olefin isomerizations, and is frequently encountered as a termination step in polymerization reactions.

β-Elimination is effectively prevented in complexes where the alkyl ligand has no β-hydrogens, e.g. CH_3, CH_2Me_3, CH_2CMe_3.

α-Elimination is a more recently characterized process. It involves transfer of hydrogen from the α-carbon atom of the alkyl ligand to the metal with concomitant formation of a hydride–metal carbene complex. The reaction can and often does occur for complexes in which β-elimination is prohibited, e.g. those containing CH_3 and CH_2SiMe_3 ligands, and it has been discussed in a review by Schrock.[13] It may be represented by the equation

$$MCH_2R \quad \rightleftharpoons \quad M{\overset{H}{\underset{CHR}{\diagdown}}}$$

What happens next depends on what other ligands are present in the complex. In Cooper and Green's initial observation on this reaction the following sequence of events was characterized, using deuterium-labelled compounds.[14]

isolated
product

The fundamental step of α-elimination, hydride transfer to metal, is exactly analogous to the 1,2-hydride shifts commonly associated with carbonium ion chemistry. Compare

$$(CH_3)_2\overset{H}{\underset{|}{C}} - \overset{+}{C}HCH_3 \quad \longrightarrow \quad (CH_3)_2\overset{+}{C} - CH_2CH_3$$

with

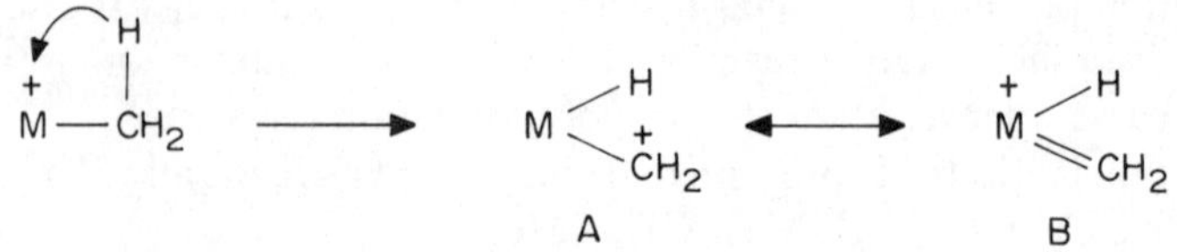

A   B

Viewed in this way, we can see that the possible driving force for the reaction is the resultant delocalization of charge, at least in the above case, over metal and carbon atoms, as shown by the canonical forms A and B.

In complexes where a hydride–metal carbene complex is produced which contains a ligand such as σ-alkyl, reductive elimination of, e.g. alkane, is often observed, e.g.

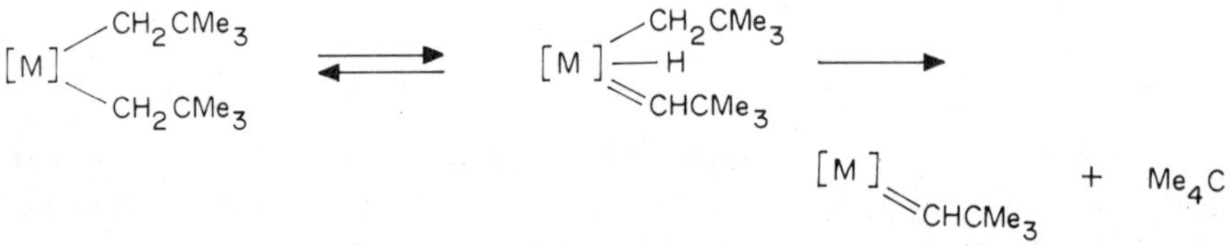

Examples of this behavior are found in Chapter 4. This therefore represents a decomposition mode for transition metal alkyl derivatives since by this mechanism an alkyl ligand departs as alkane.

OLEFIN METATHESIS

Two very good reviews on the olefin metathesis reaction have recently appeared.[15] The overall conversion in the reaction is the transformation of two olefins to two new olefins arising from cleaving the original molecules in half, through the double bond and perpendicular to it, and rejoining the separated halves to new partners. The reaction occurs in the presence of a transition metal catalyst, commonly containing W or Mo, and it can be readily appreciated that, depending on their structure, two olefins may be combined to produce at least eight new olefins together with the starting compounds, in equilibrium, e.g.

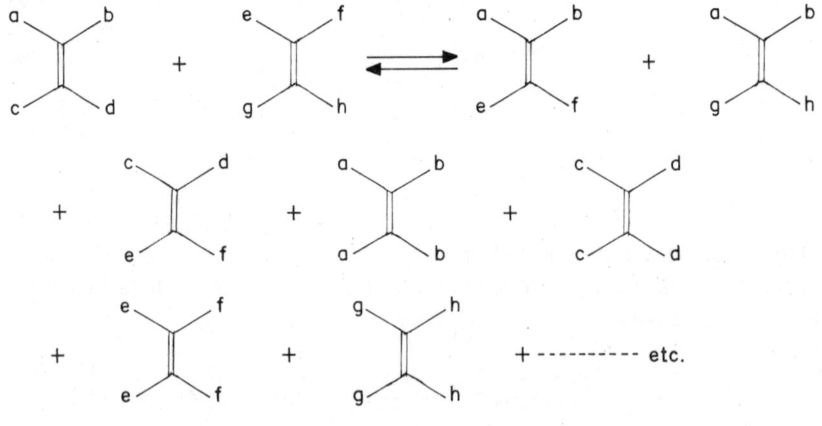

This does not include the various possible geometrical isomers of each product, so we can see that a mixture containing twenty olefins can be produced from the appropriate precursors. However, the reaction is often used with alkenes which give fewer products. For example, the first metathesis reaction, described in 1957, was the conversion of propene to a mixture of ethylene and but-2-ene:

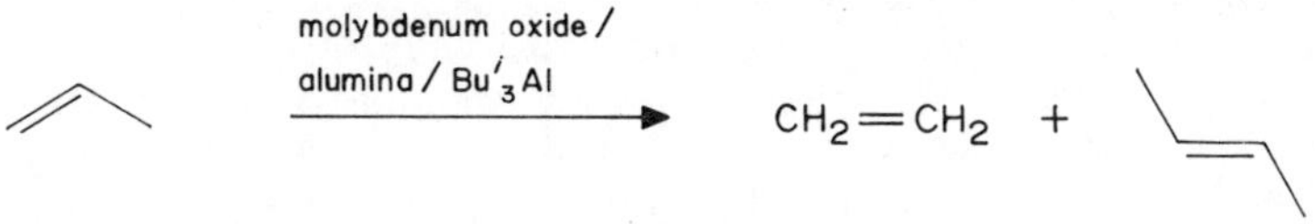

Various cyclic olefins produce polymeric olefins, having *cis* and *trans* double bonds in a ratio dependent on the catalyst used, e.g.

The product of this reaction resembles natural rubber, so the metathesis process has considerable commercial significance.

As a result of considering the overall transformation, i.e.

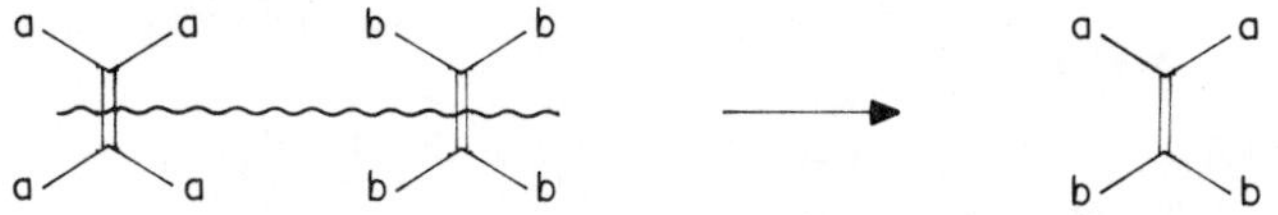

the first mechanism put forward involved the formation and decomposition of a cyclobutane–metal complex of some sort. The metal was supposed to be instrumental in circumventing orbital symmetry restrictions on the olefin cycloaddition:

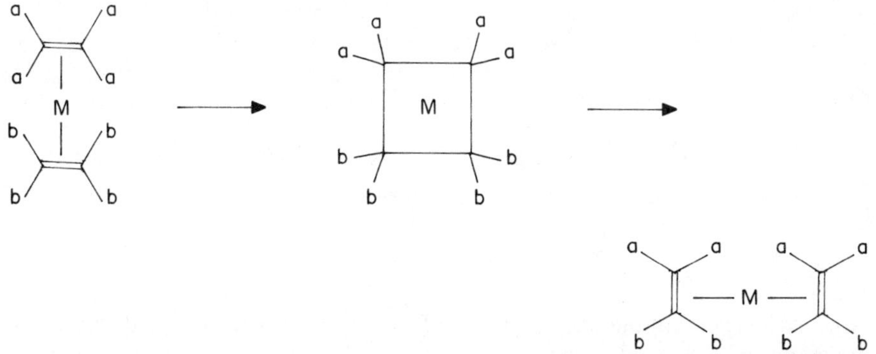

However, no cyclobutanes are observed as by-products in the reaction, and since a metal–cyclobutane complex of the type shown above is expected to have extremely weak bonding, if any, we would expect to see considerable amounts of free cyclobutanes.

The currently accepted mechanism involves prior formation of a catalytic amount of metal–carbene complex, say by α-elimination from a metal alkyl, followed by its reaction with olefin to form a metallacyclobutane derivative. This then expels olefin to regenerate a metal carbene, which continues the cycle:

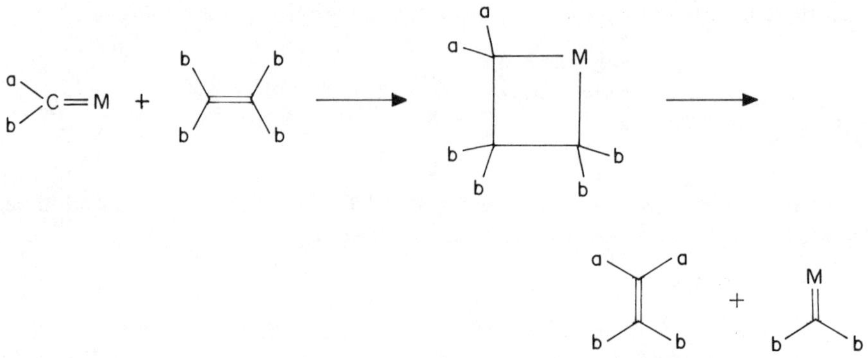

Evidence for the proposed mechanism comes mainly from single-cross and double-cross experiments, as well as from deuterium labelling studies.

Single-Cross Experiments

The product distribution at the beginning of the reaction is examined during the following metathesis:

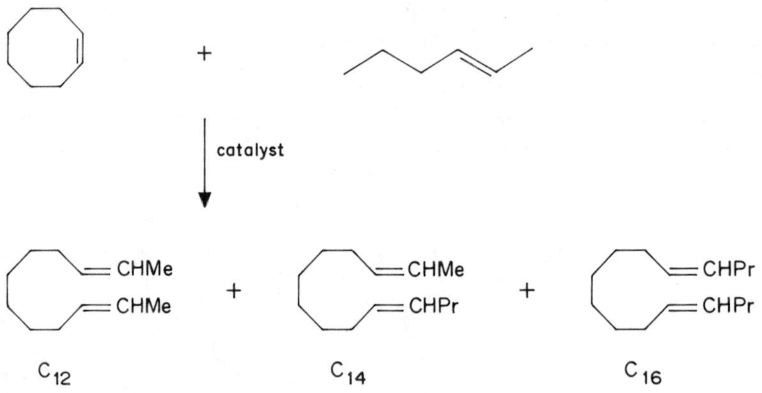

Now, if the cyclobutane–metal complex formation is operative (i.e. direct olefin dimerization), we should expect to see *only* C_{14} diene in the initial stages. On the other hand, if the mechanism is via a metallacyclobutane, we would expect to see a $C_{12}:C_{14}:C_{16}$ distribution of $1:2:1$ in the initial stages. While this ideal value is not actually observed, significant proportions of C_{12} and C_{16} compounds are obtained, suggesting the metallacyclobutane mechanism. Naturally, the actual ratio observed will not be the ideal value, since the preferred pathway for reactions of carbene complexes with olefin will depend on steric factors, e.g. the following reactions will have different rates:

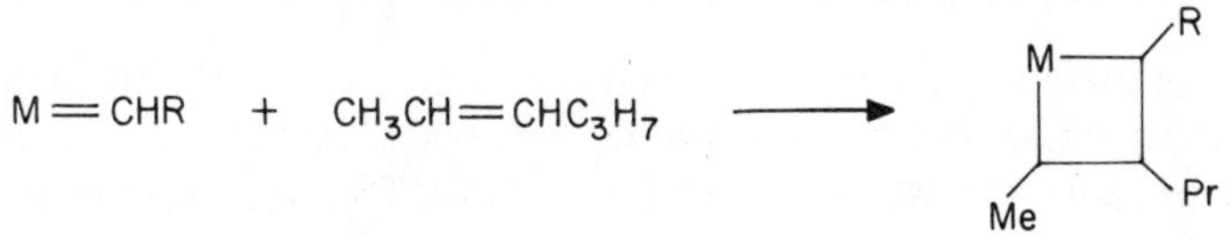

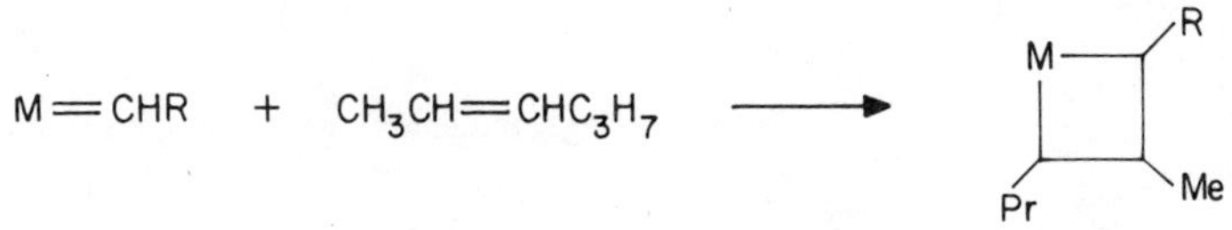

$$M=CHR \quad + \quad CH_3CH=CHC_3H_7 \longrightarrow$$

$$M=CHR \quad + \quad CH_3CH=CHC_3H_7 \longrightarrow$$

The ratio of $C_{12}:C_{14}:C_{16}$ product is also expected to be dependent on the exact nature of the acyclic olefin, i.e. whether internal or terminal alkene, and this is indeed found to be the case.

Double-Cross Experiments

In this investigation the initial product ratio is investigated for the following reaction involving the metathesis of three alkenes. Again, we would not expect to obtain ideal ratios of products, for the reasons outlined in the previous section.

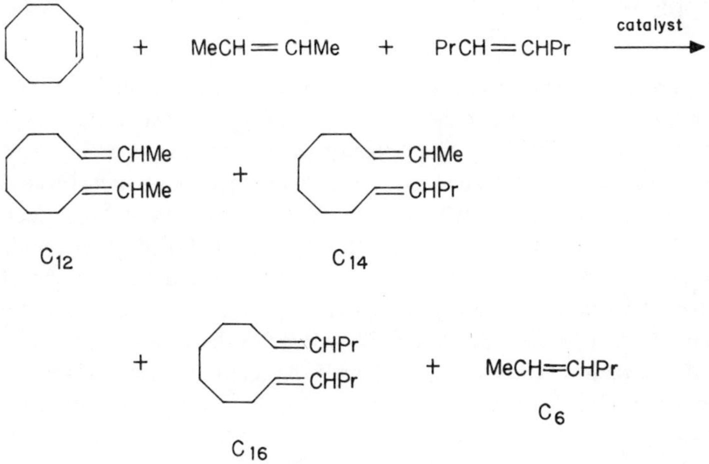

It is found that in the initial stages the C_{14} diene is formed in greater amount than C_{12} or C_{16} dienes. The ideal $C_{12}:C_{14}:C_{16}$ ratio for a metallacyclobutane mechanism is 1:2:1. However, if the first mechanism, cyclobutane–metal complex formation, is correct we should expect no formation of C_{14} diene at the start of the reaction.

These experiments therefore support (but do not prove) the metallacyclobutane mechanism.

Typical Metathesis Catalysts

Olefin Metathesis Induced by Carbene Complexes

Since the metallacyclobutane mechanism requires the initial formation of a metal–carbene complex, it is reasonable to suppose that isolated complexes of this type will act as initiators for olefin metathesis. This has been found to be the case, as shown by the following examples.[16]

high molecular weight polymer (97 % *cis*)

polynorbornene

It is interesting that *cis* to *cis* and *trans* to *trans* olefin conversions are observed with high stereoselectivity using $Ph_2C=W(CO)_5$, but the selectivity is poorer with $Ph(MeO)C=W(CO)_5$. The reasons for this are at present unknown.

In fact, the diphenylcarbene complex is an inefficient metathesis catalyst, possibly owing to the slow rate of displacement of CO ligands. This speculation is partly supported by the observation that $(CO)_5W=C(OEt)C_4H_9$, a highly stable complex, shows no catalytic activity for the metathesis of cyclopentene. However, when this complex is mixed with titanium tetrachloride in the absence of light, one equivalent of carbon monoxide is evolved. Subsequent thermal or photochemical activation leads to an extremely efficient catalyst. The role of the titanium tetrachloride is not well understood, but CO displacement appears to be essential.[17]

Catalysts Activated by Organometallic Co-catalysts

Typically a high oxidation state transition metal salt is mixed with a main group metal alkyl, e.g. WCl_6/R_mAlCl_n, WCl_6/RLi, WCl_6/R_4Sn or WCl_6/R_2Zn. These catalysts function via the formation of the transition metal alkyl σ-complex, which presumably undergoes α-elimination to give the active carbene complex.

Catalysts Without Organometallic Co-catalysts

Some transition metal salts, e.g. $MoCl_6$, sometimes combined with a Lewis acid, e.g. $AlCl_3$, will catalyze olefin metathesis in the absence of a co-catalyst. The route to carbene initiation in these systems is not well established and will not be discussed here.

Stereochemical Aspects of the Olefin Metathesis Reaction

It was noted above that, at least in relatively simple cases, there is a considerable degree of stereospecificity regarding the geometry of the product olefins. In order to explain this it has been proposed that the metallacyclobutane intermediate has a puckered ring structure. This is probably confirmed by the X-ray structure determination of the stable platinum complex

which indicated a slightly folded ring having a 12° angle between the planes.[18] It should be remembered that experimental difficulties with the measurements made this value barely significant. However, this assumption enables analogies to be made with the chemistry of cyclobutanes, and is helpful in rationalizing some of the results. Basically, reaction pathways appear to be favored which provide conformations of the metallocyclobutane possessing the fewest number of axial substituents on the α-carbon atoms in the ring, e.g.

It should be noted that the *cis → cis* olefin conversion cannot be properly explained on the basis of a flat metallocyclobutane structure. When *i*-Pr or *t*-Bu substituents are present in cyclobutanes, these groups now command an exclusively equatorial orientation. This can lead to marked changes in the stereochemical course of olefin metathesis, shown by the following example.

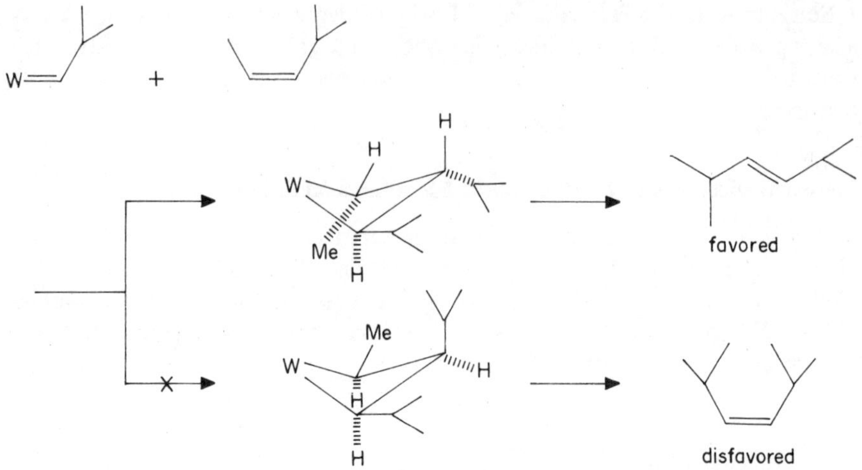

Of course, there are deficiencies in this theory, but on the whole it is a useful working model.

Metathesis of Functionalized Alkenes

This is a relatively new venture, and most metathesis catalysts are unable to withstand the poisoning effects of electronegative atoms (O, N, S). The area has been reviewed, and has significance in perfume chemistry, insect pheromone chemistry and the preparation of flame- and oil-resistant elastomers and speciality plastics. Most work has been done on readily available olefinic esters, such as methyl oleate, olive oil (mainly triglycerides of oleic acid), linseed and soybean oils (triglycerides of linoleic and linolenic acids). Examples are shown below.[19]

$$CH_3(CH_2)_7CH{=}CH(CH_2)_7CO_2Me \xrightleftharpoons[70\ ^\circ C]{WCl_6/Me_4Sn}$$

methyl
oleate

$$Me(CH_2)_7CH{=}CH(CH_2)_7Me \quad + \quad MeO_2C(CH_2)_7CH{=}CH(CH_2)_7CO_2Me$$

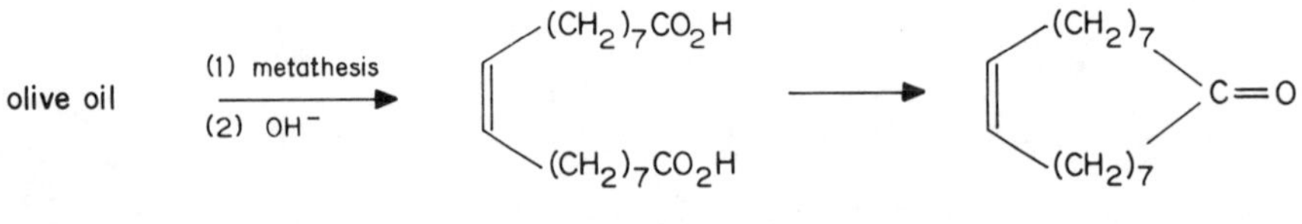

olive oil $\xrightarrow[\text{(2) OH}^-]{\text{(1) metathesis}}$

$$CH_2{=}CH(CH_2)_7CH_2OAc \xrightarrow[R_4Sn]{WCl_6}$$

$$AcOCH_2(CH_2)_7CH{=}CH(CH_2)_7CH_2OAc + CH_2{=}CH_2 + CH_3CH(Cl)(CH_2)_7CH_2OAc$$

$$\text{CH}_3(\text{CH}_2)_7\text{CH}{=}\text{CH}(\text{CH}_2)_7\text{CH}_2\text{OAc} \xrightarrow[\substack{\text{or} \\ (\text{EtO})_2\text{MoCl}_3 / \text{Et}_3\text{B} \\ 178\ °\text{C}}]{\text{WCl}_6 / \text{Et}_3\text{B}}$$

$$\text{AcOCH}_2(\text{CH}_2)_7\text{CH}{=}\text{CH}(\text{CH}_2)_7\text{CH}_2\text{OAc} \quad + \quad \text{CH}_3(\text{CH}_2)_7\text{CH}{=}\text{CH}(\text{CH}_2)_7\text{CH}_3$$

A particularly interesting application of the metathesis reaction, using a catalyst prepared from $WOCl_4$ or WCl_6 and dimethyltitanocene, (Cp_2TiMe_2), is the intramolecular reaction leading to macrolide formation. The yield is low but it is possible that further refinement of catalyst would lead to very respectable yields.[20]

It may be anticipated that development of new metathesis catalysts will allow application to highly functionalized molecules with important consequences for organic synthesis.

INSERTION REACTIONS[21]

The term 'insertion reaction' is, in most cases, an incorrect description, at least in mechanistic terms. It is simply meant to describe the overall stoichiometric result, and is amply illustrated by reference to the following general equation:

$$\text{X} \ + \ \text{M–L} \longrightarrow \text{M–X–L}$$

Thus, the species X is apparently inserted into the M—L bond. The classical example is the conversion of methylmanganesepentacarbonyl to σ-acetyl-manganesepentacarbonyl, first discovered in 1957.[22]

$$\text{MeMn(CO)}_5 \ + \ \text{CO} \ \rightleftarrows \ \text{Me}\overset{\overset{\text{O}}{\|}}{\text{C}}\text{Mn(CO)}_5$$

While the mechanism of this particular insertion reaction is now known, there are very few other examples with a comparable detailed knowledge. The most widely studied insertion reactions are those of carbon monoxide, although other unsaturated molecules, e.g. SO_2, exhibit similar behavior in many cases. Carbon monoxide insertion reactions have considerable potential for organic synthesis, since breakage of the acyl—metal bond can lead to aldehydes and acid derivatives corresponding to homologation and functionalization of an alkyl derivative. It may be noted that only a few metals are not yet known to give insertions of CO into M—C σ-bonds. Relatively few kinetically stable CO inserted products are known for Ru, Co, Rh, Ni and Pd, so it is not surprising to find that these metals are usually effective *catalysts* for the carbonylation of organic substrates. Indeed, CO insertion is a key step in many industrial-scale catalytic processes (see later). It is found that 5d metals are normally less inclined to undergo CO insertion than 3d or 4d metals of the same subgroup. This has been attributed to the greater strength of the M—C bond for the later metals (e.g. dissociation energies: Mn—Me 27.9 kcal mol^{-1}; Re—Me 53.2 kcal mol^{-1}.

In this section we shall consider in some detail the insertion and de-insertion reaction of CO and SO_2.

Carbon Monoxide Insertions

The general reaction is:

A variety of ligands L can be used, e.g. alkylphosphines, which at once gives a clue to the mechanism, e.g.

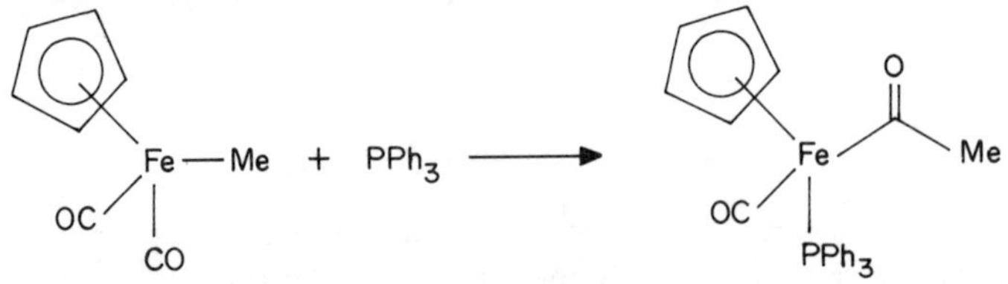

The reverse of CO insertion is termed decarbonylation and proceeds as follows:

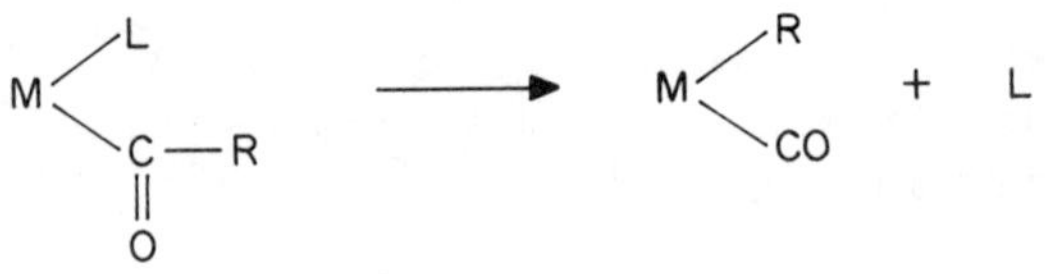

or

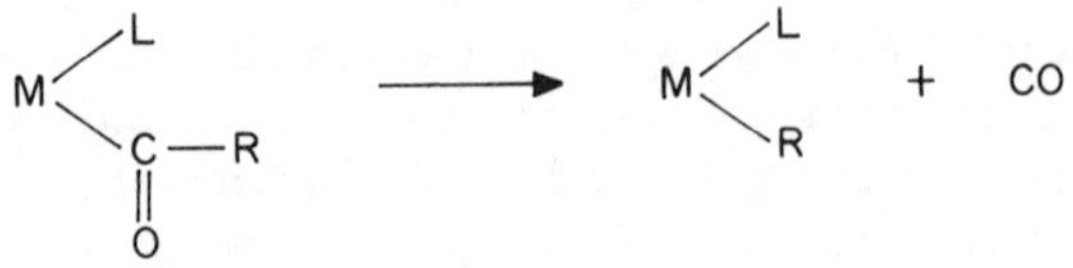

e.g.

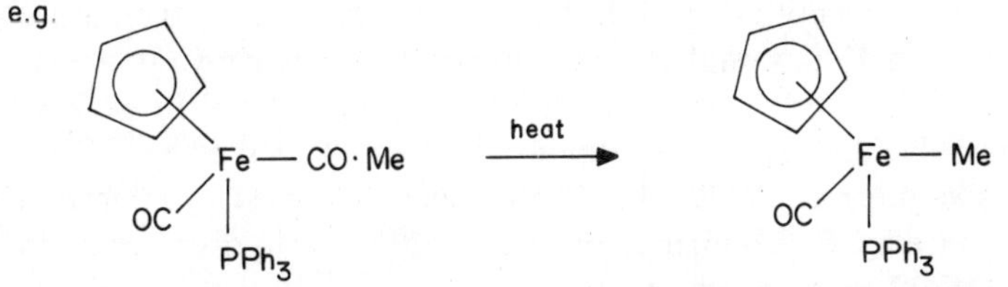

In fact, decarbonylation also proves to be a synthetically useful operation, since carbon—carbon bond formation sometimes involves the use of anion-stabilizing groups which are not required in the final product. For example, in an approach to chlorothricolide synthesis, Ireland's group produced just such a compound, which required decarbonylation of an aldehyde.[23] This was readily effected using chlorobis(triphenylphosphine)rhodium(I) (Wilkinson's catalyst):

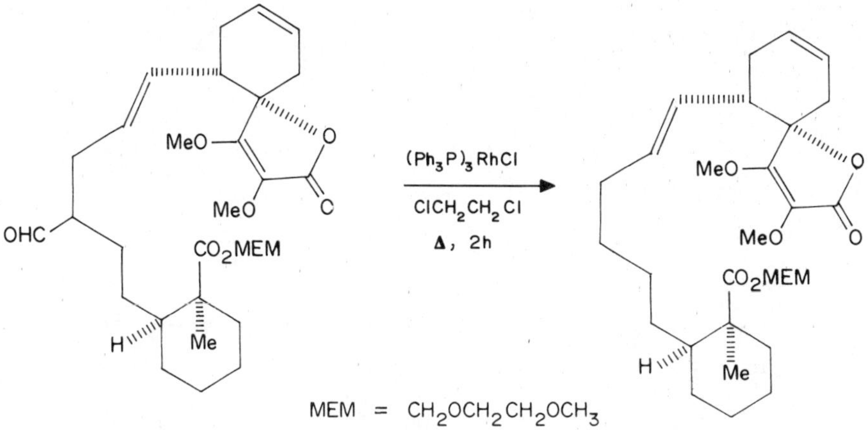

$$MEM = CH_2OCH_2CH_2OCH_3$$

Decarbonylation of aldehydes in this way involves prior oxidative addition of the aldehyde C—H bond to the metal, followed by CO deinsertion and then reductive elimination of the alkane derivatives:

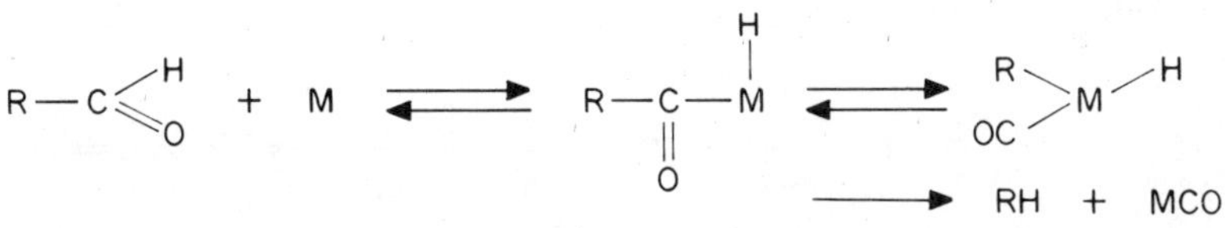

Naturally, appropriate ligand dissociations in each step ensure an adherence to the 18-electron rule. Decarbonylation can, therefore, be effected under both thermal and photochemical reaction conditions.

Kinetic studies of CO insertion indicate the following general mechanism:

Reactions which are both dependent on and independent of the nature of the solvent are known. For example, the reaction between $MeMn(CO)_5$ and aniline is 10^4 times faster in dimethylformamide than in mesitylene. In general, polar coordinating solvents are found to enhance the rate of CO insertion. This suggests that the above intermediate is a 16-electron species formed by migration of the alkyl group onto a CO ligand, in many ways analogous to organic counterparts such as the Wolff rearrangement, etc., e.g.

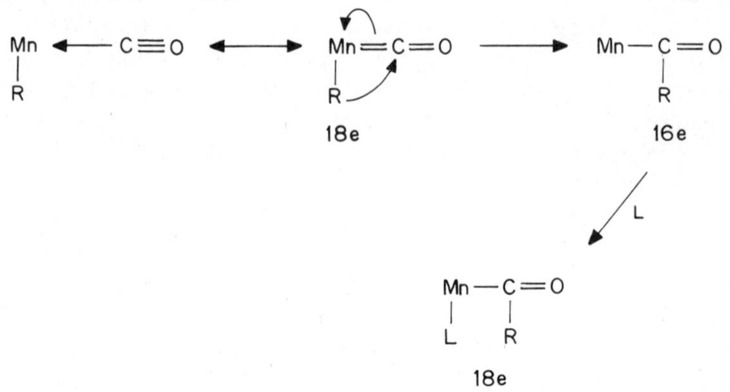

The *reverse* of this reaction, i.e.

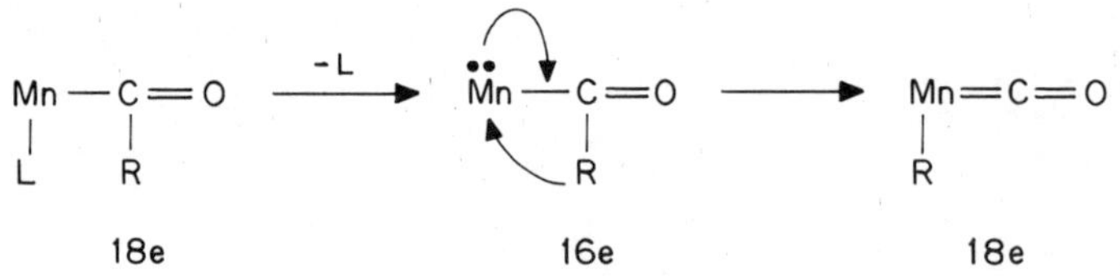

in which an electron-deficient species is produced by a dissociative process, and then undergoes alkyl migration to complete the required electron complement, compares very well with the Wolff rearrangement, and similar alkyl migrations to electron-deficient centers:

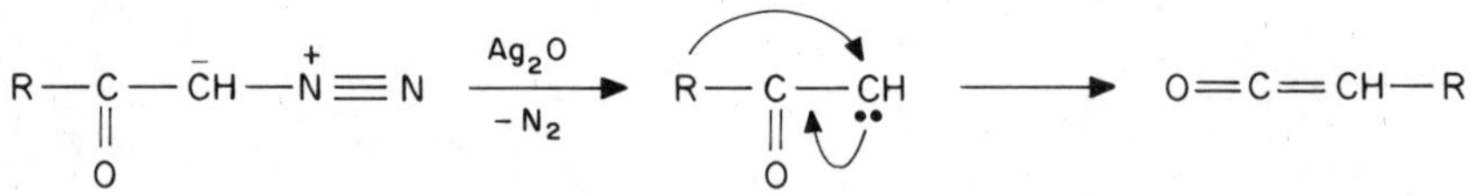

With this analogy in mind we might expect similar trends, and these are often observed. For example, the migratory aptitude of the R group usually follows the trend $i\text{-}Pr > Et > Me$ in both the CO insertion/deinsertion and the related organic reactions.

Detailed studies of the mechanism and stereochemistry of the alkylmanganese-pentacarbonyl insertion reaction have been made, and we now consider these.

The use of ^{13}C-labelled carbon monoxide gives the following result, readily determined by infrared spectroscopy:

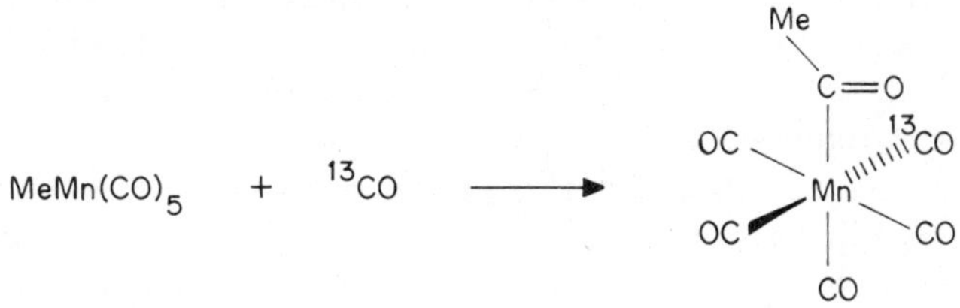

$$MeMn(CO)_5 \quad + \quad {}^{13}CO \quad \longrightarrow$$

Thus, the incoming CO occupies a position *cis* to the acetyl group. No label is found in the acetyl group, supporting the alkyl migration mechanism, and the above result indicates that the ^{13}CO occupies the coordination site vacated during the migration step. The same result is obtained when the insertion is induced by a phosphine or phosphite ligand, e.g.

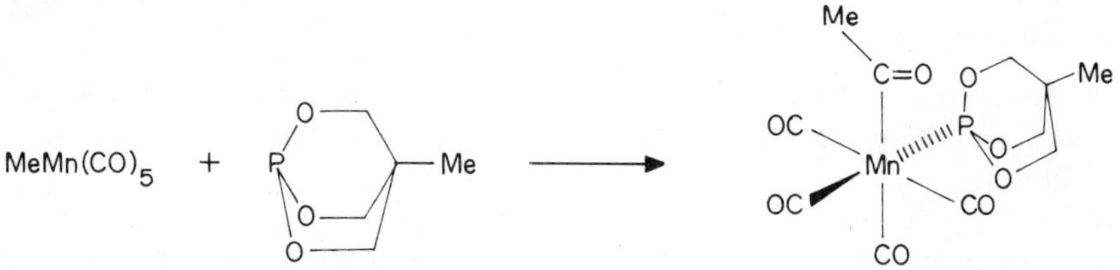

$$MeMn(CO)_5 \quad + \quad \longrightarrow$$

The alkyl migration concept is also supported by studies of both CO insertion and the reverse reaction using labelled complexes. For example, the following results are all consistent with alkyl migration, as opposed to direct CO insertion/deinsertion into the M—C σ-bond. In particular, note that during decarbonylation of the acyl complex the migration can only occur consequent upon dissociation of a ligand *cis* to the acyl group.

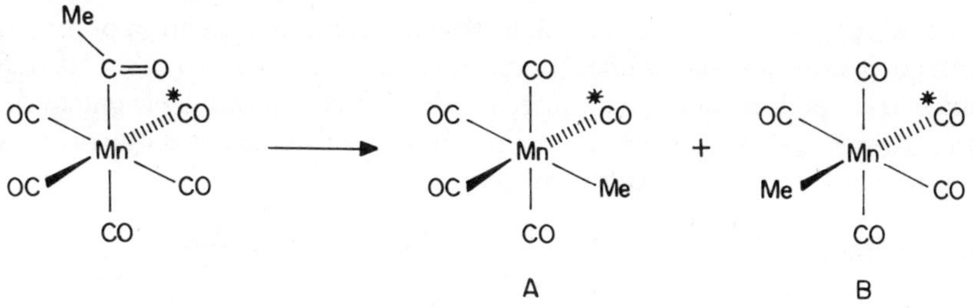

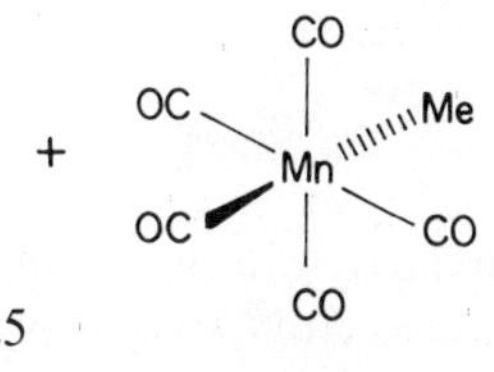

A

B

C

for alkyl migration:
 Theoretical probabilities: A 0.5; B 0.25; C 0.25
 Found: labelled product ratio A:B = 2:1

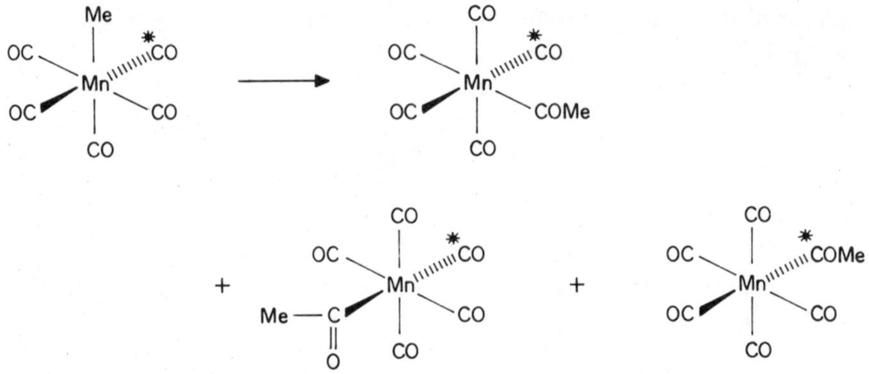

for alkyl migration:
  Theoretical ratio: 2:1:1
  Observed ratio: 2:1:1

It has also been found, using complexes containing optically active alkyl groups, that there is no loss of optical activity after prolonged time during the equilibrium:

$(+)_D - PhCH_2 \overset{*}{C}H(Me) COMn(CO)_5 \rightleftharpoons (+)_D - PhCH_2 \overset{*}{C}H(Me) Mn(CO)_5 + CO$

Also, the following decarbonylation is found to proceed without loss of optical activity:

$(-)_{546} - CpFe(CO)_2 CO\overset{*}{C}H(Me)Ph + Rh(PPh_3)_3Cl$

$\longrightarrow (+)_{546} - CpFe(CO)_2 \overset{*}{C}H(Me)Ph + Rh(CO)(PPh_3)_2Cl + Ph_3P$

However, neither of these results tell us whether complete retention or inversion occurs at the asymmetric center. Some very elegant NMR experiments in fact demonstrate that the reaction proceeds with complete *retention* of configuration at the α-carbon, shown by examining the $^1H-^1H$ coupling constants of starting material and product during the following reaction:

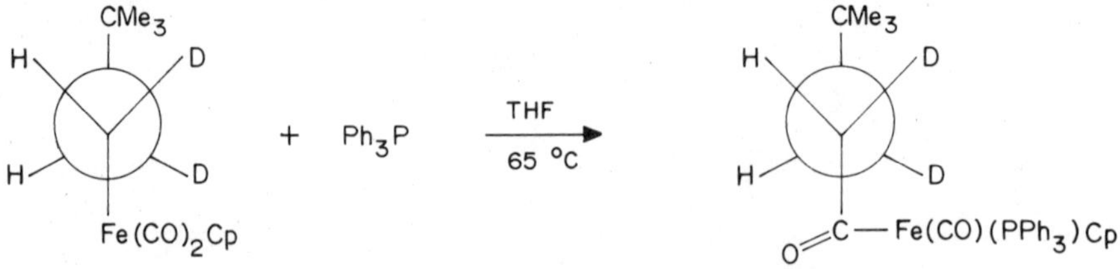

Similarly, a large number of decarbonylations of optically active aldehydes, promoted by Wilkinson's catalyst, proceed with retention of configuration. These observations are again analogous to the counterpart organic migration reactions.

Interestingly, during the following reaction, only partial epimerization occurs at the asymmetric *metal* center.

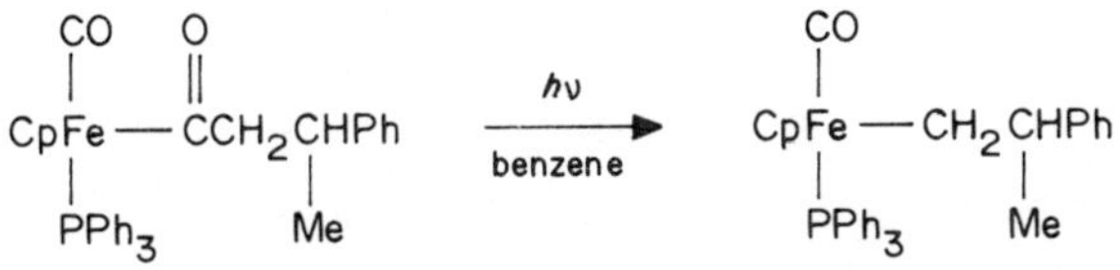

This observation indicates that the stereochemical integrity of the coordinatively unsaturated intermediate is maintained for at least the lifetime of reassociation of the incoming ligand, and any fluxional changes occur relatively slowly.

A recent illustration[24] of the potential utility of carbonyl insertion reactions involving functionalized substrates is provided by the palladium-catalyzed conversion of acetylenic alcohols to the α-methylenebutyrolactone grouping in a number of biologically important molecules (e.g. vernolepin, frulhanolide, eriolanin) so this method could prove very useful, e.g.

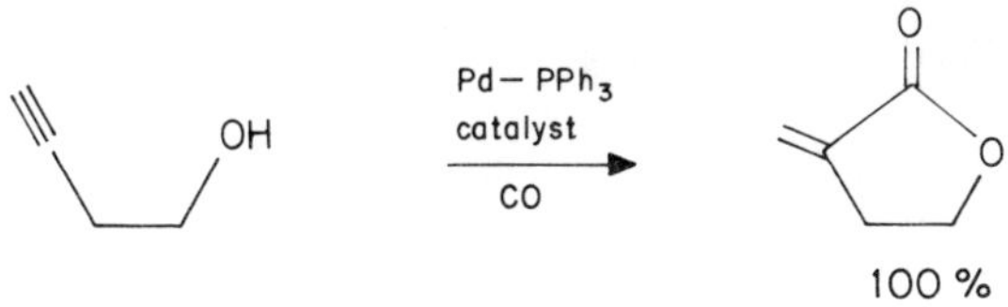

100 %

Proper development of a range of CO insertion reactions of this type might well result in a number of extremely powerful synthetic methods.

Sulphur Dioxide Insertion Reactions[25]

Insertion of sulphur dioxide into a transition metal—carbon σ-bond is now a well known process. The overall reaction is

$$M—R \ + \ SO_2 \longrightarrow M—(SO_2)—R$$

However, unlike CO insertion reactions, there are several possible structures for the final product, *viz.*

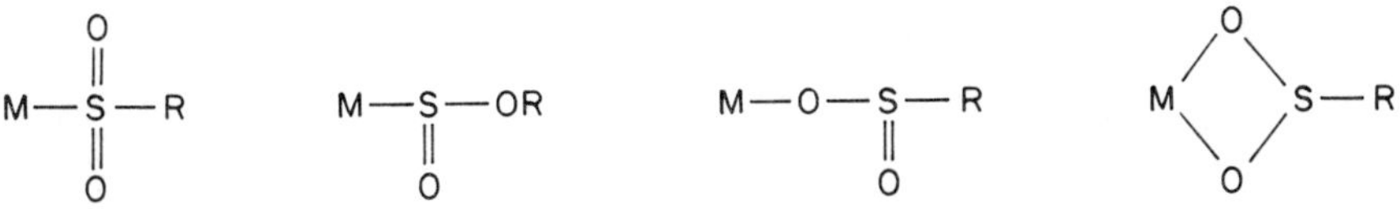

The assignment of structure to the actual product is readily achieved using infrared and NMR spectroscopy. In fact, the most common product is the S-sulphinate complex, although the O-sulphinate complex can be detected during reactions which produce the S-sulphinate, and it is widely recognized that this is an intermediate in a two-step reaction:

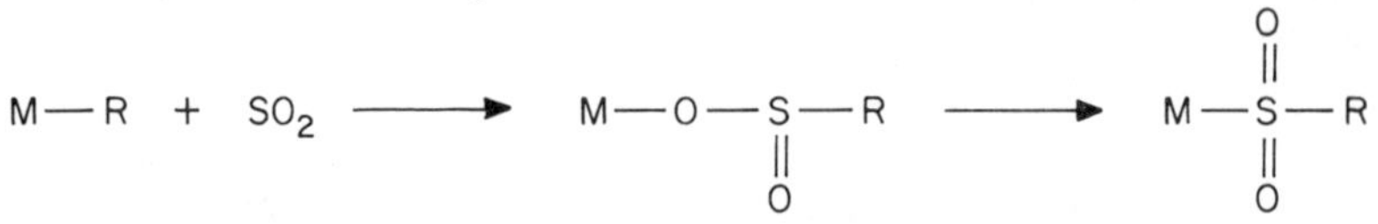

$$M-R \ + \ SO_2 \longrightarrow M-O-\underset{\underset{O}{\|}}{\overset{}{S}}-R \longrightarrow M-\underset{\underset{O}{\|}}{\overset{\overset{O}{\|}}{S}}-R$$

The last step can be thought of as a migration of the metal moiety on to sulphur. This observation, together with stereochemical considerations (see below) suggest that SO_2 insertion is mechanistically different from CO insertion, at least in some cases, in that there appears to be a direct concerted insertion into the M—C bond. In many cases, precoordination of SO_2 to the metal appears unlikely. However, there are examples where SO_2 does act as a ligand (but does not then undergo insertion). In fact, SO_2 can undergo the following reactions as well as insertion.

(i) Addition to the metal, usually occurring for coordinatively unsaturated complexes with strong M—C bonds, e.g.

$$MeC \equiv C - Ir(CO)(PPh_3)_2 \ + \ SO_2 \longrightarrow MeC \equiv C - Ir(CO)(SO_2)(PPh_3)_2$$

(ii) Addition to another coordinated ligand:

$$MePt(PPh_3)_2I \ + \ SO_2 \longrightarrow MePt(PPh_3)(ISO_2)$$

(iii) Cycloaddition to the hydrocarbon fragment:

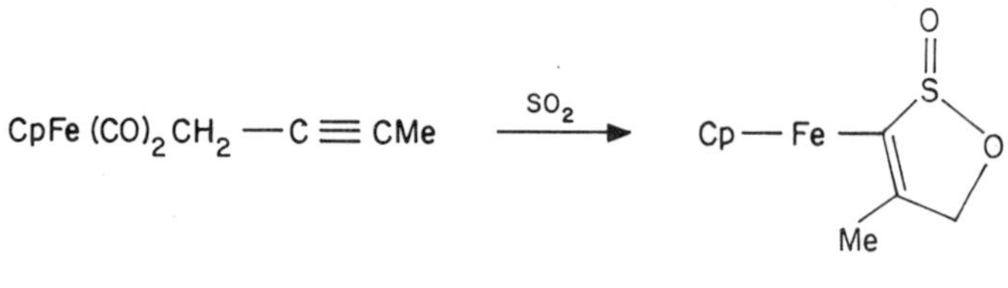

$$CpFe(CO)_2CH_2 - C \equiv CMe \xrightarrow{SO_2} Cp-Fe$$

Sulphur dioxide insertion is not limited to M—L σ-bonds (although this is the most common substrate), there being a number of examples of insertion into M—C π-bonds (also of polyhapto ligands), M—M, M—O and M—H bonds. The reverse of SO_2 insertion, desulphurization is also found, i.e.

$$M(SO_2)R \longrightarrow MR \ + \ SO_2$$

As with carbon monoxide insertion, a number of mechanistic studies have been attempted. Some of the important results are as follows. When steric effects are unimportant, the reactivity of complexes $CpFe(CO)_2CH_2X$ increases with an increase in the electron-releasing tendency of X, i.e., $X = C_6F_5 \ll Ph < C_6H_4OMe\text{-}p$, and $X = CN < OMe < i\text{-Pr}$. This appears to indicate an electrophilic nature of the insertion of SO_2 into the M—C bond. There is a dependence on the steric requirements of the R group in $CpFe(CO)_2R$, i.e. rates are $R = Me > CH_2CHMe_2 > CH_2CH_2CMe_3 > CH_2CMe_3 > CMe_3$. This is indicative of a backside approach of SO_2, a proposition which is confirmed by studies using complexes with centers of asymmetry, e.g.

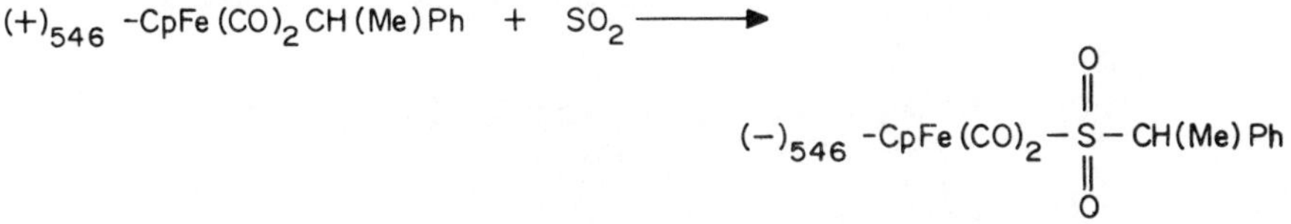

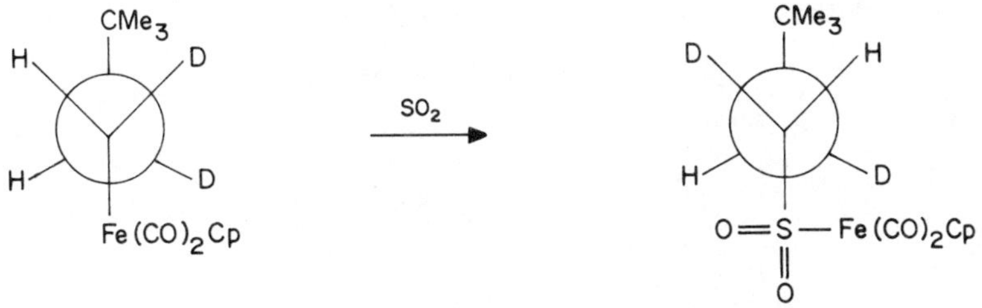

$(+)_{546}$ –CpFe$(CO)_2$CH(Me)Ph + SO$_2$ ⟶

$$(-)_{546} \text{ –CpFe}(CO)_2 - \overset{\overset{O}{\|}}{\underset{\underset{O}{\|}}{S}} - CH(Me)Ph$$

The first of these experiments indicates stereospecificity of the reactivity, but tells us nothing about whether inversion or retention occurs at the α-carbon, whilst NMR studies on the second reaction confirm that the reaction proceeds with inversion. These results have been well rationalized using frontier molecular orbital concepts by Inagaki et al.[25b]

A more sophisticated mechanism than the originally postulated intermediacy of a contact ion pair was presented. Thus, Inagaki et al. described the process as an initial interaction between the bonding orbital of the Fe—R bond and the LUMO of SO$_2$ (which has the largest amplitude at sulphur) which leads to a weakening and lengthening of the Fe—R bond. This causes charge separation between the iron and the carbon, and the resulting transition state is stabilized by sequential interaction among the HOMO of the anionic carbon, the LUMO of the cationic iron and the LUMO of SO$_2$. This can be represented diagrammatically:

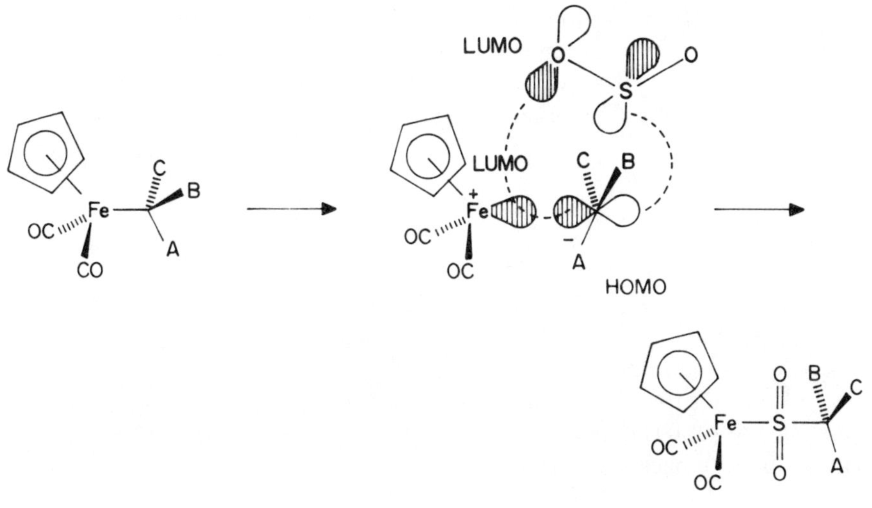

Indeed, we could even suppose that the whole process corresponds to a concerted antarafacial ($\sigma2_a + \pi2_s$) process, as is indicated by the above diagram.

IMPORTANT CATALYTIC PROCESS BASED ON CARBONYL INSERTION

Hydroformylation (the 'oxo' reaction)

This is extremely important industrially for the preparation of aldehydes from olefins. The basic reaction is

$$CH_2{=}CH_2 \;+\; H_2 \;+\; CO \xrightarrow{\text{catalyst}} CH_3CH_2CHO$$

$$CH_3CH{=}CH_2 \;+\; H_2 \;+\; CO \longrightarrow CH_3CH_2CH_2CHO$$

Butyraldehyde is an extremely versatile chemical intermediate (giving plasticizers). Originally Co catalysts were used, but more recently Rh catalysts have been investigated.

The cobalt catalyst is octacarbonyldicobalt produced by reductive carbonylaction of cobalt oxide *in situ*.

Accepted mechanism (usual temperature *ca.* 150 °C, pressure $CO + H_2 > 300$ atm).

$$Co_2(CO)_8 \;\rightleftharpoons\; Co_2(CO)_7 \;+\; CO$$

$$Co_2(CO)_7 \;+\; H_2 \;\rightleftharpoons\; \underset{16e}{HCo(CO)_3} \;+\; \underset{18e}{HCo(CO)_4}$$

Then:

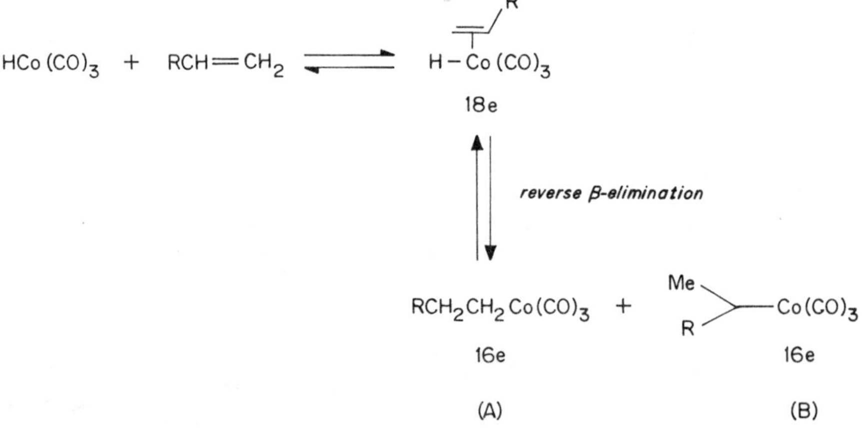

$$HCo(CO)_3 \;+\; RCH{=}CH_2 \;\rightleftharpoons\; \underset{18e}{H-Co(CO)_3}\!\!\overset{\displaystyle R}{\underset{\displaystyle}{\Vert}}$$

$$\Big\updownarrow \text{ \textit{reverse β-elimination}}$$

$$\underset{16e}{RCH_2CH_2\,Co(CO)_3} \;+\; \underset{16e}{\overset{Me}{\underset{R}{>}}\!\!-Co(CO)_3}$$

$$\text{(A)} \qquad\qquad\qquad\qquad \text{(B)}$$

Note that at this point the cycle can produce either linear (from **A**) or branched (**B**) products:

Linear:

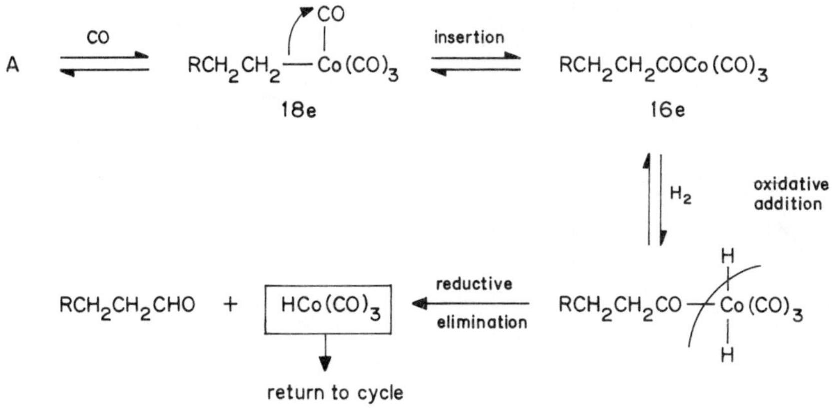

$$A \rightleftharpoons \xrightarrow{CO} RCH_2CH_2-Co(CO)_3 \underset{18e}{} \xrightarrow{insertion} RCH_2CH_2COCo(CO)_3 \underset{16e}{}$$

H_2 oxidative addition

$$RCH_2CH_2CHO + \boxed{HCo(CO)_3} \xleftarrow[elimination]{reductive} RCH_2CH_2CO-\overset{H}{\underset{H}{Co}}(CO)_3$$

return to cycle

Branched:

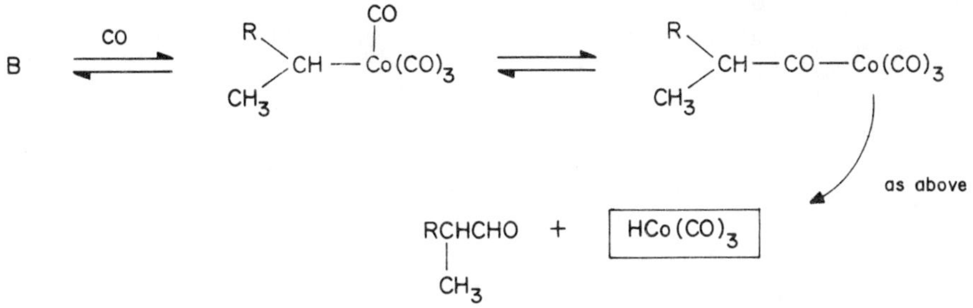

$$RCHCHO + \boxed{HCo(CO)_3}$$
$$\underset{CH_3}{|}$$

From the above, it can be seen that a better linear/branched ratio can be obtained by using bulky ligands on the Co (therefore disfavoring **B** on steric grounds).* This is observed when $HCo(CO)_3(PBu_3)$ is used as catalyst. This also leads directly to hydrogenation:

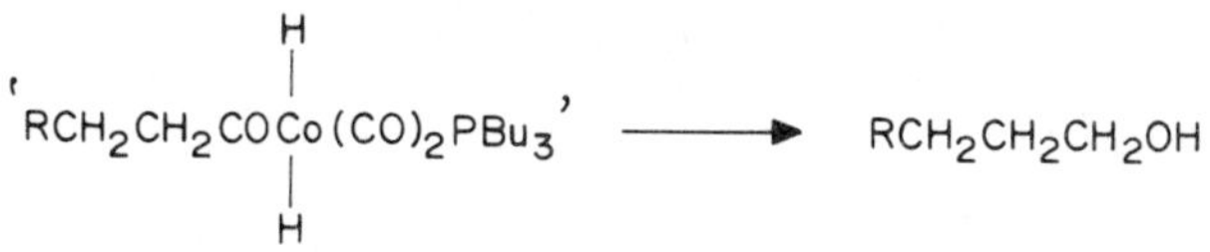

$$RCH_2CH_2CO\overset{H}{\underset{H}{Co}}(CO)_2PBu_3' \longrightarrow RCH_2CH_2CH_2OH$$

Rhodium catalysts have several advantages over the Co systems, e.g. use of $HRh(CO)(PPh_3)_3$. The mechanism has been studied by Wilkinson and co-workers.[26]

Use of this catalyst gives excellent selectivity for linear aldehydes, and is effective at 10–20 atm and *ca.* 100 °C. Little or no hydrogenation of the aldehyde is observed.

---

* This effect might also be due to the increased electron density on the metal, altering the polarization of the Co—H bond from CoH in $HCo(CO)_4$ to CoH in $HCo(PBu_3)(CO)_3$.

58

Mechanism :

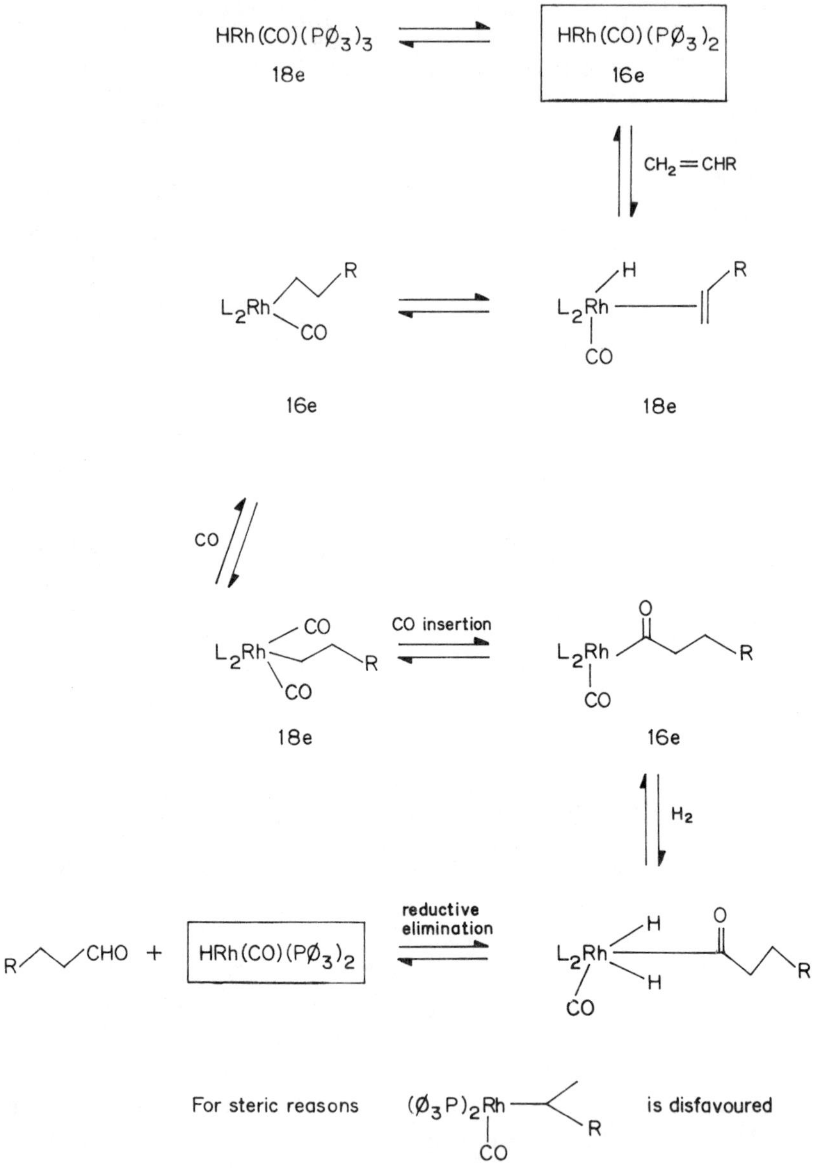

$$HRh(CO)(P\emptyset_3)_3 \rightleftharpoons \boxed{HRh(CO)(P\emptyset_3)_2}$$

18e   16e

$CH_2\!=\!CHR$

16e   18e

CO

CO insertion

18e   16e

H_2

reductive elimination

$R\!\diagup\!\!\diagdown\!\!\diagup CHO$ + $\boxed{HRh(CO)(P\emptyset_3)_2}$ $\rightleftharpoons$

For steric reasons $(\emptyset_3P)_2Rh\!-\!\!\diagup\!\!\diagdown_R$ is disfavoured

CO

Acetic Acid Synthesis via Carbonyl Insertion

Methanol is a readily available feedstock for the chemical industry, so its conversion to acetic acid and acetic anhydride, both important over a range of industry (polymers, pharmaceuticals, dyestuffs, pesticides), represents an extremely important process.

Currently, a commercial synthesis of acetic acid uses a rhodium catalyst to effect the carbonylation of methanol. Usually a rhodium iodide complex is employed, this serving as a source of HI, which converts methanol to methyl iodide, which is the actual reactant. The process occurs as shown in the following equations:

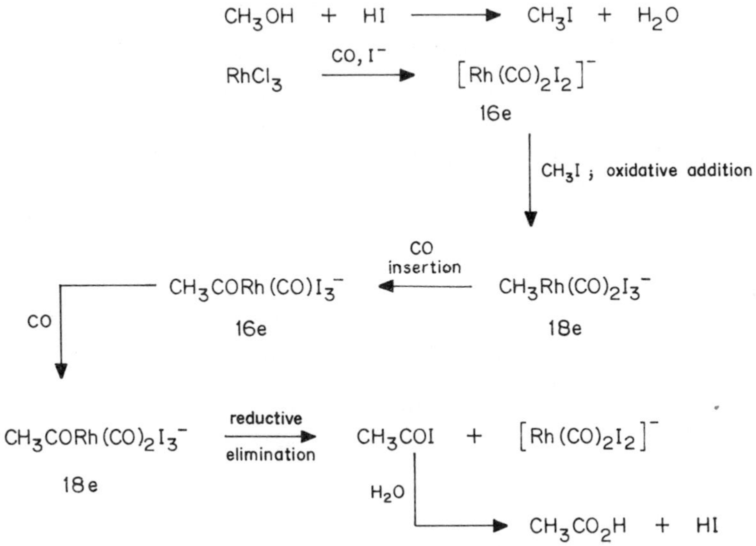

$$CH_3OH \;+\; HI \longrightarrow CH_3I \;+\; H_2O$$

$$RhCl_3 \xrightarrow{CO,\,I^-} [Rh(CO)_2I_2]^-$$
$$16e$$

CH_3I ; oxidative addition

$$CH_3CORh(CO)I_3^- \xleftarrow[\text{insertion}]{CO} CH_3Rh(CO)_2I_3^-$$
$$16e \qquad\qquad\qquad\qquad 18e$$

CO

$$CH_3CORh(CO)_2I_3^- \xrightarrow[\text{elimination}]{\text{reductive}} CH_3COI \;+\; [Rh(CO)_2I_2]^-$$
$$18e \qquad\qquad\qquad H_2O$$

$$CH_3CO_2H \;+\; HI$$

Conversion of Alkenes to Carboxylic Acids and Esters

The basic process is similar to the hydroformylation reactions (above) but the acyl–metal complex is intercepted by H_2O, ROH or R_2NH to give hydrolytic breakdown and formation of acids, esters or amides, e.g.

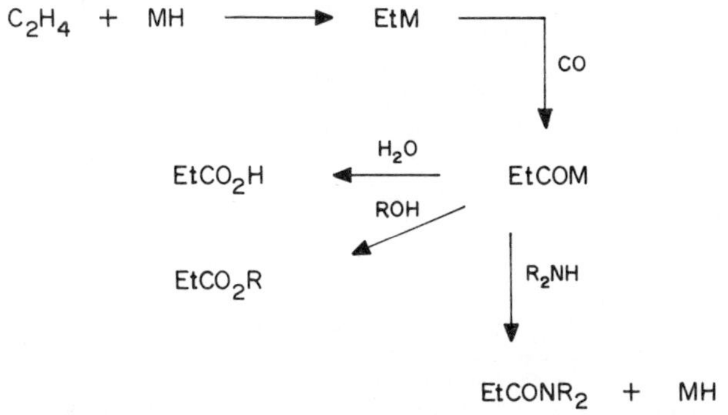

$$C_2H_4 \;+\; MH \longrightarrow EtM$$

CO

$$EtCO_2H \xleftarrow{H_2O} EtCOM$$

ROH

$$EtCO_2R$$

R_2NH

$$EtCONR_2 \;+\; MH$$

Most commercial processes use $Co_2(CO)_8$ [giving $HCo(CO)_3$] as catalyst. Palladium catalysts are useful, especially on the laboratory scale, because they can be employed at lower pressures. Similar mechanisms probably operate for both Co and Pd complexes. The mode of cleavage of the acyl–M bond is possibly interesting, e.g.

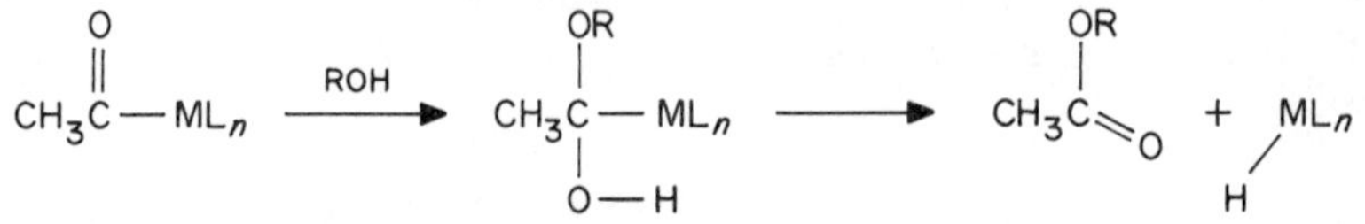

analogous to β-elimination. A similar application of Pd-catalyzed carboxylation, forming the Pd–aryl σ-complex by oxidative addition of ArI, is well known and has recently been employed by Tsuji's group in a synthesis of zearalenone, an important veterinary compound, and other interesting macrocyclic lactones,[27] e.g.

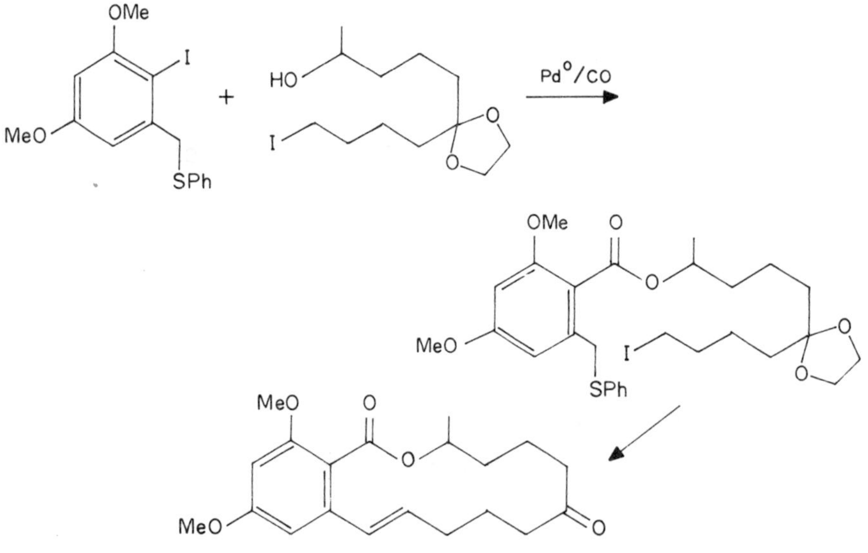

zearalenone

The carboxylation of terminal acetylenes catalyzed by $PdCl_2$ may not always proceed analogously to olefin carboxylation. The acetylene has an acidic proton which enables it to form an acetylene–Pd σ-complex directly:[28]

$$RC\equiv CH \xrightarrow{Pd^{II}} (RC\equiv CPd^{II}) \xrightarrow[R'OH]{CO} RC\equiv CCO_2R'$$

The 'normal' mode of alkyne carboxylation, however, may be obtained using Ni or Pd catalysts under appropriate conditions, e.g.

$$HC\equiv CH + H_2O \xrightarrow[150\ °C,\ 130\ atm]{Ni(CO)_4} CH_2=CHCO_2H$$

$$HC\equiv CH + CO + MeOH \xrightarrow[O_2]{PdCl_2/thiourea} MeO_2CCH=CHCO_2Me$$

STABILIZATION OF REACTIVE OR UNSTABLE MOLECULES

Whilst it is conceivable that a range of transition metals may be utilized for stabilization of formally unstable systems, by means of complexation, we shall concentrate here mainly on the application of iron(0), since this is probably the most widely used, and the complexes are readily available. The general principles are applicable to other metals.

Cyclobutadiene

Longuet-Higgins and Orgel predicted that the 'antiaromatic' molecule cyclobutadiene might be stabilized as its metal complex. This synthesis of tricarbonylcyclobutadieneiron **1** was accomplished some years later by Emerson *et al.* as outlined below.[29] Diironnonacarbonyl acts to dechlorinate a vicinal dichloride and also to complex to the resultant diene.

(1)

The analogous compound benzocyclobutadienetricarbonyliron (**2**) has also been prepared by a similar method from dibromobenzocyclobutene.

(2)

The complex **1** may be crystallized as pale yellow, low-melting prisms from pentane and shows the usual infrared absorptions at 1985 and 2055 cm^{-1} due to the Fe(CO)$_3$ group (see below), whilst the proton NMR shows a single resonance at δ 3.91. Evidence for the proposed structure, rather than the alternative bis(acetylene)iron–tricarbonyl complex, was obtained from the mass spectrum where no peaks corresponding to the loss of C$_2$H$_2$ fragments are observed.

Tricarbonylcyclobutadieneiron is a readily available source of cyclobutadiene and this, together with the intrinsic interest in the effect of Fe(CO)$_3$ complexation on the chemistry of the ligand, prompted further investigation. It appears to undergo reactions associated with aromatic molecules. Thus, electrophilic substitution, such as Friedel–Crafts acetylation and Vilsmeier formylation, occur readily at room temperature, and the resulting carbonyl compounds undergo the usual conversions with, for example, Grignard reagents, sodium borohydride and in the Wittig reaction, as outlined below.

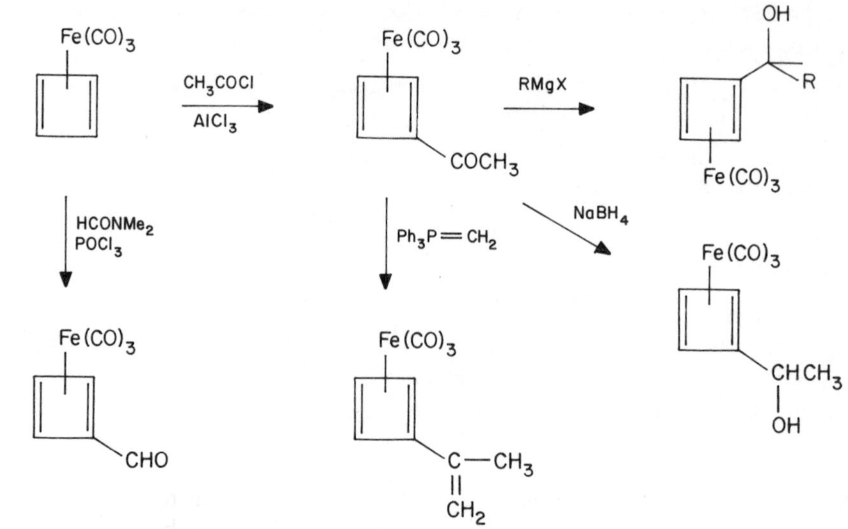

The real synthetic potential of **1** lies in its use as a source of cyclobutadiene, which is very reactive in cycloaddition reactions. Thus, treatment of **1** with cerium (IV) ammonium nitrate disengages the ligand which in the presence of a suitable dienophile, e.g. methyl propiolate, produces the 'Dewar benzene' derivative **3**.

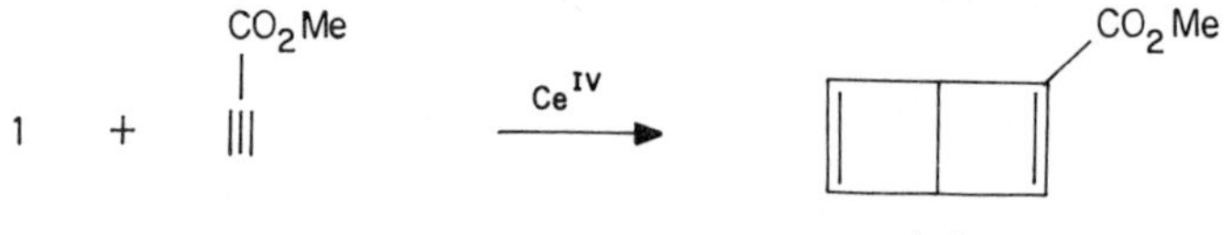

$$1 \quad + \qquad \xrightarrow{\text{Ce}^{\text{IV}}}$$

(3)

The complex has in this way been employed in an elegant synthesis of cubane derivatives, shown below.[30]

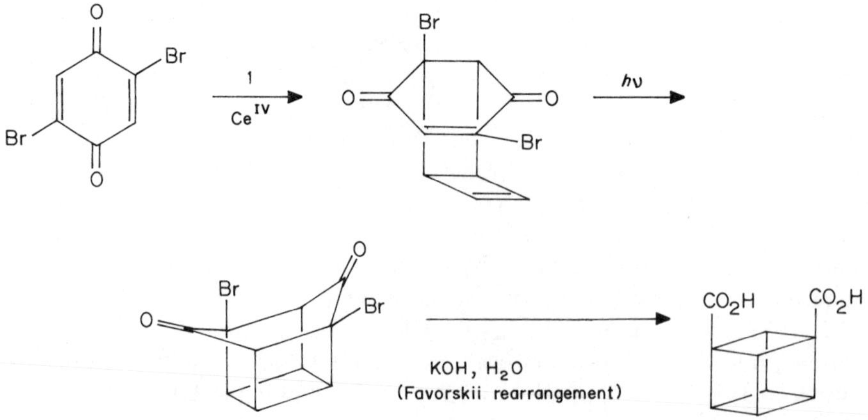

More recently, a simple synthesis of 9-hydroxyhomocubane was reported, in which the cyclobutadiene is liberated using lead tetraacetate as oxidant:[31]

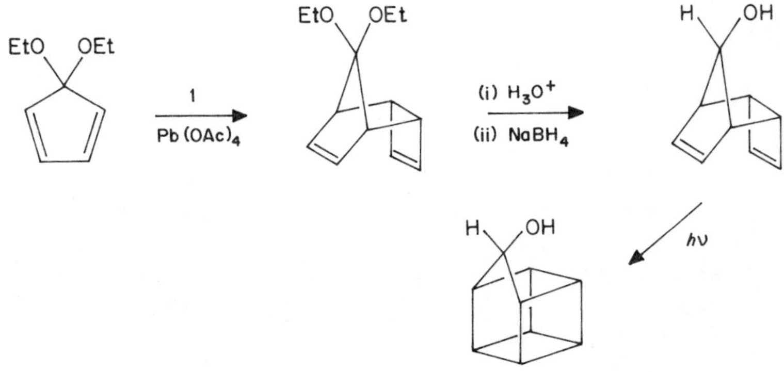

An interesting variation of the intermolecular cycloaddition reaction is the intramolecular reaction described by Grubbs et al.,[32] shown below, but no exploitation for, e.g. natural products synthesis, appears to have been undertaken.

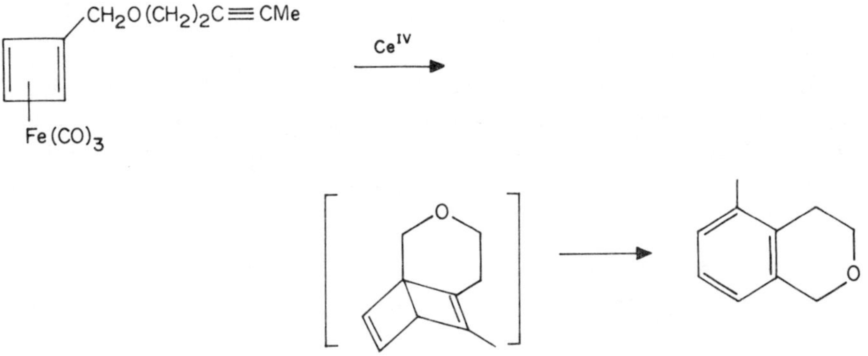

A number of spectroscopic properties of 1 and analogues have been investigated. Both carbon-13 and proton NMR spectra indicate a quadratic structure for the four-membered ring with uniform C—C bond order. Gas-phase electron diffraction studies of tricarbonylcyclobutadieneiron reveal that all Fe—C (2.063 Å) and C—C (1.456 Å) bond lengths are equal, indicating the diene to be square, but some authors consider that this is not inconsistent with a model in which alternating C—C bond lengths differ only by a few hundredths of an angstrom, owing to the errors involved.

Measurement of the Fe(CO)₃ Ir stretching frequencies and ^{13}C–H and H–H coupling constants indicates that no conjugative interaction occurs between the π-electron systems of the cyclobutadiene ligand and a phenyl substituent in complexes such as 4.[33]

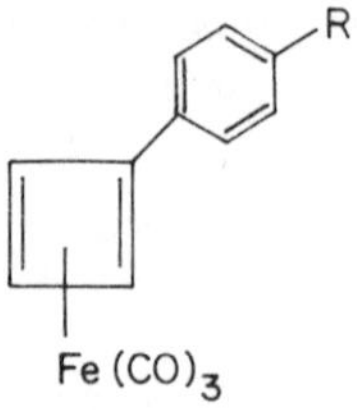

(4) 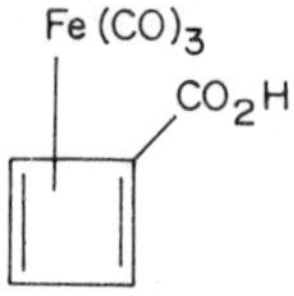 (5) pK_a 5.01 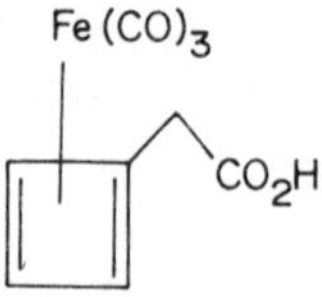 (6) pK_a 5.56

On the other hand, the pK_a values for the two carboxylic acid derivatives **5** and **6** have been interpreted in terms of electron withdrawal by induction and electron release by resonance for the $C_4H_3Fe(CO)_3$ moiety.[34] These conclusions are consistent with the ability of the group to stabilize adjacent carbenium ions as evidenced by solvolysis studies of the chloromethyl–cyclobutadiene complex **7**.

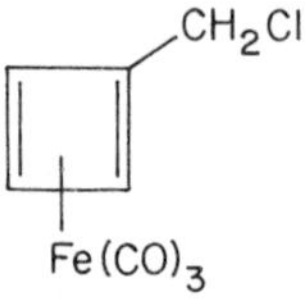

(7)

Trimethylenemethane

Trimethylenemethane is a theoretically interesting molecule, but it is unstable. Evidence for its existence at low temperatures has been obtained, but no source had been produced which might facilitate its further study until the stable tricarbonyliron derivative **8** was prepared.[36] The preparation used a similar dechlorination of chloromethallyl chloride by $Fe_2(CO)_9$ to that employed above, and gave **8** as a pale yellow solid, m.p. 28.4–29.6 °C.

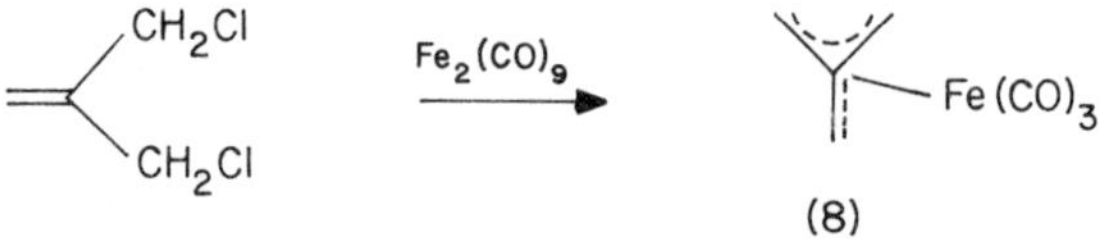

(8)

The compound obtained shows the usual infrared absorptions for the $Fe(CO)_3$ group at 2061, 1995 and 1966 cm^{-1}, these being at slightly higher energy than their identically substituted diene–$Fe(CO)_3$ complexes, indicating that a greater metal–ligand back-bonding occurs. The proton NMR spectrum consists of a sharp singlet at δ 2.00 p.p.m. and mono-substitution leads to the expected non-equivalence of all protons, together with the observation of *W* couplings of 1.9–4.5 Hz. The carbon-13 NMR spectrum shows a singlet at δ 211.6 p.p.m. for the $Fe(CO)_3$ group, a singlet at δ 105.0 p.p.m. due to the central carbon of the trimethylenemethane ligand and a triplet (J_{CH} 158 ± 10 Hz) at δ 53.0 p.p.m. for the methylene carbon atoms. Gas-phase electron diffraction studies showed that **8** has an antiprismatic C_{3v} structure with the methylene groups bent *towards* the Fe atom. Treatment of the trimethylenemethane complex with electrophiles results in addition or substitution as shown below:

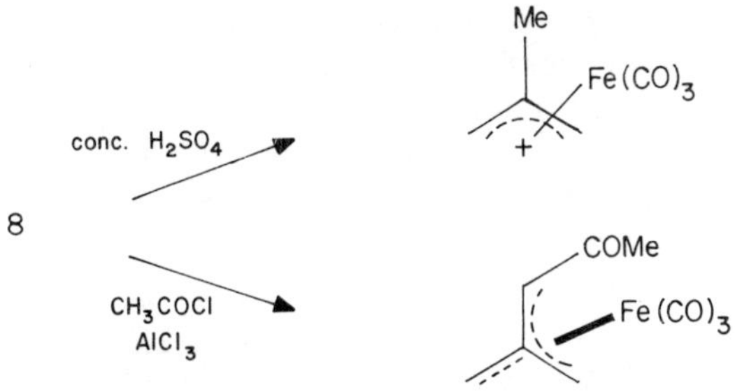

Attempts have been made to liberate the trimethylenemethane from **8** and study its chemistry. Treatment of the complex with cerium (IV) ammonium nitrate in the presence of tetracyanoethylene (tcne) gave a small amount of the corresponding Diels–Alder adduct **9**.

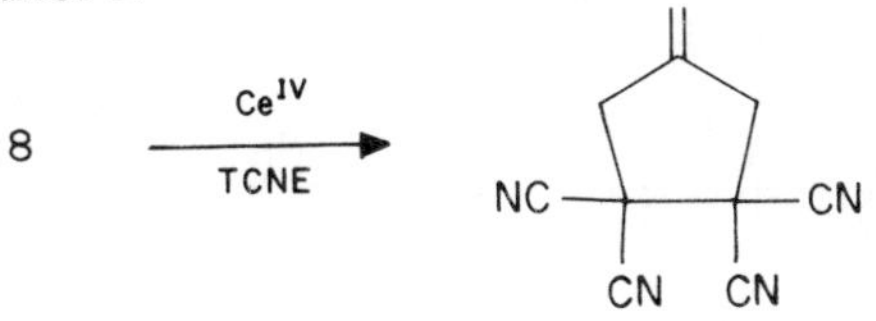

(9)

The free ligand may also be obtained photochemically and used as a diene component in a thermal cycloaddition reaction, or as a dienophile in a photochemical reaction, as outlined below. However, the low yields and high multiplicity of products indicates that a considerable amount of work is still to be done before the complexes will be synthetically useful.

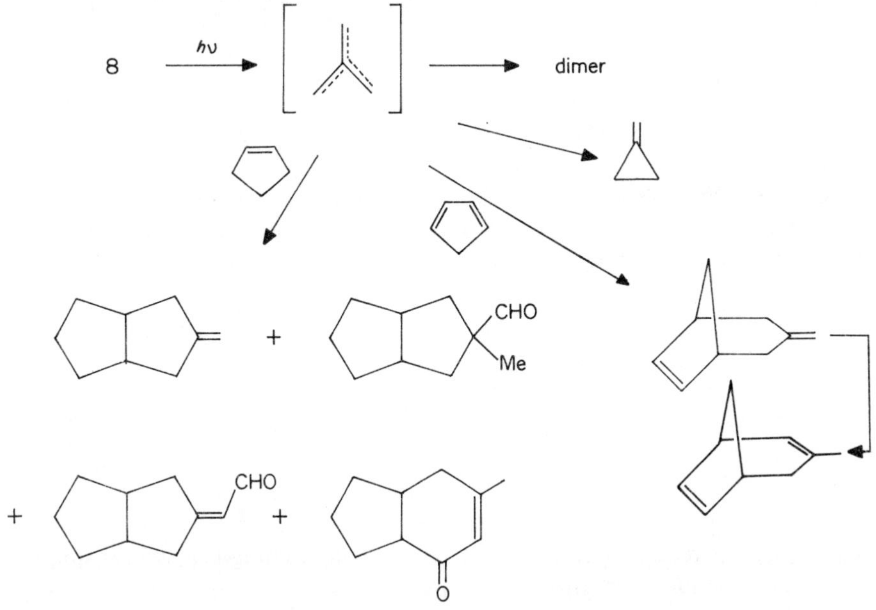

Trost and Chan[37] found that the reaction of 2-acetoxymethyl-allyltrimethyl-silane **10** with a dienophile in the presence of tetrakis(triphenyl-phosphine)palladium results in good yields of the cyclopentane derivatives **11**. The intermediacy of a trimethylenemethane–palladium complex is strongly implicated in this reaction.

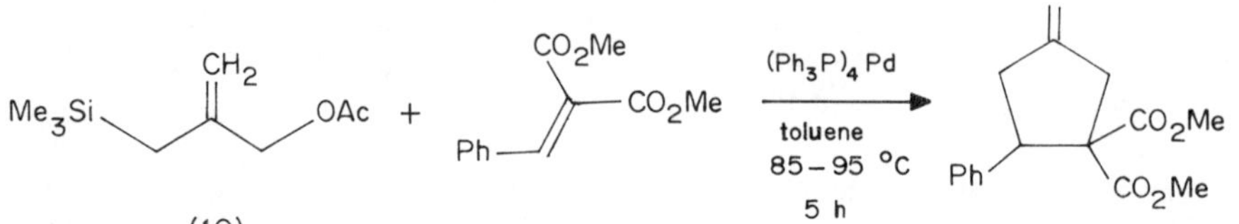

5,6-Dimethylenecyclohexa-1,3-diene

This unstable *ortho*-quinodimethane **12** has been trapped as its maleic anhydride cycloadduct during the zinc reductions of α,α'-dibromo-*o*-xylene.

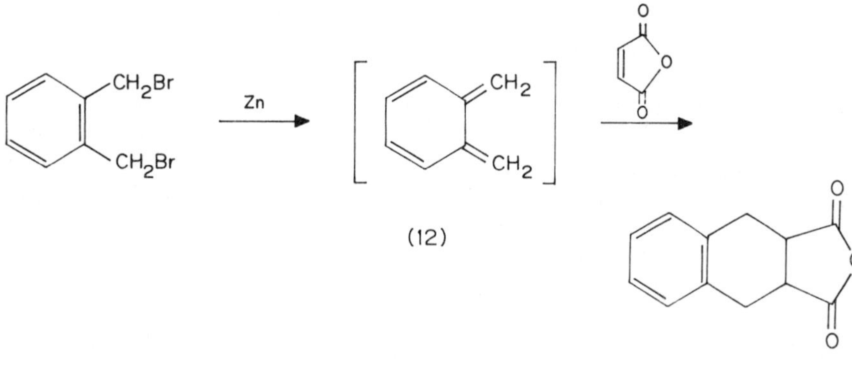

A very low yield of the stable tricarbonyliron derivative **13** has been obtained by treating the above dibromo compound with $Fe_2(CO)_9$. Use of the appropriately substituted dihalo compound leads to substituted complexes.[38] An improved yield (35%) of the parent complex may be obtained by treatment of the dibromide with disodium tetracarbonylferrate, $Na_2Fe(CO)_4$.[39]

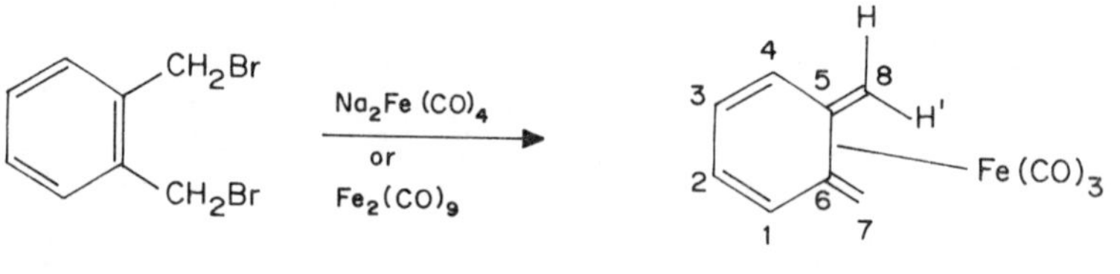

The proton NMR spectrum of **13** is similar to that of butadienetricarbonyliron (see later) showing H-7 and H-8 as a doublet ($J_{gem} = 2.5$ Hz) at δ 2.2 p.p.m., whilst H'-7 and H'-8 occur as a doublet ($J = 2.5$ Hz) at $\delta -0.05$ p.p.m., thus confirming the structure. The remaining protons are found in the aromatic region at δ 7.2 p.p.m.. The complex appears to behave as a normal aromatic compound rather than a diene, undergoing facile acetylation to give **14**, although it behaves as a diene in its reactions with pentacarbonyliron (see chart below). Treatment with aluminum chloride results in a carbonyl insertion reaction to give 2-indenone. The aromatic character of the ring in this complex is understandable in view of the synergic interaction of the complexed diene with Fe leading to fairly high π-bond order in the 5,6-bond (see Chapter 1).

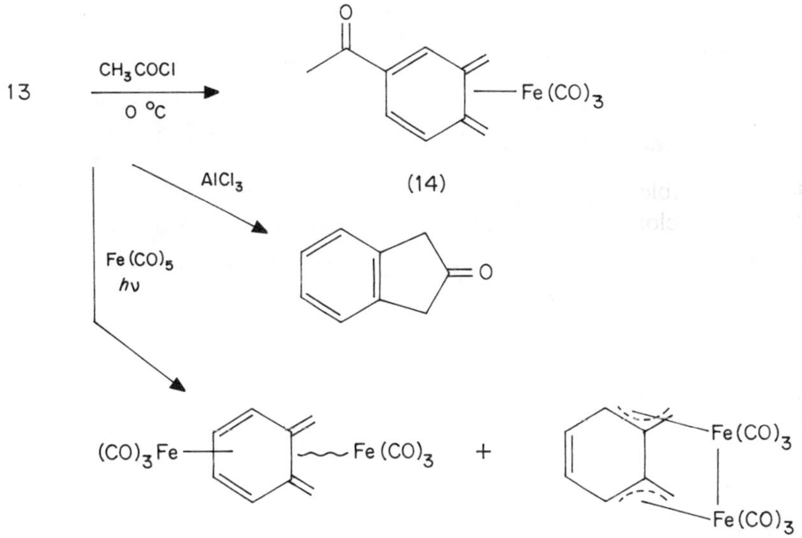

There does not appear to have been any use of the liberated ligand in Diels–Alder reactions. A more rigorous study of the chemistry of this system would be useful in view of the current applications of similar (uncomplexed) systems in steroid synthesis.[40] Usually the o-quinodimethane derivative, e.g. **15** carrying an olefinic group to act as dienophile, is generated by thermolysis of a benzocyclobutene derivative, and undergoes stereoselective cyclization to give an estrone derivative. This has become a very popular method of constructing aromatic A-ring steroids.

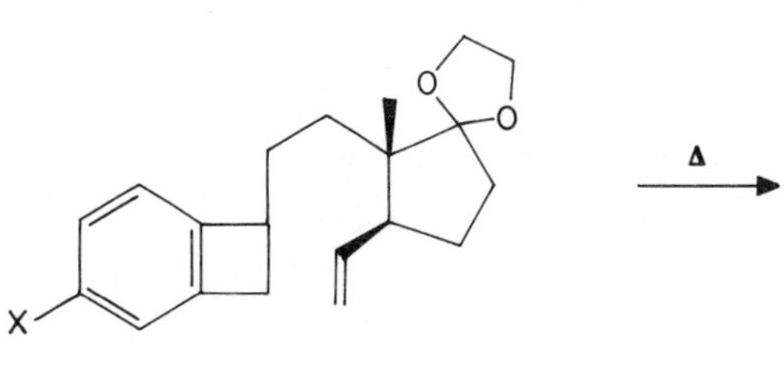

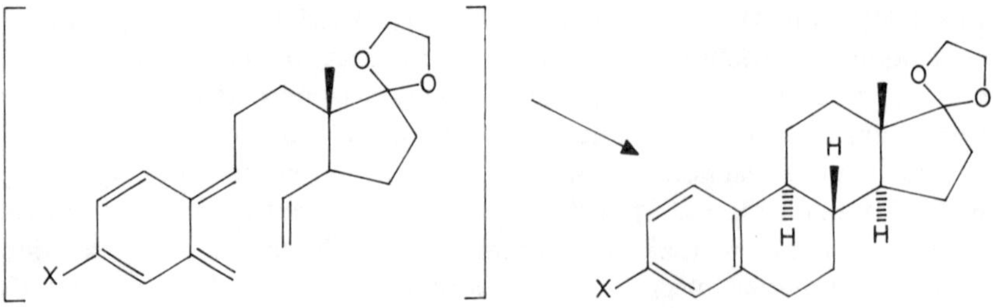

(15)

Cyclopentadienone

Owing to polarization of the carbonyl group, cyclopentadienone is analogous to the cyclopentadienyl cation and therefore shows antiaromatic character. The free dienone cannot be isolated at normal temperatures. Tricarbonylcyclopentadienoneiron **16** is obtained as a yellow crystalline air-sensitive solid, m.p. 114–115 °C, by reaction of acetylene with $Fe(CO)_5$ under pressure in light petroleum or benzene solution.[41] The complex is readily soluble in polar organic solvents and water.

(16)

Substituted acetylenes react with $Fe(CO)_5$ to give substituted derivatives of **16**. Thus, reaction of $Fe(CO)_5$ with phenylacetylene gives **17**. Similarly, a number of other acetylenes, e.g. PhC_2Me, PhC_2SiMe_3, Me_3SiC_2H, $BrC_6H_4C_2H$, $CF_3C_2CF_3$ and PhC_2Ph, give the appropriate complexes.

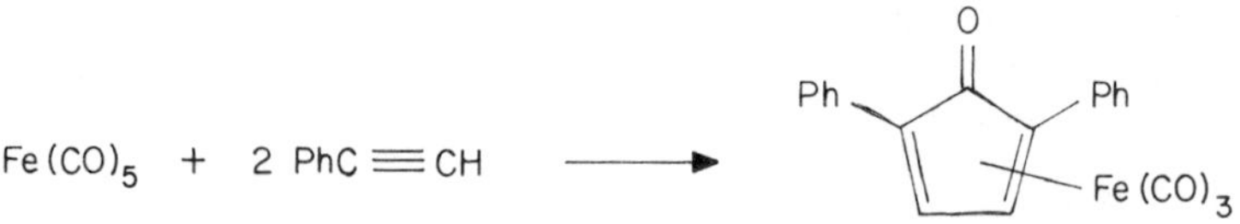

(17)

The ketone carbonyl group of these complexes tends to be very unreactive in comparison with normal ketones. Thus, 2,4-dinitrophenylhydrazones cannot be prepared by reaction with the hydrazine. This is not due to problems of stability of the desired products, since cyclopentadienone phenylhydrazones react readily with $Fe(CO)_5$ to give the appropriate complexes.[42]

Undoubtedly the stability of cyclopentadienone complexes is due to synergic interaction with the metal which leads to electronic population of the LUMO, presumably leading to an electronic structure resembling the cyclopentadienyl anion, which is well known to possess aromatic stability.

Fulvene Complexes[43]

The ready polarizability of fulvenes leads to facile formation of complexes formally containing cyclopentadienyl anions. This usually can arise by synergic interaction between canonical form **B** with zerovalent metal, in which an extra pair of electrons is pushed into **B**.

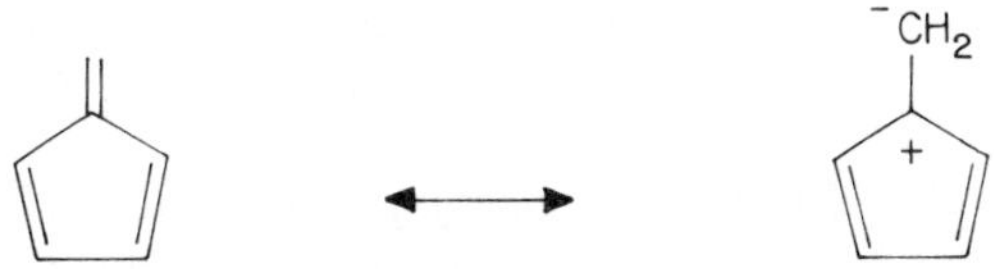

(A)   (B)

Reaction with metal carbonyls in general gives complexes with symmetrical five-membered ring ligands, e.g.

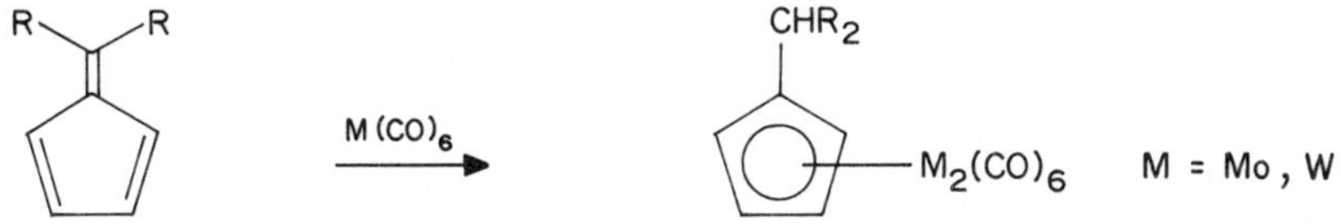

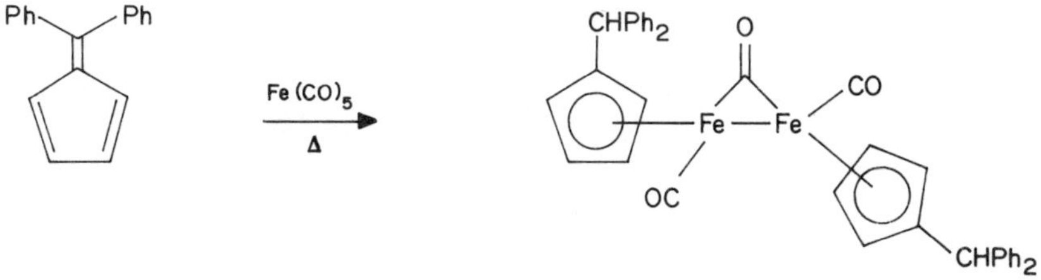

$M = Mo, W$

These are not fulvene complexes. It is possible to prepare iron carbonyl derivatives containing unchanged fulvene as a ligand.

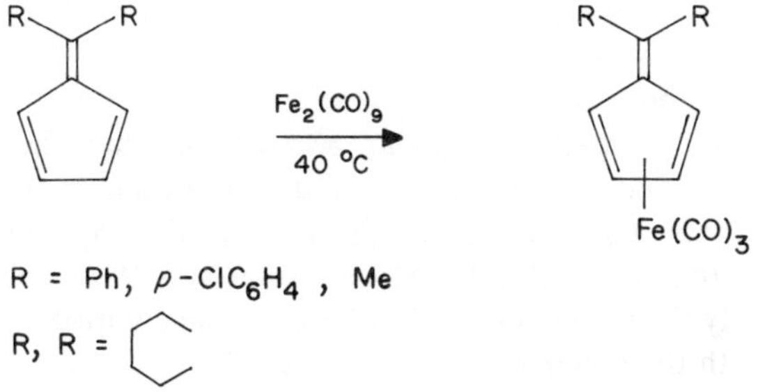

$R = Ph, p-ClC_6H_4, Me$

$R, R =$

The products show IR spectra fairly typical of diene–Fe(CO)$_3$ complexes, e.g. 2049, 1988, 1972 cm^{-1} for R = Ph. The complexes show appreciable dipole moments, suggesting the importance of the canonical forms shown below:

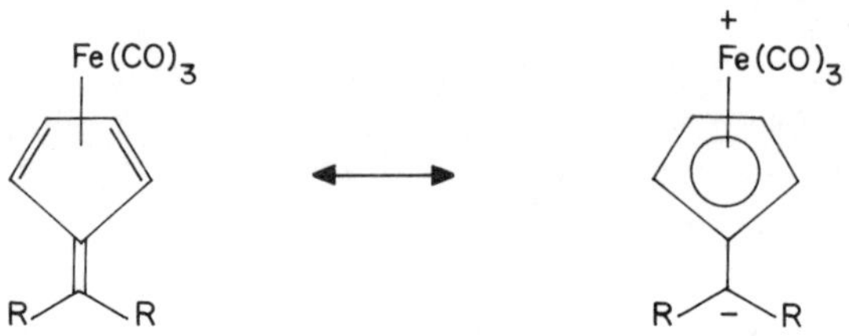

Indeed, treatment of the diphenylfulvene complex (R = Ph) with HCl in benzene leads to protonation exactly as expected from this formalism to give

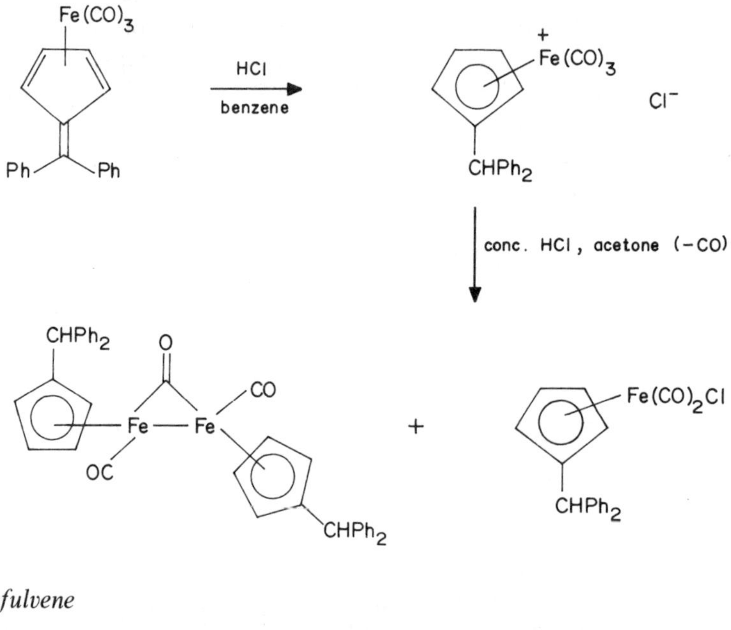

Heptafulvene

Von Doering and Wiley[44] first prepared heptafulvene and found it to be unstable even at $-180\,°C$, although the Diels–Alder adduct with dimethylacetylenedicarboxylate could be obtained. The molecule may be stabilized by coordination of a tricarbonyliron group, but close inspection of the conjugated π-system reveals that there are two ways in which the metal can bond, leading to either a trimethylenemethane (TMM)-like complex or a typical diene complex. Both types of compound have been successfully synthesized, and some of their chemistry studied. The TMM-like derivative 19 was prepared in 25% yield by reaction of 7-hydroxymethylcycloheptatriene with Fe$_2$(CO)$_9$, and was obtained as a red crystalline solid, m.p. 38–41 °C.

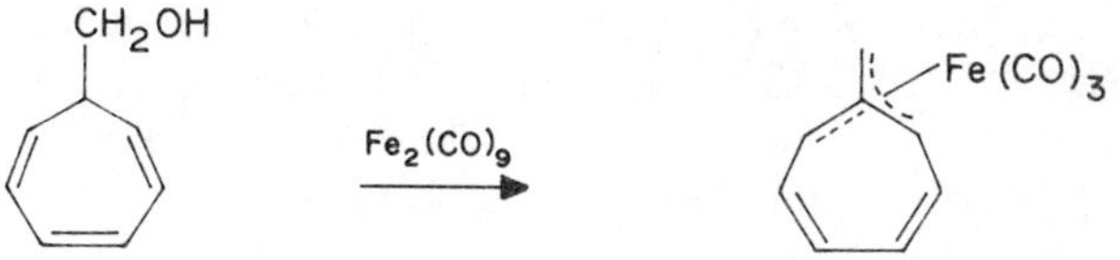

(19)

The proton NMR spectrum of **19** is in agreement with the TMM-type structure [δ 5.87 (4H, m, free diene), 3.70 (2H, dd, $J = 6$, 1 Hz), 1.40 p.p.m. (2H, s)]. The complex shows low reactivity, the free diene moiety undergoing no cycloaddition reaction with tetracyanoethylene. Thermolysis in the presence of dimethyl-acetylenedicarboxylate gave rearranged tricarbonyliron complexes and an organic product which, on dehydrogenation with Pd/C, gave a purple product with a UV spectrum similar to that reported for dimethyl-azulene-1,2-dicarboxylate. If a tricarbonyliron complex of an appropriately substituted cycloheptatriene is used as a source of the heptafulvene system, substituted analogues of the diene-like complex **20** can be obtained, some examples being shown below.

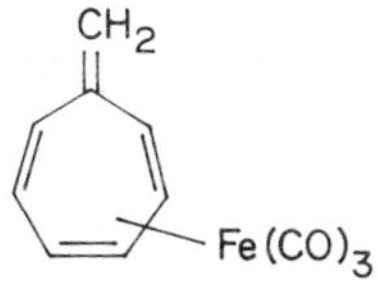

(20)

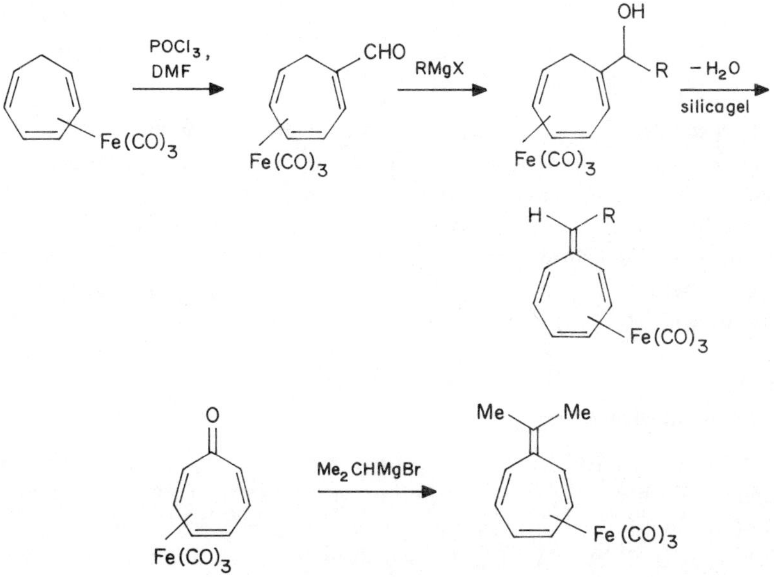

Attempts to prepare the parent complex **20** by this route led mainly to a dimeric species $[C_8H_8Fe(CO)_3]_2$, the monomer being produced in only 1% yield. Similarly, attempts to deprotonate tricarbonyl(methyltropylium)iron tetrafluoroborate **21** with triethylamine gave only the dimer assigned as **22**, which contains two *trans* double bonds in an eight-membered ring!

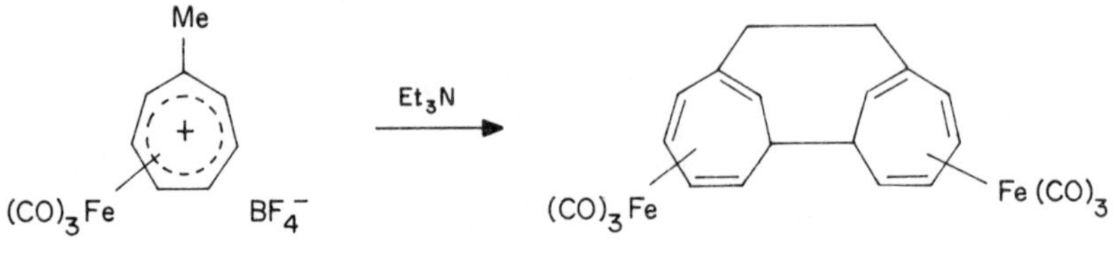

(21)   (22)

A proposed mechanism for dimer formation is by addition across the exocyclic double bond of **20**, the formally non-allowed [8 + 8] cycloaddition being promoted by the $Fe(CO)_3$ fragment. This proposal is not consistent with the fact that the complex **20**, prepared from tricarbonyl(6-methylenecyclohepta-dienyl)iron tetrafluoroborate as shown below, undergoes dimerization extremely slowly.

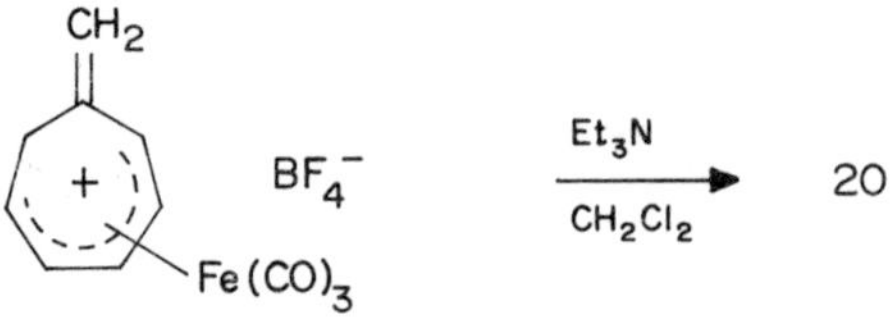

The proton NMR spectra of **20** and its substituted analogues are different from that of **19** and are consistent with the diene–$Fe(CO)_3$ structure proposed, e.g. **20** showing a multiplet (4H) at δ 5.44 p.p.m., two singlets (1 H each) at δ 5.19 and 4.89 p.p.m., a doublet (1 H) at δ 3.63 p.p.m. (7 Hz) and a triplet (1 H) at δ 2.93 p.p.m. Deuterium labelling shows the signals at δ 5.19 and 4.89 p.p.m. to be due to the exocyclic methylene group, the remainder of the spectrum being entirely consistent with the presence of a cyclic diene–$Fe(CO)_3$ group (see later).[45]

Cyclohexa-2,4-dien-1-one

Cyclohexadienone is a non-isolable, unstable tautomer of phenol, stabilization of which is readily achieved by complex formation with $Fe(CO)_3$. This has the effect of locking the diene so that tautomerization is prevented. The complex **25** is readily prepared by the acidic hydrolysis of tricarbonyl(1-methoxycyclohexadienyl)iron tetrafluoroborate **23** (see below), or by the chromium trioxide oxidation of tricarbonyl(5-hydroxycyclohexa-1,3-diene)iron **24**.

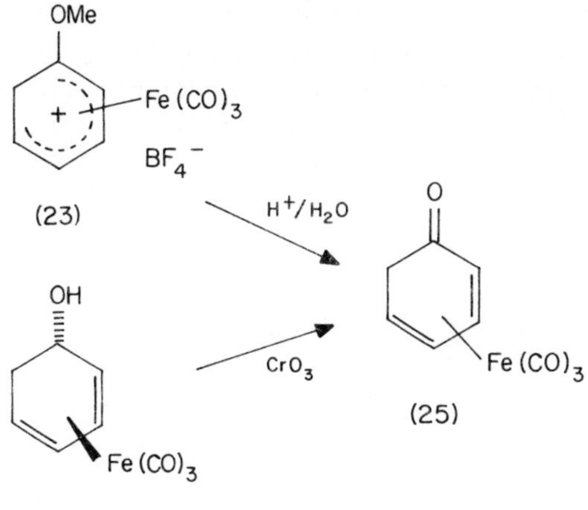

There has been only a limited amount of work reported on this complex and consequently few useful synthetic applications. The carbonyl group may be reduced with sodium borohydride stereospecifically from the face opposite that occupied by the Fe(CO)$_3$ group, undoubtedly due to the steric effects, to give **26**.

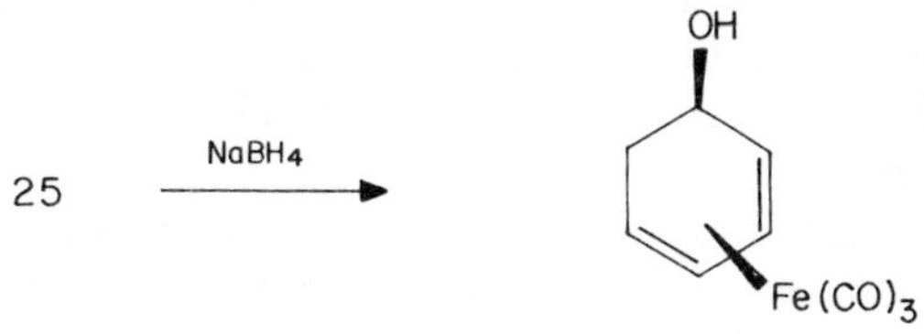

The ketone group also reacts under Reformatsky conditions to give the ester **27**, and reaction with amines results in *N*-phenylation.

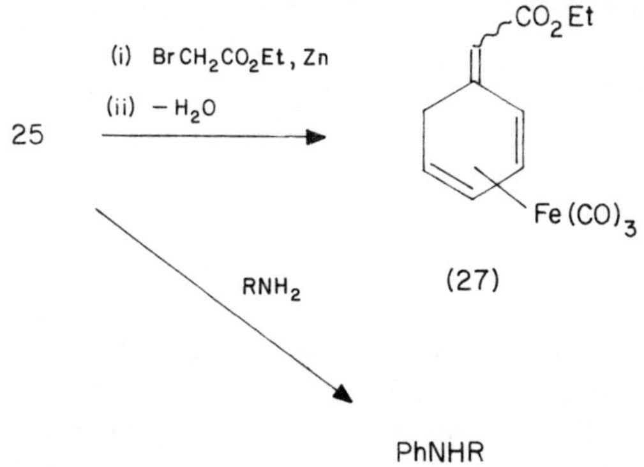

POLYMERIZATION AND OLIGOMERIZATION[46]

Ziegler–Natta Polymerization of Alkenes

Ziegler and Natta found that rapid polymerization of ethene and α-olefins occurs using catalysts prepared from, e.g. titanium trichloride and an alkyl aluminum derivative, AlR_3. Using α-olefins, such as propylene, there are many possible stereoisomeric products, but in practice regular polymers are obtained:

isotactic:

Me H Me H Me H

Syndiotactic:

H Me Me H H Me Me H

Some catalysts are found to produce isotactic and some syndiotactic polymers. A typical laboratory catalyst is a slurry of violet crystalline $TiCl_3 \cdot 1/3AlCl_3$ which is treated with diethylaluminum chloride in the presence of propylene at room temperature and 3–4 atm pressure. Rapid polymerization occurs to give highly isotactic polypropylene. Ziegler–Natta polymerization now represents a major industrial process (more than 2 million tons per year of polyethene and polypropene). Most of the catalyst systems are heterogeneous, but some homogeneous ones are known.

In the original Ziegler–Natta process ethene, a monomer which is normally difficult to activate, was polymerized at atmospheric pressure and room temperature in a glass vessel using petrol as the solvent and $TiCl_4/Al_2Et_3Cl_3$ as catalyst. This gives a product containing appreciable catalyst residues, and modern polymer research is aimed at developing catalysts which produce so much polyolefin that the amount of such residue is negligible. This then avoids difficult purification steps.

In very rough terms the basic mechanism is as follows:

Chain propagation:

Possible chain terminations:

(*a*) *β-elimination / reverse β-elimination*:

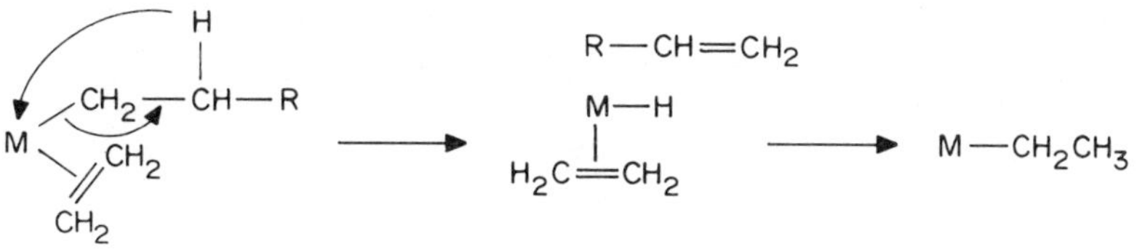
H
CH2
CH—R
M
CH2
CH2
R—CH=CH2
M—H
H2C=CH2
M—CH2CH3

(*b*) *hydrogenation*.

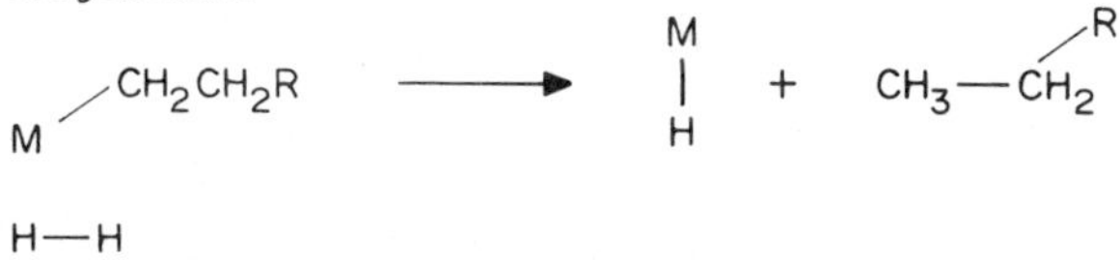
M
CH2CH2R
H—H
M
H
+
R
CH3—CH2

The propagation step shown above appears to be an insertion of alkene into the M–C σ-bond. If such a step does occur directly, then we must somehow account for the stereoregularity of the product polymer. It has been proposed that approach of the monomer to the reactive metal carbon bond, or vacant coordination site, should occur from the less hindered side. If the *metal* has very bulky substituents the second insertion step is impossible, but with smaller substituents the monomer always approaches the reactive M—C bond from the same side, resulting in an isotactic polymer. Consider the following:

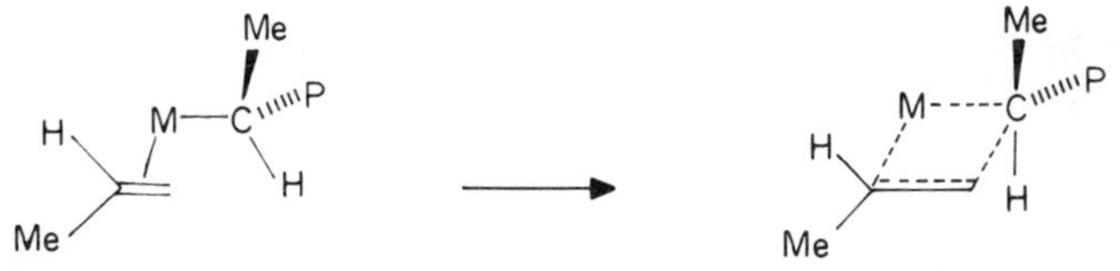
Me
H
M—C
P
H
Me

(A)

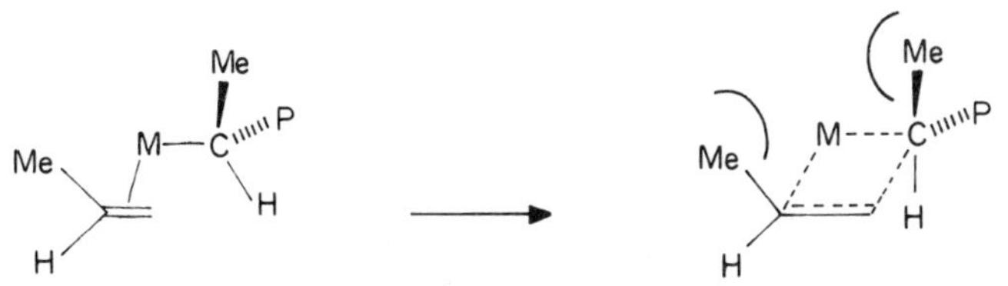
Me
Me
M—C
P
H
H

(B)

It can be seen that the insertion step requires a four-membered ring transition state, and the *trans* system (**A**) would appear to be favored on steric grounds over the *cis* arrangement; this would lead to a syndiotactic polymer.

Isotactic polymerization requires an intermediate **C** shown below, in which the metal—polymer σ-bond is formed at the unsubstituted carbon. It is difficult to rationalize the stereoregularity using this model, but presumably conformational effects in the four-center transition state are important.

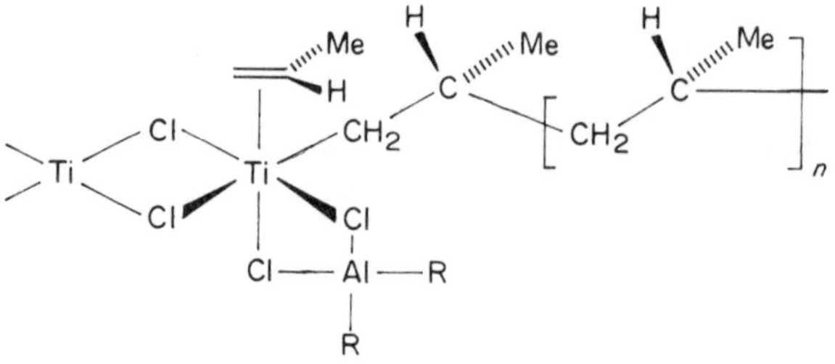

(C)

Green and co-workers have proposed an alternative mechanism for stereoregular polymerization which involves the formation of a carbenoid intermediate and its reaction with alkene monomer to give a metallocyclobutane intermediate, as encountered in olefin metathesis.[47] It was pointed out that, at that time, there were no unambiguous examples where a characterized metal–alkyl–olefin compound could be induced to give the desired insertion product. The following alternative mechanism was proposed.

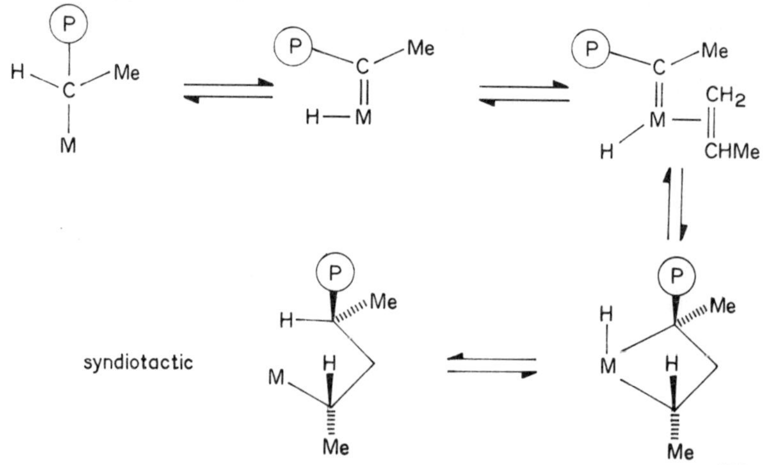

(D)

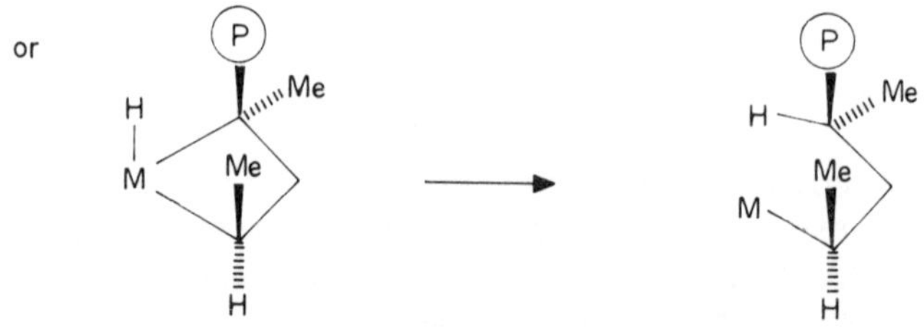

(E)   isotactic

The factors influencing the stability of metallacyclobutanes **D** and **E** are expected to be the same as those found in the metathesis reaction. In a buckled ring we might expect the largest substituents to occupy equatorial positions, i.e.

favoured over

We can then see that **D** in fact has a diaxial Me–Me interaction, which is absent in **E**, so that isotactic polymerization is favored.

More recently, however, an unambiguous example of olefin insertion has been found.[48] The following reaction was observed:

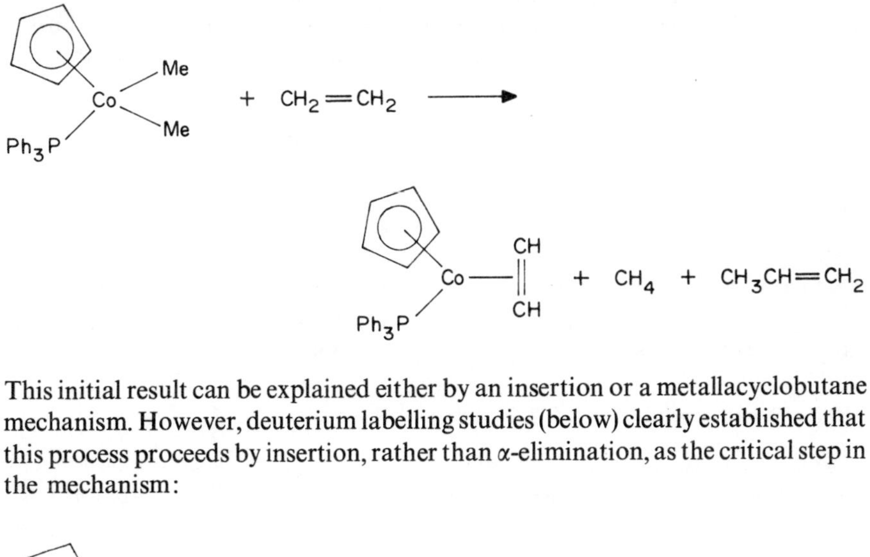

This initial result can be explained either by an insertion or a metallacyclobutane mechanism. However, deuterium labelling studies (below) clearly established that this process proceeds by insertion, rather than α-elimination, as the critical step in the mechanism:

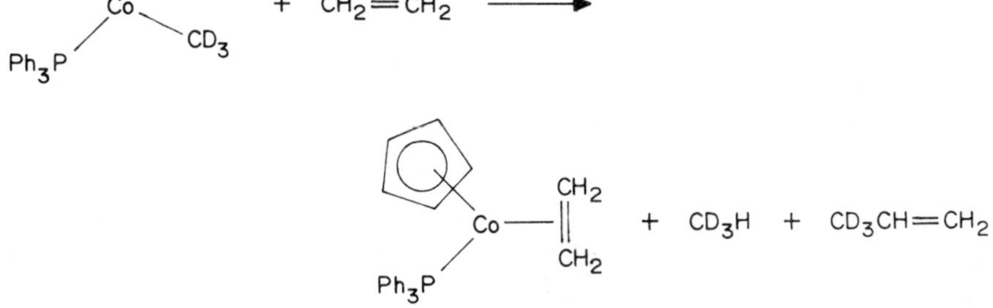

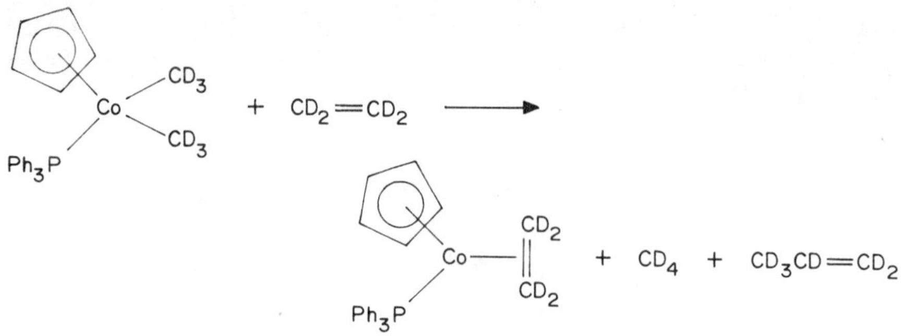

Whilst it is still possible that Ziegler–Natta polymerization takes place by the metallacyclobutane mechanism, 'those in favor of this mechanism must now shoulder the burden of proof for establishing it.'

1,3-Diene Polymerization

Butadiene and isoprene have long been feedstocks for synthetic rubbers. Natural rubber, an isoprene polymer, has the structure given below:

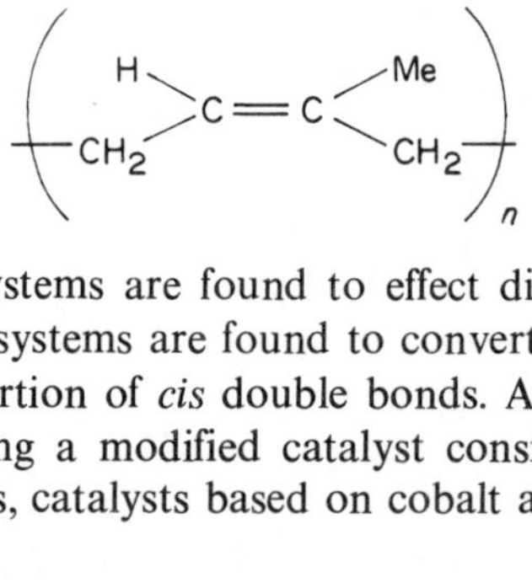

A number of catalyst systems are found to effect diene polymerization. The standard Ziegler–Natta systems are found to convert butadiene to a polymer containing a high proportion of *cis* double bonds. A better proportion of *cis* product is obtained using a modified catalyst consisting of TiI_4/R_xAlI_y. In addition to these systems, catalysts based on cobalt and nickel are effective.

Organocobalt Catalysts

These seem to be true homogeneous catalysts. Used in combination with alkylaluminum halides they give 96–98% *cis*-polybutadiene. The mechanism is as follows:

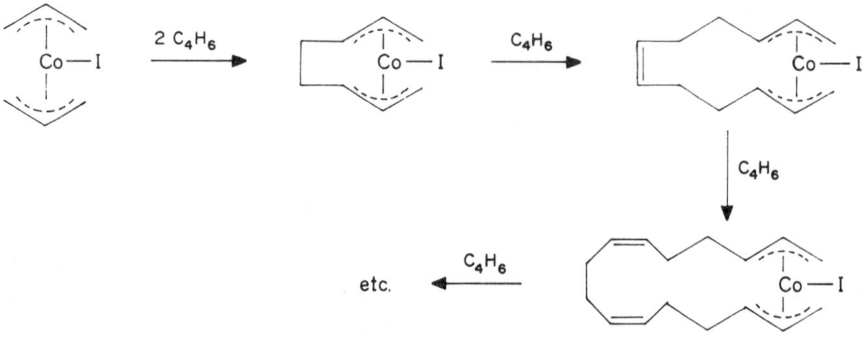

Nickel Catalysts

Catalysts based on π-allyl nickel complexes, although less used, have been well studied. Usually a π-allyl complex is formed *in situ* as follows:

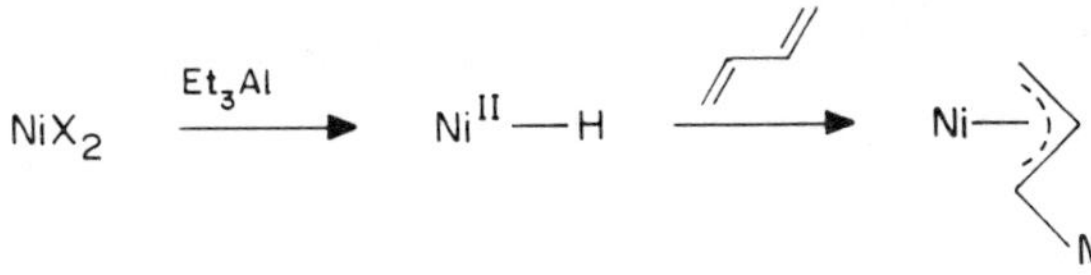

The double bond geometry of polybutadienes produced using these catalysts is largely dependent on the nature of the ligand attached to the metal. This is explained as being due to coordination of butadiene as either η^4 or η^2 ligands, as shown by the following examples.

(a)

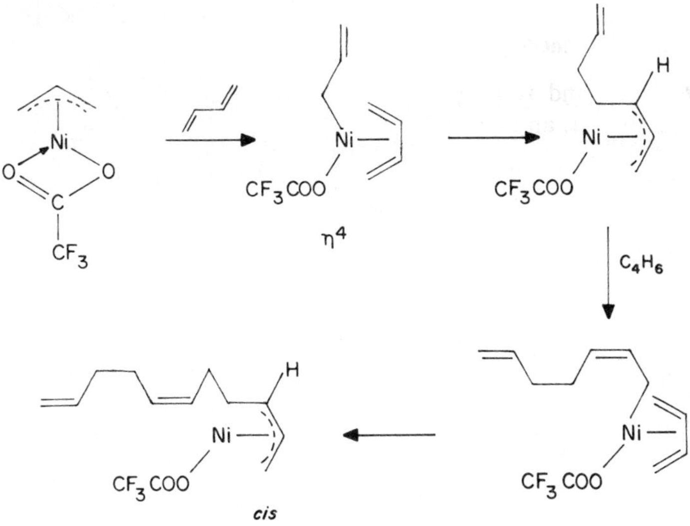

(b)

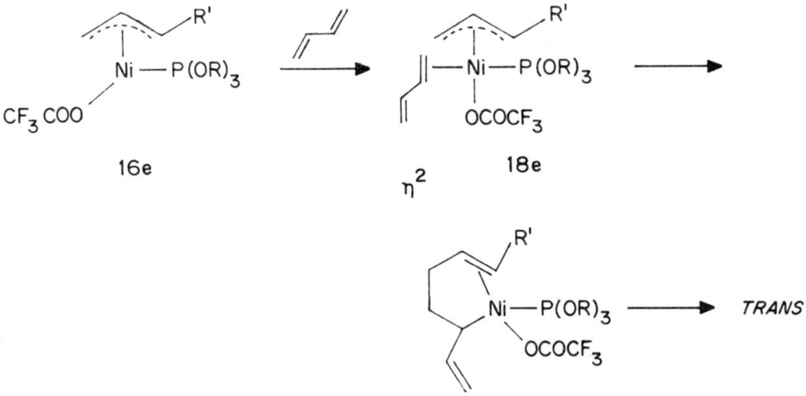

Oligomerization

The self-addition of alkenes, dienes or alkynes to form dimers, trimers and low polymers is termed oligomerization. The mechanisms are essentially the same as for polymerization except that chain termination occurs much more frequently. We shall consider in this section the dimerization and trimerization of olefinic and acetylenic compounds.

Olefin Oligomerizations

During Ziegler's studies of ethene oligomerization to long-chain terminal olefins using alkylaluminum compounds, it was found that traces of nickel from reactor corrosion caused a change from oligomerization to formation of but-1-ene, and this led to the invention of catalysts for a number of processes. Effective catalysts for olefin dimerization are based on nickel and rhodium, and a reaction of alkylaluminum compounds with cobalt salts and titanium(IV) complexes. Combinations of π-allylnickel halides with phosphines and Lewis acids (e.g. $AlCl_3$, $EtAlCl_2$) or $Ni(cod)_2$, $Ni[P(OPh)_3]_4$ and Lewis acids are also commonly employed. In the case of nickel, and probably other metals, the active catalyst is thought to be a metal hydride species formed by β-elimination, a mechanism being as follows (ancillary ligands omitted):

$$Ni-CH_2CH_2R \longrightarrow Ni-H + RCH=CH_2$$

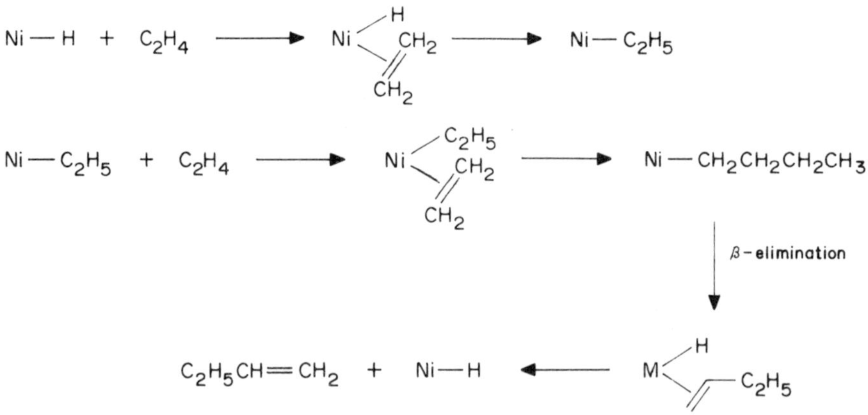

With propene a number of products are possible, the major one being determined to a large extent by the types of ligand (steric crowding) on the metal:

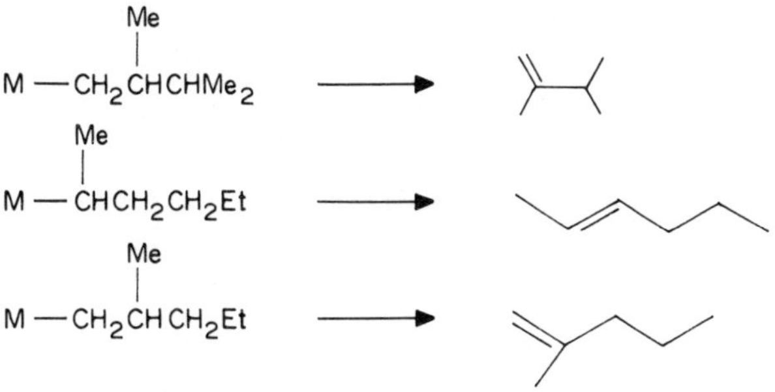

Me
|
M—CH$_2$CHCHMe$_2$ ⟶

Me
|
M—CHCH$_2$CH$_2$Et   ⟶

Me
|
M—CH$_2$CHCH$_2$Et   ⟶

Co-dimerization of ethene and butadiene to give hexa-1,4-dienes is also successfully achieved in the presence of a number of Ziegler–Natta catalysts based on rhodium, nickel, iron and cobalt. The *trans*-hexa-1,4-diene isomer, used in synthetic elastomer production, is obtained using Rh or Ni catalysts, whilst Fe and Co give *cis* product:

Again, the mechanism of the reaction appears to involve prior formation of a metal hydride (ancillary ligands omitted):

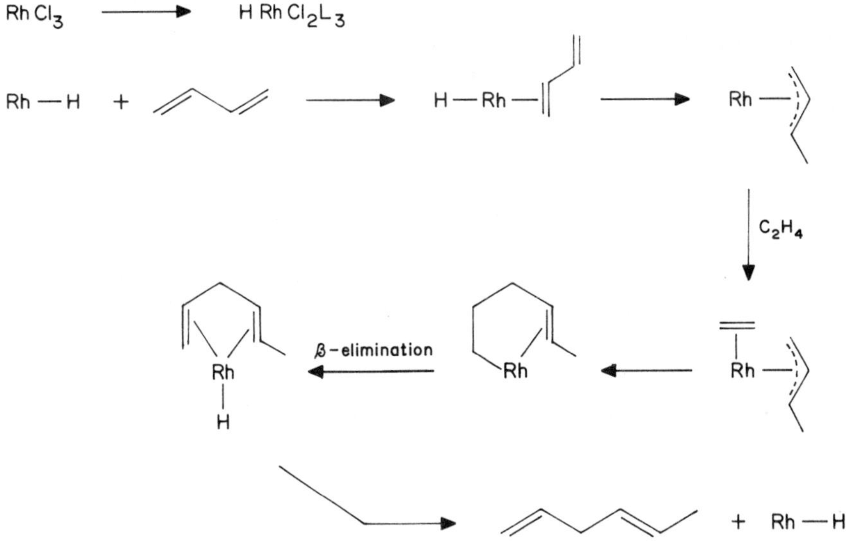

RhCl$_3$ ⟶   HRhCl$_2$L$_3$

Rh—H +   ⟶   H—Rh—   ⟶   Rh—

↓ C$_2$H$_4$

β-elimination

+ Rh—H

With cobalt catalyst an *anti*-allyl complex is formed, leading to *cis*-hexa-1,4-diene.

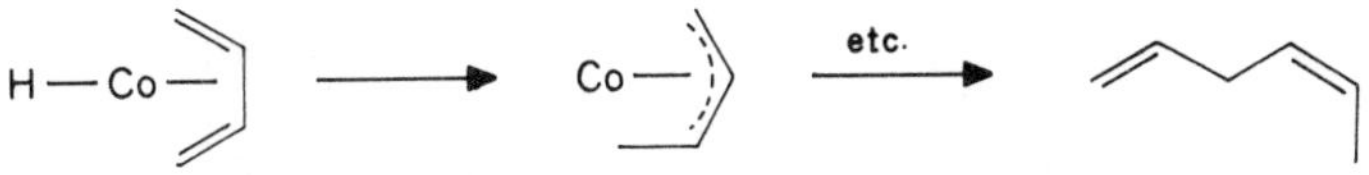

H—Co—   ⟶   Co—   etc.   ⟶

It is not certain whether the rhodium catalyst forms the *syn*-allyl derivative directly via an η^2-butadiene to σ-allyl transformation or whether an initially formed *anti*-allyl complex rapidly rearranges to the *syn* complex:

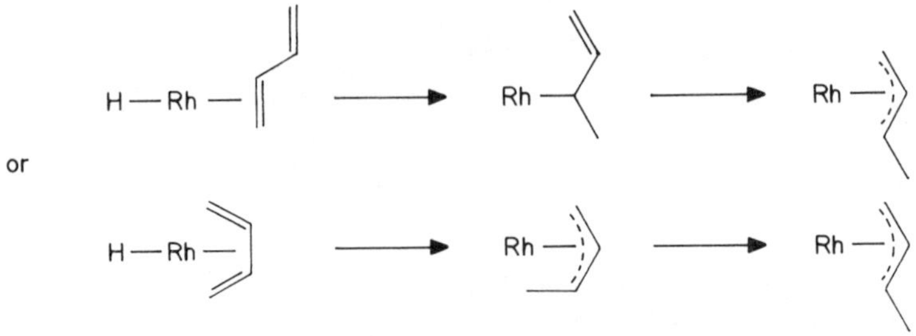

or

The dimerization of butadiene, using a modified Ziegler catalyst, usually nickel-based, can be used to produce cycloocta-1,5-diene (cod) or octa-1,3,7-triene, depending on the ligands and reaction conditions chosen, e.g.

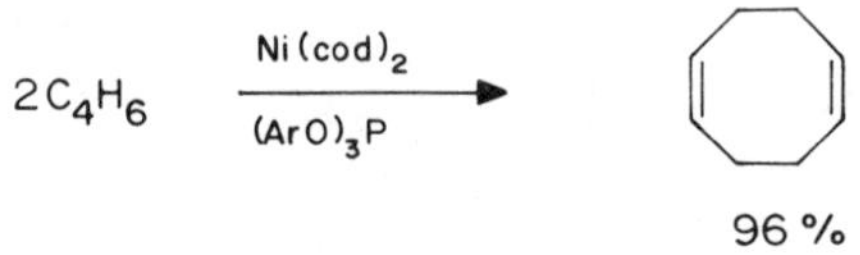

96 %

Addition of an acidic component, such as ROH, R_2NH or HCN, results in the production of linear dimers. However, linear dimers are best obtained using palladium catalysts, e.g.

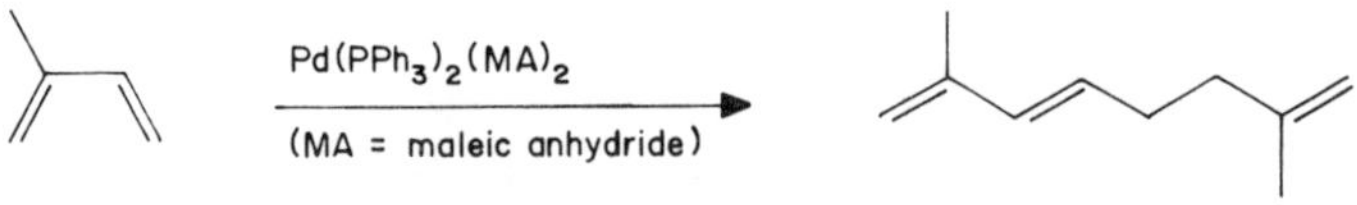

The mechanism of the nickel-catalyzed reaction reveals the conditions necessary for preferential formation of linear or cyclic dimers.

(a) *Linear:*

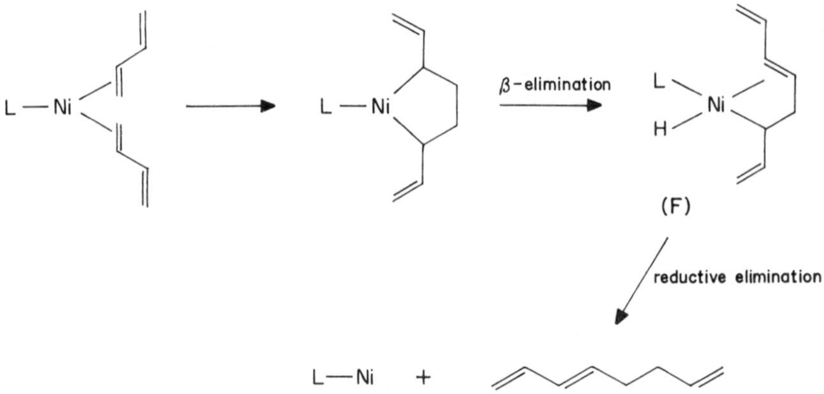

(b) *Cyclic:*

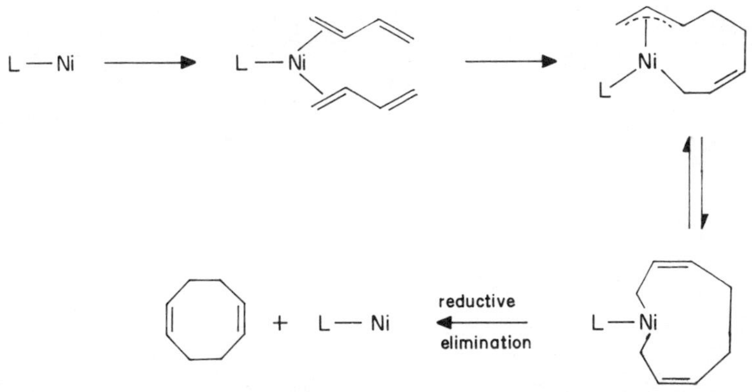

(G)

Thus, the formation of linear dimers requires the formation of the secondary σ-allyl complex **F**, whilst cyclodimerization involves production of a primary σ-allyl derivative **G**. Consequently, we might expect that bulky ligands **L** will disfavor the first pathway and lead to cyclization, as is indeed observed.

A useful dimerization of butadiene, in which an acetoxy group is also introduced, has been developed by Tsuji's research group.[49] The product is used as source of octa-1,7-dien-3-one, which is useful as a building block for steroid synthesis

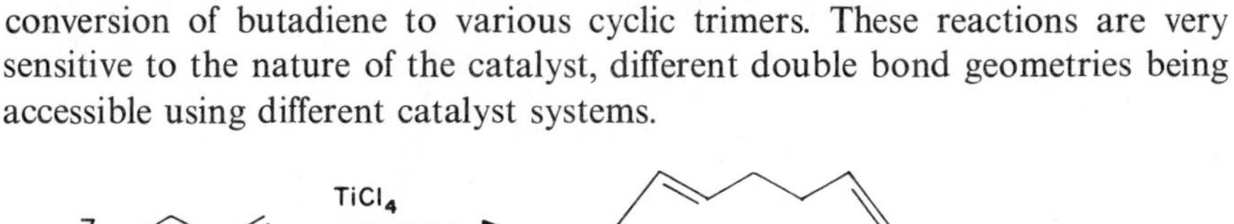

Yet further variation of the Ziegler catalyst, already so useful, allows the conversion of butadiene to various cyclic trimers. These reactions are very sensitive to the nature of the catalyst, different double bond geometries being accessible using different catalyst systems.

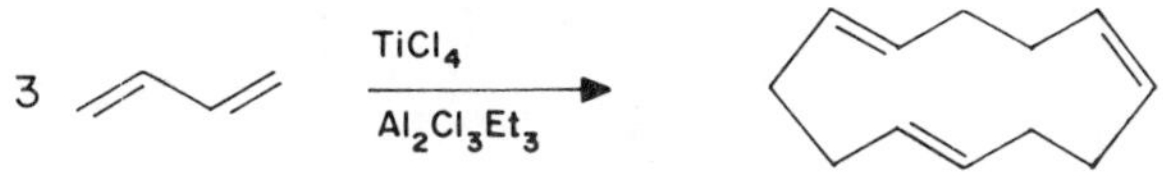

cis, trans, trans - cyclododecatriene

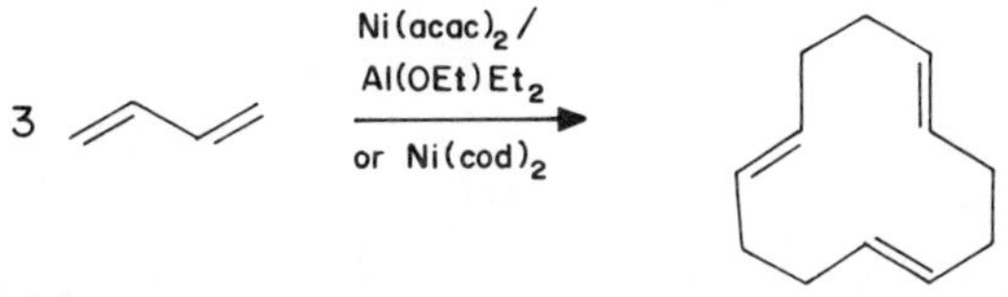

trans, trans, trans - cyclododecatriene

Alkyene Oligomerization

Conversion of acetylene to cyclooctetratraene is readily achieved using nickel salts as catalysts, e.g. $Ni(CN)_2$ in anhydrous THF at 60–70 °C and 15–20 atm.[50]

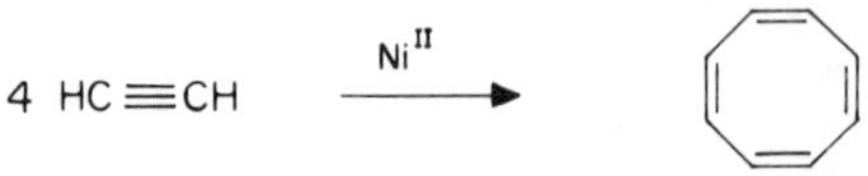
4 HC≡CH
Ni^II

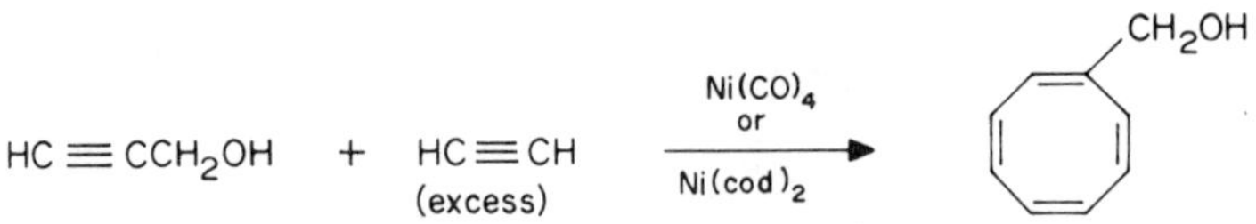
HC≡CCH2OH + HC≡CH
(excess)
Ni(CO)4
or
Ni(cod)2
CH2OH

Trimerization

A useful method for the preparation of aromatic compounds involves catalytic acetylene trimerization:

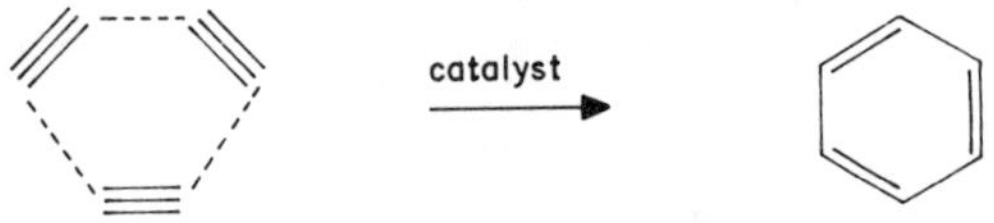
catalyst

e.g.

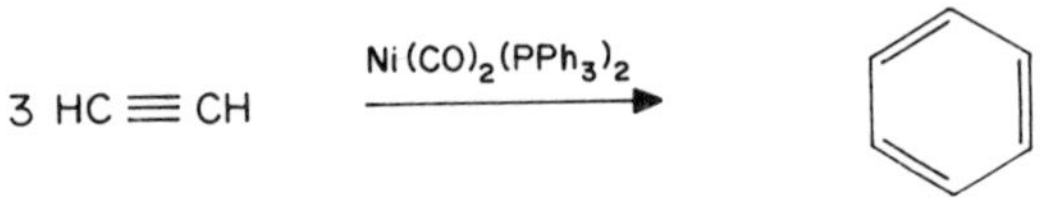
3 HC≡CH
Ni(CO)2(PPh3)2

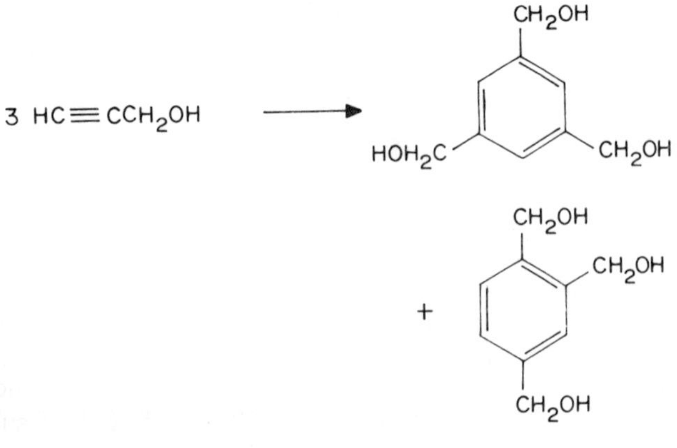
3 HC≡CCH2OH
CH2OH
HOH2C
CH2OH
+
CH2OH
CH2OH
CH2OH

Cobalt catalysts have more recently been found extremely useful,[51] e.g.

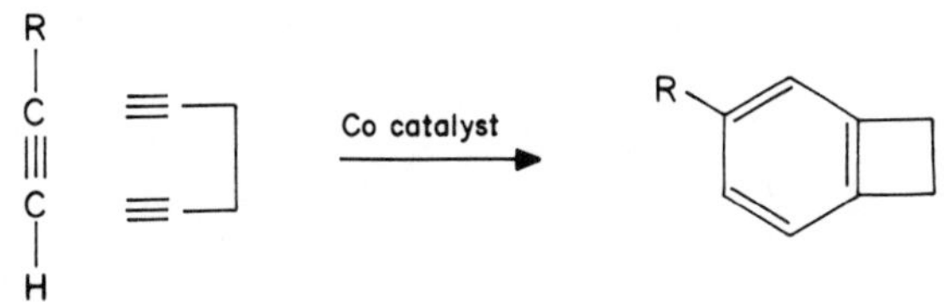

This can be applied to the synthesis of polycyclic compounds by means of further reactions of the benzocyclobutene:[52]

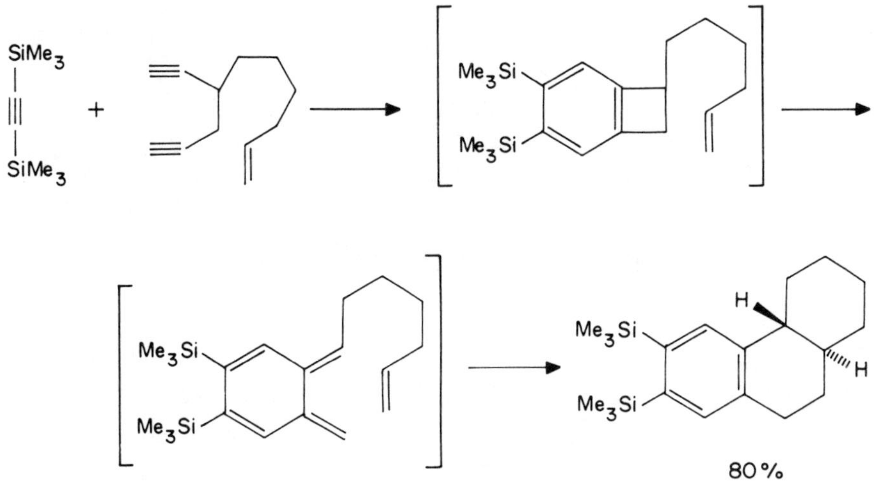

80%

This has been applied to the synthesis of (±)-oestrone:[53]

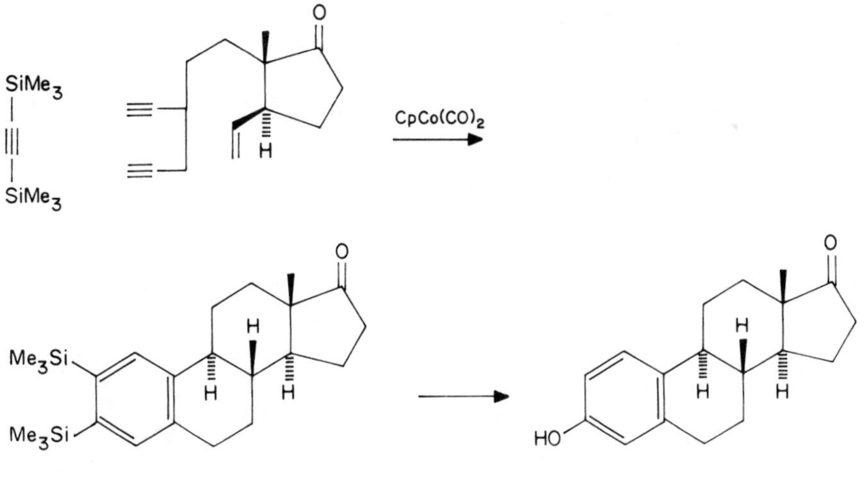

The mechanism of these reactions is via a metallacyclopentadiene inter-mediate,[54] e.g.

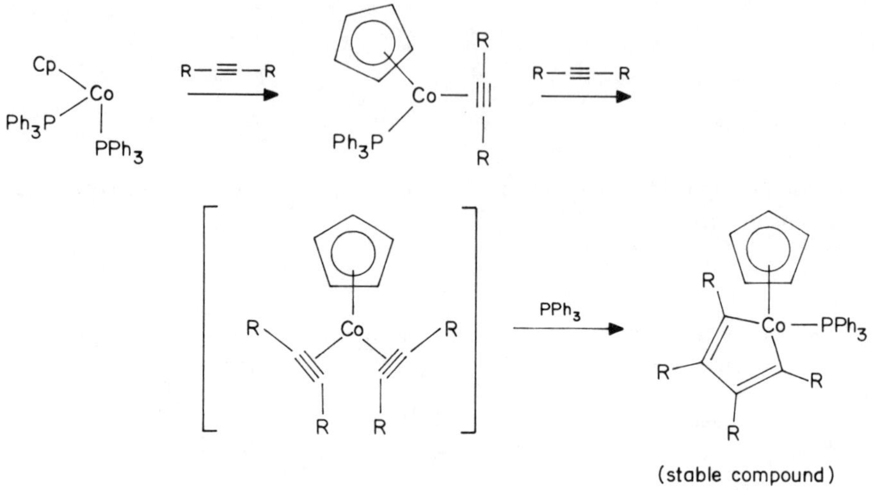

(stable compound)

The third acetylene may add in a number of possible ways:

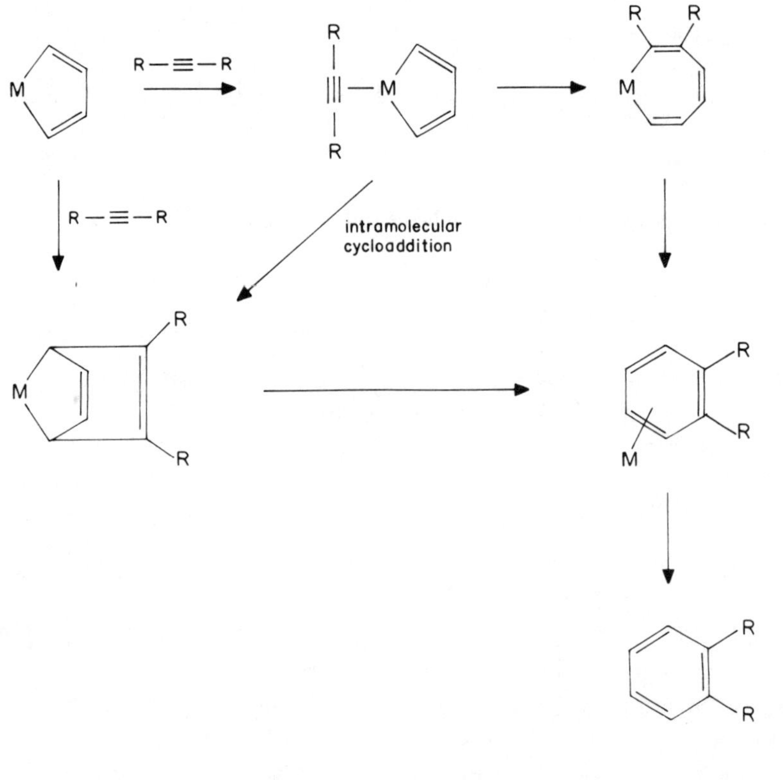

intramolecular
cycloaddition

Dimerization of acetylenes to give cyclobutadiene–metal complexes may also be obtained, e.g. with Pd catalysts:

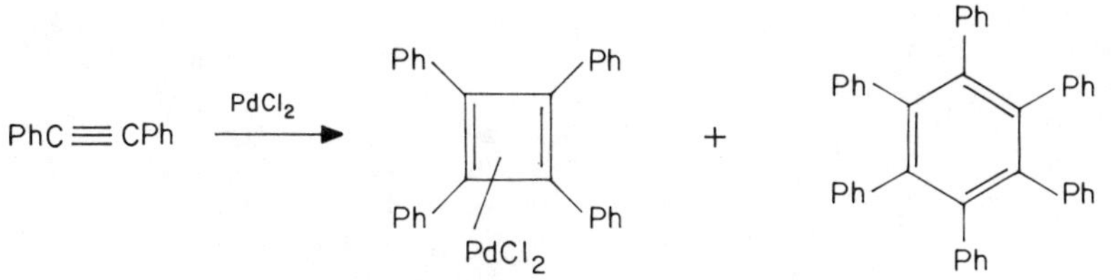

The aromatic product is major in non-hydroxylic solvents.

THE WACKER PROCESS

This reaction represented a major contribution to manufacturing organic chemistry, and is even now becoming obsolete. Its discovery meant that acetaldehyde could be obtained from ethene, thus displacing the dependence of industry on acetylene as a feedstock. Very similar processes are used to produce related compounds, e.g. vinyl acetate, but only the basic Wacker oxidation will be presented here.

The basic process is

$$CH_2\!=\!CH_2 \ + \ PdCl_2 \ + \ H_2O \ \longrightarrow \ CH_3CHO \ + \ Pd^0 \ + \ 2\ HCl$$

Of course the Pd^0 produced is useless and so for an efficient catalytic cycle it must be re-oxidized to Pd^{II}. This is readily achieved by incorporation of $CuCl_2$ in the reaction mixture. In fact, only a catalytic amount of $CuCl_2$ is required, since the product of its reduction, $CuCl$, is reoxidized by atmospheric oxygen in the system. Thus the overall process is

$$C_2H_4 \ + \ \tfrac{1}{2}O_2 \ \xrightarrow{\ catalyst\ } \ CH_3CHO$$

The overall mechanism (schematic) is

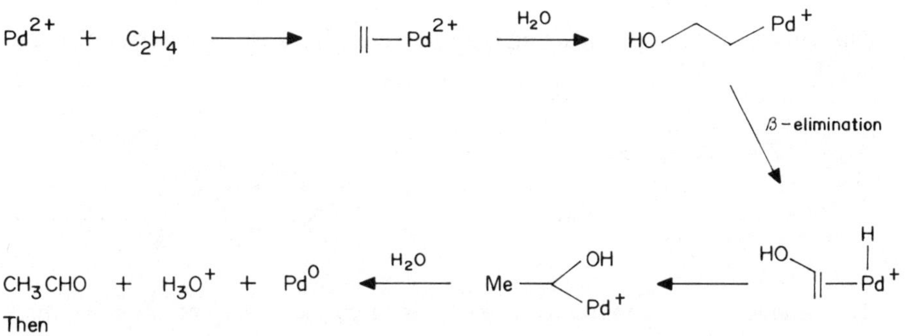

Then

$$Pd^0 \ + \ 2\ Cu^{II} \ \longrightarrow \ Pd^{II} \ + \ 2\ Cu^I$$

Terminal olefins may be subjected to Wacker oxidation to give methyl ketones. Good use is made of this in Tsuji's work on steroid synthesis:[49]

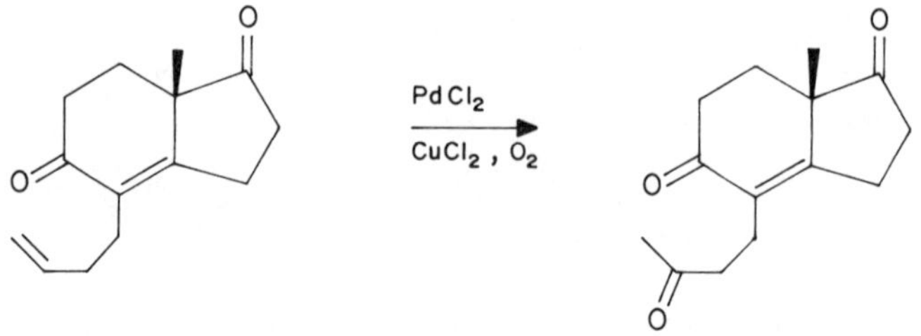

A modified catalyst system, replacing oxygen with t-BuOOH or H_2O_2 has been found useful for the conversion of α,β-unsaturated esters to β-keto esters:[55]

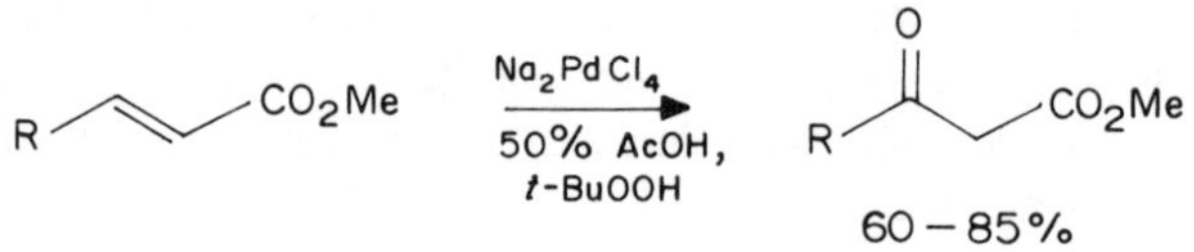

$$60 - 85\%$$

OLEFIN EPOXIDATION

Fairly recently it has been discovered that alkenes may be converted to epoxides in good yield using t-Butyl hydroperoxide in the presence of a transition metal catalyst. Simple olefins, e.g. propylene, can be converted using $Mo(CO)_6$ as catalyst. This represents a potentially important industrial synthesis of the important propylene oxide. Furthermore, allylic alcohols can be epoxidized stereospecifically and selectively using $VO(acac)_2$ as catalyst, e.g.

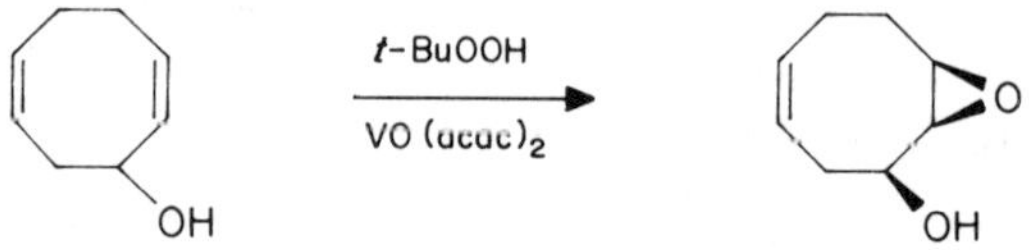

The vanadium-catalyzed reaction is also effective for some homoallylic alcohols, e.g.[56]

The reagent, now commonly termed the Sharpless reagent after its main developer, is one of the choice for such epoxidations. More recently, Sharpless has developed reagents based on titanium tetra-*iso*-propoxide in the presence of tartrate esters, which effect asymmetric epoxidation, e.g.

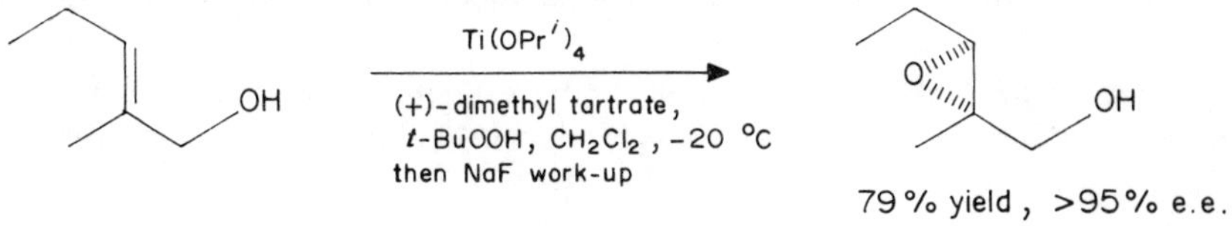

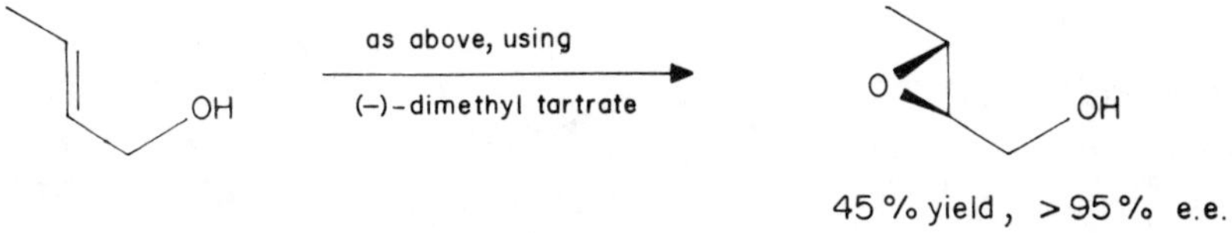

79% yield, >95% e.e.

45% yield, >95% e.e.

The products are precursors for the total synthesis of a range of interesting and important natural products, e.g. erythromycins and leukotrienes, and the availability of the optically active epoxy alcohols leads to a useful entry into asymmetric synthesis.

CATALYTIC HYDROGENATION

This subject is extensively treated in a number of authoritative texts,[58] so we shall content ourselves with a brief discussion of the salient features of homogeneously catalyzed reactions, and an introduction to asymmetric hydrogenation.

There are basically three modes of activation of hydrogen and three mechanisms for hydrogenation:

1. Oxidative addition of H_2 to complex:

2. Heterolytic addition:

$$M + H_2 \longrightarrow M-H + H^+$$

3. Homolytic addition:

$$2M + H_2 \longrightarrow 2M-H$$

These three processes lead to different overall mechanisms and also differences in the nature of the substrate which can be reduced.

Activation by Oxidative Addition—Wilkinson's Catalyst

Two independent reports appeared in 1965 that the complex tris(triphenylphosphine)chlororhodium(I), $RhCl(PPh_3)_3$, in organic solvents is an active catalyst for reduction of alkynes and alkenes at ambient temperature and atmospheric pressure (H_2).[59] This compound has subsequently become known as Wilkinson's catalyst. The currently favored mechanistic schemes are shown below.

There is a subtle difference between the two possible mechanisms, being the point at which solvent acts as a ligand. This depends, of course, on the presence of a suitable solvent.

(a) *Not incorporating solvent until a late stage*

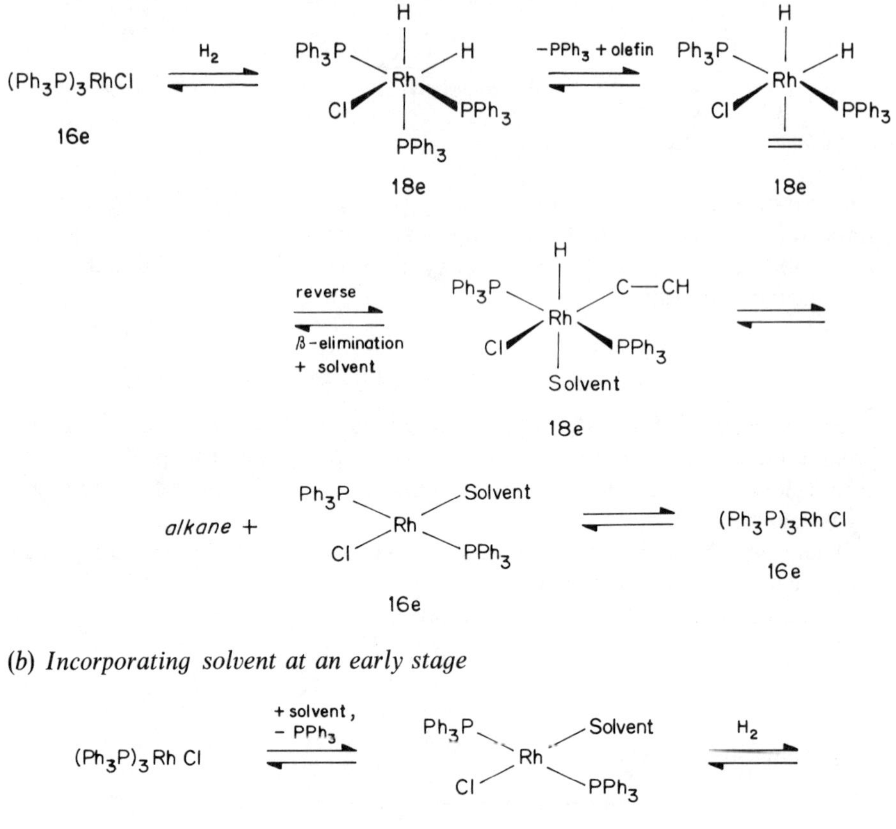

$(Ph_3P)_3RhCl$ 16e $\xrightleftharpoons{H_2}$

Ph₃P, Rh, Cl, PPh₃, PPh₃, H, H — 18e $\xrightleftharpoons{-PPh_3 + olefin}$ Ph₃P, Rh, Cl, PPh₃, H, H — 18e

$\xrightarrow[\text{+ solvent}]{\text{reverse } \beta\text{-elimination}}$ Ph₃P, Rh, Cl, PPh₃, Solvent, H, C—CH — 18e $\rightleftharpoons$

alkane + Ph₃P, Rh, Cl, PPh₃, Solvent — 16e $\rightleftharpoons$ $(Ph_3P)_3Rh\,Cl$ 16e

(b) *Incorporating solvent at an early stage*

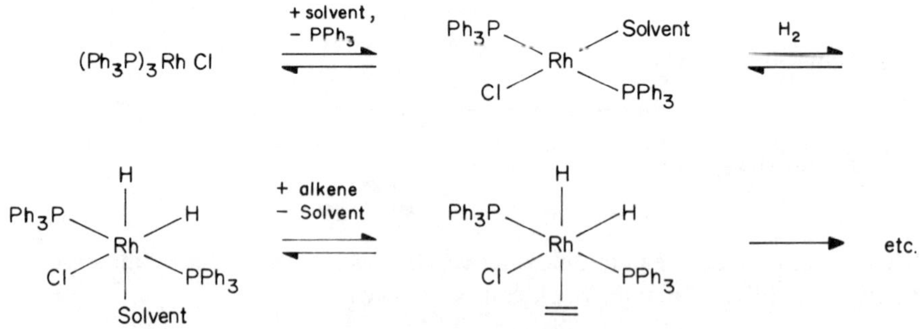

$(Ph_3P)_3Rh\,Cl$ $\xrightleftharpoons{\substack{+ \text{ solvent,} \\ - PPh_3}}$ Ph₃P, Rh, Cl, PPh₃, Solvent $\xrightleftharpoons{H_2}$

Ph₃P, Rh, Cl, PPh₃, Solvent, H, H $\xrightleftharpoons{\substack{+ \text{ alkene} \\ - \text{ Solvent}}}$ Ph₃P, Rh, Cl, PPh₃, H, H $\longrightarrow$ etc.

The rate-determining step might be expected to be the oxidative addition of H_2 or ligand migration (reverse β-elimination) and the importance of path (b) is shown by the observation that the rate of hydrogenation of cyclohexene in benzene–ethanol is *ca.* twice that in pure benzene. In any case, there is no prior dissociation of Ph_3P from the 16e $(Ph_3P)_3RhCl$ to give a 14e complex—either solvent adds first (18e) followed by Ph_3P dissociation, or oxidative addition occurs directly to the trisphosphine complex. The more usual pathway, starting with $(Ph_3P)_3RhCl$, appears to be (*a*).

It can be readily appreciated that since the intermediates along either mechanistic pathway contain triarylphosphine ligands, these tend to be sterically congested. Consequently, we should expect the reaction to be susceptible to steric effects in the olefinic substrate. This is indeed the case, and it can be put to good use in achieving selective reduction of less hindered double bonds in a polyolefinic substrate. Some examples are given below.

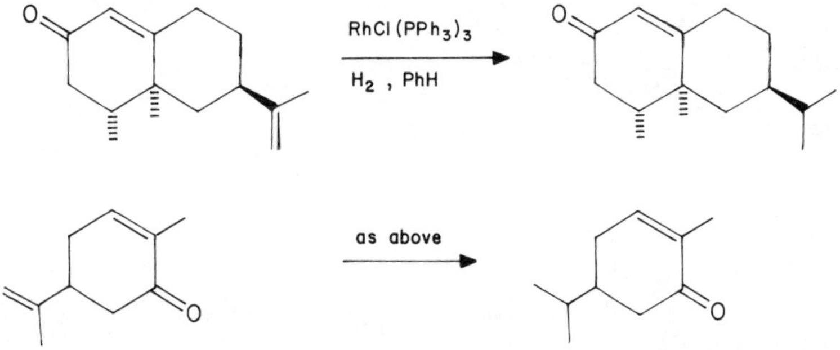

The general trend in reduction rates is alk-1-enes > *cis*-alk-2-enes ≫ *trans*-alk-2-enes > *trans*-alk-3-enes. Exocyclic double bonds are generally reduced more rapidly than endocyclic ones. A number of functional groups, e.g. $C\!=\!O$, $C\!=\!N$, NO_2, Ar, CO_2R, are unaffected. Thus, Wilkinson's catalyst is an extremely selective hydrogenation catalyst with excellent potential for use with complex organic molecules.

Heterolytic Addition of Hydrogen

Hydrogenation of olefinic double bonds conjugated with carbonyl groups, e.g. α,β-unsaturated carboxylic acids or ketones, is readily achieved using ruthenium(II) salts. Unactivated olefins are not so readily reduced by these systems. The basic mechanism, outlined below, involves heterolytic addition of H_2 to a Ru complex, typically ruthenium(II) chloride in dilute hydrochloric acid.

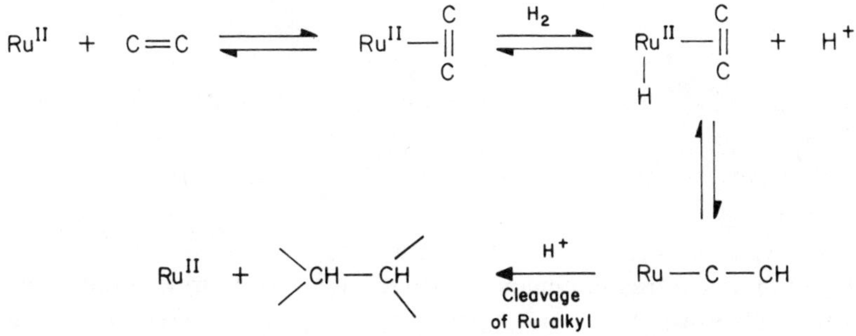

The last two steps of this sequence can occur only with olefins that are highly activated towards nucleophilic attack.

A very useful catalyst[60] for a range of hydrogenations, including terminal alkenes, nitro compounds ($\rightarrow NH_2$) and aldehydes ($\rightarrow$ alcohols), is the triphenylphosphine derivative $Ru^{II}Cl_2(PPh_3)_3$. This complex reacts by the mechanism shown below; it should be pointed out that heterolytic addition of H_2 is formally identical with oxidative addition, except that subsequent loss of H^+ occurs from the $M\overset{H}{\underset{H}{<}}$ complex.

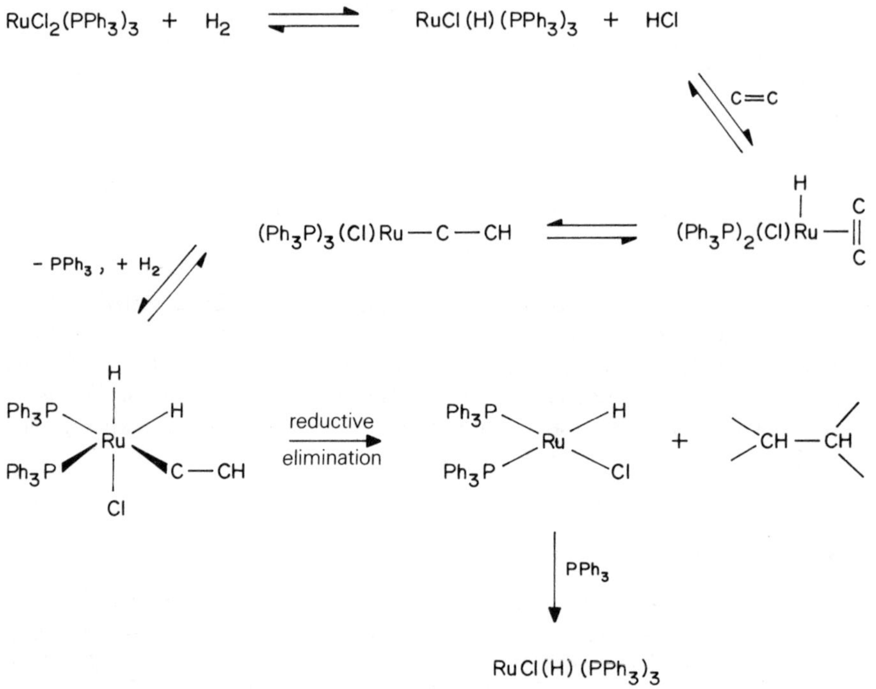

This reagent has been used to reduce selectively the less substituted double bond of, e.g., steroidal dienones:[61]

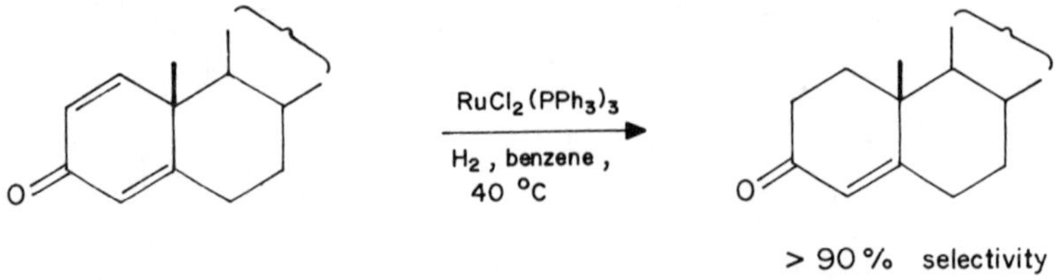

> 90 % selectivity

The anionic platinum complex $Pt(SnCl_3)_5{}^{3-}$ also appears to react by heterolytic cleavage of H_2:

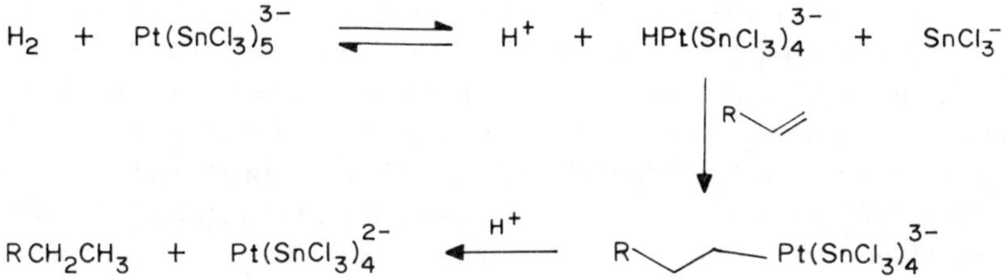

$$H_2 + Pt(SnCl_3)_5^{3-} \rightleftharpoons H^+ + HPt(SnCl_3)_4^{3-} + SnCl_3^-$$

$$\downarrow R\diagdown$$

$$RCH_2CH_3 + Pt(SnCl_3)_4^{2-} \xleftarrow{H^+} R\diagup\diagdown - Pt(SnCl_3)_4^{3-}$$

Homolytic Addition of H_2 to Metal Catalyst

This type of behavior is typically ascribed to pentacyanocobalt(II), which readily absorbs hydrogen, and will catalyze the hydrogenation of conjugated double bonds, but *not* isolated $C{=}C$ double bonds, e.g.

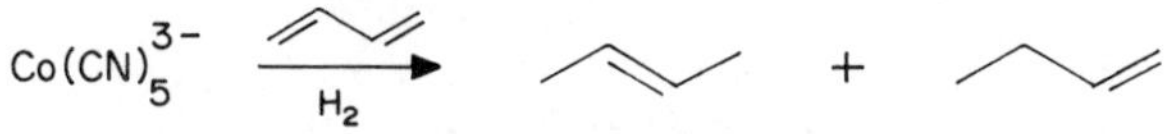

$$Co(CN)_5^{3-} \xrightarrow[H_2]{} \quad + \quad$$

There are two catalytic mechanisms for the action of this system, differing in whether or not homolytic dissociation of an intermediate alkyl complex occurs. Both involve prior reactions of the cobalt salt with H_2:

$$2\, Co(CN)_5^{3-} + H_2 \longrightarrow 2\, HCo(CN)_5^{3-}$$

Mechanism 1:

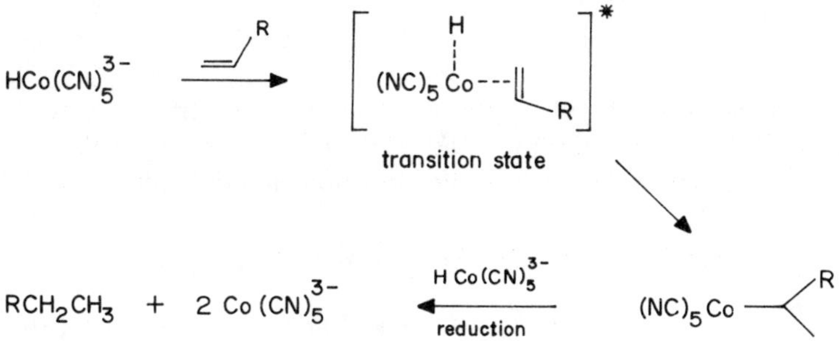

$$HCo(CN)_5^{3-} \quad \diagup R \longrightarrow \left[(NC)_5 \overset{H}{Co} {-}{-}{-} \Big|\Big|_{R} \right]^*$$

transition state

$$RCH_2CH_3 + 2\, Co(CN)_5^{3-} \xleftarrow[\text{reduction}]{H\,Co(CN)_5^{3-}} (NC)_5 Co -\!\!\!\diagup^{R}$$

Mechanism 2:

$$HCo(CN)_5^{3-} \quad \diagup R \longrightarrow \left[(NC)_5 \overset{H}{Co} {-}{-}{-} \Big|\Big|_{R} \right]^*$$

$$RCH_2CH_3 + Co(CN)_5^{3-} \xleftarrow{HCo(CN)_5^{3-}} R\diagdown_{\overset{\bullet}{CH}}\diagup CH_3 + Co(CN)_5^{3-}$$

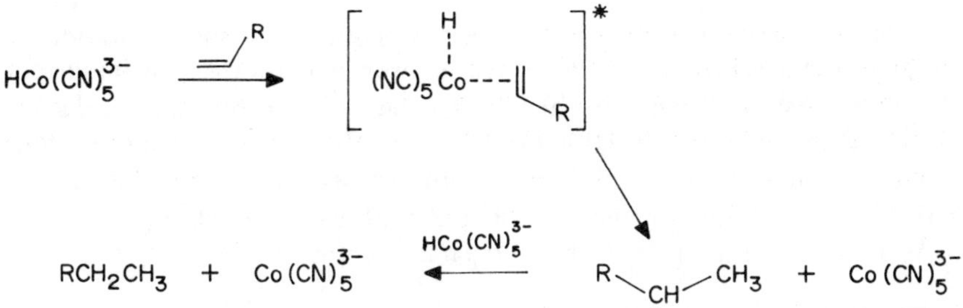

Buta-1,3-diene is reduced according to mechanism 1 and in fact the product ratio of but-2-ene to but-1-ene is changed according to the reaction conditions. For example, in the presence of excess of cyanide ion, but-1-ene is the major product. However, when excess of CN^- is absent, *trans*-but-2-ene becomes the major product. This is caused by dissociation of CN^- from the intermediate secondary σ-allyl complex, giving a π-allyl which can convert into the primary σ-allyl, as shown below:

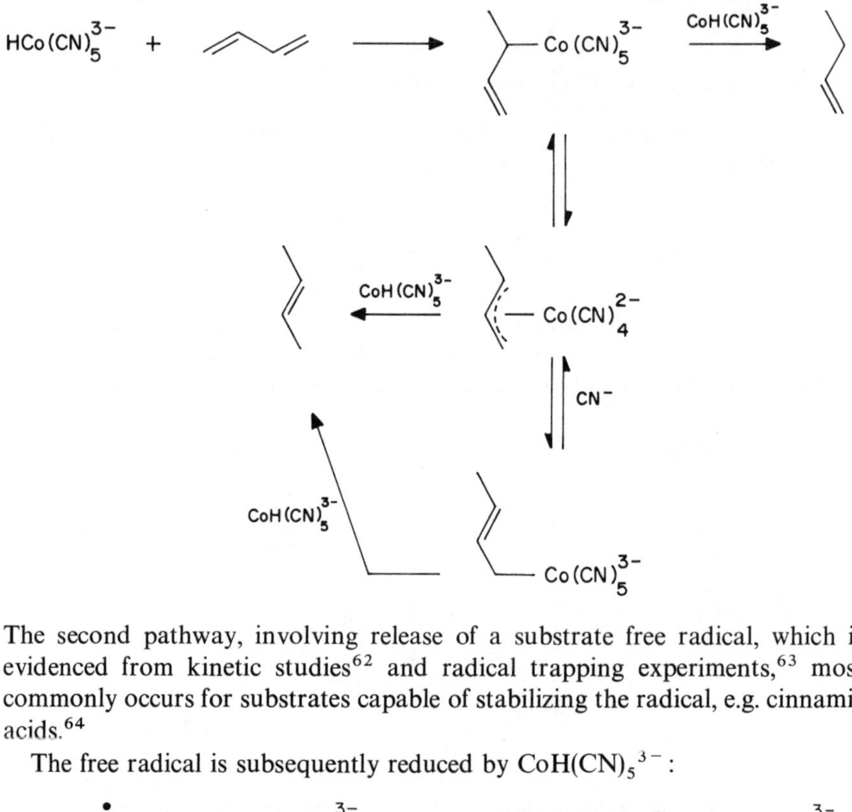

The second pathway, involving release of a substrate free radical, which is evidenced from kinetic studies[62] and radical trapping experiments,[63] most commonly occurs for substrates capable of stabilizing the radical, e.g. cinnamic acids.[64]

The free radical is subsequently reduced by $CoH(CN)_5{}^{3-}$:

$$PhCH_2\overset{\bullet}{C}HCO_2^- \; + \; CoH(CN)_5^{3-} \; \longrightarrow \; PhCH_2CH_2CO_2^- \; + \; Co(CN)_5^{3-}$$

Asymmetric Hydrogenation*

In general terms asymmetric synthesis, whereby only one enantiomer of a chiral molecule is produced from an achiral precursor, is extremely important in industries such as pharmaceutical and agrochemicals, since two optical isomers of the same molecule usually exhibit markedly different pharmacological properties. Indeed, often only one enantiomer shows the required biological activity (e.g. L-DOPA for the treatment of Parkinson's disease).

Asymmetric catalytic hydrogenation has become an extremely elegant and

*Definition: ee = $\%R - \%S$ (enantiomeric excess).

important means of producing, e.g., α-amino acids with the natural handedness. Normally one aims to utilize an optically active catalyst, based on the assumption that coordination with a prochiral olefin favors a particular orientation of the olefin, thereby leading to the formation of only one diastereoisomeric σ-alkyl complex during the reduction step. An optically active transition metal complex used for this purpose is commonly based on the use of some form of optically active phosphine ligand, which in turn can be derived either from asymmetry at the central phosphorus atoms, or asymmetry in the groups attached to the phosphorous. To date the latter approach has been the more successful.

Rhodium complexes, $RhCl_3L_3$, using S-(+)-methyl-phenyl-n-propyl-phosphine, PMePhPr, as ligands were initially used to effect asymmetric hydrogenation.[65] This complex effected the reduction of saturated acids, but in low optical yield.

15 – 28 % e.e.

Better results have been obtained using the chiral ligand 2-methoxyphenylcyclo-hexylmethylphosphine, although the mechanism of its action is still uncertain,[66] e.g.

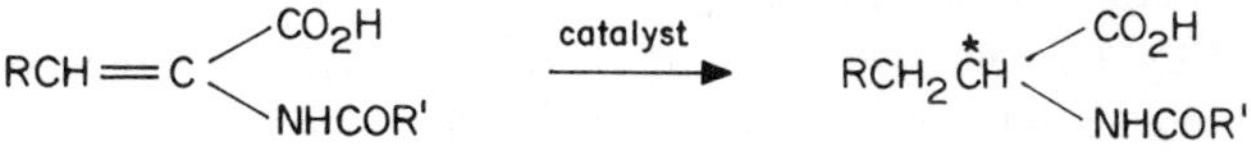

77 – 90 % e.e.

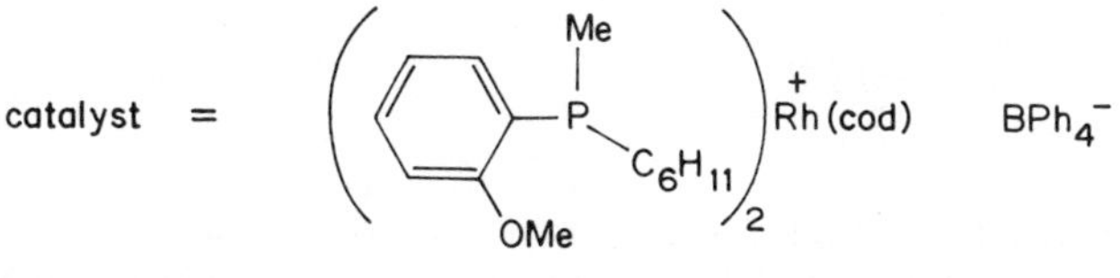

The main drawback of this approach is in the preparation of the optically active phosphine derivatives, usually only available by multistep processes.

Use of phosphines containing optically active alkyl substituents has proved to be extremely useful in recent years. Some examples are given below.

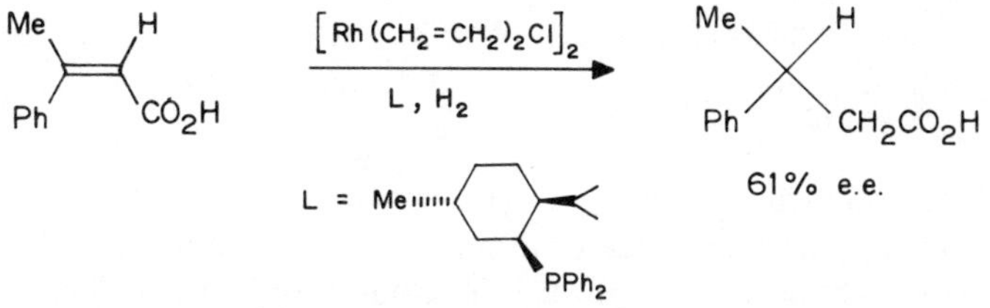

61 % e.e.

L = Me⠀⠀

(+)-neomenthyl diphenylphosphine

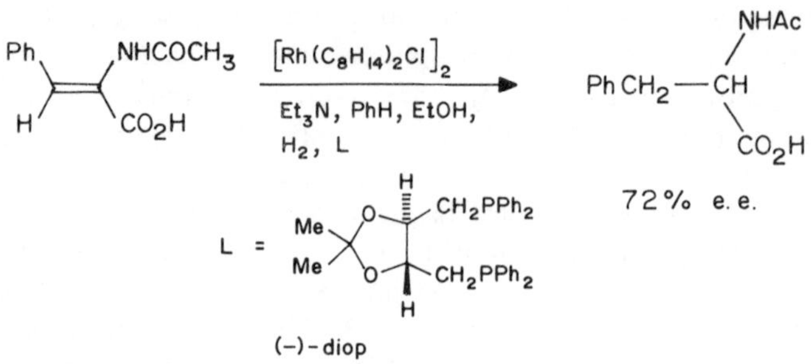

(−)-diop

The chelating diphosphine systems are extremely useful and easy to prepare, and their use as ligands with rhodium catalysts is very common.

EFFECT OF TRANSITION METAL COMPLEXATION ON THE STEREOCHEMICAL COURSE OF ELECTROCYCLIC REACTIONS

At this point it is worth discussing briefly the effect of complexation on the outcome of pericyclic reactions involving coordinated polyenes, since some understanding of the frontier orbital controlling factors is currently being obtained. One example is provided by the ring closure of cyclooctatetraene–$Fe(CO)_3$ which occurs by formal disrotatory closure of a butadiene.[67] The thermal ring closure of butadiene (to give cyclobutene) is normally conrotatory (Woodward–Hoffman rules) so the synergic interactions between cot (two double bonds) and Fe is in some way causing the 'free' diene to behave as the excited state. The exact details are not understood at present.

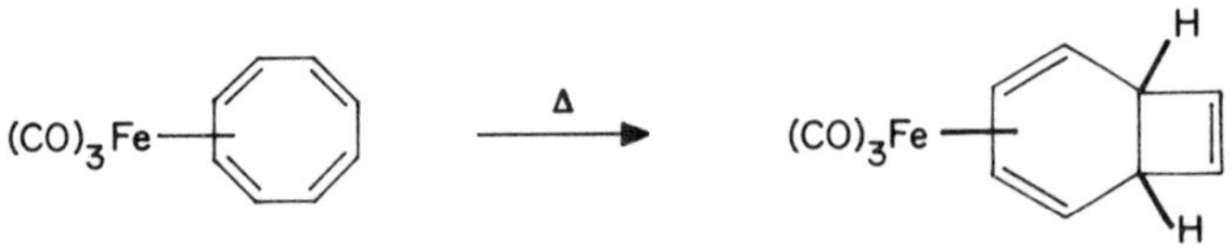

An attempt has been made to rationalize the disrotatory ring opening of a 16-electron cyclobutene–$Fe(CO)_3$ complex using frontier molecular orbital approach, e.g.

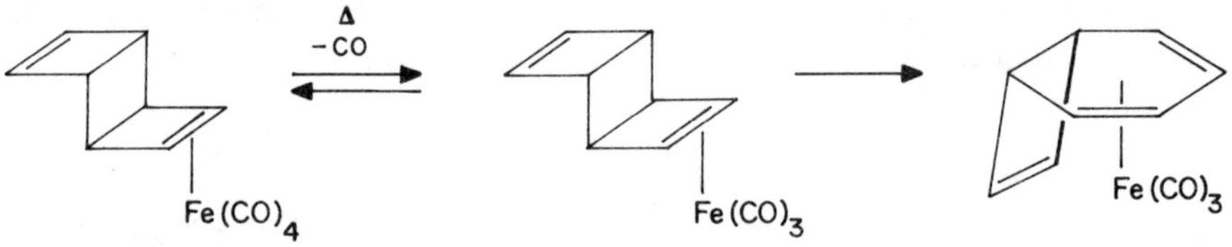

The highest occupied MO (HOMO) of the complex is described as a bonding interaction between an antisymmetric metal orbital (or hybrid combination) and the alkene π^* orbital. The Woodward–Hoffman rules for ring opening demand that the antibonding σ^* orbital mixes in a bonding manner with this HOMO. For maximum overlap to occur (including metal orbitals) the σ-bond breaking must occur in a disrotatory sense and *towards* the metal, as is indeed observed:[68]

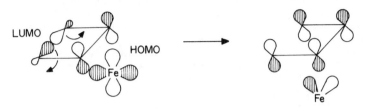

REFERENCES

1. E. Cramer, *J. Am. Chem. Soc.*, 1964, **86**, 217.
2. S. Winstein and H. J. Lucas, *J. Am. Chem. Soc.*, 1938, **60**, 836.
3. J. W. Faller and A. M. Rosan, *J. Am. Chem. Soc.*, 1976, **98**, 3388; J. W. Faller, C. C. Chen, M. J. Mattina and A. Jakubowski, *J. Organomet. Chem.*, 1973, **52**, 361.
4. J. K. Becconsall and S. O'Brien, *Chem. Commun.*, 1966, 302.
5. J. K. Becconsall and S. O'Brien, *Chem. Commun.*, 1966, 720.
6. P. Corradini, G. Maglio, A. Musco and G. Paiaro, *Chem. Commun.*, 1966, 618.
7. J. W. Faller, D. A. Haitko, R. D. Adams and D. F. Chodosh, *J. Am. Chem. Soc.*, 1977, **99**, 1654.
8. J. P. Collman, *Adv. Organomet. Chem.*, 1968, **7**, 53; *Acc. Chem. Res.*, 1968, **1**, 136. J. Halpern, *Acc. Chem. Res.*, 1970, **3**, 386; M. F. Lappert and P. W. Lednor, *Adv. Organomet. Chem.*, 1976, **14**, 345; R. D. W. Kemmit, *Inorg. React. Mech.*, 1972, **2**, 350; A. J. Deeming, *MTP Int. Rev. Sci., Inorg. Chem. Ser.*, 1972, **19**, 117; L. Vaska, *Acc. Chem. Res.*, 1968, **1**, 335; R. Ugo, *Coord. Chem. Rev.*, 1968, **3**, 319.
9. P. S. Braterman and R. J. Cross, *Chem. Soc. Rev.*, 1973, **2**, 271.
10. Review: J. R. Norton, *Acc. Chem. Res.*, 1979, **12**, 139.
11. R. R. Schrock and G. W. Parshall, *Chem. Rev.*, 1976, **76**, 263.
12. G. M. Whitesides, J. F. Gaasch and E. R. Stedronsky, *J. Am. Chem. Soc.*, 1972, **94**, 5258.
13. R. R. Schrock, *Acc. Chem. Res.*, 1977, **12**, 98.
14. W. J. Cooper and M. L. H. Green, *J. Chem. Soc., Chem. Commun.*, 1974, 761.
15. T. J. Katz, *Adv. Organomet. Chem.*, 1977, **16**, 283; N. Calderon, J. P. Lawrence and E. A. Ofstead, *Adv. Organomet. Chem.*, 1979, **17**, 447.
16. T. J. Katz, S. J. Lee and N. Acton, *Tetrahedron Lett.*, 1976, 4247.
17. J. Chauvin and D. Commereuc, *Pap. Int. Symp. Metathesis, Mainz*, 1976, p. 116.
18. C. P. Casey, L. D. Albin and T. J. Burkhardt, *J. Am. Chem. Soc.*, 1977, **99**, 2533.
19. R. Nakamura, S. Furuhare, S. Matsumoto and K. Komatsu, *Chem. Lett.*, 1976, 253; review: R. Streck, *Chem. Ztg.*, 1975, **99**, 397.
20. J. Tsuji and S. Hashiguchi, *Tetrahedron Lett.*, 1980, **21**, 2955.
21. Reviews: A. Wojcicki, *Adv. Organomet. Chem.*, 1972, **11**, 88; 1974, **12**, 33; F. Calderazzo, *Angew. Chem., Int. Ed. Engl.*, 1977, **16**, 299.
22. T. H. Coffield, J. Kozikowski and R. D. Clossom, *J. Org. Chem.*, 1957, **22**, 598.
23. R. E. Ireland and W. J. Thompson, *Tetrahedron Lett.*, 1979, 4705.
24. T. F. Murray, E. G. Samsel, V. Varma and J. R. Norton, *J. Am. Chem. Soc.*, 1981, **103**, 7520.
25. (a) C. K. Brown, D. Georgiou and G. Wilkinson, *J. Chem. Soc. A*, 1971, 3120; M. R. Snow, J. McDonald, F. Basolo and J. Ibers, *J. Am. Chem. Soc.*, 1972, **94**, 2526; M. R. Churchill, J. Wormald, D. A. Ross, J. E. Thomasson and A. Wojcicki, *J. Am. Chem. Soc.*, 1970, **92**, 1795; J. E. Thomasson, P. W. Robinson, D. A. Ross and A. Wojcicki,

Inorg. Chem., 1971, **10**, 2130; (b) S. Inagaki, H. Fujimoto and K. Fukui, *J. Am. Chem. Soc.*, 1976, **98**, 4693.

26. D. Evans, A. Yagupsky and G. Wilkinson, *J. Chem. Soc. A*, 1968, 2660; C. Yagupsky, C. K. Brown and G. Wilkinson, *J. Chem. Soc. A*, 1970, 1392, 2753.
27. T. Takahashi, T. Nagashima and J. Tsuji, *Chem. Lett.*, 1980, 369; T. Takahashi, H. Ikada and J. Tsuji, *Tetrahedron Lett.*, 1980, 3885.
28. J. Tsuji, M. Takahashi and T. Takahashi, *Tetrahedron Lett.*, 1980, 849.
29. G. F. Emerson, L. Watts and R. Pettit, *J. Am. Chem. Soc.*, 1965, **87**, 131.
30. R. Pettit, J. C. Barborak, and L. Watts, *J. Am. Chem. Soc.*, 1966, **88**, 1328.
31. P. Dowd, *J. Am. Chem. Soc.*, 1966, **88**, 2587.
32. R. H. Grubbs, T. A. Pancoast and R. A. Grey, *Tetrahedron Lett.*, 1974, 2425.
33. H. A. Brune, G. Horlbeck, H. Rottele and U. Tanger, *Z. Naturforsch., Teil B*, 1973, **28**, 68.
34. D. Stierle, E. R. Biehl and P. C. Reevs, *J. Organomet. Chem.*, 1976, **72**, 221.
35. R. E. Davis, H. D. Simpson, N. Grice and R. Pettit, *J. Am. Chem. Soc.*, 1971, **93**, 6688.
36. G. F. Emerson, K. Ehrlich, W. P. Giering and P. C. Lauterbur, *J. Am. Chem. Soc.*, 1966, **88**, 3172; K. Erhlich and G. F. Emerson, *J. Am. Chem. Soc.*, 1972, **94**, 2462.
37. B. M. Trost and D. M. T. Chan, *J. Am. Chem. Soc.*, 1979, **101**, 6429, 6432; 1980, **102**, 6359.
38. W. R. Roth and J. D. Meier, *Tetrahedron Lett.*, 1967, 2053.
39. B. F. G. Johnson, J. Lewis and D. J. Thompson, *Tetrahedron Lett.*, 1974, 3789.
40. Review: T. Kametani and H. Nemoto, *Tetrahedron*, 1981, **37**, 3.
41. E. Weiss, W. Hubel and R. Merenyi, *Chem. Ind. (London)*, 1960, 407.
42. E. Weiss and W. Hubel, *J. Inorg. Nucl. Chem.*, 1959, **11**, 42.
43. E. O. Fischer and H. Werner, *Metal π-Complexes*, Elsevier, Amsterdam, 1966, Vol. 1; E. Weiss and W. Hubel, *Angew. Chem.*, 1961, **73**, 298; *Chem. Ber.*, 1962, **95**, 1186.
44. W. von Doering and D. W. Wiley, *Tetrahedron*, 1960, **11**, 183.
45. G. T. Rodeheaver, G. C. Farrant and D. F. Hunt, *J. Organomet. Chem.*, 1971, **30**, C22.
46. Reviews: H. Sinn and W. Kaminsky, *Adv. Organomet. Chem.*, 1980, **18**, 99; P. W. Jolly and G. Wilke. *The Organic Chemistry of Nickel*, Academic Press, New York, 1975, Vol. 2; A. C. L. Su, *Adv. Organomet. Chem.*, 1978, **17**, 269; G. W. Parshall, *Homogeneous Catalysis*, Wiley, New York, 1980.
47. K. J. Irvin, J. J. Rooney, C. D. Stewart, M. L. H. Green and R. Mahtab, *J. Chem. Soc., Chem. Commun.*, 1978, 604.
48. E. R. Evitt and R. G. Bergman, *J. Am. Chem. Soc.*, 1979, **101**, 3973.
49. J. Tsuji, I. Shimizu, H. Suzuki and Y. Naito, *J. Am. Chem. Soc.*, 1979, **101**, 5070.
50. W. Reppe, O. Schlichting, K. Klager and T. Toepel, *Justus Liebigs Ann. Chem.*, 1948, **560**, 104.
51. K. P. C. Vollhardt and R. G. Bergman, *J. Am. Chem. Soc.*, 1976, **96**, 4996.
52. R. L. Funk and K. P. C. Vollhardt, *J. Am. Chem. Soc.*, 1979, **101**, 215.
53. R. L. Funk and K. P. C. Vollhardt, *J. Am. Chem. Soc.*, 1980, **102**, 5253.
54. H. Yamazaki and Y. Wakatsuki, *J. Organomet. Chem.*, 1977, **139**, 157, 169; J. P. Collman, J. W. Kang, W. F. Little and M. F. Sullivan, *Inorg. Chem.*, 1968, **7**, 1298.
55. J. Tsuji, H. Nagashima and K. Hori, *Chem. Lett.*, 1980, 257.
56. W. C. Still and M. T. Tsai, *J. Am. Chem. Soc.*, 1980, **102**, 3656.
57. T. Katsuki and K. B. Sharpless, *J. Am. Chem. Soc.*, 1980, **102**, 5974; B. E. Rossiter, T. Katsuki and K. B. Sharpless, *J. Am. Chem. Soc.*, 1981, **103**, 464.
58. C. Masters, *Homogeneous Transition-metal Catalysis*, Chapman and Hall, London, 1981; B. R. James, *Homogeneous Hydrogenation*, Wiley, New York, 1974; F. J. McQuillin, *Homogeneous Hydrogenation in Organic Chemistry*, Reidel, Dordrecht, 1976; A. P. G. Kieboom and F. van Rantwijk, *Hydrogenation and Hydrogenolysis in Synthetic Organic Chemistry*, Delft University press, Delft, 1979; P. N. Rylander, *Organic Synthesis with Noble Metal Catalysts*, Academic Press, New York, 1973, Ch. 2.

59. J. F. Young, J. A. Osborne, F. H. Jardine and G. Wilkonson, *J. Chem. Soc., Chem. Commun.*, 1965, 131; R. S. Coffey, ICI Ltd., *Br. Pat.*, 1,121,642 (1965).
60. J. F. Knifton, *J. Org. Chem.*, 1975, **40**, 519; 1976, **41**, 1200; *Tetrahedron Lett.*, 1975, 2163; J. Tsuji and H. Suzuki, *Chem. Lett.*, 1977, 1085.
61. S. Nishimura and K. Tsuneda, *Bull. Chem. Soc. Jpn.*, 1969, **62**, 852.
62. J. Halpern and L. Y. Wong, *J. Am. Chem. Soc.*, 1968, **90**, 6665; L. Simandi and F. Nagy, *Acta Chim. Acad. Sci. Hung.*, 1965, **46**, 137.
63. J. Kwiatek and J. K. Seyler, *Adv. Chem.*, 1968, **70**, 2073.
64. J. Kwiatek, I. L. Mador and J. K. Seyler, *Adv. Chem.*, 1963, **37**, 201; *J. Am. Chem. Soc.*, 1962, **84**, 304.
65. W. S. Knowles and M. J. Sabacky, *J. Chem. Soc., Chem. Commun.*, 1968, 1445; L. Horner, H. Siegel and H. Butte, *Angew. Chem., Int. Ed. Engl.*, 1968, **7**, 942.
66. W. S. Knowles, M. H. Sabacky and B. D. Vineyard, *J. Chem. Soc., Chem. Commun.*, 1972, 10; *Ann. N.Y. Acad. Sci.*, 1973, **214**, 119.
67. M. Cooke, J. A. K. Howard, C. R. Russ, F. G. A. Stone and P. Woodward, *J. Organomet. Chem.*, 1974, **78**, C43; *J. Chem. Soc., Dalton Trans.*, 1976, 70.
68. A. R. Pinhas and B. K. Carpenter, *J. Chem. Soc., Chem., Commun.*, 1980, 15.

CHAPTER 3

σ-Alkyl Complexes

The most progress has been made in preparing σ-alkyl complexes which do not contain β-hydrogen, thereby preventing decomposition by a β-elimination pathway.[1] It was originally thought that all transition metal alkyls were unstable, until it became generally recognized that the apparent instability was kinetic rather than thermodynamic, by virtue of the low activation energy pathways available for their decomposition. It was also thought that stable metal alkyls must be diamagnetic or else be coordinatively saturated. Some striking examples now known are WMe_6, containing an obvious abundance of M—C σ bonds, $Ti(CH_2SiMe_3)_4$, which grossly violates the 18-electron rule, and the paramagnetic derivative $V(CH_2Ph)_4$. The groups Me, CH_2SiMe_3 and CH_2Ph are therefore alkyl ligands which do not undergo β-elimination. Other obvious choices are t-butyl and (bridgehead) norbornyl, the latter giving rise to anti-Bredt olefins from elimination. Bulky ligands are also useful in suppressing bimolecular decomposition pathways, although these are sometimes prone to α-elimination.

The most common synthetic approach to transition metal alkyls involves reaction of a Grignard or organolithium reagent with a transition metal halide, although the experimental conditions employed are often critical, e.g.

$$4 \ PhCH_2MgCl \ + \ ZrCl_4 \ \xrightarrow{Et_2O} \ Zr(CH_2Ph)_4$$

The complexes may be subdivided into neutral, anionic and cationic complexes of both homoleptic alkyls, and those containing other ligands, e.g. nitrogen ligands and η^5-cyclopentadienyl.

HOMOLEPTIC COMPLEXES—PREPARATION AND REACTIONS

These complexes are prepared almost exclusively by the Grignard method.

Neutral Complexes.

The bright yellow complex $TiMe_4$ was first prepared in 1959 by Clauss and Beermann:[2]

$$\text{TiCl}_4 \quad + \quad 4 \text{ MeLi} \quad \xrightarrow[-78\ ^\circ\text{C}]{\text{Et}_2\text{O}} \quad \text{TiMe}_4 \quad + \quad 4 \text{ LiCl}$$

The desired product and ether are co-distilled from the reaction mixture at ca. $-30\ ^\circ$C, and decomposition occurs near room temperature to give mainly methane, together with traces of ethene and ethane. The complex may be obtained crystalline by removing ether from an ether–hexane solution at $-78\ ^\circ$C, but it is not stable above this temperature. Preparation of other methyl complexes are summarized below:

$$\text{ZrCl}_4 \quad + \quad 4 \text{ MeLi} \quad \xrightarrow[-45\ ^\circ\text{C}]{\text{ether}-\text{toluene}} \quad \text{ZrMe}_4 \quad + \quad 4 \text{ LiCl}$$

$$\text{TaMe}_3\text{Cl}_2 \quad \xrightarrow[-78\ \rightarrow\ 0\ ^\circ\text{C}]{\text{MeLi}/\text{Et}_2\text{O}} \quad \text{TaMe}_5$$

$$\text{WCl}_6 \quad \xrightarrow[\substack{\text{Et}_2\text{O} \\ -20\ ^\circ\text{C}}]{3\ \text{MeLi}} \quad \substack{\text{WMe}_6 \\ ca\ 40\%}$$

$$\text{MnI}_2 \quad \xrightarrow[\text{Et}_2\text{O}]{\text{MeLi}} \quad \text{MnMe}_2$$

The stability characteristics of these complexes show wide variation. For example, TaMe_5 can be isolated as volatile, pale yellow needles which melt ca. $10\ ^\circ$C to a yellow oil, decomposing smoothly at $25\ ^\circ$C to give methane. The methyl groups occur at $\delta\ 0.82$ in the NMR spectrum (toluene-d_8, $-10\ ^\circ$C). The complex of TiMe_4 with dioxane, TiMe_4 (diox), explodes at $0\ ^\circ$C, while other TiMe_4L complexes are more stable in the order $\text{L} = \text{diox} < \text{NMe}_3 < \text{PMe}_3 < \text{py}$. A number of complexes of other metals, e.g. TiMe_3 and CrMe_4 are only apparently formed in solution at low temperature and are too unstable to allow isolation.

Whilst phenyl complexes were believed for sometime to be inherently more stable than alkyl complexes, this in fact applies only to those alkyls possessing β-hydrogen. Neutral phenyl complexes are, however, difficult to prepare, since reduction of the metal and biphenyl formation are common, and the difficulties may also be ascribed to a low-energy decomposition route involving ortho-hydrogen abstraction, analogous to β-hydride elimination. Some examples of preparation and decomposition of these complexes are given below.

$$4 \text{ PhLi} \quad + \quad \text{TiCl}_4 \quad \xrightarrow[-70\ ^\circ\text{C}]{\text{Et}_2\text{O}} \quad \text{Ph}_4\text{Ti} \quad \xrightarrow{25\ ^\circ\text{C}} \quad \text{Ph}_2\text{Ti} \quad + \quad \text{PhPh}$$

A convenient method of testing for the persistence of any unreacted phenyllithium in this reaction is to add carbon dioxide. Phenyllithium gives benzoic acid, whereas Ph_4Ti is unreactive.[3] The Ph_2Ti complex is a black, easily characterized crystalline material. Similarly:

$$ZrCl_4 + 4PhLi \xrightarrow[-40\ °C]{Et_2O} Ph_4Zr + 4LiCl$$

$$TiCl_4(py)_2 + 4PhMgBr \xrightarrow[-16\ °C]{Et_2O} TiPh_4$$

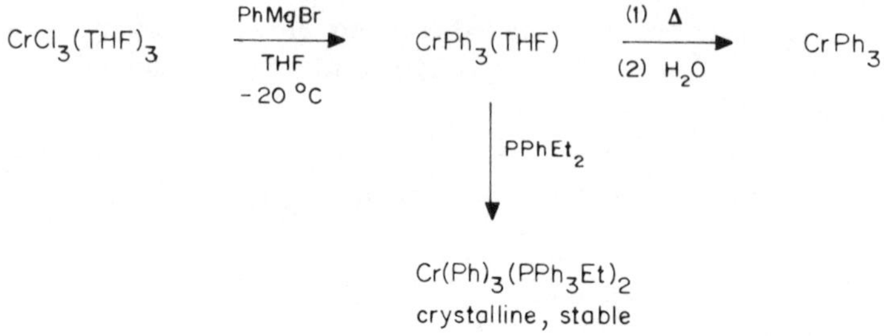

$$CrCl_3(THF)_3 \xrightarrow[\substack{THF \\ -20\ °C}]{PhMgBr} CrPh_3(THF) \xrightarrow[(2)\ H_2O]{(1)\ \Delta} CrPh_3$$

$$\downarrow PPhEt_2$$

Cr(Ph)$_3$(PPh$_3$Et)$_2$
crystalline, stable

$$Cr(CHPh_2)Cl_2(THF)_2 \xrightarrow[THF]{PhLi} CrPh_2(CHPh_2)(THF)_2$$

Alkyl ligands such as neopentyl, trimethylsilylmethyl and benzyl are similar in that they have no β-hydrogens; the former two are also rather bulky ligands. The benzyl complexes, in addition to their scientific interest, also have practical potential as olefin polymerization catalysts. Examples are shown below, and it may be noted that Ti(CH$_2$Ph)$_4$, a red solid, m.p. 70 °C, is much easier to prepare than TiPh$_4$.

$$MCl_4 \xrightarrow[Et_2O]{PhCH_2MgCl} M(CH_2Ph)_4$$
$$M = Ti,\ Zr,\ Hf$$

$$VCl_4 \xrightarrow[pentane,\ ether]{PhCH_2MgCl} V(CH_2Ph)_4(Et_2O)$$

Some reactions of the titanium group tetrabenzyl derivatives are given below.

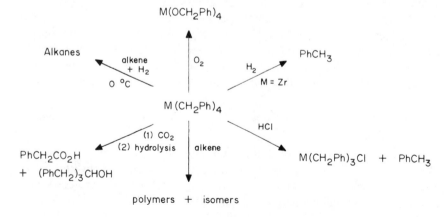

Examples of neopentyl, Me_3SiCH_2—, and other bulky alkyl group complexes are shown in the following examples:

$$MCl_4 \ + \ 4Me_3CCH_2Li \quad \xrightarrow{\quad Et_2O \quad} \quad M(CH_2CMe_3)_4$$

$$MCl_4 \quad \xrightarrow{\quad Me_3SiCH_2Li \quad} \quad M(CH_2SiMe_3)_4$$

$$M = Ti, Zr, Hf$$

Usually the neopentyl derivatives of Group IV are pale yellow to colorless, low-melting solids which sublime without decomposition *in vacuo*, whereas the corresponding trimethylsilylmethyl complexes are all liquids at room temperature.

$$MCl_4 \ + \quad \text{[structure: norbornyl-Li]} \quad \xrightarrow{\quad pentane \quad} \quad (1-norbornyl)_4\,M$$

$$M = Ti, Zr, Hf, V, Cr \text{ from } CrCl_3$$

All of these 1-norbornyl complexes are pentane-soluble, monomeric and can be passed through an alumina column unchanged. Their decomposition by the normal β-elimination is disfavored since this would produce a bridgehead double bond (anti-Bredt). The *t*-butyl ligand is normally very susceptible to β-elimination but $Cr(CMe_3)_4$ can be prepared in 60% yield from $Cr(OCMe_3)_4$ and $LiCMe_3$ in pentane. Its stability is probably due to the tight packing of the methyl groups, thereby preventing a proper orientation for β-elimination.

Anionic Complexes

Most anionic metal alkyls contain one or more alkali metal counter ions, usually lithium, owing to the method of preparation. In fact, Li^+ usually attaches itself in some way to the 'bridging' alkyl group, resulting in an Li—R—M arrangement. Perhaps the best known examples of anionic alkyl complexes, from the organic chemist's standpoint, are the lithium dialkyl cuprates. These compounds are discussed more fully later. Some examples of preparation of anionic methyl complexes are given in the following equations.

$$TiMe_4 \quad \xrightarrow[\substack{Et_2O \\ \text{then dioxane}}]{RLi} \quad Li[TiMe_4R] \cdot 2diox$$

$$R = Me, Ph, CH_2Ph$$

$$ZrCl_4 \quad \xrightarrow[Et_2O-toluene]{>6 \ MeLi} \quad Li_2[ZrMe_6]$$

$$CrCl_3 \quad \xrightarrow[Et_2O, \ -18\,°C]{6 \ MeLi} \quad Li_3[CrMe_6]$$

Again, there is a variety of stability associated with these complexes. The titanium derivatives $Li[TiR_5] \cdot 2diox$ all decompose at -20 to $0\,°C$, while $Li_2[ZrMe_6]$ may be isolated as bright yellow crystals which are stable under nitrogen at $0\,°C$ for several hours. The chromium complex $Li_3(CrMe_6) \cdot 3diox$ is isolated as blood-red crystals on addition of dioxane to the complex shown above. It is insoluble in hexane and benzene, but soluble and stable in ether and dioxane, in which it is monomeric and undissociated. The crystal and molecular structure of this compound have been determined by X-ray analysis and low-temperature IR spectroscopy. The six methyl groups are arranged in an octahedron at a distance of 2.30 Å around the chromium atom. Weak distortion of the octahedron occurs by the interaction of three Li atoms ($Li-Me = 2.17$ Å) at three edges and each Li atom is surrounded tetrahedrally by two oxygen atoms of two different dioxane molecules and by two methyl groups. The dioxane molecules connect the chromium complexes in the solid state, so the overall structure can be written as $\{Cr[(Me)_2-Li-2(diox)_{1/2}]_3\}_n$. This structure is shown below:[4]

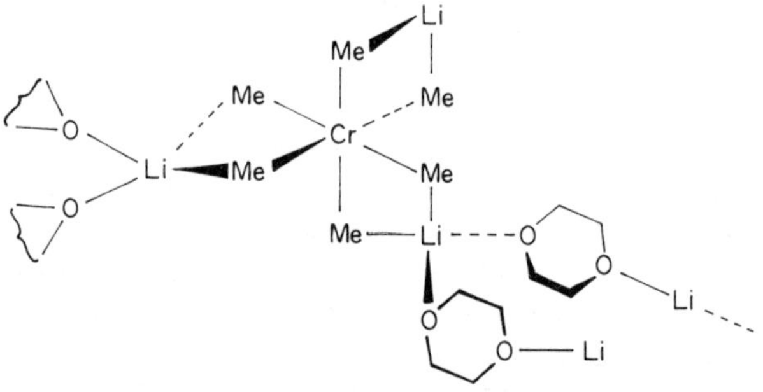

Slow decomposition of Li_3CrMe_6 in ethers gives a dimeric Cr(II) complex whose reduced paramagnetism ($\mu = 0.57$) suggests a Cr—Cr bond. The X-ray structure has been determined and is shown (partial) below.[5]

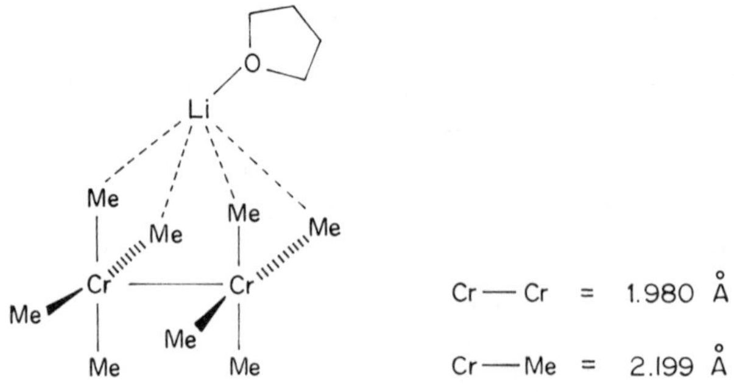

Cr — Cr = 1.980 Å

Cr — Me = 2.199 Å

A number of anionic σ-phenyl transition metal complexes are also known. Treatment of $CrCl_3$ with Grignard reagents normally leads to reduction to Cr(0), but reaction with PhLi produces a complex of formula $(LiPh)_3Cr(Ph)_3(Et_2O)_{2.5}$. It is obtained as an orange crystalline material in high yield. Preparations of other phenyl complexes are shown below; again the types of complex produced depend on the reaction conditions.

$$NaPh \ + \ CrCl_5 \ \longrightarrow \ Na_2CrPh_5(Et_2O)_3$$

$$4 \ Na_2CrPh_5 \ + \ CrCl_3 \ \longrightarrow \ 5 \ NaCrPh_4 \ + \ 3 \ NaCl$$

$$2 \ Li_3CrPh_6 \ + \ CrCl_3 \ \longrightarrow \ 3 \ LiCrPh_4 \ + \ 3 \ LiCl$$

$$2 \ Li_3Cr^{III}Ph_6(Et_2O)_{2.5} \ + \ CrCl_3(THF)_3 \ \longrightarrow \ 3 \ Li \ Cr^{II}Ph_3(Et_2O)_{1.5} \ +$$
$$1.5 \ Ph_2 \ + \ 2 \ LiCl$$

$$MoCl_5 \ \xrightarrow[\ Et_2O, \ -30 \ °C\ ]{\ 10 \ equiv. \ PhLi\ } \ (LiPh)_3MoPh_3(Et_2O)_3$$
$$20\%$$

$$VCl_3(THF)_3 \ \xrightarrow[\ THF\ ]{\ 10 \ PhLi\ } \ (LiPh)_4VPh_2(THF)_4$$

$$Cp_2V \ \xrightarrow[\ Et_2O\ ]{\ 7 \ PhLi\ } \ (LiPh)_5VPh_2(Et_2O)_5$$

$$NbBr_5 \ \xrightarrow[\ Et_2O\ ]{\ 8 \ PhLi\ } \ (LiPh)_3NbPh_4(Et_2O)_3$$

$$\xrightarrow[\ Et_2O\ ]{\ 9 \ PhLi\ } \ (LiPh)_4NbPh_2(Et_2O)_{3.5}$$

$$TaCl_5 \ \xrightarrow[\ (2) \ THF\ ]{\ (1) \ 5 \ PhLi, \ benzene, \ 20 \ °C\ } \ \left[Li(THF)_4\right]^+ \left[TaPh_6\right]^-$$

As with the electroneutral alkyl complexes discussed above, the preparation of anionic derivatives containing neopentyl and trimethylsilylmethyl groups has been popular. Some examples are

$$CrCl_3(THF)_3 \ \xrightarrow[\ THF\ ]{\ LiCH_2CMe_3\ } \ Li^+\left[Cr(CH_2CMe_3)_4\right]^-$$

$$\xrightarrow[\ THF\ ]{\ LiCH_2SiMe_3\ } \ Li^+\left[Cr(CH_2SiMe_3)_4\right]^-$$

The neopentyl complex is dark blue and the silyl derivative is blue–green. Both are stable only in thf, and attempts to isolate crystalline salts by addition of large cations have been unsuccessful.

NON-HOMOLEPTIC COMPLEXES

Most commonly, transition metal alkyls containing halide, CO, olefins and η^5-cyclopentadienyl ancillary ligands are encountered. There is a greater range of methods for the preparation of these complexes than for the corresponding homoleptic alkyls. Thus, preparations based on the Grignard method, reactions of alkyl halides with metal anions, oxidative addition, insertion and elimination (or de-insertion) reactions are all well characterized. The following discussion is subdivided according to these methods.

Preparation by the Grignard Method

Halides.

Modification of the Grignard approach described above results in the preparation of halide–alkyl complexes MR_yX_z. Examples are given below

$$TiCl_4 + AlMe_2Cl \rightleftharpoons TiMeCl_3 + AlMeCl_2 \xrightarrow{NaCl} Na^+ \; AlMeCl_3$$

$$TiCl_4 \xrightarrow[\text{or } Me_2Zn]{MeMgCl} TiMeCl_3$$
(deep violet solid)

As with the foregoing complexes, much lower stability is found for complexes containing alkyl groups other than methyl. $TiRCl_3$ complexes have been prepared where $R = Et$, Pr, $i\text{-}Pr$, C_5H_{11}, using the alkylaluminum or zinc reagents, but they are stable only at low temperatures. Pure $TiEtCl_3$ can be obtained as a violet low-melting solid by vacuum distillation.

$$NbCl_5 \xrightarrow[\text{pentane}]{Me_2Zn} NbMe_2Cl_3$$

$$TaCl_5 \xrightarrow{Me_2Zn} TaMe_3Cl_2$$

$$VCl_4 \xrightarrow{BMe_3} VMe_2Cl_2 + VMeCl_3$$

$$\xrightarrow{BEt_3} VEt_2Cl_2$$

$$TaCl_5 \xrightarrow[\text{Et}_2O]{3 \; Me_3CCH_2MgCl} Ta(CH_2CMe_3)_3Cl_2$$

$$MCl_5 \xrightarrow{Zn(CH_2SiMe_3)_2} M(CH_2SiMe_3)_3Cl_2$$
$$+ \; M(CH_2SiMe_3)_3Cl_3$$
$$+ \; M(CH_2SiMe_3) \; Cl_4$$
e.g. M = Ta, Mo

$$CrCl_3(THF)_3 + AlR_3 \xrightarrow{THF} CrRCl_2(THF)_3$$
R = Me, Et, Pr, *i*-Bu

A number of phenyl complexes have also been prepared:

$$TiCl_4 \xrightarrow[\text{pentane, } -20 \text{ °C}]{Ph_2Zn} TiPhCl_3$$

$$WCl_6 \xrightarrow[\text{pentane, reflux}]{SnPh_4} WPhCl_3$$

The above types of alkyl metal halide complexes offer some possibilities for synthetic application. For example, it has been shown that tertiary halides are successfully methylated using $MeTiCl_3$. The same type of conversion may also be achieved using $Me_2Zn–ZnCl_2$, e.g.[6]

$$\xrightarrow[\substack{\text{or } Me_2Zn-ZnCl_2 \\ -40 \text{ °C, } 0.5 \text{ h}}]{MeTiCl_3}$$

80 – 90 %

η^5-Cyclopentadienyl and Group V donor complexes

It is widely accepted that η^5-cyclopentadienyl (Cp) and phosphine ligands stabilize the M—L σ-bond by occupying coordination sites which would otherwise be used in decomposition pathways. We shall concentrate on the Cp complexes, for which there are similar methods of preparation to those discussed above, e.g.

$$Cp_2ZrCl_2 \xrightarrow[\text{or } MeMgX]{MeLi} Cp_2ZrMe_2$$

$$Cp_2ZrCl_2 \xrightarrow{PhLi} Cp_2ZrPh_2$$

Thermal stabilities vary, usually $R = Me < CH_2Ph < Ph < CH_2SiMe_3$ and $Ti < Zr$. Whilst the colorless complex Cp_2ZrMe_2 sublimes with little decomposition at 100 °C, the analogous orange–yellow Cp_2TiMe_2 blackens in the solid state at 40 °C and decomposes in solution at 0 °C.

Other complexes are as follows:

$$Cp_2ZrMe_2 \xrightarrow{PbCl_2} Cp_2ZrMe(Cl) \xrightarrow{LiAlH_4} Cp_2Zr(H)Me$$

$$Cp_2ZrCl_2 \xrightarrow[\text{CH}_2\text{Cl}_2, \, 0 \text{ °C}]{Et_3Al} Cp_2ZrEt(Cl)$$

$$Cp_2TiCl_2 \xrightarrow{1 \text{ mol } RMgX} Cp_2TiR(Cl)$$

$$Cp_2TiCl \xrightarrow{RMgX} Cp_2TiR$$

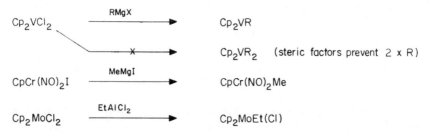

$$Cp_2VCl_2 \xrightarrow{\text{RMgX}} Cp_2VR$$

$$\xrightarrow{\text{x}} Cp_2VR_2 \quad \text{(steric factors prevent 2 x R)}$$

$$CpCr(NO)_2I \xrightarrow{\text{MeMgI}} CpCr(NO)_2Me$$

$$Cp_2MoCl_2 \xrightarrow{\text{EtAlCl}_2} Cp_2MoEt(Cl)$$

A range of alkyl complexes containing auxiliary ligands other than those presented above are also known. A number of isopropoxytitanium alkyls have been prepared. For example, the methyl complex $(i\text{-PrO})_3TiMe$ is a distillable viscous yellow liquid. Preparations of some of these complexes are shown below.

$$Ti(OPr^i)_4 \xrightarrow[\text{LiBr, Et}_2O]{\text{LiPh}} \xrightarrow{0.25 \ TiCl_4} PhTi(OPr^i)_3$$

$$Ti(OPr^i)_3Cl \xrightarrow{\text{MeLi}} MeTi(OPr^i)_3$$

$$2 \ TiMe(OR)_3 + TiMeCl_3 \xrightarrow{CH_2Cl_2} 3 \ TiMe(OR)_2Cl$$

$$ZrPh_2(Et_2O)_2 \xrightarrow{\Delta} ZrPh(OEt)$$

Other Ancillary Ligands

Another counter ligand which has been used is the oxo group, an example being $V(O)PhCl_2$, which is a red crystalline compound. The reaction conditions are often critical.[7]

$$V(O)Cl_3 \xrightarrow[\substack{\text{or } Ph_2Zn \\ -25 \ ^\circ C}]{\text{Ph}_2Hg} V(O)PhCl_2$$

$$V(O)(OPr^i)_3 \xrightarrow{\text{Me}_2Zn} V(O)Me(OPr^i)_2$$

$$Nb(O)Cl_3 \xrightarrow[\substack{\text{toluene} - Et_2O \\ -15 \ ^\circ C}]{\text{MeMgI}} \left[Nb(O)MeCl_2\right] \xrightarrow{OPMe_3} NbMe(O)Cl_2(OPMe_3)_2$$

not isolable

$$Cp_2W(O)Cl_2 \xrightarrow[\text{Et}_2O \text{ or THF}]{\text{RMgI}} Cp_2W(O)R_2$$

$$R = Me, Et, Ph, CH_2Ph$$

$$Re(O)Cl_4 \xrightarrow{\text{MeLi}} Re(O)Me_4$$

Various other similar complexes are summarized below.

$$Ti(NR_2)_3X \xrightarrow{R'MgX} Ti(NR_2)_3(R')$$

$$(COT)ZrCl_2 \xrightarrow{RMgX} (COT)ZrR_2$$

$$R = Me, Et$$

$$NbCl_5 \xrightarrow[\text{(2) LiR}]{\text{(1) } K_2^{2+}(COT)^{2-}} (COT)_2NbR$$

$$Me_3TaCl_2 \xrightarrow{K_2^{2+}(COT^{2-})} (COT)TaMe_3$$

Preparation Using Insertion Reactions.

Formation of a transition metal—carbon σ-bond by this approach most commonly involves the insertion of some carbon-containing species into a metal—hydrogen bond. No mechanistic significance is meant to be implied by the name 'insertions reaction,' which merely describes the outcome of the reaction. The reaction is of considerable significance in that it is often implicated in transition metal-catalyzed reactions of olefins such as Ziegler–Natta polymerizations, carbonylations, dimerization or hydrogenation (see Chapter 2). The inserting species may be a carbene, olefin, acetylene or epoxide. Examples of these follow. The first sub-section deals with hydrozirconation, a reaction which allows the isolation and characterization of a range of interesting and useful insertion products.

Hydrozirconation and its synthetic applications

The above M—H insertion reaction, using zirconium complexes and resulting in hydrozirconation, has been extensively studied by Hart and Schwartz, e.g.[8]

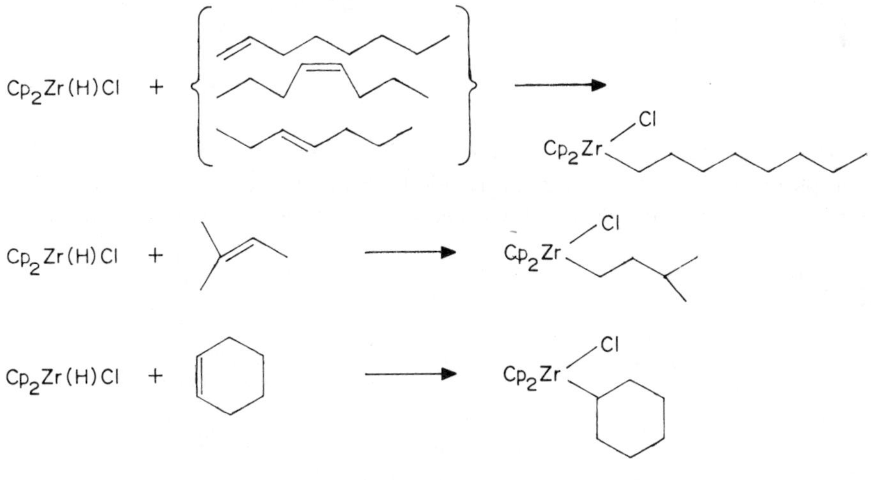

As can be seen, hydrozirconation of alkenes always occurs to give the alkylzirconium complexes in which the metal is attached to the sterically least hindered, accessible position of the olefin *as a whole*, in many cases via double bond migration. The reaction is of some potential for organic synthesis, since the product σ-alkyl complexes react with a range of electrophiles. Also, the reagents are inexpensive, easy to prepare, and require only moderate care in their handling. Some reactions are as follows:[9]

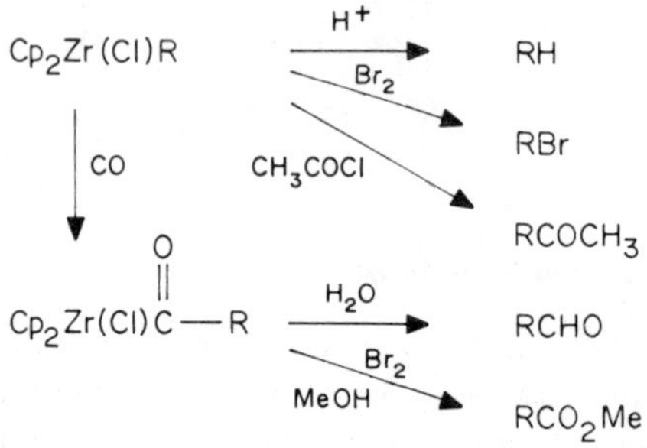

The reaction and subsequent manipulation also proceed smoothly with conjugated dienes:[10]

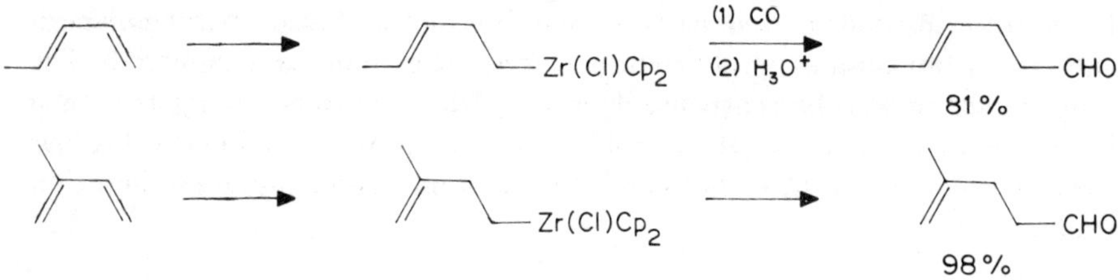

The alkylzirconium derivatives can be converted to alkylaluminum compounds by transmetallation:[11]

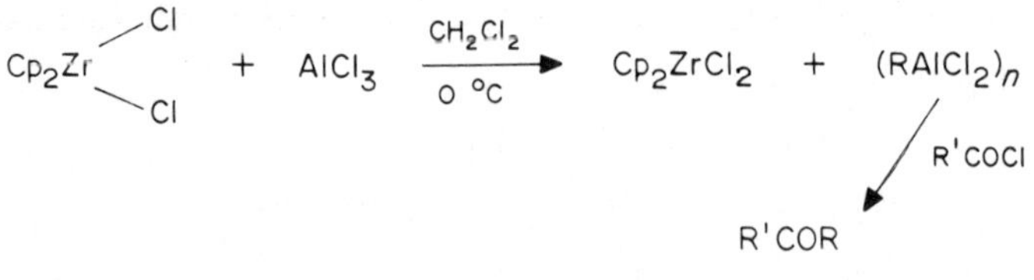

The stereochemistry of cleavage of the Zr—C bond in the product σ-alkyl complexes, by a number of electrophilic reagents, has been studied using the diastereoisomeric complexes shown below, coupled with NMR spectroscopy.

Cleavage with Br_2, I_2 and *N*-bromosuccinimide (nbs) gives the halide with retention at the carbon atom, CO insertion proceeds with retention, and reaction with molecular oxygen to give an alcohol proceeds with *some* loss of stereochemical integrity.

Hydrozirconation also proceeds using acetylenes to give σ-vinyl zirconium complexes, this time without movement of the double bond. Treatment of these with various electrophiles leads to a convenient preparation of, e.g., vinyl halides:[12]

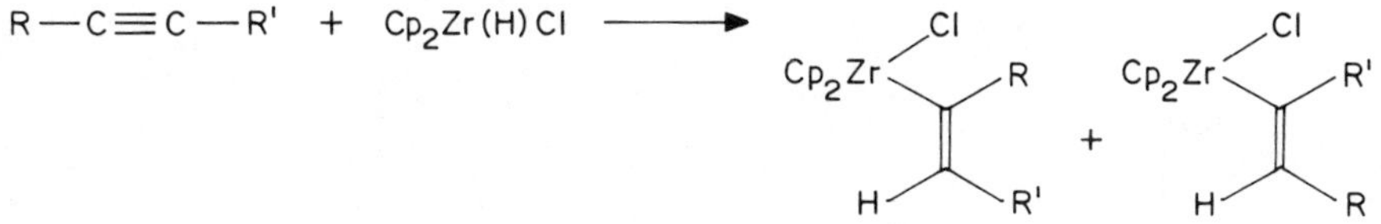

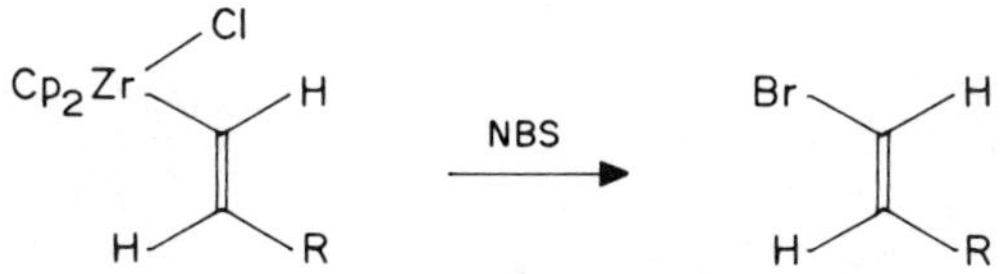

Terminal acetylenes invariably give a single complex by attachment of metal to the unsubstituted carbon, and this gives rise to a number of useful synthetic applications, such as the preparation of a vinyl iodide required in Grieco *et al.*'s synthesis of the macrolide antibiotic methynolide:[13]

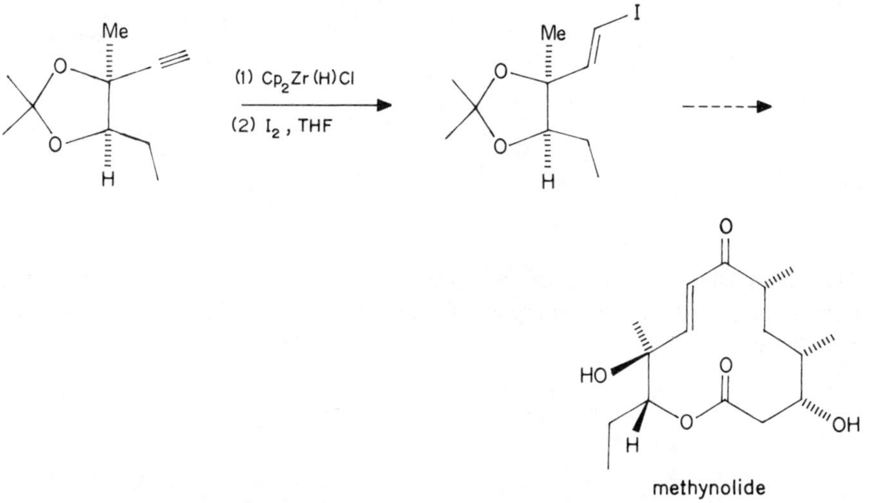

methynolide

Whilst the alkenyl zirconium complexes obtained in this way do not undergo direct reaction with α,β-unsaturated ketones, it is found that prior treatment with a catalytic amount of $Ni(acac)_2$ and diisobutylaluminum hydride (DIBAl) results in transmetallation to give an alkenyl nickel species, which undergoes very efficient conjugate addition. This has been employed in an approach to prostaglandin synthesis:[14]

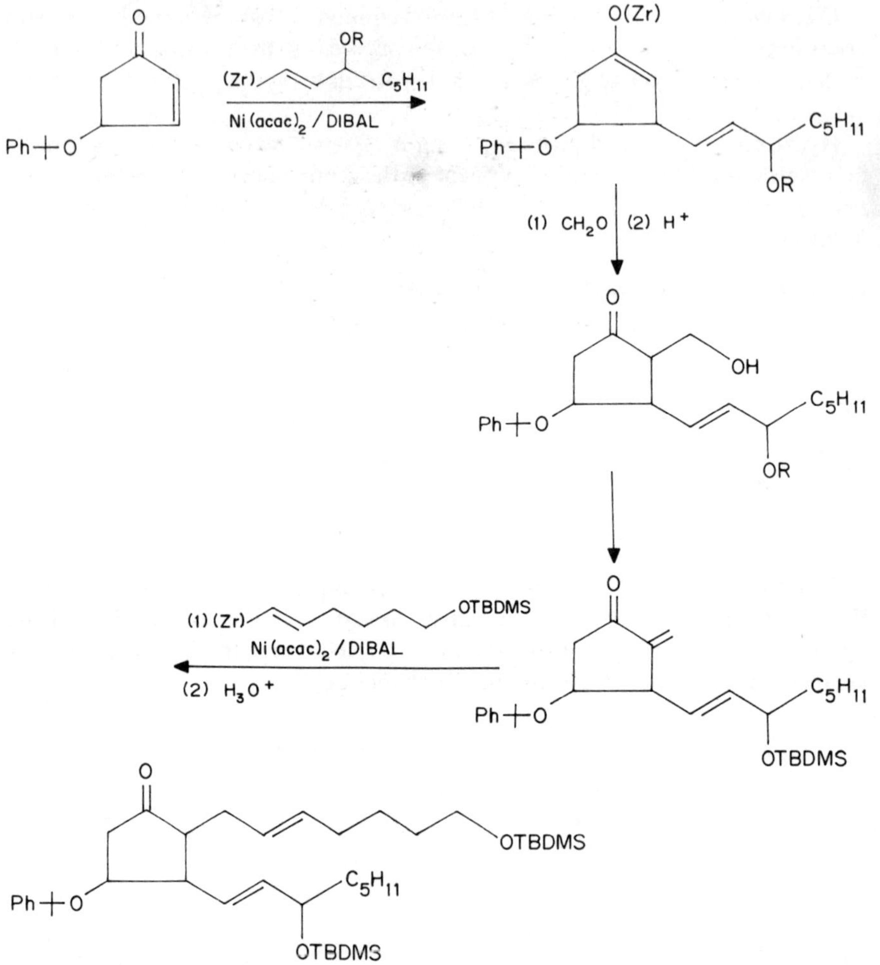

The alkenylzirconium derivatives have also been found to react with π-allyl palladium complexes, providing a novel method for introducing steroidal ring side chains, e.g.[15]

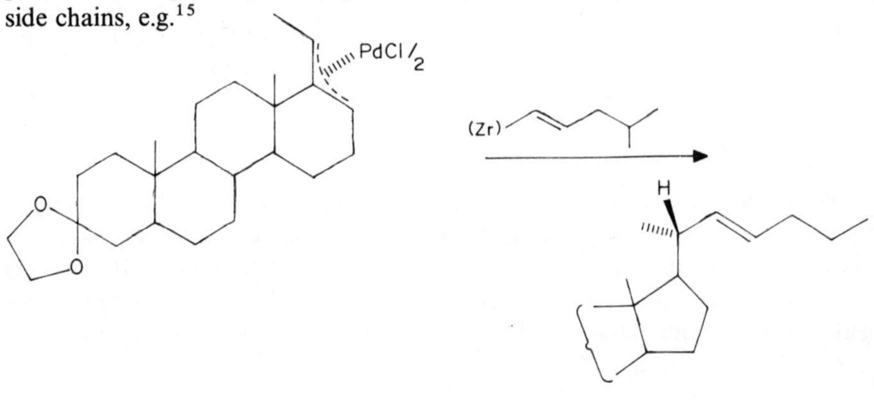

Other additions of metal hydride species to olefinic double bonds are summarized below.

$$HPtCl(PEt_3)_2 \ + \ CH_2{=}CH_2 \ \longrightarrow \ CH_3CH_2PtCl(PEt_3)_2$$

$$HRhCl_2(PPh_3)_2 \ + \ CH{\equiv}CH \ \longrightarrow \ CH_2{=}CHRhCl_2(PPh_3)_2$$

$$(\eta^5{-}Cp)FeH(CO)_2 \ + \ \text{/\!\!=\!\!\textbackslash} \ \longrightarrow \ CH_3CH{=}CHCH_2Fe(CO)_2Cp$$

What may be regarded formally as insertion of a carbene into M—H is the reaction of diazomethane, resulting in the formation of σ-methyl complexes:

$$HMo(CO)_3(\eta^5{-}Cp) \ \xrightarrow{CH_2N_2} \ CH_3Mo(CO)_3Cp$$

$$HMn(CO)_5 \ \xrightarrow{CH_2N_2} \ CH_3Mn(CO)_5$$

This reaction also occurs with metal halides:

$$ClIr(CO)(PPh_3)_2 \ \xrightarrow{CH_2N_2} \ ClCH_2Ir(CO)(PPh_3)_2$$

Reaction of metal hydride derivatives with epoxides gives rise to β-hydroxyalkyl complexes:

$$HCo(CO)_4 \ + \ \text{(epoxide)} \ \longrightarrow \ HOCH_2CH_2Co(CO)_4$$

There appears to be very little success in the synthesis of alkyl metal complexes by insertion of olefins, etc., into metal—carbon bonds, probably owing to further reactions of the immediate insertion products to give polymerization or decomposition.

Preparation Using Elimination or De-insertion Reactions

The formation of σ-alkyl complexes by expulsion of a group separating the carbon and metal bonds is termed elimination. The process may involve loss of, e.g., CO, SO_2 or N_2, and may be achieved either thermally or photochemically. It is the reverse of insertion, and has been discussed (for CO insertion/elimination) in Chapter 2. Some examples of preparation of σ-alkyl and aryl complexes by this method are given below.

$$NaMn(CO)_5 \ + \ RCOCl \ \longrightarrow \ R{-}\overset{\overset{\displaystyle O}{\displaystyle \|}}{C}{-}Mn(CO)_5 \ \xrightarrow{\Delta} \ R\,Mn(CO)_5$$

$$R = Me, Ph$$

$$O_2N{-}\!\!\!\bigcirc\!\!\!{-}N{=}NPtCl(PEt_3)_2 \ \xrightarrow{Al_2O_3} \ O_2N{-}\!\!\!\bigcirc\!\!\!{-}PtCl(PEt_3)_2$$

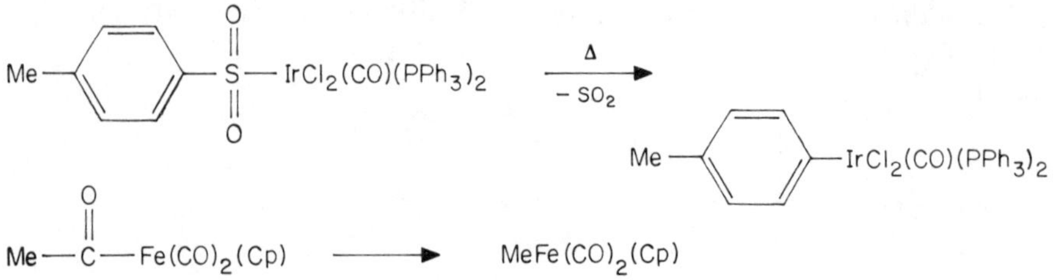

This provides a useful method for the preparation of complexes not available by the Grignard method with the corresponding metal halides, even though these may be readily available.

Preparation by Reaction of Metal Anions with RX

Formally the inverse of the Grignard method is the reaction of a transition metal anionic species with an appropriate alkyl derivative RX, where X is a good leaving group.

$$L_nM^- \quad R\!-\!X \longrightarrow L_nM\!-\!R \ + \ X^-$$

When it is possible to prepare the desired complex by either this or the Grignard method, then the transition metal anion approach is usually better, resulting in higher yield and fewer side-reactions. This method may also be used in conjunction with deinsertion (above), by replacing the alkyl halide with acyl halide:

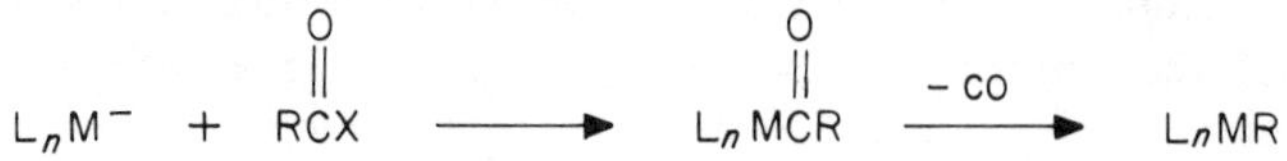

$$L_nM^- \ + \ RCX \longrightarrow L_nMCR \xrightarrow{-CO} L_nMR$$

The most commonly used metal anions are the carbonyl derivatives which may or may not contain other ligands, e.g. R_3P, η^5-Cp. Some general examples are given below.

$$\left[CpMo(CO)_3\right]_2 \xrightarrow{NaHg} Na^+\left[CpMo(CO)_3\right]^-$$

$$\downarrow RX$$

$$CpMo(CO)_3R$$

(Also for Cr and W)

$$CpM(CO)_2^- \xrightarrow{RX} CpM(CO)_2R$$

$$M = Fe, \ Ru$$

$$(CO)_5M^- \xrightarrow{RX} (CO)_5MR$$

$$M = Mn, Re$$

$$(CO)_4Co^- \xrightarrow{RX} (CO)_4CoR$$

$$L_nCo^- + \triangle\!\!\!\!O \longrightarrow \xrightarrow{H^+} L_nCoCH_2CH_2OH$$

$$CpFe(CO)_2^- + \triangle\!\!\!\!O \longrightarrow \xrightarrow[\text{(mild)}]{H^+} CpFe(CO)_2CH_2CH_2OH$$

$$\Big\downarrow H^+$$

$$CpFe(CO)_2(\eta^2\text{-}CH_2\!\!=\!\!CH_2)^+$$

The relative nucleophilicities of a series of transition metal anions has been estimated semi-quantitatively by measuring the rates of their reactions with alkyl halides. The observed order is: $CpFe(CO)_2^- > CpRu(CO)_2^- > CpNi(CO)^- > (CO)_5Re^- > CpW(CO)_3^- > (CO)_5Mn^- > CpMo(CO)_3^- > CpCr(CO)_3^- > (CO)_4Co^- > (CO)_5(CN)Cr^- \approx (CO)_5(CN)Mo^- \approx (CO)_5(CN)W^-$.

There are a number of organic compounds RX that are useful in this reaction, including alkyl, aryl, alkenyl, benzyl, allyl and propargyl halides. As expected from parallel organic reactions (S_N2, etc.), phenyl and vinyl halides react considerably more slowly than the corresponding alkyl, benzyl, allyl or propargyl derivatives, and the reaction is also influenced by the nature of the leaving group X.

Preparation Using Oxidative Addition Reactions

Oxidative addition reactions of transition metal complexes have already been discussed in detail (Chapter 2). Suffice it to say here that reaction of an appropriate organic species (e.g. alkyl halide) with a metal complex often results in the oxidation of the metal from a low to a higher valence state with concomitant metal—carbon bond formation. Both one- and two-electron oxidations are known:

One electron:

$$2 L_nM + RX \longrightarrow L_nMR + L_nMX$$

Examples:

$$2 Cp_2V + RX \longrightarrow Cp_2VR + Cp_2VX$$

$$R = Me, Et, PhCH_2$$

$$2 \left[Co(CN)_5 \right]^{3-} + PhCH_2Br \longrightarrow \left[PhCH_2Co(CN)_5 \right]^{3-} + \left[BrCo(CN)_5 \right]^{3-}$$

These reactions mostly appear to proceed by a free radical mechanism.

Two-electron:

$$L_nM + RX \longrightarrow L_nM \overset{R}{\underset{X}{\diagdown}}$$

Examples:

$$Rh^I Cl(CO)(PBu_3)_2 \xrightarrow{MeI} MeRh^{III} ICl(CO)(PBu_3)_2$$

$$Ir^I Cl(CO)(PPh_3)_2 + \text{/=\\/}I \longrightarrow CH_2{=}CHCH_2 Ir^{III} ICl(CO)(PPh_3)_2$$

$$(Ph_3P)_3 Pt^0 \xrightarrow{MeI} MePt^{II} I(PPh_3)_2 \xrightarrow{MeI} Me_2 Pt^{IV}(I)_2(PPh_3)_2$$

Oxidative addition methods also provide useful syntheses of σ-aryl and σ-vinyl complexes. An interesting variation is the intramolecular oxidative addition of a C—H bond, commonly referred to as cyclometallation,[16] e.g.

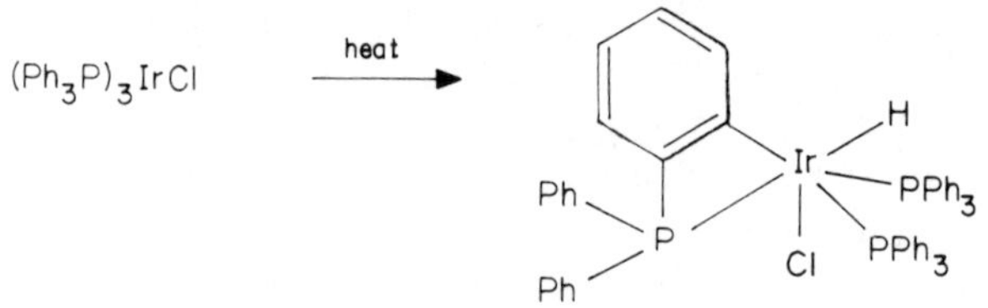

$$(Ph_3P)_3 IrCl \xrightarrow{heat}$$

PHYSICAL AND SPECTRAL PROPERTIES OF TRANSITION METAL ALKYLS

Alkyl and aryl ligands are strong σ-donors in the context of crystal-field theory, and consequently lie high in the 'spectroscopic series' in electronic spectroscopy. They have strong *trans* effects in terms of weakening a *trans* metal —ligand bond in a square-planar or octahedral complex. This is manifested in low force constants of *trans* M—X bonds and the observation from X-ray data that M—X bonds *trans* to organic groups are longer than *cis* M—X bonds.

Proton and ^{13}C NMR spectroscopy are, of course, useful tools in studies of these complexes. However, a wide variation of chemical shifts is observed, depending on the metal involved and the ancillary ligands. Some examples are shown below ($\delta\ ^1H$).[17]

2.78 (J_{Rh-H} = 2.4 Hz)
CH_3—$RhICl(PPh_3)_2(CH_3I)$

−2.1
CH_3—Ni

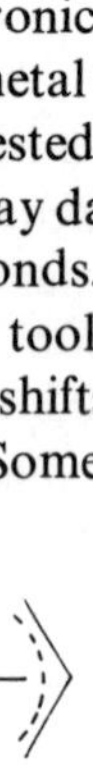

1.31   1.31
CH_3—CH_2—$Mn(CO)_5$

1.05   0.80
$(CH_3$—$CH_2)_2 Ni(Bipy)$

In the infrared spectra of transition metal alkyl complexes the alkyl C—H frequencies are little changed in comparison with organic compounds. In contrast, and perhaps expectedly so, C=O stretching frequencies of acyl metal complexes may be considerably different from those of aldehydes or ketones. There is also a dependence on the nature of the metal and counter ligands, e.g.

$$CH_3CO \bullet Fe(CO)(PBu_3)(\eta^5\text{-}Cp) \qquad\qquad 1590 \ cm^{-1}$$

$$CH_3CO \bullet Co(salen)(H_2O) \qquad\qquad 1728 \ cm^{-1}$$

REACTIONS OF σ-ALKYL COMPLEXES

The more common decomposition pathways, α- and β-hydride elimination and reductive elimination, have already been discussed in Chapter 2.

Reactivity towards various electrophiles, some of which have been met in the preceding discussion, is possibly the most common property. Further examples are given in the following summary.

The most simple electrophile, H^+, generally cleaves the metal—alkyl bond to produce the corresponding alkane. There are two possible mechanisms for the process: (a) protonation of the metal followed by reductive elimination of alkane or (b) direct protonation of the alkyl carbon atom. A means of distinction is not yet available (see later). It may be noted that some complexes are fairly stable and require strong acids to effect this transformation, e.g. $Cr(CMe_3)_4$ and $Ti(1\text{-}adamantyl)_4$.

Reactions with various other reagents which may be regarded as electrophiles are listed below.

(a) *Halogens :*

$$MR \ + \ X_2 \ \longrightarrow \ MX \ + \ RX$$

(b) *Carbon Dioxide :*

$$Ti(CH_2Ph)_4 \ \xrightarrow[\substack{(2) \ H_2O}]{\substack{(1) \ CO_2 \\ 20\ °C}} \ PhCH_2CO_2H \ + \ (PhCH_2)_3COH$$

Most transition metal alkyls are less reactive towards CO_2 than are Grignard reagents, so that it is possible to destroy excess of the alkylmagnesium reagent used in preparation of such alkyls by treatment of the reaction mixture with CO_2 at low temperature.

(c) *Oxygen:*

$$MR_4 \ \xrightarrow{\ O_2\ } \ M(OR)_4$$

$$M \ = \ Ti, \ Zr, \ etc. \ ; \ R \ = \ CH_2SiMe_3$$

There is some evidence to suggest that this reaction occurs by a free radical chain mechanism involving peroxide intermediates.

(d) *Ketones:* These are fairly mild electrophiles, being very reactive towards alkyl-magnesium and -lithium reagents and unreactive towards zinc reagents, etc. A broad spectrum of activity of transition metal alkyls is observed, and qualitatively it is anticipated that the more powerful is the alkylating agent which is required to prepare a given alkyl, the more readily will the alkyl react with the carbonyl function.

Some interesting observations which might have mechanistic implications are summarized below.[18]

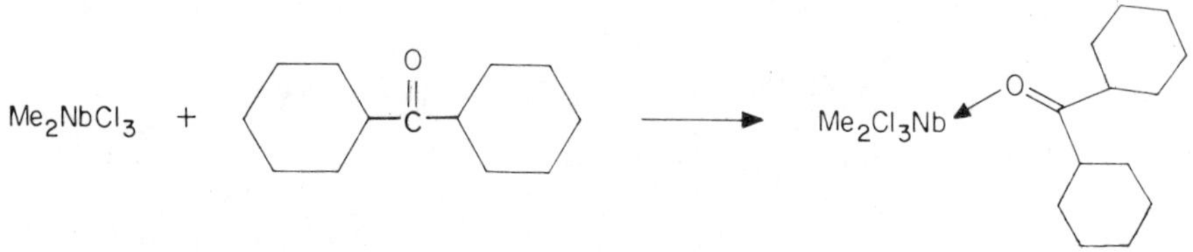

$$Me_2NbCl_3 + CH_3COCH_3 \longrightarrow 2\left[(Me_3CO)NbOCl_2\right]Me_2CO$$

Thus a bulky ketone simply adds to the metal of the complex as a donor ligand, whilst a less sterically demanding ketone undergoes further reaction, i.e. transfer of a methyl group to the ketone carbonyl carbon atom, with formation of the expected alkoxide complex.

The synthetic potential offered by the range of reactivity of transition metal alkyls does not yet appear to have received much attention, with the exception of cuprate reagents.

Mechanistic Aspects of Reactions with Electrophiles

Mechanistic studies of the above reactions are mainly limited to those complexes which are relatively stable, usually diamagnetic. This subject has been reviewed in some detail.[19]

In general terms, the reaction between a nucleophilic and an electrophilic species may be regarded as an interaction between the highest occupied molecular orbital (HOMO) of the former and the lowest unoccupied molecular orbital (LUMO) of the latter species.[20] For such a reaction attack may occur on *either* side of the carbon atom of the nucleophilic species under attack, e.g.

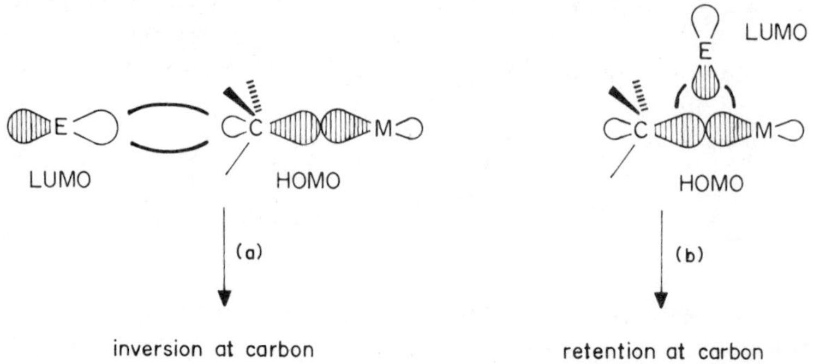

Clearly, the mode of reaction will depend on many factors, e.g. orbital overlap in path (a) or (b), steric effects for paths (a) or (b), the nature of the electrophile and/or the nucleophile (M—C bond). In fact, we shall see that both stereochemical modes of reaction may occur for the same metal complex (nucleophile) depending on the electrophile employed, and that the above picture, whilst well suited to main group alkyls, is not necessarily appropriate for transition metal alkyls. The reactions of σ-alkyl transition metal complexes is complicated by other factors. For example, the HOMO, which according to the above scheme plays a vital role in these reactions, may well be located on the metal, on one of the ligands, or in metal—ligand bonds, depending on the system involved. Consequently, reaction between an organotransition metal complex and an electrophile may not be a concerted cleavage of the carbon—metal bond, though such cleavage frequently occurs in later stages of the reaction.

The following examples are of reactions of a σ-alkyl complex with an electrophile, which are believed to proceed with substantial retention of configuration at the α-carbon atom:

(a) $(-)$-MeCHBrCO$_2$Et + Mn(CO)$_5^-$ $\xrightarrow[\text{inversion}]{\text{THF}}$ MeCH(CO$_2$Et)Mn(CO)$_5$

$\downarrow$ Br$_2$, retention

$(+)$-MeCHBrCO$_2$Et

(b) t-BuCHDCHDZr(Cp)$_2$Cl $\xrightarrow{\text{Br}_2}$ t-BuCHDCHDBr

single diastereoisomer
with retention

(c) $\xrightarrow[\text{PhH}]{\text{HgCl}_2}$

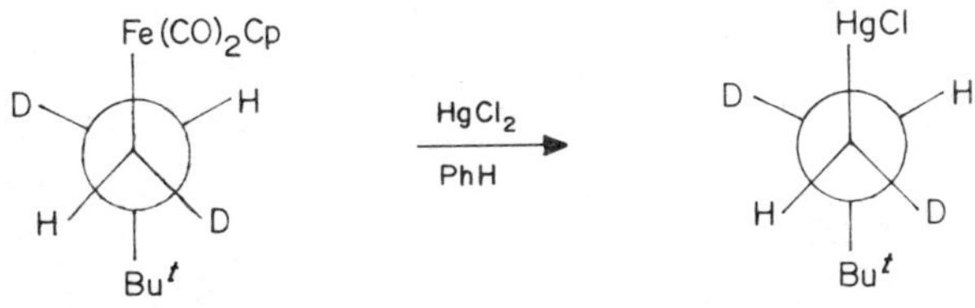

The last example is particularly interesting, the entity HgCl being, in a formal sense, the electrophile, because when the electrophile is changed to Br$_2$ (in aprotic solvents) reaction occurs with *inversion* of configuration:

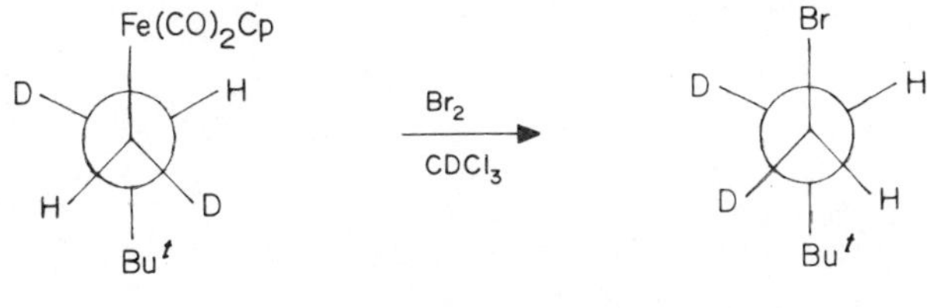

$\xrightarrow[\text{CDCl}_3]{\text{Br}_2}$

This poses a particular problem in terms of detailed mechanism. The reaction may be a direct interaction of the electrophile LUMO with the 'backside' of the Fe—C HOMO [path (a) above]. Alternatively, the electrophile LUMO might now give a better match in energy (and therefore better overlap) with a complex orbital which is metal-centered (e.g. t_{2g} orbital). This can then lead to two possibilities, subsequent addition of electrophile to the metal:

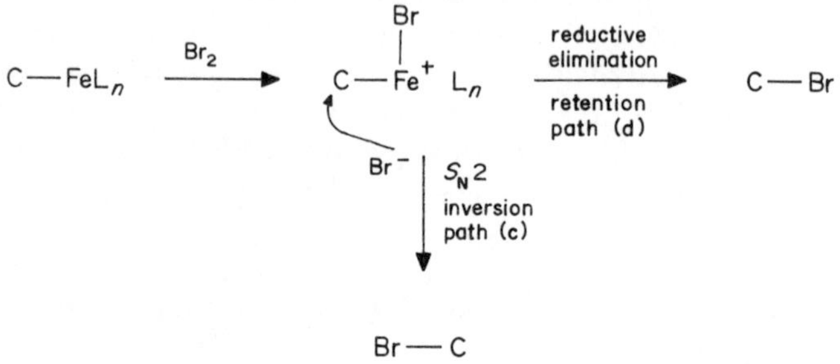

Thus the metal–bromide species now bears a positive charge and is activated towards nucleophilic attack. An S_N2 reaction would thereby lead to inversion. Alternatively, reductive elimination would lead to retention. Clearly, then, this reaction follows either path (a) or (c). As a further example, the complex **1** below gives a number of reactions with Br_2 under a range of conditions which may be interpreted in terms of an anchimeric assistance from the phenyl group, and proceeding via a phenonium ion. Undoubtedly, the first step is some kind of oxidation at the metal to create a good leaving group.

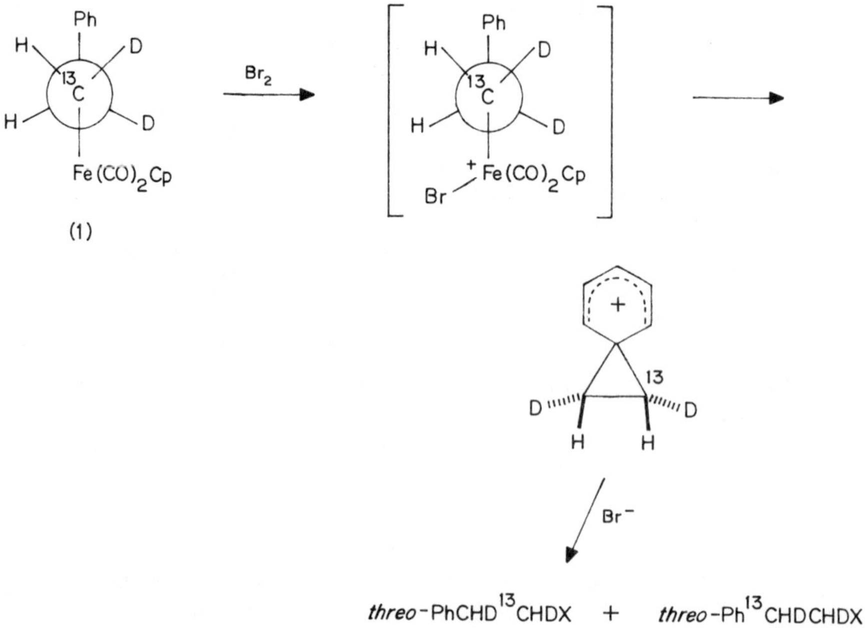

$$ threo\text{-}PhCHD^{13}CHDX \quad + \quad threo\text{-}Ph^{13}CHDCHDX $$

This is evidenced also by competition studies in the presence of other nucleophiles:

$$\text{PhCHDCHD} - \text{Fe}(\text{CO})_2\text{Cp} \xrightarrow[\text{MeOH}]{\text{Cl}_2} \text{PhCHDCHDOMe} \; (\geqslant 90\% \text{ retention})$$

$$\xrightarrow[\text{excess Cl}^-]{\text{Br}_2} \text{PhCHDCHDBr} \; + \; \text{PhCHDCHDCl}$$

However, when the bromination of PhCHDCHD–Fe is carried out in the presence of fluoride ion, under conditions where F^- is known to be as powerful a nucleophile toward carbon-centered electrophiles as Br^-, no phenethyl fluoride is produced. Consequently, Slack and Baird[21] postulated that the reaction involves nucleophilic attack on an iron species, which would not readily react with F^-. It was proposed that the phenomium ion is not free but is somehow complexed to Fe, and may be represented as the rapidly equilibrating species.[21]

Whilst the situation is still not absolutely clear, most of the evidence points to an (oxidative) addition of electrophile to the metal as the general first step, followed by either nucleophilic displacement of the metal leaving group [path (c) → inversion] or reductive elimination [path (d) → retention] of the substituted alkyl group. Clearly, whether path (c) or (d) is followed will depend on the electrophile, e.g. HX and HgX_2 react with retention, whilst Br_2 may react with inversion. Any preference for prior additions of electrophilic species at the metal itself is entirely consistent with available molecular orbital data. For example, MO treatments of transition metal complexes of high symmetry predict that nonbonding d orbitals lie at higher energy than do the filled bonding orbitals.[22] Consequently, the filled orbital nearest in energy to the halogen LUMO is a nonbonding d orbital, so that this is the favored interaction rather than a concerted attack on the metal—carbon bond. It may be noted in this context that photoelectron spectroscopic studies of $CpFe(CO)_2CH_3$ suggest that the $Fe-CH_3$ σ-bonding orbital is a least 0.6 eV lower in energy than the nonbonding metal-centered d orbitals.[23]

SOME SPECIFIC TRANSITION METAL ALKYL σ-COMPLEXES

Organocopper Reagents

We mention these reagents here because, whilst copper is left out of discussion of transition metal organometallics to a large extent, the alkyl copper reagents are exceedingly important to the synthetic organic chemist, since many strategic bond formations can be effected by their use. There are a number of reviews available.[24]

A large number of organocuprate reagents have been prepared and used synthetically. The reagents can either be generated by stoichiometric reaction or catalytically, the former usually being more appropriate for preparation from alkyllithium precursors, whilst the catalytic method is often employed using Grignard reagents. The resultant dialkyl cuprates undergo 1,4-addition to α,β-unsaturated ketones:

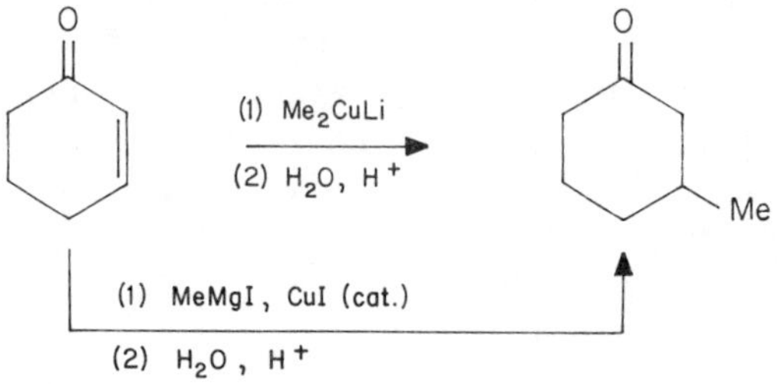

The reaction proceeds with the intermediacy of an enolate derivative, which may be trapped with an appropriate electrophile e.g.

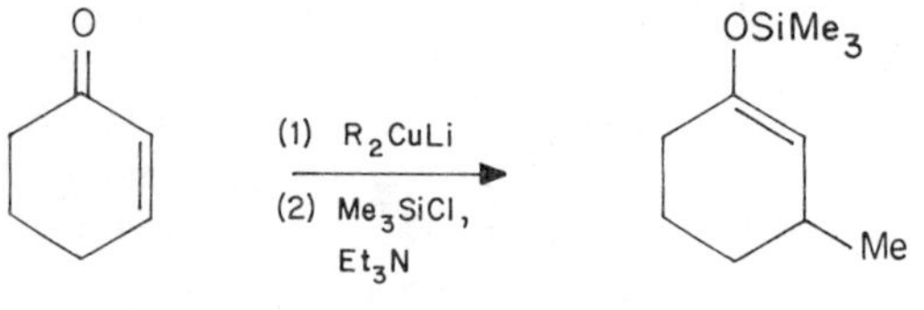

The trapped derivative may be subjected to further transformation, e.g.

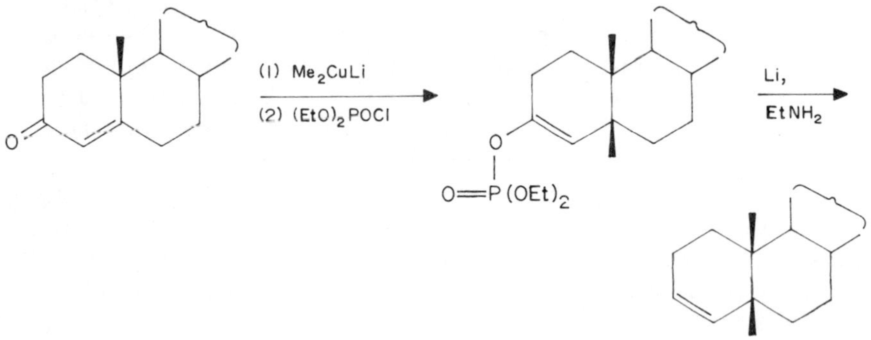

There are many syntheses of natural products involving the use of alkyl cuprates in this manner. Organocuprates are also effective for the formation of carbon—carbon bonds by reaction with alkyl halides, tosylates and various other substrates. Some examples are given below. The copper reagents tend to avoid some of the unwanted side reaction experienced with e.g., alkyl lithium and Grignard reagents.

(i) 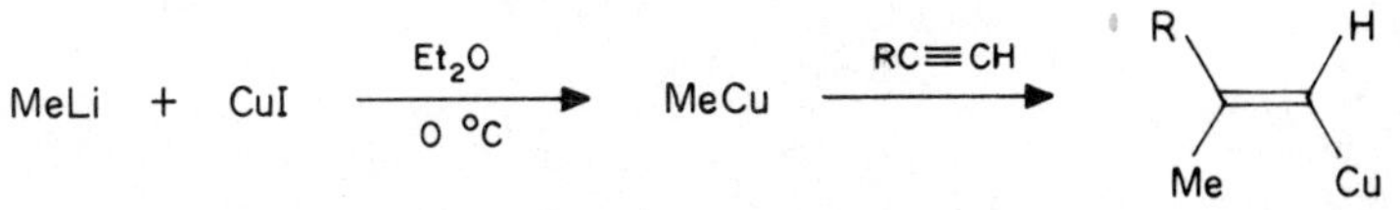 $\quad$ MeLi $\ +\ $ CuI $\ \xrightarrow[0\ °C]{Et_2O}\ $ MeCu $\ \xrightarrow{RC≡CH}$

(ii)

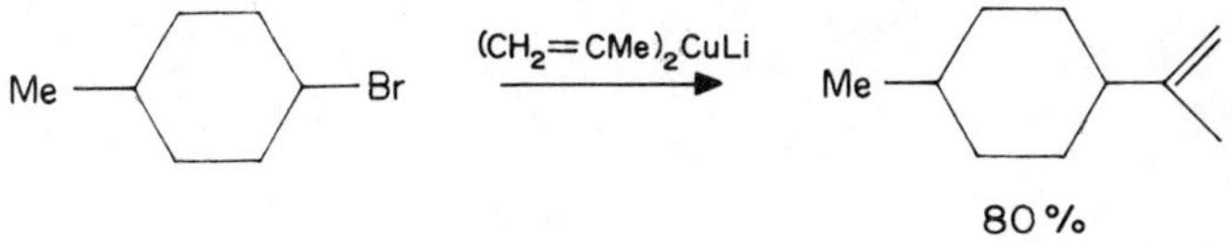

$(CH_2{=}CMe)_2CuLi$

80%

(iii)

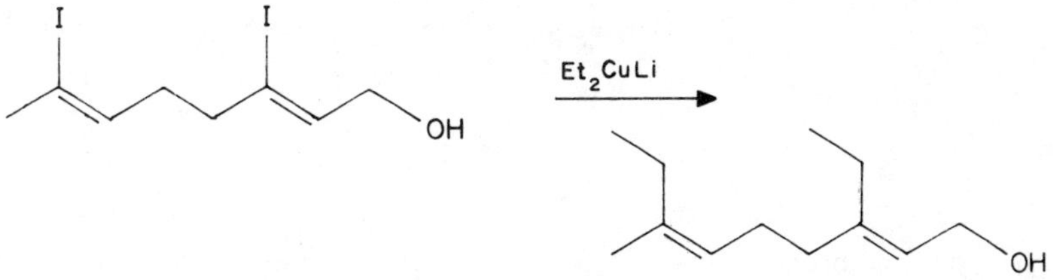

Et_2CuLi

(iv)

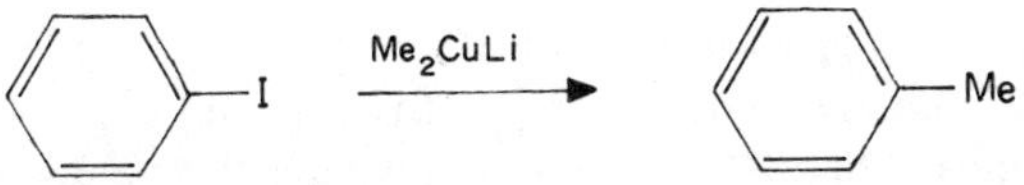

Me_2CuLi

(v)

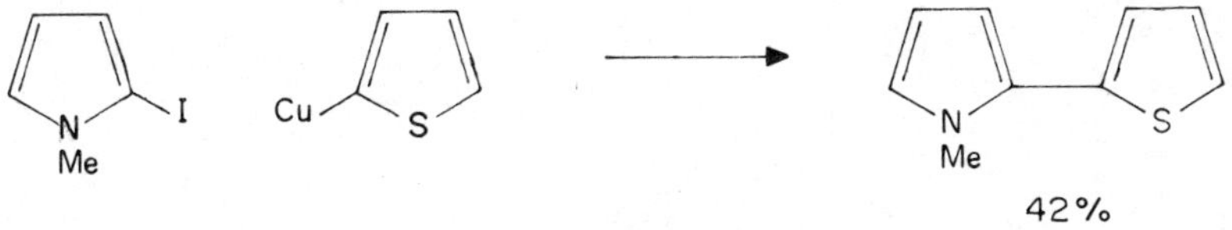

42%

(vi)

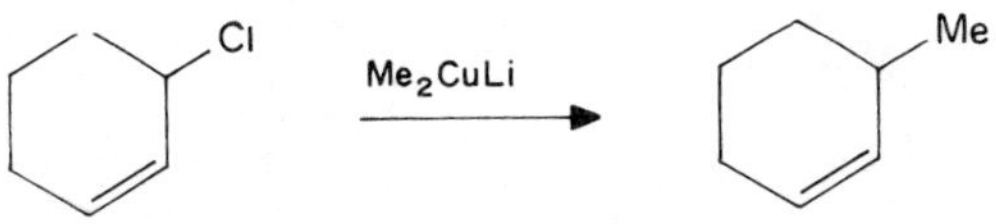

Me_2CuLi

(vii)

$(CH_2{=}CMe)_2CuLi$

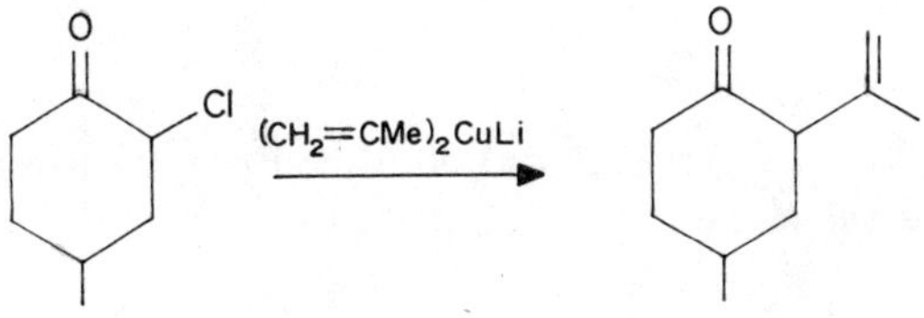

(viii) $\quad$ RCOCl $\ \xrightarrow{R'_2CuLi}\ $ RCOR'

(ix)

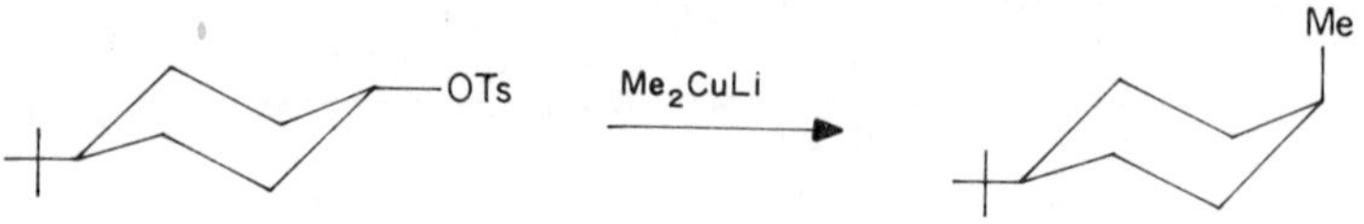

(x)

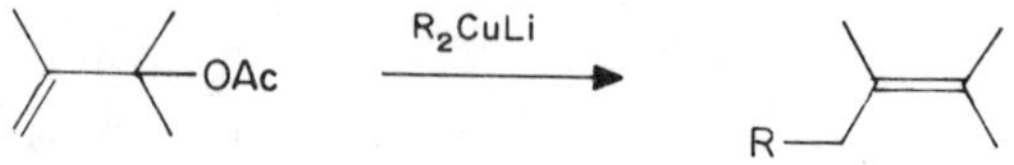

(xi)

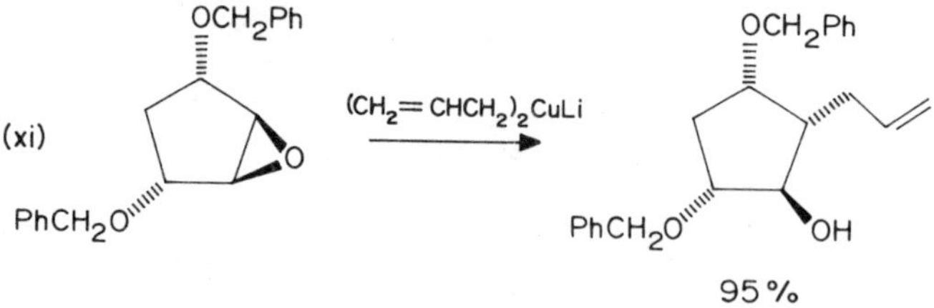

95%

Allyl–Fp σ-complexes

A range of Fp–R alkyl complexes are available by reaction of the anionic derivative Na–Fp with alkyl halides. These include allyl–Fp complexes, in which the allyl group is σ-bonded to the metal, and which we discuss here since they show some promise as synthetic reagents. An early review dealt with developments in this area.[25] The general scheme for the preparation of the complexes is shown below.

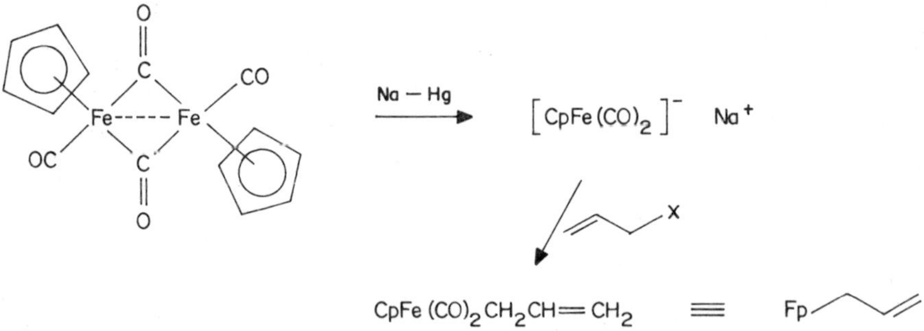

An alternative method for the preparation of allyl–Fp derivatives is by deprotonation of the cationic η^2-olefin–Fp complexes, e.g.

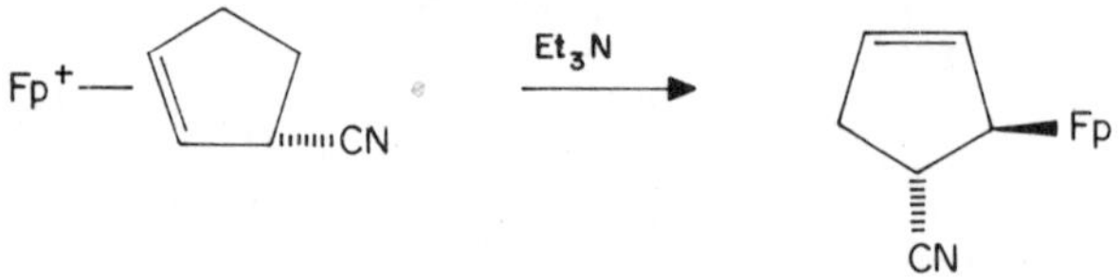

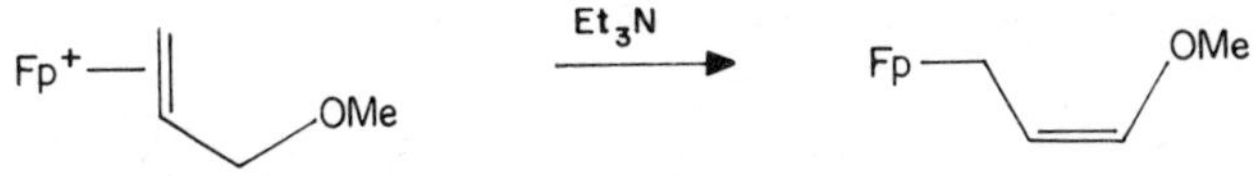

Allyl–Fp complexes behave as nucleophiles. The immediate product of such reaction is the cationic η^2-olefin complex, and since this is itself a reactive entity (see Chapter 5), there would appear to be some scope for synthetic applications of such compounds, although this has not been fully explored. Some typical nucleophilic behavior is summarized below.

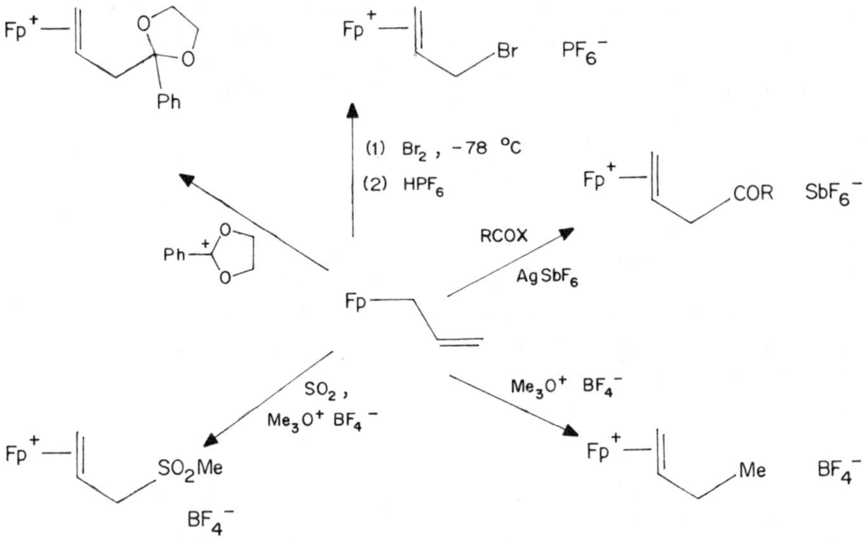

Production of the electrophilic olefin complex during these reactions has occasionally been exploited for further reaction, e.g.[26]

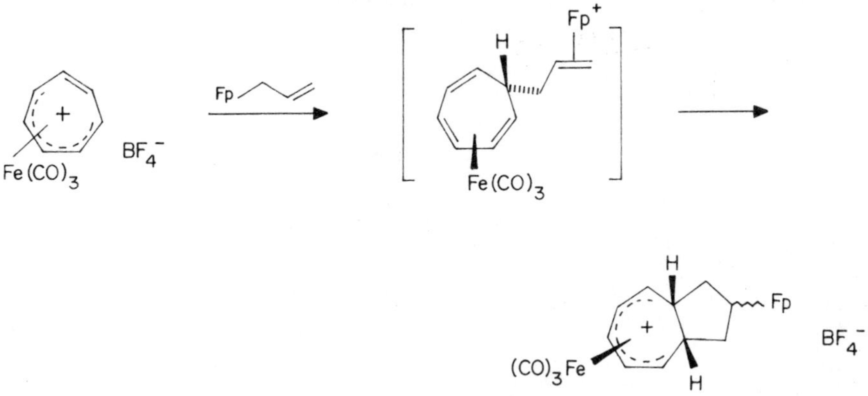

In this reaction the cycloheptatriene–Fe(CO)$_3$ complex behaves as a nucleophile, the driving force being the formation of the dienyl–Fe(CO)$_3$ complex (see also Chapter 8). The product may be reacted even further, since the dienyl–Fe(CO)$_3$

system is reactive towards a number of nucleophiles:

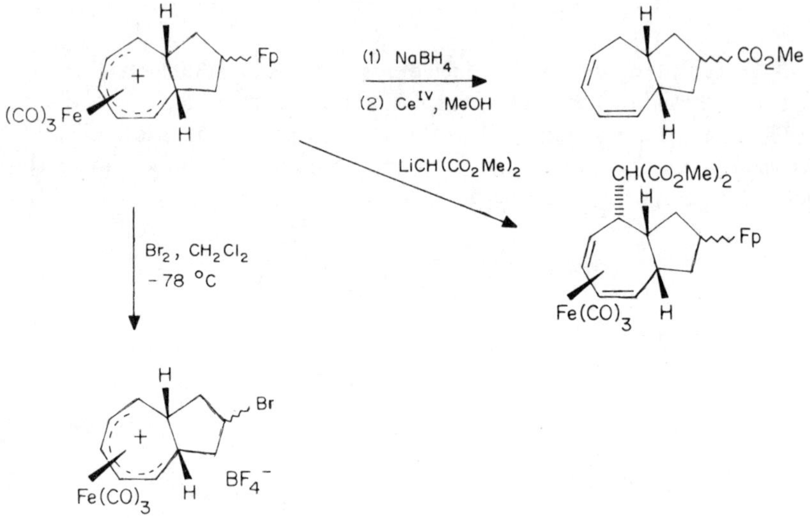

Allyl–Fp complexes also react with a number of powerful dienophiles, although it seems unlikely that this is a concerted cycloaddition reaction, but rather a stepwise process, e.g.

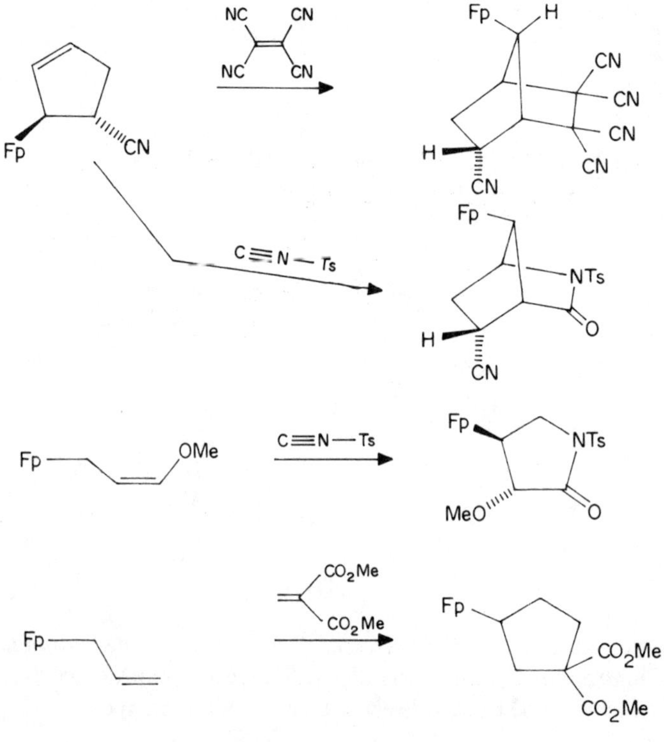

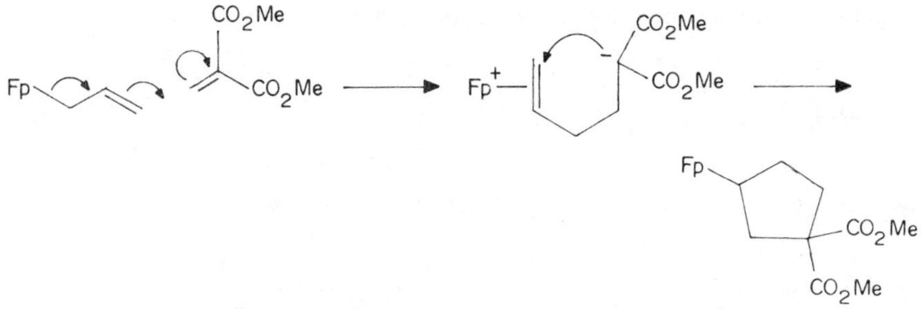

The non-concerted nature of these reactions is evidenced by the fact that they are non-stereospecific, e.g.

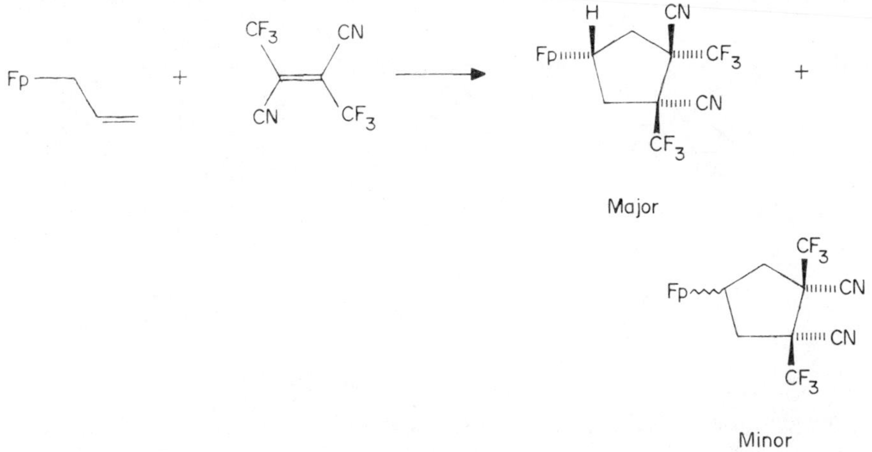

Major

Minor

Aryl and vinyl Palladium σ-complexes

The so-called Heck reaction is a process in which a carbon—carbon bond is created by reaction between aryl or vinyl halides and olefinic compounds in the presence of an amine base and catalytic amounts of palladium(0).[27] Whilst the detailed mechanism is not known, it is fairly clear from the chemistry of isolated organopalladium complexes that the reaction involves the formation of σ-aryl or vinyl palladium complexes. Thus, the following stoichiometric reaction is obtained:

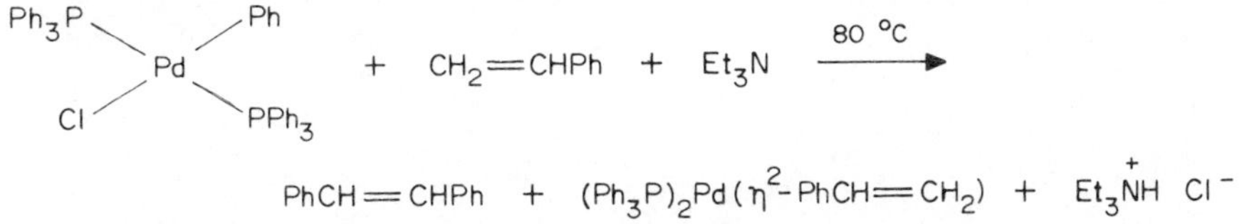

$$PhCH\!=\!CHPh \;+\; (Ph_3P)_2Pd(\eta^2\!-\!PhCH\!=\!CH_2) \;+\; Et_3\overset{+}{N}H \; Cl^-$$

Currently, the reaction is carried out using a catalytic amount of $Pd(PPh_3)_4$, a development which followed the discovery that organopalladium species could be generated by an oxidative addition of organic bromides to this complex:

$$RBr \quad + \quad Pd(PPh_3)_4 \quad \longrightarrow \quad RPd(PPh_3)_2Br \quad + \quad 2\,PPh_3$$

A typical reaction is given below.

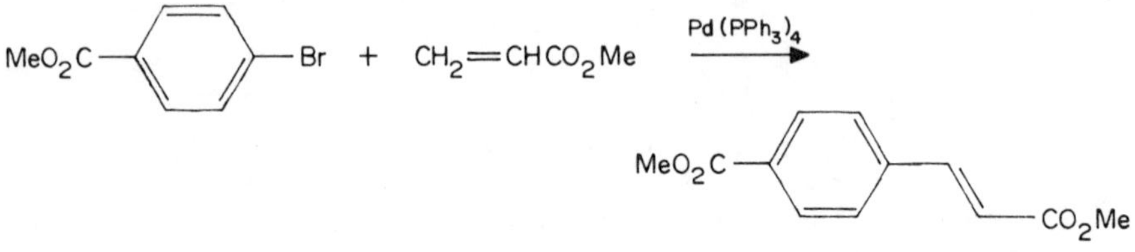

The mechanism is

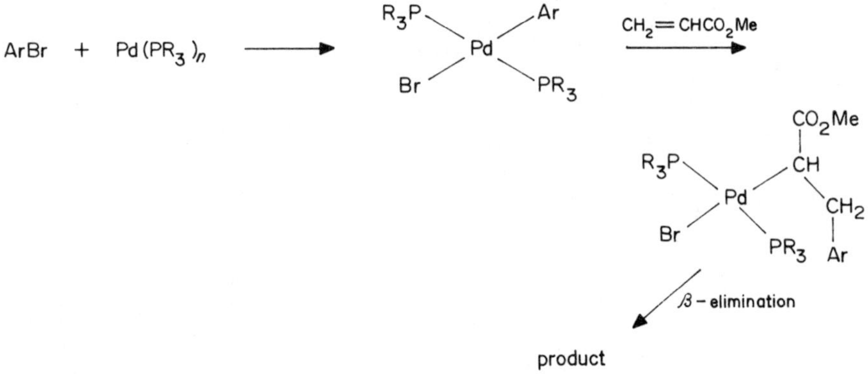

$$ArBr \quad + \quad Pd(PR_3)_n \quad \longrightarrow$$

product

A wide variety of substituents may be present in the aryl halide for the reaction with methyl acrylate. A variety of other organic halides, including benzyl, heterocyclic and vinyl halides, may be employed, as indicated in the following examples.

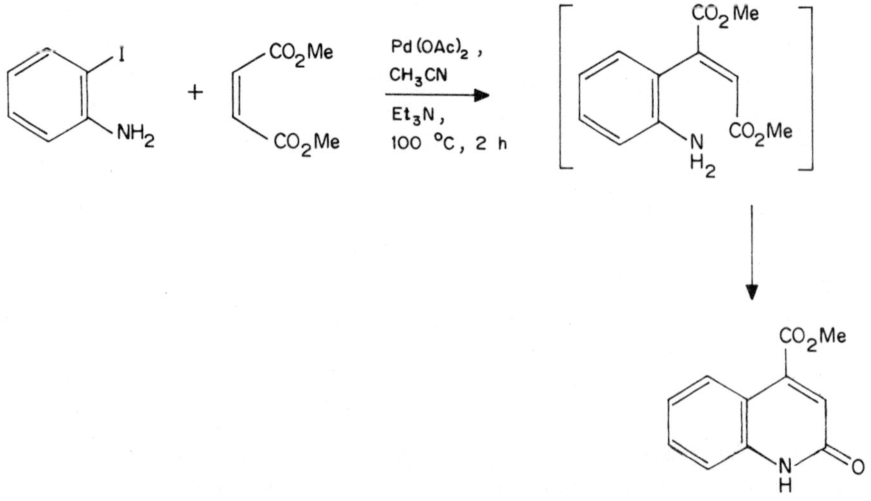

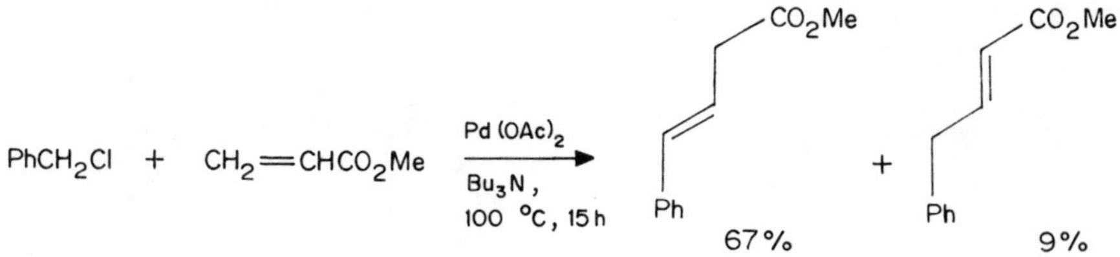

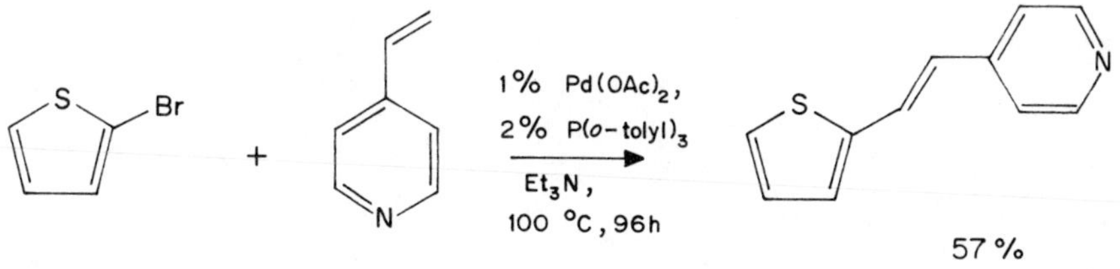

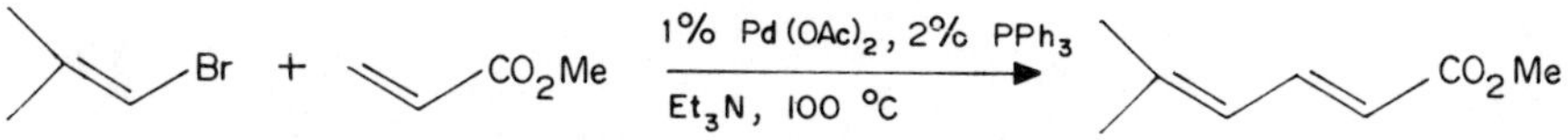

The last vinyl halide reaction is limited to olefins which contain electron-withdrawing groups. For example, reaction of 2-bromopropene and but-3-en-2-ol fails under the usual conditions. However, when the reaction is carried out in the presence of a nucleophilic secondary amine it leads to a product corresponding to amine addition to an intermediate electrophilic π-allyl palladium complex (see also chapter 6).

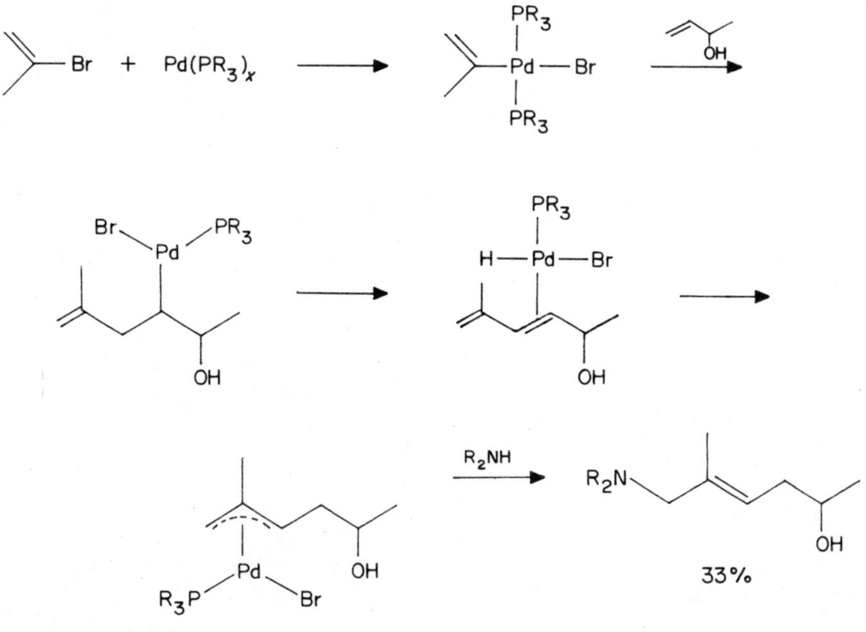

also

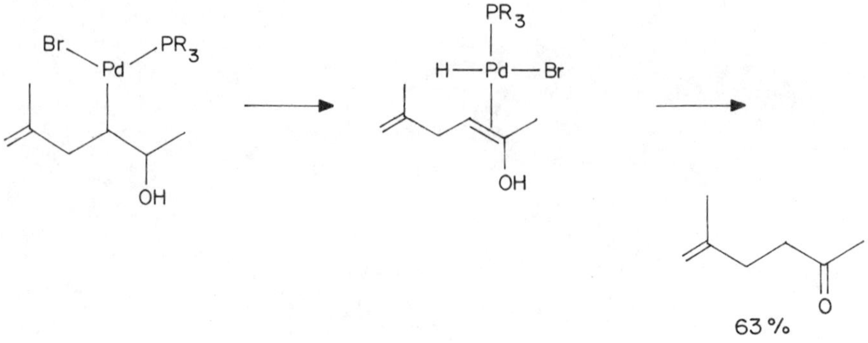

63%

Similar reactions are shown below.

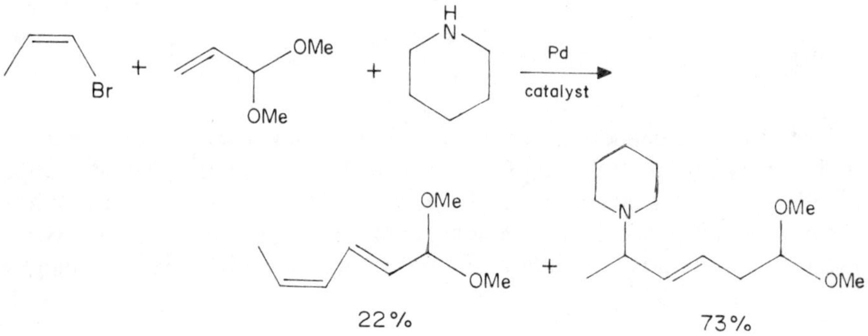

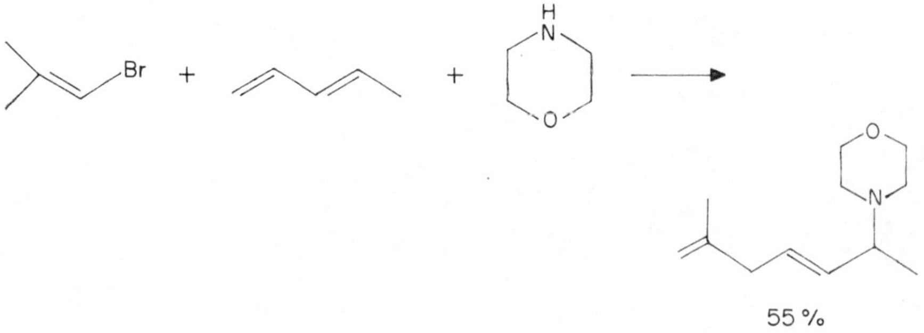

22%   73%

55%

These reactions appear to present useful synthetic possibilities but they have not been pursued in detail.

Metallacycles

A review on metallacycles as novel organic reagents has been published.[28] We have already met metallacyclobutanes and metallacyclopentadienes in connection with olefin metathesis and acetylene oligomerization, respectively (Chapter 2). This section presents briefly these and other metallacycle complexes.

Metallacyclobutanes are now recognized as key intermediates in olefin metathesis reactions. Stable metallacyclobutanes have been prepared for a number of Group VIII metals, although these are not metathesis catalysts. Oxidative addition reactions of Pt(II) complexes with cyclopropane derivatives results in opening of the carbocycle with formation of the Pt(IV) metallacycle:[29]

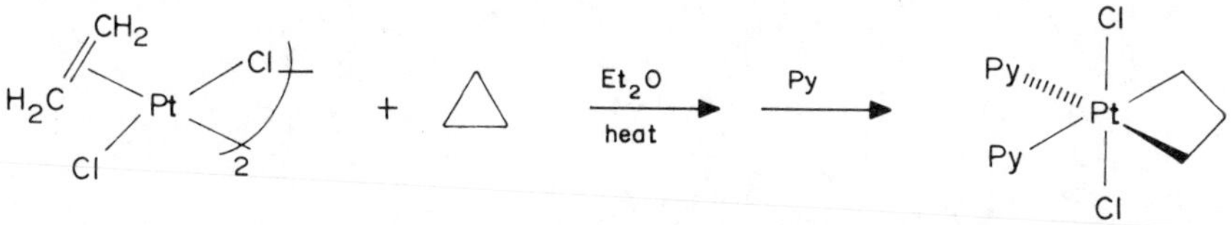

Stable metallacyclobutane derivatives of nickel, palladium and platinum are prepared by cyclometallation of bis-neopentyl complexes, e.g.[30]

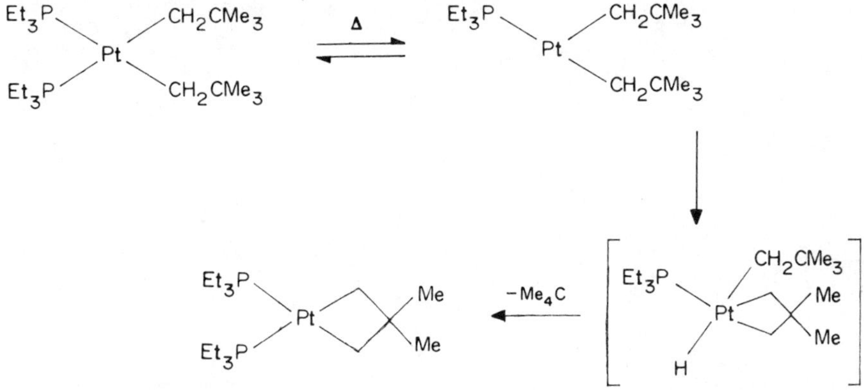

An unusual attack by nucleophiles at the central carbon atom of certain π-allyl metal complexes has also been found to yield stable metallacyclobutanes, but this is by no means a general method (see also Chapter 6), e.g.

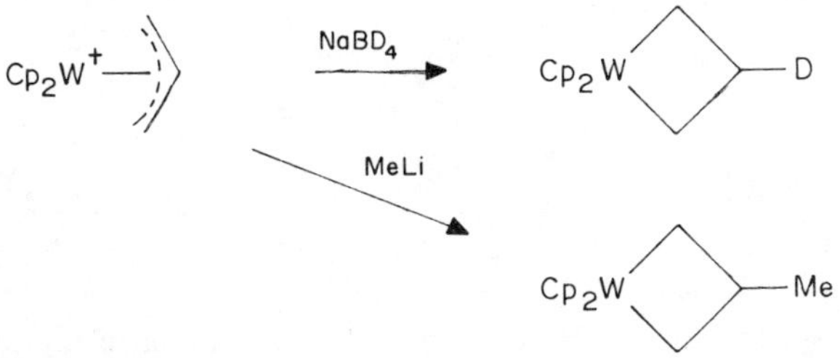

Metallacyclopentanes may be prepared in basically three ways: (a) reaction of a metal dihalide with a 1,4-dilithiobutane; (b) reaction of an appropriate metal complex with two molecules of olefin; (c) reaction of a metal complex with a cyclobutane. Examples are give below.[31-36]

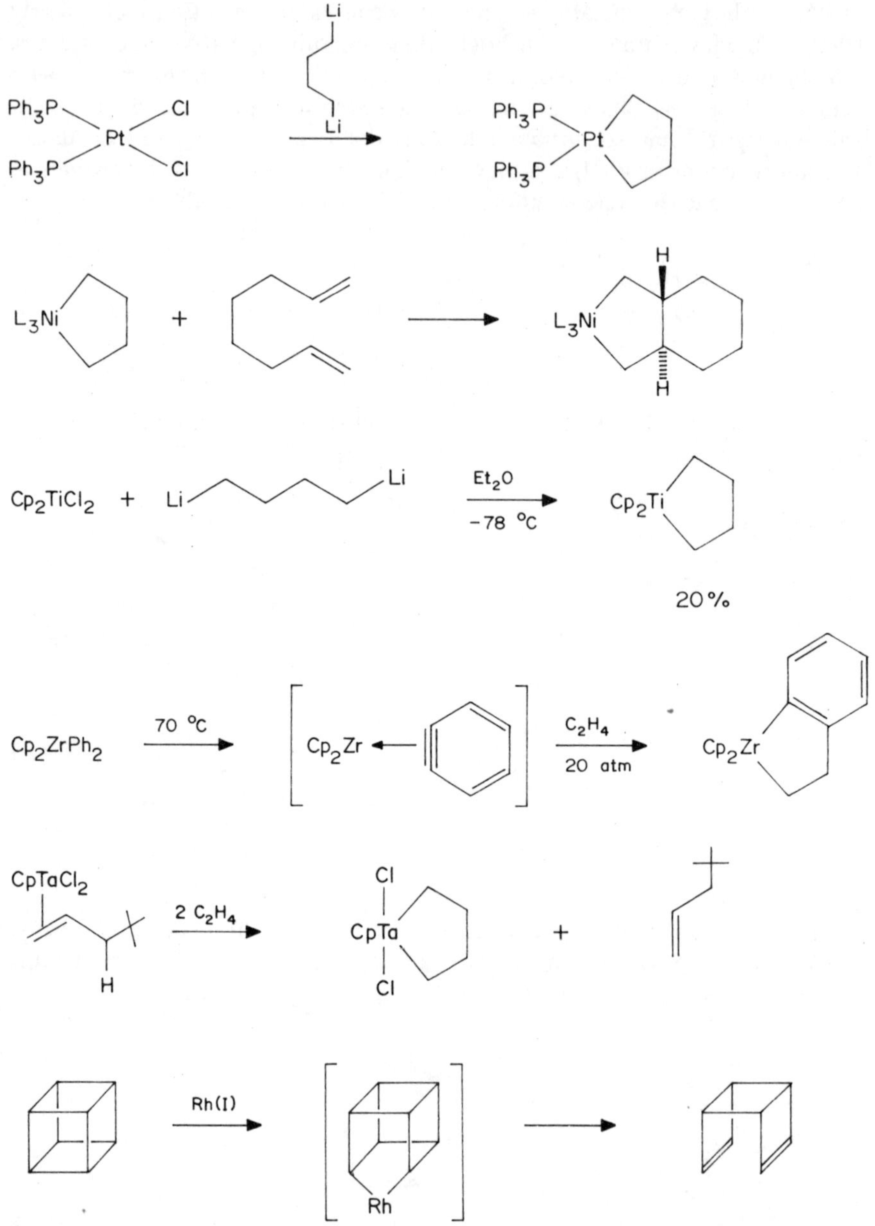

Six- and seven-membered metallacycles of Pt and Ni have been characterized, e.g.

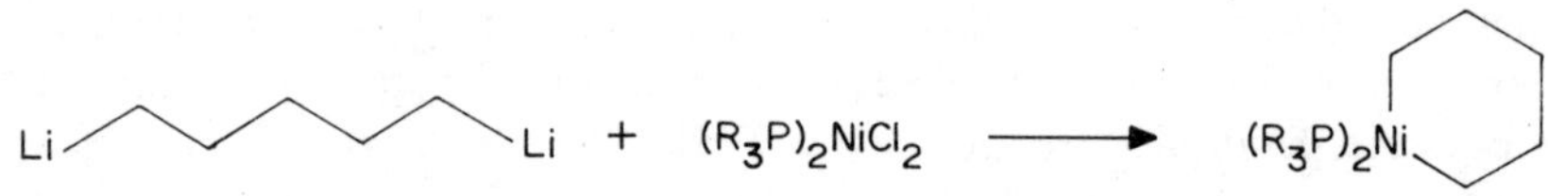

With regard to the stability of all the above complexes, it is found that whereas the smaller rings (up to six-membered) are relatively stable to β-elimination, the analogous metallacycloheptane derivatives undergo this mode of decomposition much more easily. The reason generally accepted is that the inflexibility of the smaller rings prevents arrangement of the β-C—H bonds in a manner necessary to allow transfer of the β-hydrogen to the metal. The more flexible seven-membered ring allows appropriate alignment and is consequently less stable.

Metallacycles may undergo reactions analogous to the related σ-alkyl complexes, e.g.

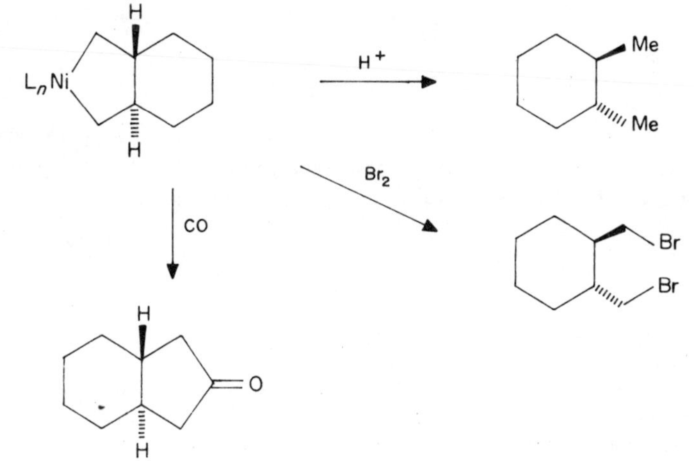

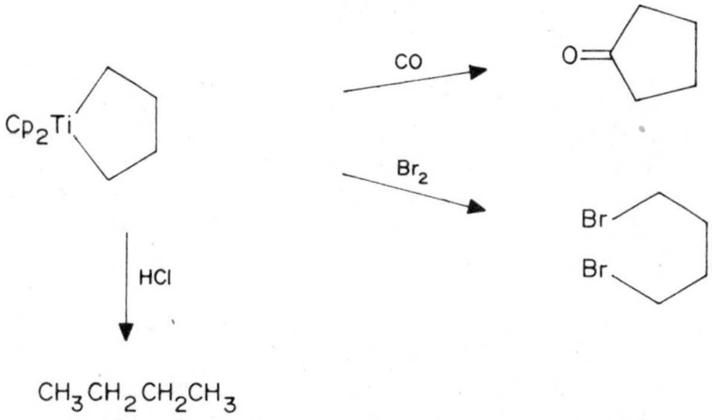

Other less obvious reactivity is also found, some of which still remains to be properly explained, e.g.

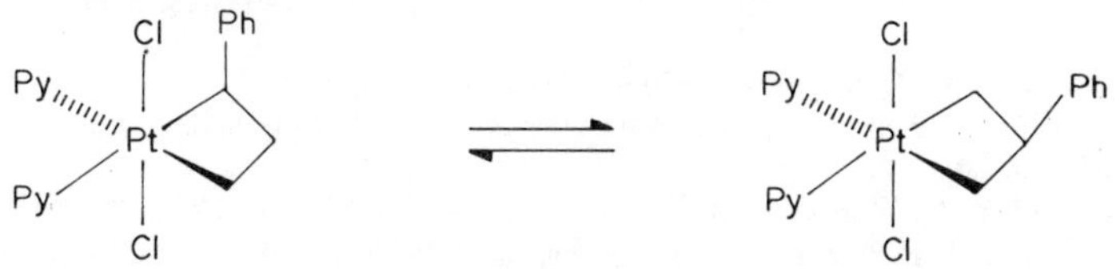

This chapter has attempted to provide a glossary of methods of synthesis of various transition metal alkyls, their reactions and some consideration of demonstrated or potential utility for organic synthesis.

REFERENCES

1. Review: R. R. Schrock and G. W. Parshall, *Chem. Rev.*, 1976, **76**, 243.
2. K. Clauss and C. Beermann, *Angew. Chem.*, 1959, **71**, 627.
3. V. N. Latjaeva, G. A. Razuvaev, A. V. Malisheva and G. A. Kiljakova, *J. Organomet. Chem.*, 1964, **2**, 388.
4. J. Krausse and G. Marx, *J. Organomet. Chem.*, 1974, **65**, 215.
5. J. Krausse, G. Marx and G. Schodl, *J. Organomet. Chem.*, 1970, **21**, 159.
6. M. Reetz, B. Wenderoth, R. Peter, R. Steinbach and J. Westerman, *J. Chem. Soc., Chem. Commun.*, 1980, 1202.
7. C. Santini-Scampucci and J. G. Reiss, *J. Chem. Soc., Dalton Trans.*, 1974, 1433.
8. D. W. Hart and J. Schwartz, *J. Am. Chem. Soc.*, 1974, **96**, 8115.
9. C. A. Bertelo and J. Schwartz, *J. Am. Chem. Soc.*, 1975, **97**, 228.
10. C. A. Bertelo and J. Schwartz, *J. Am. Chem. Soc.*, 1976, **98**, 262.
11. D. B. Carr and J. Schwartz, *J. Am. Chem. Soc.*, 1977, **99**, 638; 1979, **101**, 3521.
12. D. W. Hart, T. F. Blackburn and J. Schwartz, *J. Am. Chem. Soc.*, 1975, **97**, 679.
13. P. A. Grieco, Y. Ohfune, Y. Yokoyama and W. Owens, *J. Am. Chem. Soc.*, 1979, **101**, 4749.
14. J. Schwartz, M. J. Loots and H. Kosugi, *J. Am. Chem. Soc.*, 1980, **102**, 1333.
15. J. S. Temple, M. Riediker and J. Schwartz, *J. Am. Chem. Soc.*, 1982, **104**, 1310.
16. Review: M. I. Bruce, *Angew. Chem., Int. Ed. Engl.*, 1977, **16**, 73.
17. See, for example, H. C. Beachell, S. A. Butter, *Inorg. Chem.*, 1965, **4**, 1133; K. Kite, J. A. S. Smith and E. J. Wilkins, *J. Chem. Soc. A*, 1966, 1744.
18. J. D. Wilkins, *J. Organomet. Chem.*, 1974, **80**, 357.
19. M. D. Johnson, *Acc. Chem. Res.*, 1978, **11**, 57.
20. I. Fleming, *Frontier Orbitals and Organic Chemical Reactions*, Wiley, Chichester, 1976, p. 74.
21. D. A. Slack and M. C. Baird, *J. Am. Chem. Soc.*, 1976, **98**, 5539.
22. F. A. Cotton and G. Wilkinson, *Advanced Inorganic Chemistry*, Wiley, New York, 1972, 3rd ed., Ch. 20.
23. D. A. Symon and T. C. Waddington, *J. Chem. Soc., Dalton Trans.*, 1975, 2140; D. L. Lichtenberger and R. F. Fenske, *J. Am. Chem. Soc.*, 1976, **98**, 50.

24. See, for example, G. H. Posner, *Org. React.*, 1972, **19**, 1; 1975, **22**, 253.
25. M. Rosenblum, *Acc. Chem. Res.*, 1974, **7**, 122.
26. N. Genco, D. Marten, S. Raghu and M. Rosenblum, *J. Am. Chem. Soc.*, 1976, **98**, 848.
27. R. F. Heck, *Acc. Chem. Res.*, 1979, **12**, 146.
28. A. Nakamura and H. Yasuda, *Yuki Gosei Kagaku Kyokaishi*, 1980, **38**, 975.
29. J. Rajaram and J. A. Ibers, *J. Am. Chem. Soc.*, 1978, **100**, 829.
30. P. Foley and G. M. Whitesides, *J. Am. Chem. Soc.*, 1979, **101**, 2732.
31. J. X. McDermott, J. F. White and G. M. Whitesides, *J. Am. Chem. Soc.*, 1976, **98**, 6521.
32. R. H. Grubbs and A. Miyashita, *J. Organomet. Chem.*, 1978, **161**, 371.
33. J. X. McDermott, M. E. Wilson and G. M. Whitesides, *J. Am. Chem. Soc.*, 1976, **98**, 6529; R. H. Grubbs and A. Miyashita, *J. Chem. Soc., Chem. Commun.*, 1977, 864.
34. G. Erker and K. Kropp, *J. Am. Chem. Soc.*, 1979, **101**, 3659.
35. S. J. McLain, C. D. Wood and R. R. Schrock, *J. Am. Chem. Soc.*, 1979, **101**, 4558.
36. L. Cassar, P. E. Eaton and J. Halpern, *J. Am. Chem. Soc.*, 1970, **92**, 3515.

CHAPTER 4

Carbene and Carbyne Complexes

CARBENE COMPLEXES

The implication of transition metal–carbene complexes in a number of reactions involving transformations of, e.g., alkenes, provided impetus for a search for stable complexes. Since the first report of such a compound in 1964 by Fischer and Maasbol[1] there have been hundreds isolated. There have been a number of authoritative reviews on the subject in the intervening years,[2] as well as an account of Schrock's work on alkylidene complexes.[3]

Synthesis of Stable Carbene Complexes

While the isolation and full characterization of carbene complexes is a fairly recent event, it has been recognised[4] that the Chugaev salts, prepared as long ago as 1915,[5] contain carbene complexes. The prototypical preparation of stable complexes depended on the reactivity of certain metal carbonyls towards strong nucleophiles, in particular organolithium compounds. Thus, treatment of hexacarbonyl tungsten with methyl- or phenyllithium in ether produced an acyl pentacarbonyl tungstate **1** which could be converted into pentacarbonyl[hydroxy(organo)carbene]tungsten(0) complexes **2** acidification.

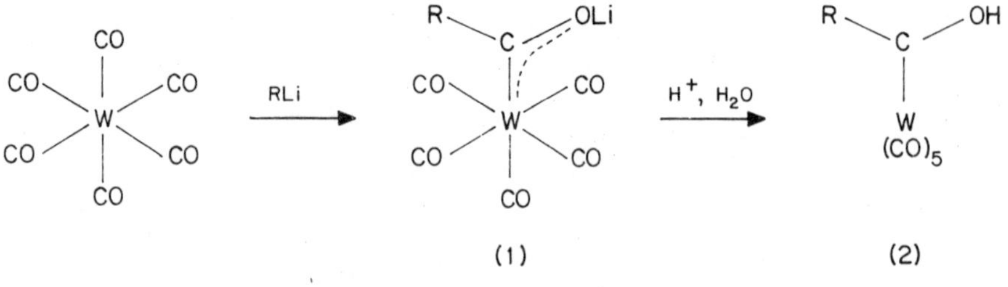

Complexes of type **2** were found to be unstable, but they could be converted into the more stable methoxy carbene derivatives **3** by treatment with diazomethane. The same compounds can be prepared by direct alkylation of the lithio derivatives **2** with Meerwein reagents, a high-yielding synthetic route.

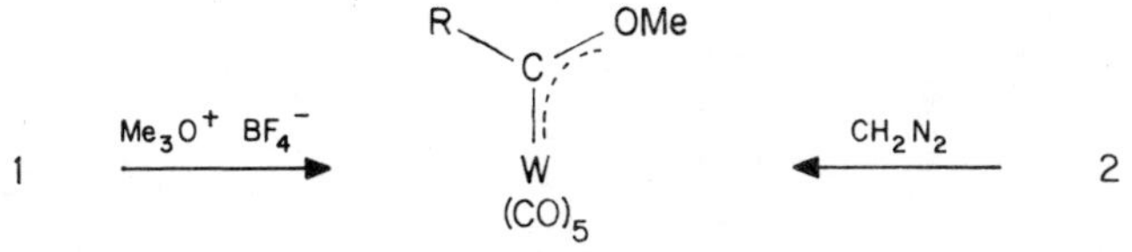

$$1 \xrightarrow{Me_3O^+ \, BF_4^-} \quad \underset{\substack{W \\ (CO)_5}}{\overset{\displaystyle R\diagdown\!\!\underset{\|}{C}\diagup\!\!OMe}{}} \quad \xleftarrow{CH_2N_2} \quad 2 \qquad (3)$$

This same method may be applied, using a range of organolithium reagents, to the preparation of analogous methoxycarbene complexes derived from hexacarbonylchromium, hexacarbonylmolybdenum, decacarbonyldimanganese, $Tc_2(CO)_{10}$, $Re_2(CO)_{10}$, $Fe(CO)_5$ and $Ni(CO)_4$. The product carbene complexes become increasingly labile in that order.

A number of other methods now exist for the preparation of carbene complexes, all of which are stabilized by the presence of an electron-releasing group capable of reducing the electron deficiency of the carbene-like carbon atom. These are summarized in the following equations:

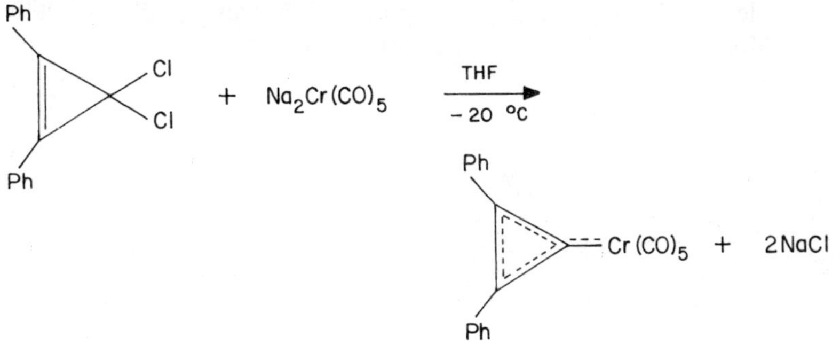

$$(4.1)$$

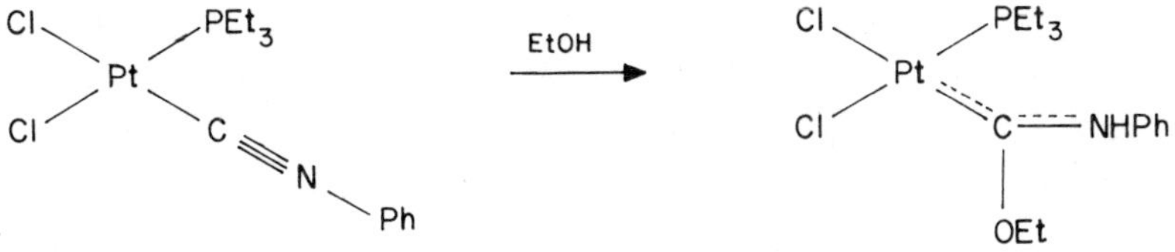

$$(4.2)$$

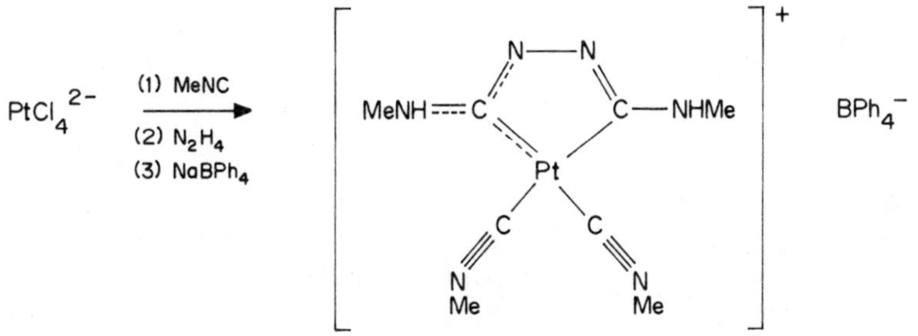

$$(4.3)$$

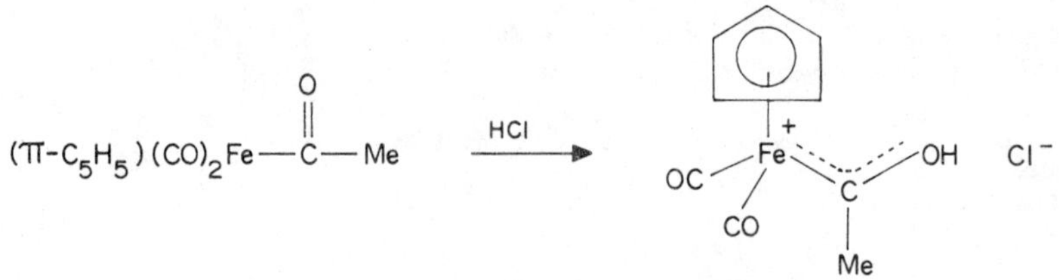

$$(\pi\text{-}C_5H_5)(CO)_2Fe\text{---}\overset{O}{\overset{\|}{C}}\text{---}Me \xrightarrow{\text{HCl}} \quad Cl^- \quad (4.4)$$

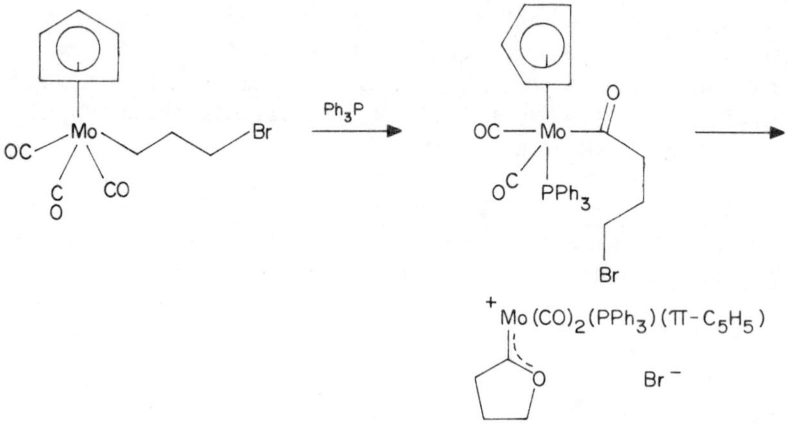

$$\xrightarrow{Ph_3P} \quad (4.5)$$

$$Mo(CO)_2(PPh_3)(\pi\text{-}C_5H_5) \quad Br^-$$

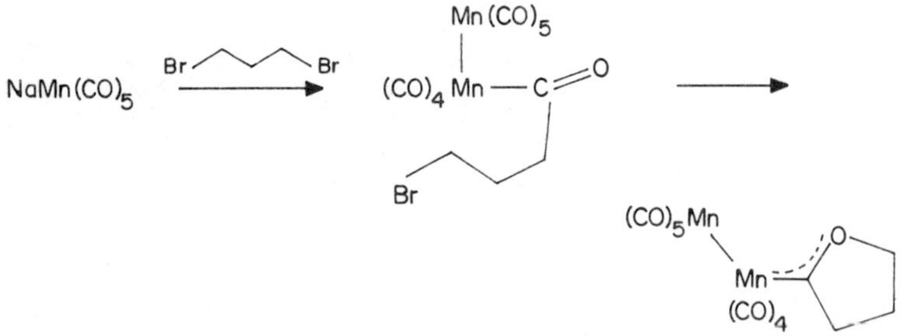

$$NaMn(CO)_5 \xrightarrow{Br \diagup\diagdown Br} (CO)_4Mn\text{---}C\!\!=\!\!O \longrightarrow \quad (4.6)$$

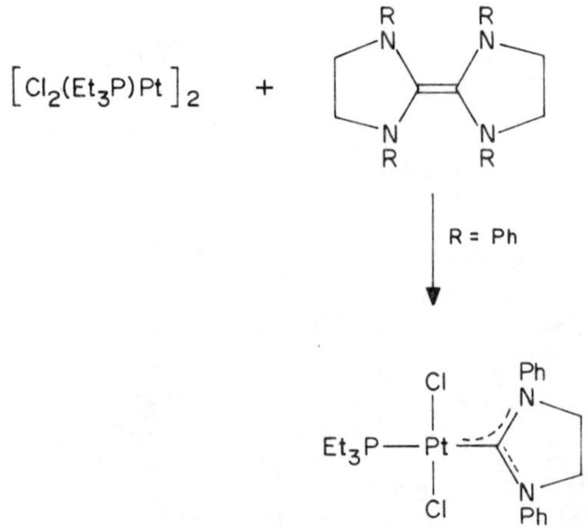

$$\left[Cl_2(Et_3P)Pt\right]_2 \quad + \qquad\qquad\qquad (4.7)$$

$$\downarrow R = Ph$$

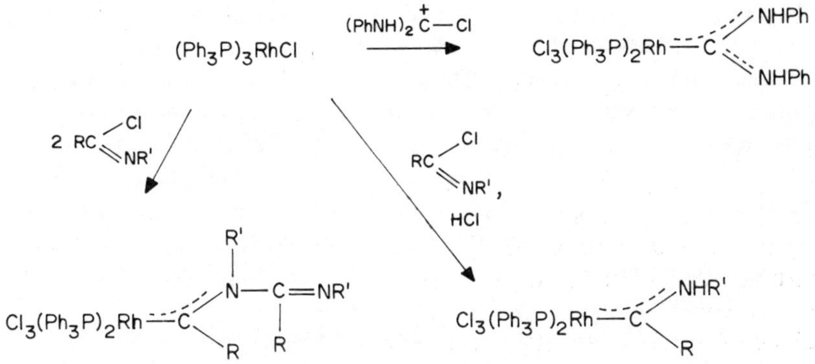

$$(4.8)$$

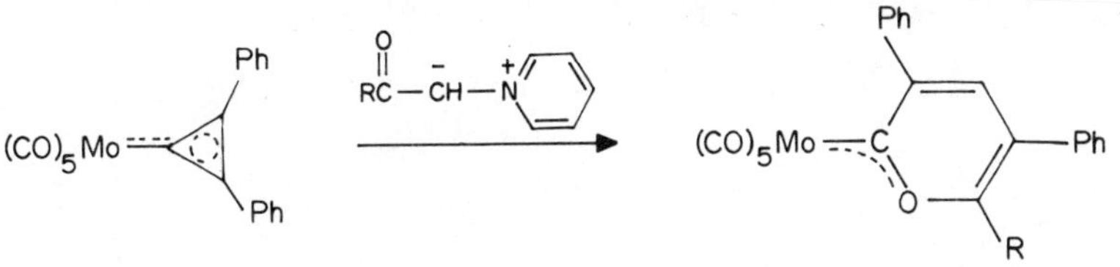

$$(4.9)$$

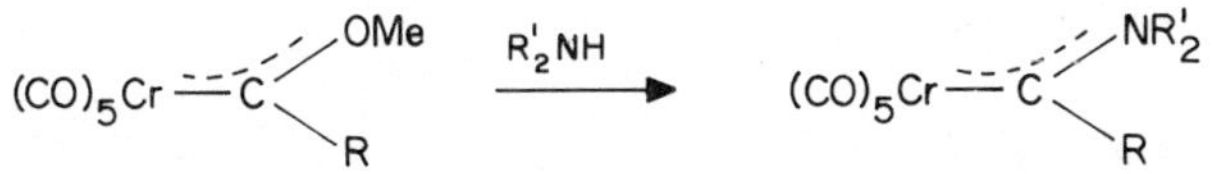

$$(4.10)$$

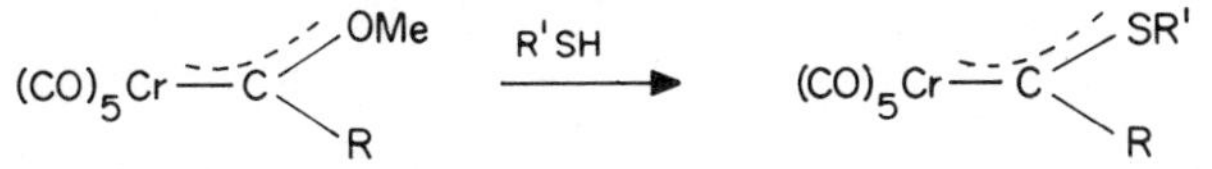

$$(4.11)$$

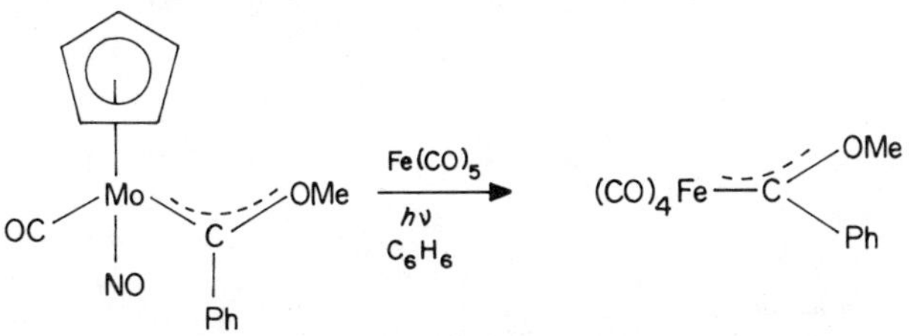

$$(4.12)$$

Hence there exist, among others, methods based on the ability of a metal to split an activated olefin into carbene components (method 4.7), the ability of metal anions to displace two halogens from a *gem*-dihalo compound (method 4.1), the conversion of the carbene by nucleophilic substitution (methods 4.9, 4.10, 4.11) and the ability of some complexes to transfer their carbene ligand to other metals (method 4.12). It may be readily seen that all of the above complexes have some means of stabilization from a substituent attached to the carbene carbon atom. These complexes are the ones commonly referred to as carbene complexes, while those compounds which have only hydrogens or alkyl groups attached to the carbene carbon are commonly referred to as alkylidene complexes. Examples of the latter complexes have only recently been prepared as stable compounds, although they have been implicated in a number of interesting organometallic reactions, e.g. the decomposition of certain transition metal alkyl complexes, Cu(I)-catalysed formation of a cyclopropane from diazomethane and an olefin, olefin metathesis, reduction of CO by H_2 and the rearrangement of small-ring hydrocarbons. A number of alkylidene derivatives of tantalum and niobium have in recent years been prepared by Schrock's group.

Treatment of $Ta(CH_2CMe_3)_3Cl_2$ with 2 mol of $LiCH_2CMe_3$ in pentane does *not* give the expected $Ta(CH_2CMe_3)_5$ (cf. preparation of $TaMe_4$). Instead, one obtains the orange crystalline complex 4 in quantitative yield:

$$M(CH_2CMe_3)Cl_2 \longrightarrow (Me_3CCH_2)_3M=CHCMe_3$$

$$M = Ta, Nb$$

$$(4)$$

A plausible mechanism for the formation of 4 is by α-hydride elimination from the intermediate 5 to give neopentane and 6, which is attacked by a molecule of neopentyl lithium to give 4.

$$M(CH_2CMe_3)_4Cl \xrightarrow{-Me_4C} (Me_3CCH_2)_2M=CHCMe_3 \xrightarrow{Me_3CCH_2Li} 4$$

$$(5) \qquad\qquad (6)$$

The corresponding η^5-cyclopentadienyl complex, readily prepared from $M(CH_2CMe_3)_2X_3$, also undergoes α-hydride elimination to give the neopentylidene complex 7. This and some other preparations of alkylidene complexes are given below.

$$M(CH_2CMe_3)_2X_3 \xrightarrow[\substack{toluene \\ 25\ °C}]{TlCp} \quad X\text{\tiny\textbackslash\textbackslash\textbackslash\textbackslash}M\diagdown CHCMe_3 \xrightarrow[\substack{toluene \\ 25\ °C}]{TlCp}$$

$$(7)$$

$$Cp_2M\diagup^{CHCMe_3}_{\diagdown X} \qquad (4.13)$$

$$M = Nb, Ta$$

$$\text{Ta(CH}_2\text{Ph)}_3\text{Cl}_2 \xrightarrow[\text{toluene}]{\text{LiC}_5\text{Me}_5} \quad \begin{array}{c} \text{Cp}^*\\ \text{Cl}^{\text{\tiny III}}\text{Ta}=\text{CHPh}\\ |\\ \text{CH}_2\text{Ph} \end{array} \qquad (4.14)$$

$$\downarrow \text{2 LiCp}$$

$$\text{Cp}_2\text{Ta}\begin{array}{c} \diagup\text{CHPh}\\ \diagdown\text{CH}_2\text{Ph} \end{array}$$

$$\text{TaMe}_3\text{Cl}_2 \xrightarrow{\text{TlCp}} \text{TaCpMe}_3\text{Cl} \xrightarrow{\text{TlCp}} \text{Cp}_2\text{TaMe}_3$$

$$\downarrow \text{Ph}_3\text{C}^+\ \text{BF}_4^- \qquad (4.15)$$

$$\begin{array}{c} \text{Cp}_2\text{Ta}\begin{array}{c}=\text{CH}_2\\ \diagdown\text{Me}\end{array} \end{array} \xleftarrow[\text{e.g. Ph}_3\text{P}=\text{CH}_2]{\text{base}} \left[\text{Cp}_2\text{TaMe}_2\right]^+ \quad \text{BF}_4^-$$

(8)

$$\text{Cp}_2\text{Ta(PMe}_3)\text{Me} \rightleftharpoons \text{Cp}_2\text{TaMe}$$

$$\downarrow \text{Et}_3\text{P}=\text{CHMe} \qquad (4.16)$$

$$\text{Cp}_2\text{Ta}\begin{array}{c}\diagup\text{CHMe}\\ \diagdown\text{Me}\end{array} \xleftarrow{-\text{PEt}_3} \left[\text{Cp}_2\text{Ta}\begin{array}{c}\diagup\text{CHMe PEt}_3\\ \diagdown\text{Me}\end{array}\right]$$

(9)

It may be noted that the parent complex **8** is readily prepared by a three- to four-step sequence involving deprotonation of a cationic dimethyltantalum complex by base (the best results are obtained with phosphorus ylide). Of particular interest is the preparation of the ethylidene complex **9**, in which techniques other than α-hydride abstraction or deprotonation must be used, owing to the propensity with which transition metal alkyls undergo β-elimination (see Chapter 2). Generally, Nb and Ta alkylidene complexes react with even traces of oxygen or water (or other protic solvents), although sterically crowded 18-electron derivatives, e.g. $Cp_2Ta(CHCMe_3)Cl$, can be handled as solids briefly in air. These are also fairly stable thermally, while the least stable are the methylene complexes **8**, and relatively uncrowded molecules, e.g. **6**. A number of bis(alkylidene) complexes are also preparable, e.g. treatment of $Ta(CH_2CMe_3)_3(CHCMe_3)$ with PMe_3 causes loss of neopentane:

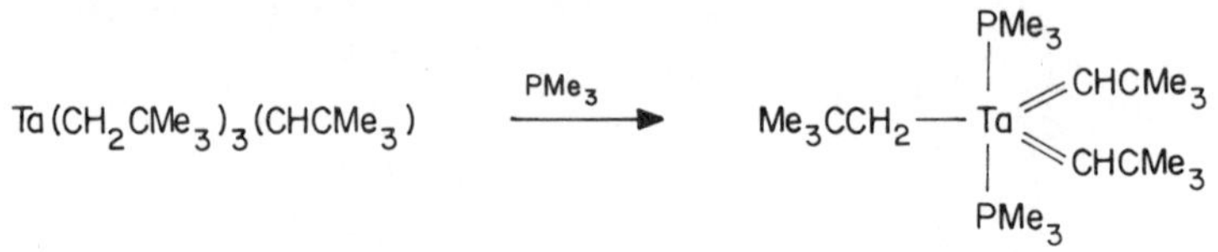

Also:

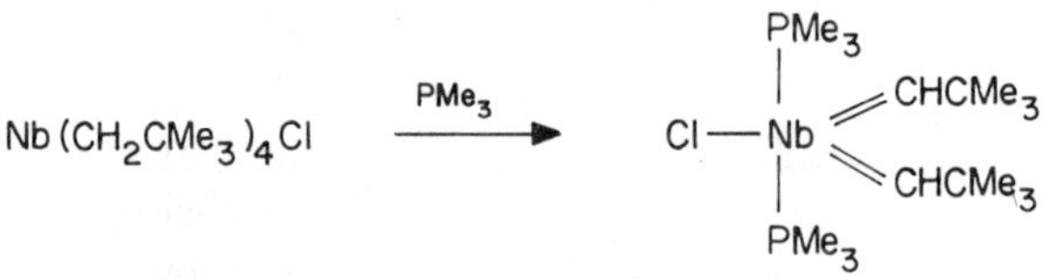

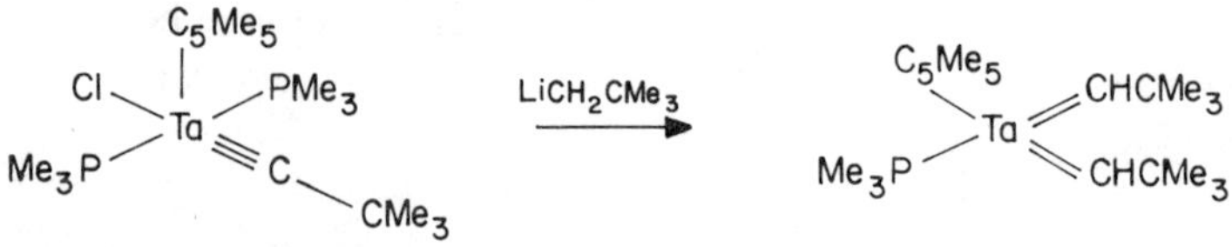

Such complexes are of practical and theoretical interest, since they might undergo transformation to an olefin complex. In the above examples this might be prevented by the bulk of the groups on the alkylidene ligands. A transformation of a less hindered derivative to an olefin complex could mean that the reverse reaction is possible, which may be important in, e.g., olefin metathesis reactions (see Chapter 2), i.e.

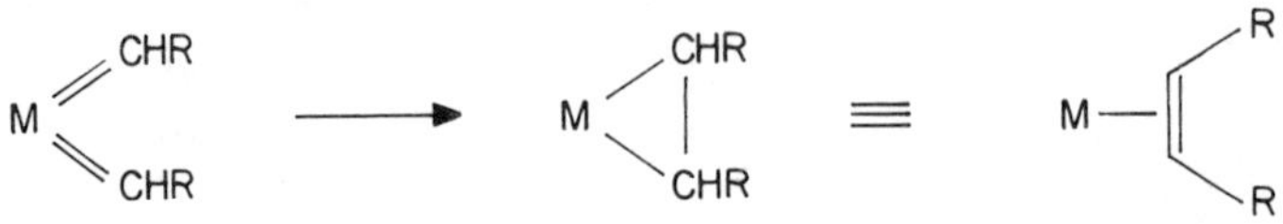

Bonding in carbene complexes has already been briefly discussed. In the most stable metal carbene complexes, it is found that C(carb) is highly electrophilic. This results in conjugative interaction with the heteroatoms of the ligand, and it might also therefore be expected to lead to a degree of $d\pi$–$p\pi$ interaction with the metal. Generally, it is found that the coordinated carbene is a strong σ-donor but a weak π-acceptor. The principal canonical forms involved in bonding are

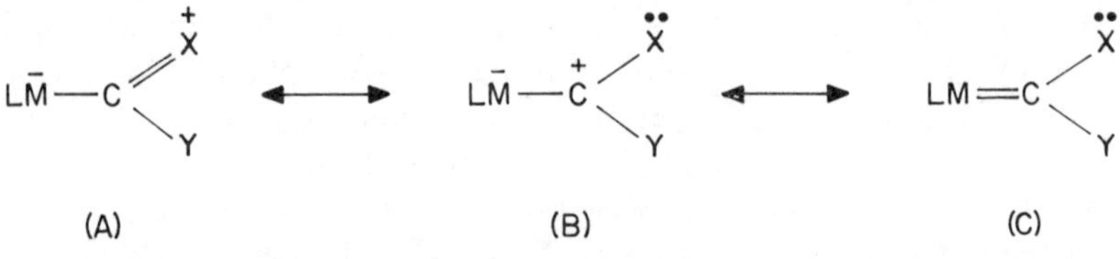

(A)   (B)   (C)

This implies strong polarization towards the metal, so that the dipole moments of the complexes are relatively large [$ca.$ 5 D for $(CO)_5Cr$–carbene]. Evidence for the substantially larger σ-donor/π-acceptor ratio than those found for CO derives from infrared spectral data. While the symmetrical Raman-active v_{co} absorption in $Cr(CO)_6$ is found at 2108 cm^{-1}, the frequency for the CO ligand *trans* to the carbene ligand of $(CO)_5Cr{=}C(OMe)Ph$ occurs at 1953 cm^{-1}, a very much lower frequency. This is due to the lower π-acceptor property of the carbene, which leads to expansion of metal t_{2g} orbitals and therefore their greater availability for interactions with the CO π^* orbital.

Carbon-13 NMR spectra are also useful. For example, the observed chemical shift of the carbene carbon in $(CO)_5CrC(OMe)Ph$ is at 351.42 p.p.m. This lies well within the range of values found for carbo-cations of organic chemistry, again implying substantial positive charge on the carbene carbon, due to the importance of canonical forms **A** and **B**.

X-ray crystal structure analysis reveals a number of interesting features. The carbene carbon atom is essentially planar, therefore sp^2 hybridized, and possessing an empty p orbital suitably disposed for interaction with the metal in $(CO)_5CrC(OMe)Ph$. The $C_{carbene}$–O distance (*ca.* 1.33 Å) lies between the values for a single C—O bond (1.43 Å in Et_2O) and a C$=$O double bond (1.23 Å in acetone), indicating substantial interaction with the heteroatom (oxygen in this case) lone pair. The average $Cr{=}C_{co}$ distance (1.87 Å) is shorter than the Cr—$C_{carbene}$ distance (2.04 Å), consistent with the greater multiple bond character of the former (but note that CO has two orthogonal π^* orbitals for overlap with metal d orbitals). However, the bond length is significantly shorter than the distance of 2.21 Å expected for a pure chromium—carbon σ-bond.[6] This implies some degree of metal–carbene $d\pi$–$p\pi$ interaction and a contribution from canonical form **C**. The phenyl group in this particular complex is at *ca.* 90° to the plane of the sp^2-C_{carb}, indicating no conjugation with this substituent. That canonical form **C** makes only a minor contribution, however, is implied by the chemical reactivity of the complexes, since scarcely any reactions are found, apart from insertion of PhSeH and RNC, which suggest $M{=}C_{carb}$ double-bond character.

Proton NMR studies of rotation about C_{carb}—NR_2 or C_{carb}—OR bonds indicate substantial energy barriers for this process, in many cases higher than carboxylic esters or amides, again implying substantial double bond character (canonical form **A**).

Turning to the alkylidene complexes, where the opportunity does not exist for canonical form **A**, we find that there is definite evidence for multiple Ta—carbene bond character. Some X-ray data are shown below, indicating significantly shorter Ta—C_{carb} than Ta—C single bonds.

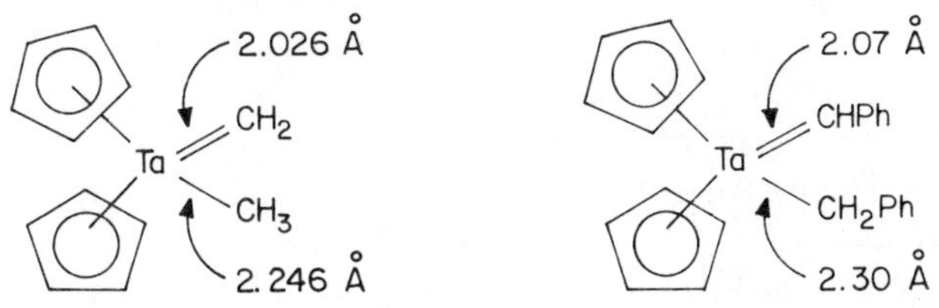

This could be a consequence of the Ta—C bond being derived from carbon sp^2 or sp^3 hybrids, the former being 'shorter,' but it is generally taken as being evidence for $d\pi$–$p\pi$ interaction, particularly in view of the fact that the Ta=CH$_2$ moiety shows restricted rotation of the methylene group. However, the overlap between metal and carbon π-type orbitals is rather poor, an *approximate* measure of the metal—alkylidene π-bond being 25 kcal mol^{-1}, compared with the Ta—C single bond strength, which is *ca.* 40–60 kcal mol^{-1}.

Carbene Complexes as Reactive Intermediates

Metal–carbene complexes of various types have been implicated in a number of chemical reactions, and only rarely characterized. In 1966, Jolly and Pettit[7] observed that reactions of chloromethyl methyl ether with CpFe(CO)$_2$Na (or NaFp) yielded the σ-bonded complex **10**. Treatment of this complex with acid in the presence of an olefin led to the formation of cyclopropane derivatives corresponding to carbene addition to the olefin double bond. Thus, it was proposed that a cationic carbene complex was formed by protonation of **10** the subsequent loss of methanol.[7]

(10)

The observation that the geometry of the double bond is maintained during formation of the cyclopropane implies a concertedness to the reaction rather than an electrophilic attack on the olefin, to form a cation, followed by further reactions. Supporting evidence for the intermediacy of a cationic carbene complex was provided by Brookhart and Nelson,[8] who were able to isolate and characterize the related phenyl carbenium derivatives **11** and **12** by NMR spectroscopy. These gave *trans*-stilbene when thermally decomposed, typically a carbene reaction.

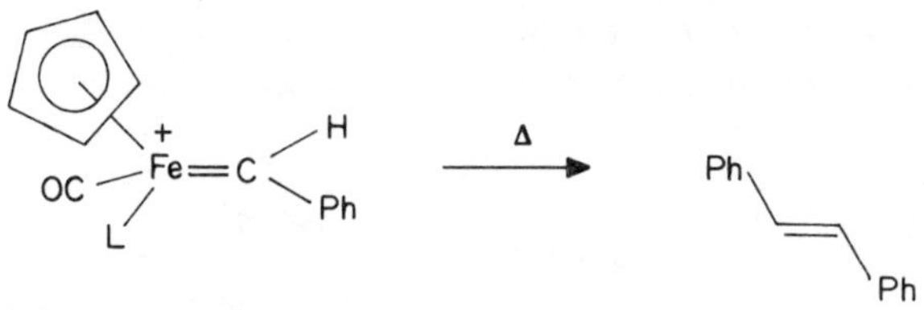

(11) L = CO

(12) L = Ph₃P

The cycloproponation reactions were potentially valuable, but gave low yields. The reaction has recently been improved by Brandt and Helquist,[9] who converted the thioether complex **13** into the stable sulphonium compound **14**. When this complex was heated in refluxing dioxane in the presence of olefin, good yields of cyclopropanes were obtained.

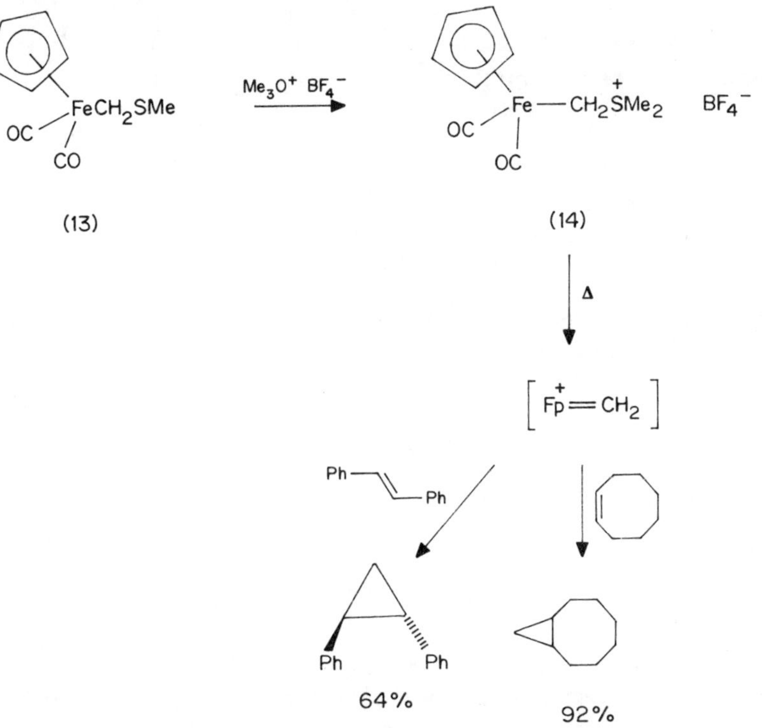

(13)

(14)

$$\left[\overset{+}{Fp} = CH_2 \right]$$

64%

92%

Metal-catalysed carbene generation from diazoalkanes has also been recognized as proceeding via the formation of metal–carbene complexes which are often probably the species involved in the reaction. Interest in the mechanism of cycloproponation using, e.g., the Simmons–Smith reagent $IZnCH_2I$, is largely responsible for the extensive studies in this area. Evidence for the involvement of metal–carbene complexes in diazoalkane decomposition is as follows:

(a) The metal-promoted reaction of diazoacetic ester with cyclohexene gives a different ratio of *exo/endo* product than the corresponding photochemical reactions. Also, changing the phosphite ligand $(RO)_3P$ affects the ratio.

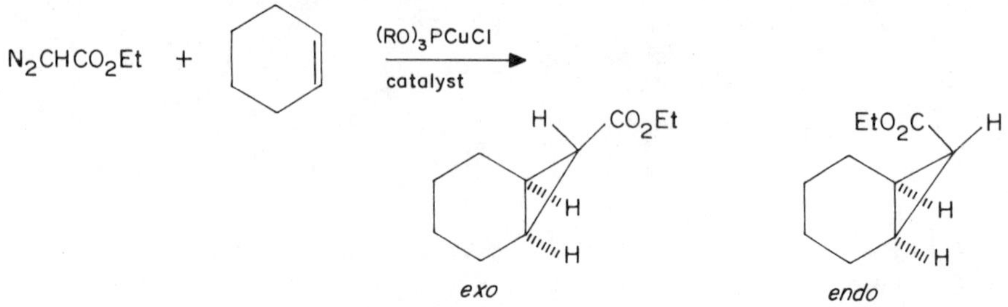

N_2CHCO_2Et + (cyclohexene) $\xrightarrow[\text{catalyst}]{(RO)_3PCuCl}$

exo   endo

(b) Small but significant degrees of asymmetric induction are obtained when optically active metal complexes are used, e.g.

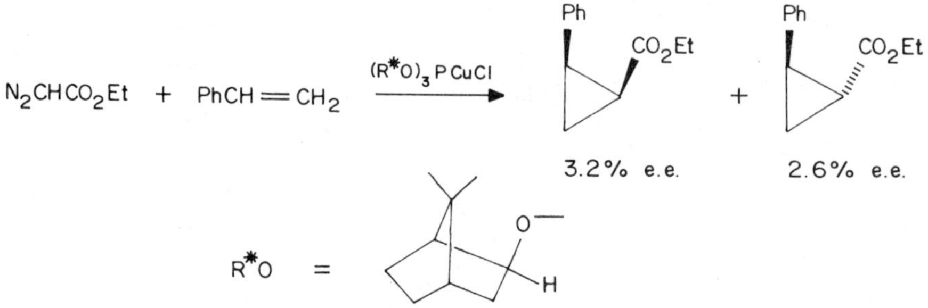

N_2CHCO_2Et + $PhCH\!=\!CH_2$ $\xrightarrow{(R^*O)_3PCuCl}$

3.2% e.e.   2.6% e.e.

R^*O =

On the basis of these results a transition state, e.g. **15**, for cyclopropanation involving metal–carbene derivatives has been proposed:

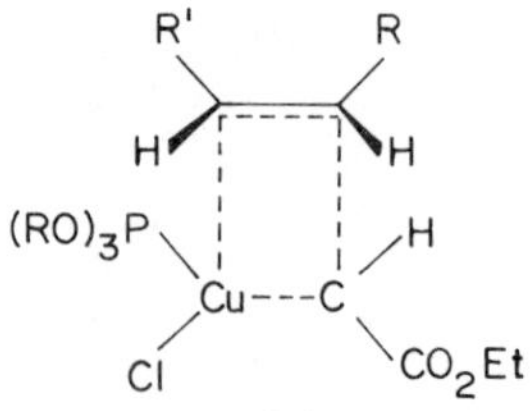

(15)

Metal–carbene-like complexes are also clearly involved in the synthesis of organic carbonyl compounds from metal carbonyls, as exemplified below.

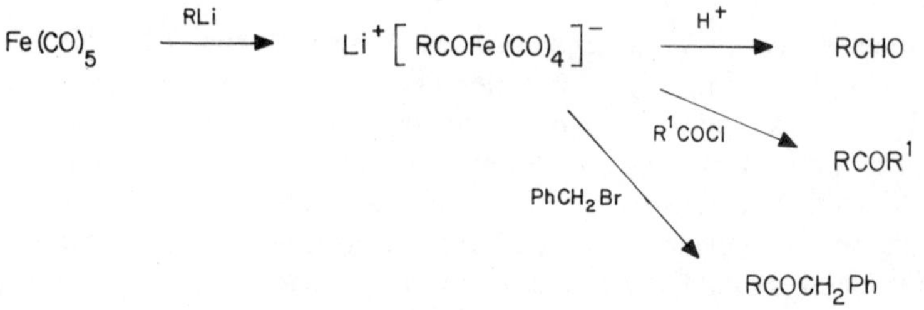

$Fe(CO)_5$ $\xrightarrow{RLi}$ $Li^+\left[RCOFe(CO)_4\right]^-$ $\xrightarrow{H^+}$ $RCHO$

$\xrightarrow{R^1COCl}$ $RCOR^1$

$\xrightarrow{PhCH_2Br}$ $RCOCH_2Ph$

In contrast to the aforementioned chromium complexes, protonation or reaction of the intermediate lithium salt with electrophiles does not appear to occur on oxygen, but rather on the metal, followed by reductive elimination, e.g.

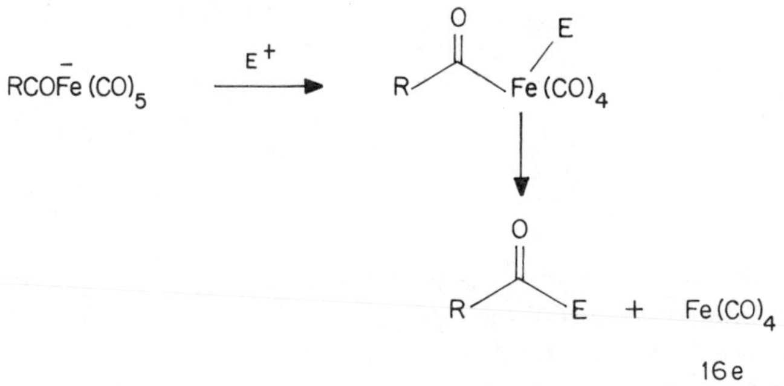

16e

Clearly these are examples of the anionic metal complex behaving as an ambient nucleophile, owing to the involvement to the following canonical forms, the second being carbene-like.

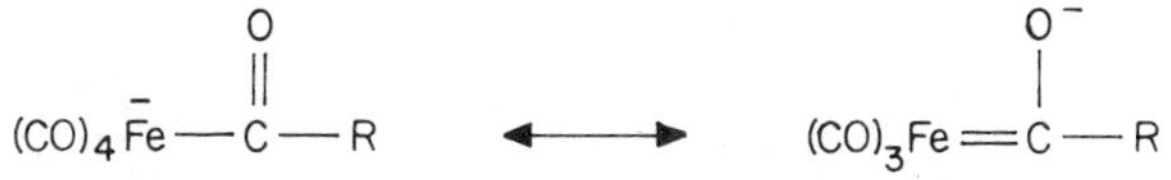

Presumably, reaction with an appropriate (hard ?) electrophile would lead to *O*-alkylation.

Metal–carbene complexes are also widely accepted as being involved in metal-promoted valence isomerizations of strained-ring carbocyclic species. The course of such reactions, in terms of product distribution and product type, is often altered markedly in the presence of transition metals. Some examples are given below.

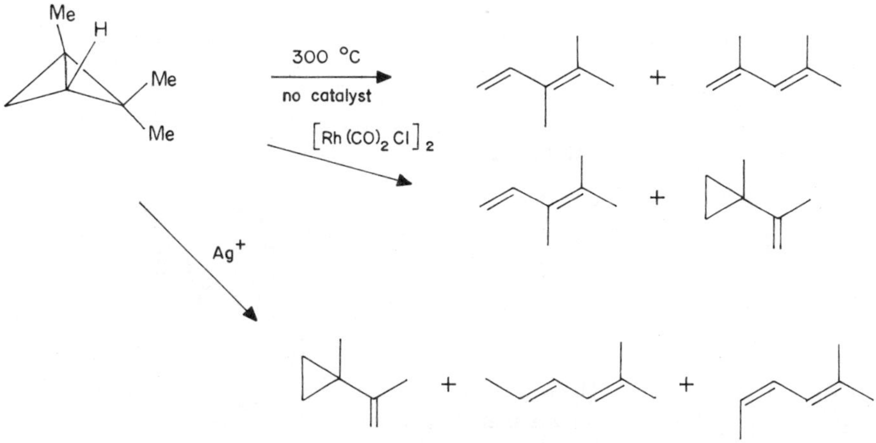

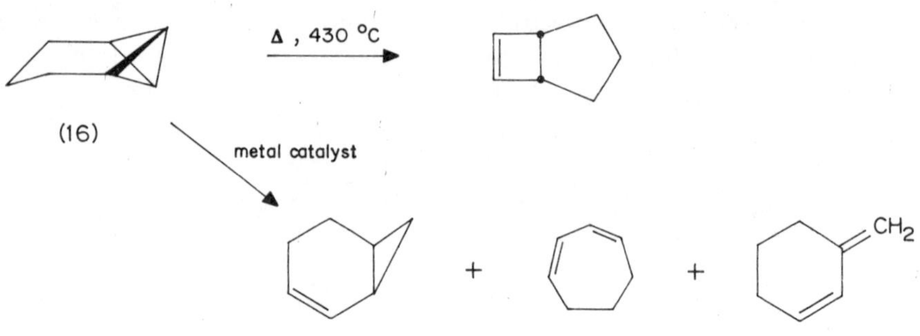

The latter reaction provides a useful discussion example. A possible reaction pathway for the formation of the methylenecyclohexene product is given below. It may be noted that the intermediate **18** may be written as a cationic metal carbene (alkylidene) complex or a metallo-carbonium ion which undergoes 1,2-hydride shift and loss of the metal.

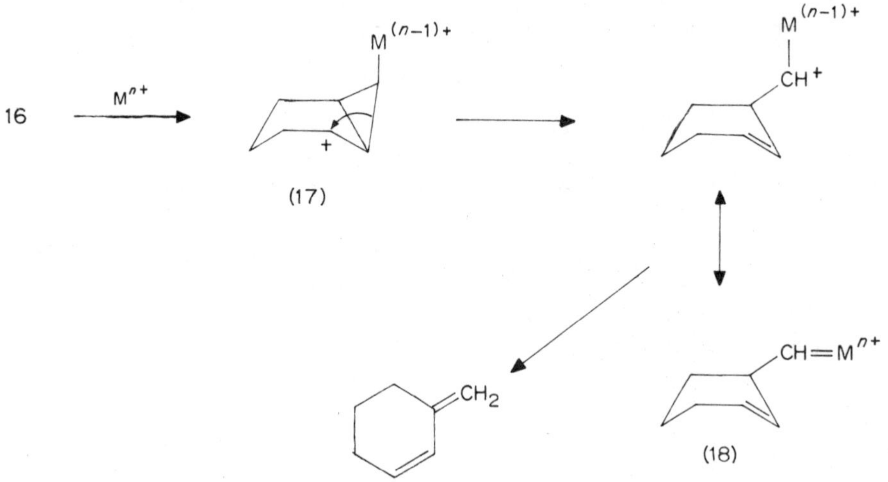

The intermediacy of carbonium ion **17** is supported by the formation of ethers when the reaction is carried out in methanol:

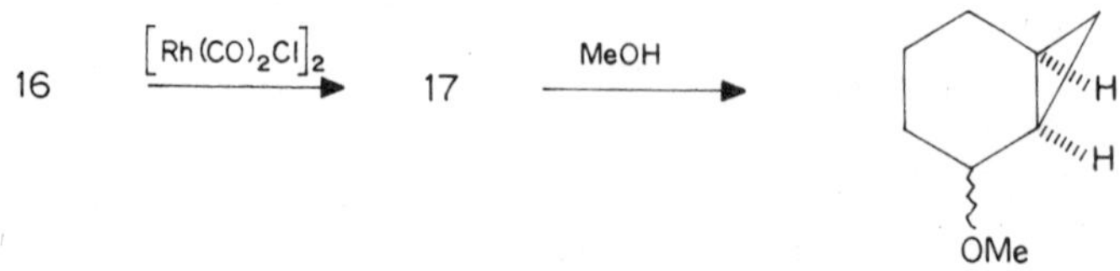

Significantly, reaction of **16** with sulphuric acid in methanol produces the same stereoisomer ratio of the product methyl ethers.

The opening of bicyclobutanes shown above also must proceed via carbonium ion intermediates, leading to cyclopropanes which do not react further, in contrast to the thermal reaction, which is a concerted process:

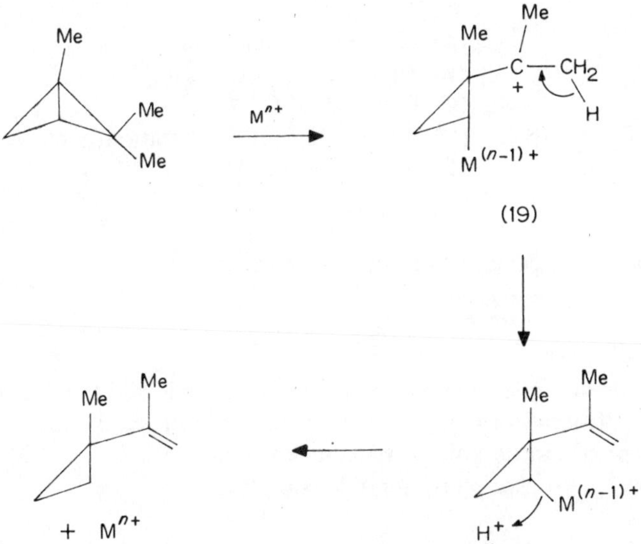

(19)

The formation of a single cyclopropane product is probably due to steric factors preventing metal attack at the alternative cyclopropane bond which is flanked by three methyl groups instead of two. Formation of the butadiene product probably involves breakdown of carbonium ion species **19** to form the carbene intermediate:

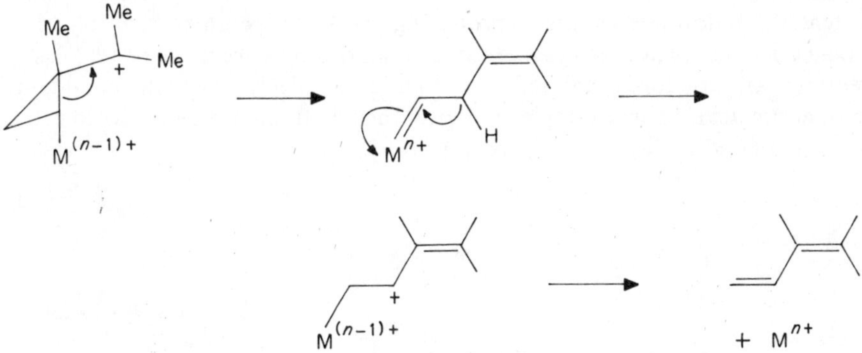

Carbene complexes are also implicated in olefin metathesis reactions, which have been fully discussed in Chapter 2.

Reactions of Carbene Complexes and Applicability to Organic Synthesis

Apart from cycloproponation, there are also a number of other features of metal–carbene complexes which might well be exploited for organic synthesis. The reactivity of heteroatom-substituted carbene complexes may be summarized diagrammatically:

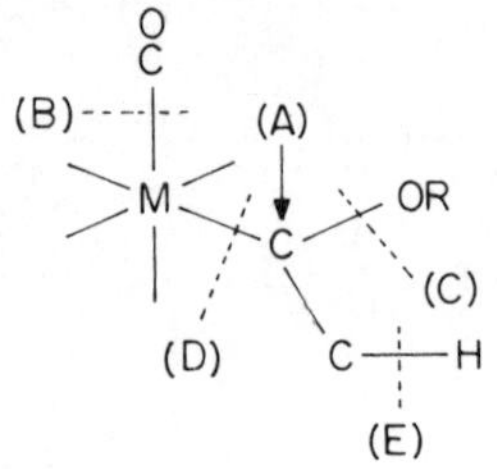

(A) Nucleophilic addition at C_{carb}
(B) Substitution of CO with L
(C) Cleavage of C—O
(D) Rearrangement or liberation of carbene ligand
(E) Substitution of H

We shall discuss these aspects in the above order.

A. Addition at C_{carb}

We have already met examples of amine nucleophile addition. Treatment of pentacarbonyl[alkoxy(organo)carbene] complexes of Cr(0) and W(0) with trialkylphosphines in organic solvents at low temperature ($-30\,^\circ$C) results in addition to C_{carb} to give phosphorylide complexes:

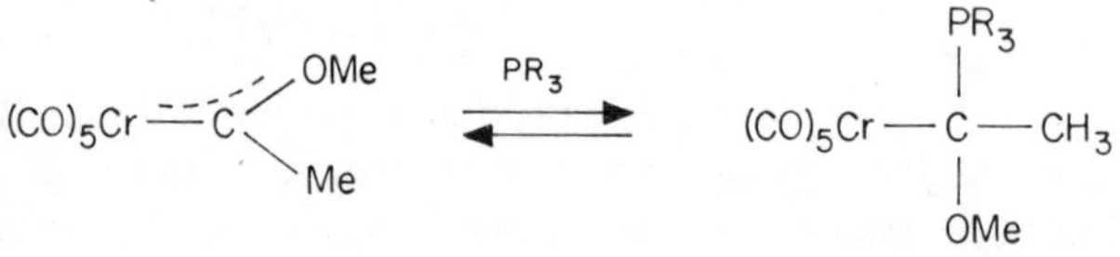

$$(20)$$

Reactions of the alkoxycarbene derivatives with amines lead to several potential applications as amino-protecting groups in peptide chemistry. The organometallic residue offers some scope in this respect since it can be very easily removed with trifluoroacetic acid, or in some cases under even milder conditions with acetic acid. Some examples are given below. It should be noted that the C_{carb}—O bond is cleaved [reaction type (C)].

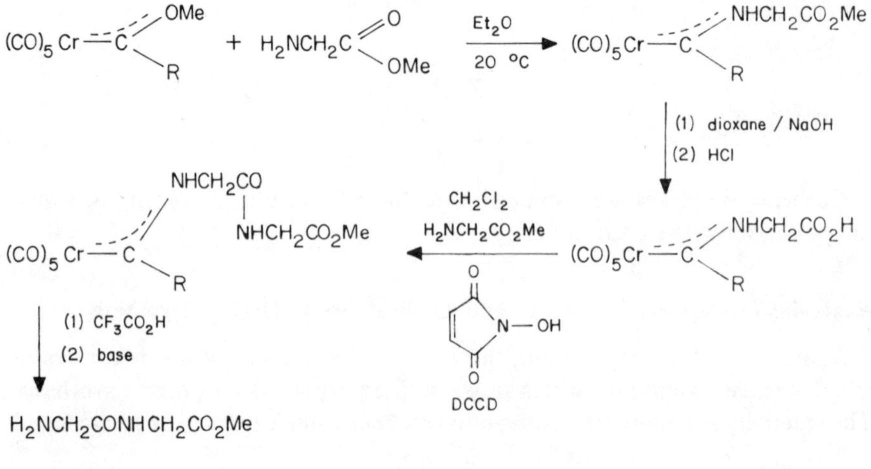

A number of advantages attach to the use of the organometallic protecting group: (a) the complexes are usually yellow and easily made visible for chromatography, etc.; (b) the protecting group is removed under mild conditions and side-products are easily separated; and (c) most of the complexes are sufficiently volatile to allow mass spectral analysis.

B. *Substitution of CO*

If the product of alkylphosphine addition **20** is irradiated in hexane–toluene at $-15\,^{\circ}\text{C}$, a CO ligand is eliminated and phosphine is introduced into the *cis* position. In contrast, if the starting carbene complex is simply heated in solution with the alkylphosphine both *cis* and *trans* substitution occur:

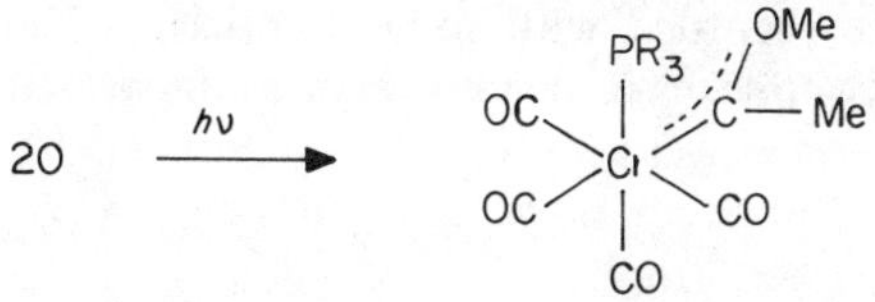

(21)

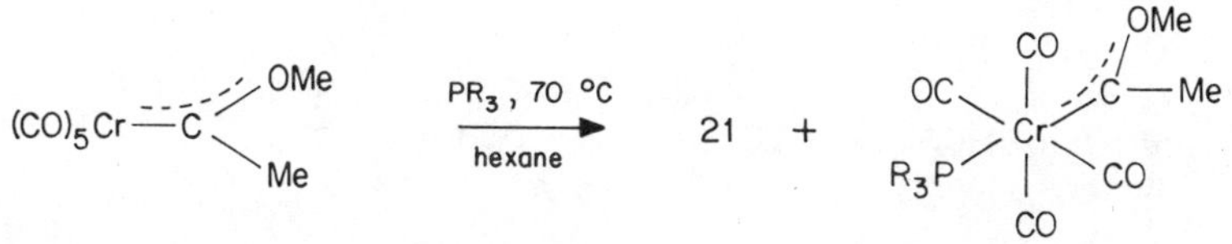

(22)

If either pure complex **21** or **22** is heated then the mixture is regenerated.

D. *Rearrangement or liberation of carbene ligand*

Two examples in which rearrangement of the carbene ligand occurs are given below.

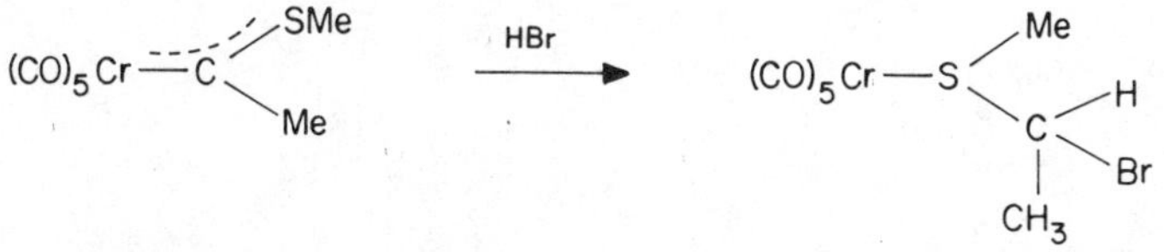

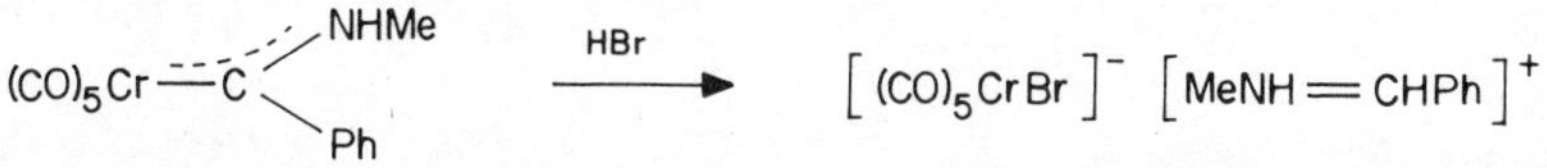

The apparent insertion of PhSeH into the metal—C_{carb} bond referred to earlier undoubtedly takes place by this type of mechanism, but occurs so readily that methanol instead of HBr can act as the source of proton and nucleophile:

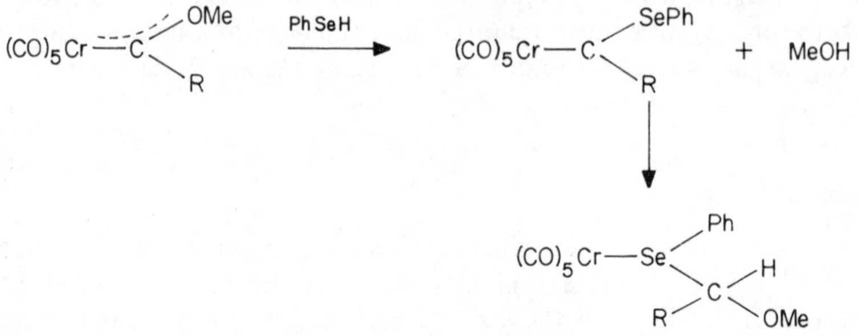

$$(CO)_5Cr=C{\overset{OMe}{\underset{R}{}}} \quad \xrightarrow{\text{Ph SeH}} \quad (CO)_5Cr-C{\overset{SePh}{\underset{R}{}}} \quad + \quad MeOH$$

$$(CO)_5Cr-Se{\overset{Ph}{\underset{}{}}}C{\overset{H}{\underset{OMe}{}}} \quad ;\; R$$

Therefore, the inference that the overall 'insertion' can be taken as evidence for M—C double bond character must be regarded with some scepticism. The carbene ligand can be cleaved from the complexes in various ways, as illustrated below:

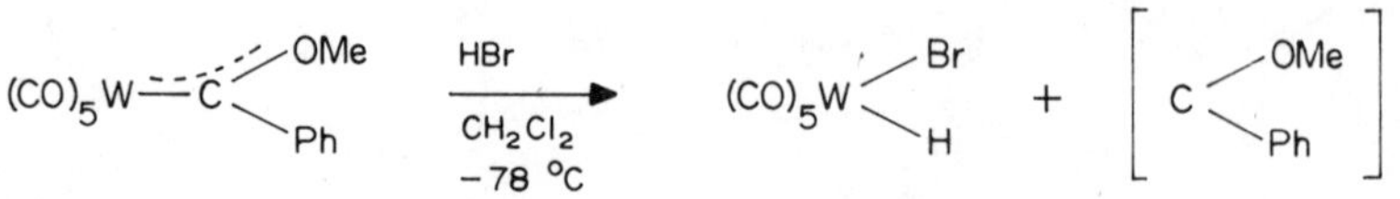

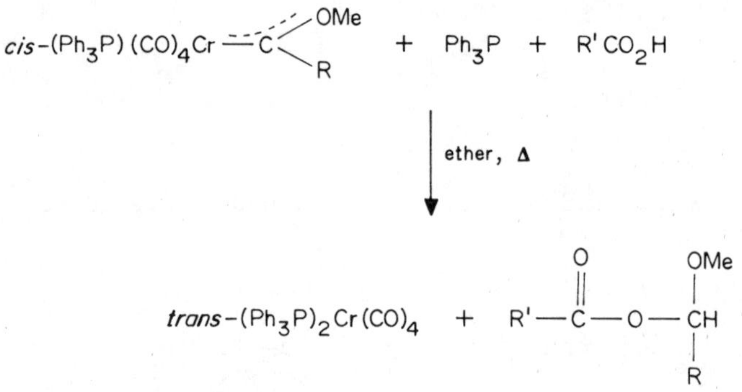

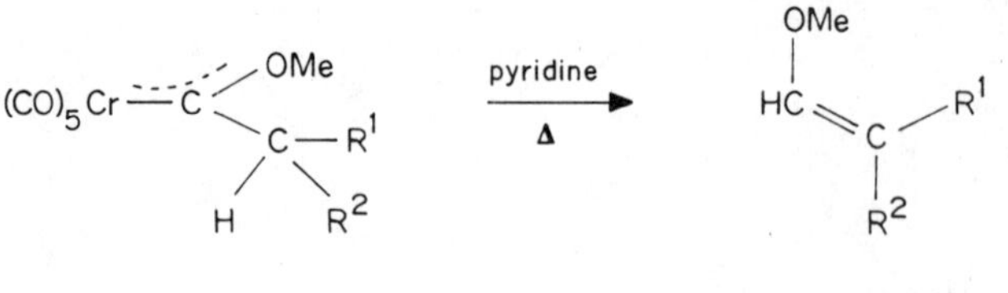

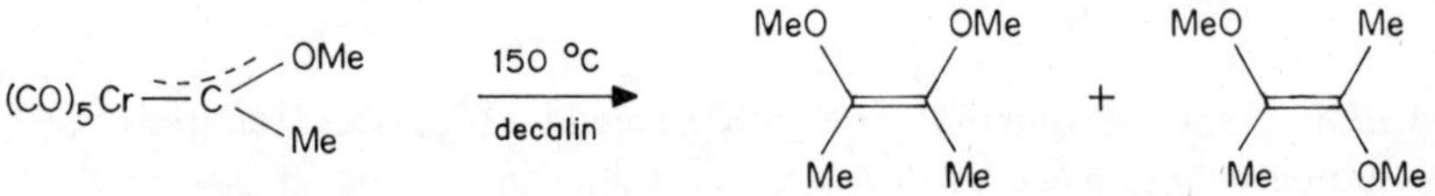

$$(CO)_5W=C{\overset{OMe}{\underset{Ph}{}}} \quad \xrightarrow[-78\ ^\circ C]{\underset{CH_2Cl_2}{HBr}} \quad (CO)_5W{\overset{Br}{\underset{H}{}}} \quad + \quad \left[C{\overset{OMe}{\underset{Ph}{}}} \right]$$

$$cis-(Ph_3P)(CO)_4Cr=C{\overset{OMe}{\underset{R}{}}} \quad + \quad Ph_3P \quad + \quad R'CO_2H$$

$$\downarrow \text{ether},\ \Delta$$

$$trans-(Ph_3P)_2Cr(CO)_4 \quad + \quad R'-\overset{O}{\overset{\|}{C}}-O-\overset{OMe}{\underset{R}{\overset{|}{C}}H}$$

$$(CO)_5Cr=C{\overset{OMe}{\underset{\underset{H}{C}-R^1}{}}}{\underset{R^2}{}} \quad \xrightarrow[\Delta]{\text{pyridine}} \quad HC{\overset{OMe}{\underset{}{}}}=C{\overset{R^1}{\underset{R^2}{}}}$$

$$(CO)_5Cr=C{\overset{OMe}{\underset{Me}{}}} \quad \xrightarrow[\text{decalin}]{150\ ^\circ C} \quad {\overset{MeO \quad OMe}{\underset{Me \quad Me}{}}} \quad + \quad {\overset{MeO \quad Me}{\underset{Me \quad OMe}{}}}$$

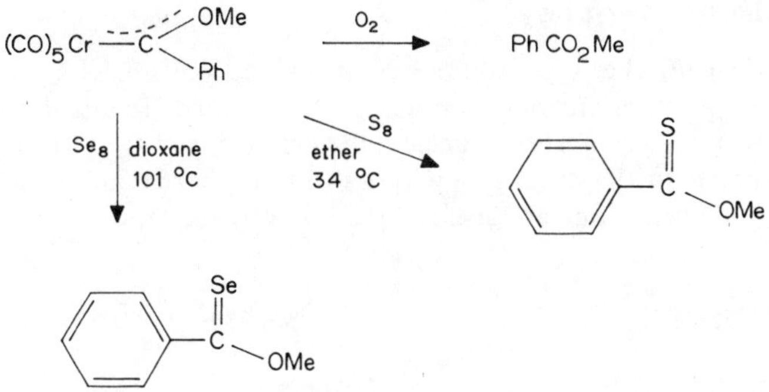

$(CO)_5Cr=C$ with OMe, Ph substituents $\xrightarrow{O_2}$ $Ph\,CO_2Me$

Se_8 | dioxane 101 °C

$\xrightarrow[\text{ether } 34\,°C]{S_8}$

(structure: Ph–C(=S)–OMe)

(structure: Ph–C(=Se)–OMe)

Decomposition of the carbene complex under suitable conditions in the presence of a reactive olefin gives cyclopropanes:

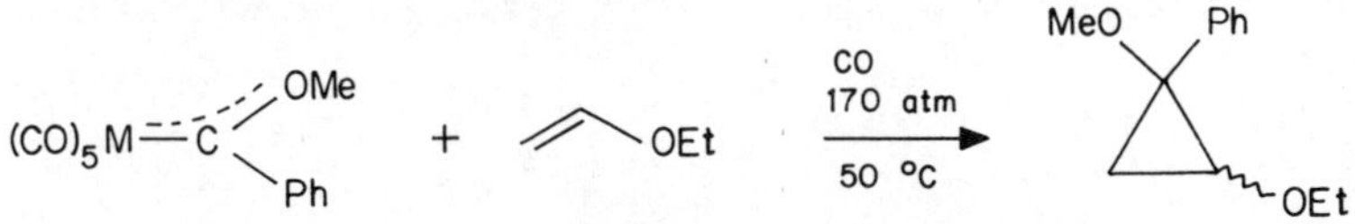

$(CO)_5M=C$ with OMe, Ph substituents $+$ (CH$_2$=CH–OEt) $\xrightarrow[50\ °C]{\substack{CO \\ 170\ atm}}$ (cyclopropane: MeO, Ph, OEt substituents)

The ratio of stereoisomers obtained varies with the metal (M = Cr, Mo, W), indicating that free carbene is probably not involved.

E. Substitution of α-Hydrogen

The hydrogen atoms which are bonded to the α-carbon atom of alkoxy(alkyl) carbene complexes have been shown to be quite acidic by ^1H NMR studies. Deuterium exchange is observed in CH_3OD solution in the presence of catalytic amounts of sodium methoxide. This reflects the ability of the metal carbonyl unit to act as an electron acceptor:

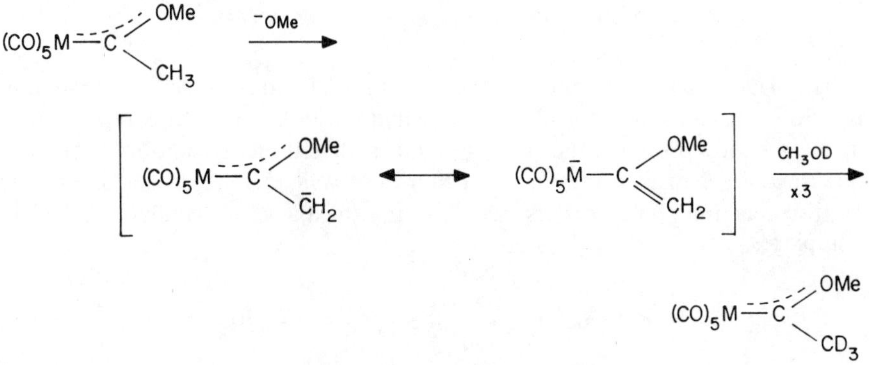

$(CO)_5M=C$ with OMe, CH$_3$ substituents $\xrightarrow{^-OMe}$

$\left[(CO)_5M=C \text{ (OMe, } \bar{C}H_2) \longleftrightarrow (CO)_5\bar{M}-C \text{ (OMe, }=CH_2) \right] \xrightarrow[\times 3]{CH_3OD}$

$(CO)_5M=C$ with OMe, CD_3 substituents

Some potential exists here for synthetic application, for example if the mononion could be generated selectively and reacted with appropriate electrophiles.

Reactions of Alkylidene Complexes

Whereas the above Fischer-type complexes, including $W(CO)_5CPh_2$ and $W(CO)_5CHPh$, behave as electrophiles at C_{carb}, the Nb and Ta alkylidene complexes prepared by Schrock show nucleophilic behaviour of C_{carb}. In this respect they are similar to the main group ylides of phosphorus and arsenic. Some examples of this behaviour are given in the following equations:

$$Nb(CH_2CMe_3)_3(CHCMe_3) \xrightarrow{HCl} Nb(CH_2CMe_3)_4Cl$$

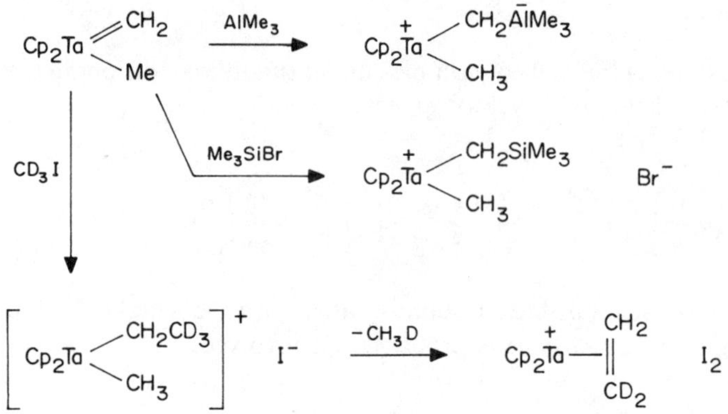

$$(Me_3CCH_2)_3M{=}CHCMe_3 \quad + \quad R'COR''$$

$$\downarrow \text{pentane, 25 °C}$$

$$cis- \quad + \quad trans-R'R''\,C{=}CHCMe_3 \quad + \quad M \text{ products}$$

The last reaction resembles the Wittig olefination reaction, involving reaction of a phosphorus ylide with aldehydes and ketones. In comparison, it appears highly probable that the titanium-mediated methylenation of ketones described by Nozaki's group,[10] which works well in many cases where the Wittig reaction fails, occurs by the intermediacy of similar $Ti{=}CH_2$ complexes, e.g.

$$R'R''C{=}O \xrightarrow[\text{TiCl}_4,\ \text{Zn}]{\text{CH}_2\text{Br}_2} R'R''C{=}CH_2$$

Methylene–titanium carbene complexes are proving to have considerable potential as reagents for organic synthesis. Treatment of Cp_2TiCl_2 with 2 equiv. of $AlMe_3$ gives the bridged methylene complex **23** as reddish orange crystals.[11]

$$Cp_2TiCl_2 \ + \ 2 \ AlMe_3 \ \longrightarrow \ CH_4 \ + \ Cp_2Ti\underset{Cl}{\overset{CH_2}{\diagdown}}AlMe_2 \ + \ AlMe_2Cl$$

(23)

In most of its reactions complex **23** behaves like an alkylidene complex polarized similar to the tantalum derivatives above, i.e. $\overset{+}{Ti}-\overset{-}{C}H_2$. It is probably formed by partial dissociation of **23**:

$$Cp_2Ti\underset{Cl}{\overset{CH_2}{\diagdown}}AlMe_2 \quad \rightleftharpoons \quad \underset{ClAlMe_2}{\overset{Cp_2Ti=CH_2}{|}}$$

The reactivity is probably better understood by reference to the following mechanistic scheme:

$$Cp_2Ti\underset{\overset{+}{Cl}}{\overset{CH_2}{\diagdown}}\bar{A}lMe_2 \ \longrightarrow \ Cp_2Ti\overset{\bar{C}H_2}{\diagdown}_{\overset{+}{Cl}-AlMe_2} \ \longrightarrow \ Cp_2\overset{+}{Ti}-\overset{-}{C}H_2$$

The carbene complex reacts with certain olefins in a homologation reaction, e.g.

$$\mathbf{23} \quad \overset{CH_2=CH_2}{\nearrow} \quad \overset{CH_3CH=CH_2}{32\%}$$

$$\overset{CH_3CH=CH_2}{\searrow} \quad Me_2C=CH_2$$

The reaction probably proceeds via formation of a metallacyclobutane intermediate, which undergoes β-elimination followed by reductive elimination:

$$Cp_2Ti\underset{CH_2}{\overset{CH_2}{<}}CH-H \ \longrightarrow \ Cp_2Ti\underset{H}{\overset{CH_2}{<}}\overset{CH}{\underset{CH_2}{\diagdown}} \ \longrightarrow \ CH_3CH=CH_2$$

Evidence for this mechanism comes from the observation that isobutene does not undergo the reaction, since the metallacyclobutane cannot undergo β-elimination:

$$Cp_2Ti\overset{Me}{\underset{Me}{\diagup}} \quad \overset{\times}{\longrightarrow} \quad no \ \beta-elimination$$

Perhaps the most potentially useful reactivity of the complex **23** is its reaction with ketones and related carbonyl compounds. Treatment of ketones with **23** results in methylenation, which appears to give higher yields that the standard Wittig olefination. The complex also reacts with esters to produce vinyl ethers. Examples are given below.

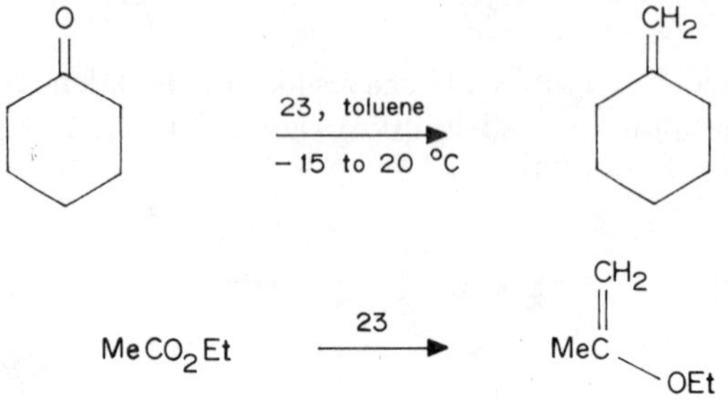

Ethene, propene and styrene react with electron-deficient alkylidene complexes with the probable intermediacy of metallacyclobutanes to give olefins as the final products, formed by β-elimination mechanisms.

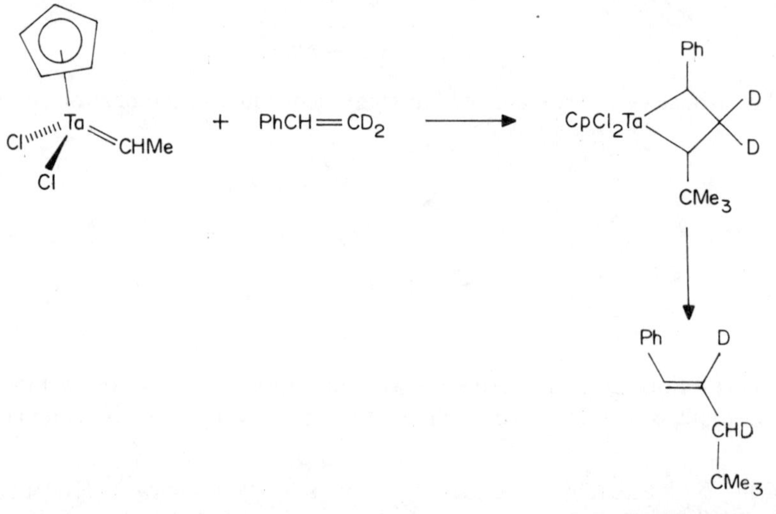

Acetylenes 'insert' in the M=CHR bond, to give a new alkylidene complex which is probably formed by rearrangement of an intermediate metallacyclobutene complex:

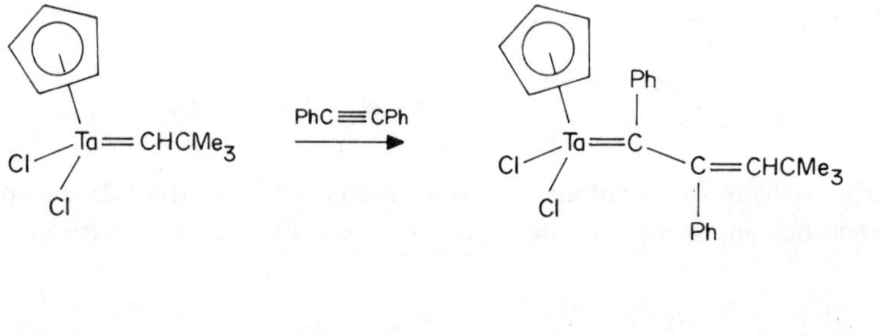

In conclusion, it may be noted that although there is some potential for synthetic application of these complexes, no concerted effort in this direction has been made. In this connection, the recent preparation of β-lactam derivatives by Hegedus' group,[12] shown in the following equations, may be noted.

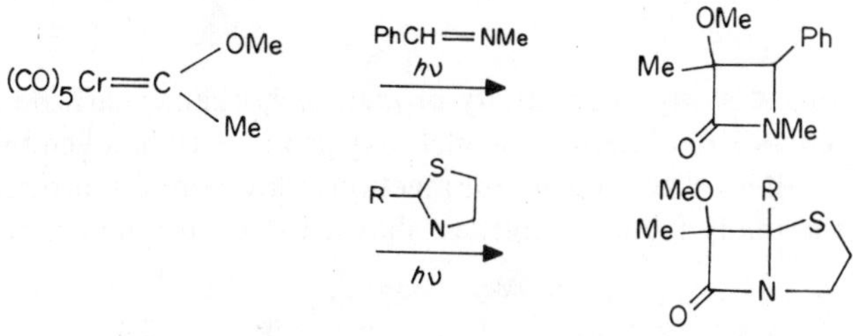

The photochemical reaction was readily brought about by exposure 'to bright sunlight on the roof of the chemistry building for 1–3 h at ambient temperature (10–20 °C),' assuming the availability of an appropriate climate. A plausible mechanism is shown below.

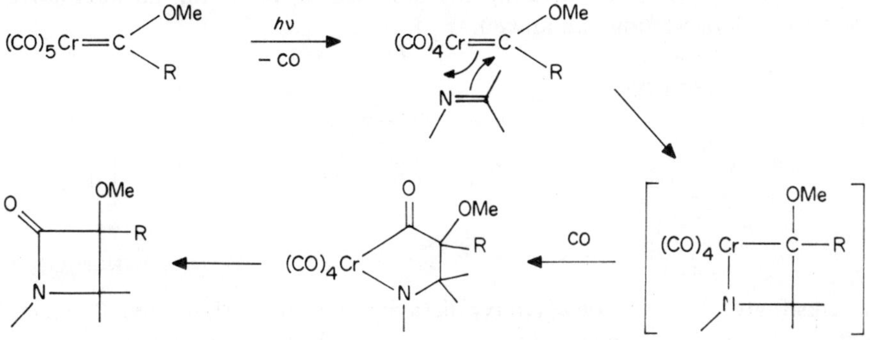

CARBYNE COMPLEXES

In this section our attention is confined to complexes which may be crudely represented as containing the $M \equiv C$ triple bond, and we omit any discussion of alkylidyne cluster complexes of type

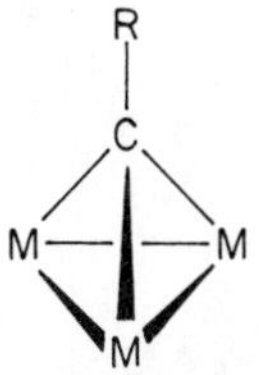

The discovery of stable carbyne complexes, of general structure **24**, by Fischer's group was largely accidental.

$$OC \quad CO$$

X—M≡C—R

$$OC \quad CO$$

(24)

M = Cr, Mo, W

X = Cl, Br, I

R = Me, Et, Ph

During a broad study of reactivity of methoxy(organo) carbene complexes, attempts were made to exchange the methoxy group with halogen by treatment with boron trihalides. Carrying out the reaction at low temperature permitted the isolation of thermally labile complexes, shown to be carbyne derivatives.

$$(CO)_5 M = C \begin{smallmatrix} OMe \\ \\ R \end{smallmatrix} \quad \xrightarrow[\text{pentane}]{BX_3} \quad 24$$

These complexes may be similarly prepared by treatment of the carbene complex with $AlCl_3$, $AlBr_3$ and $GaCl_3$ instead of boron trihalides. Reaction of α-amino ester carbene derivatives with boron tribromide serves a dual purpose of forming a carbyne complex *and* liberating the amino acid, providing an alternative method of deprotection (see above):

$$(CO)_5 W = C \begin{smallmatrix} NHCH_2CO_2Me \\ \\ Ph \end{smallmatrix} \quad \xrightarrow[\substack{CH_2Cl_2 \\ -25\,°C}]{BBr_3} \quad Br—W≡C—Ph \; + \; BBr_2NHCH_2CO_2Me$$

$$\downarrow H_2O$$

$$HBr.\,H_2NCH_2CO_2H$$

Interestingly, pentacarbonyl[ethoxy(diethylamino)carbene]tungsten (0) reacts with boron trihalides to give *trans*-(bromo)tetracarbonyl(diethyl-aminocarbyne)tungsten (0), resulting from exclusive loss of the alkoxy group. The relative stability found for this carbyne derivative can be attributed to interaction of the metal—carbon bond with the free electron pair of the nitrogen:

$$(CO)_5 W = C \begin{smallmatrix} OEt \\ \\ NEt_2 \end{smallmatrix} \quad \xrightarrow[\substack{\text{pentane} \\ -10\,°C}]{BBr_3} \quad Br—W≡C—NEt_2$$

$$\updownarrow$$

$$Br—W \overset{-}{=} C = \overset{+}{N}Et_2$$

It may be noted that while carbyne complexes of chromium are also accessible by treatment of hydroxycarbene derivatives with boron tribromide, similar reaction of the related manganese hydroxycarbene derivative merely results in the formation of an alkoxyborane derivative:

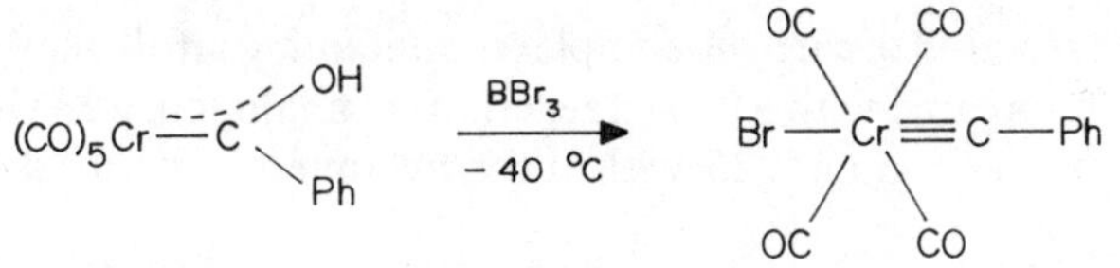

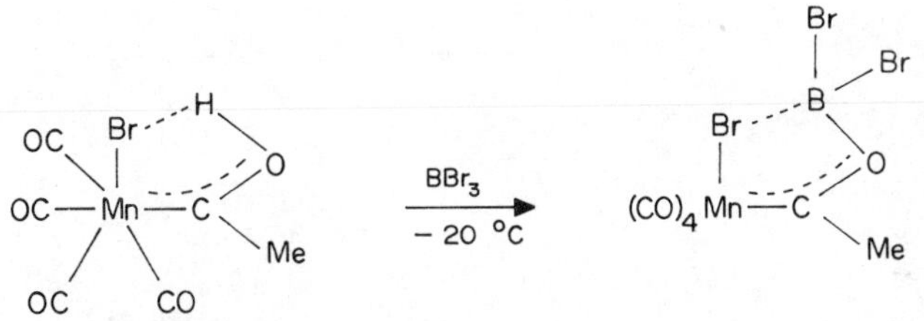

A novel and interesting synthetic route to carbyne complexes is found in the treatment of lithium benzoylpentacarbonyl tungstate with triphenyldibromophosphorane in ether at low temperature:

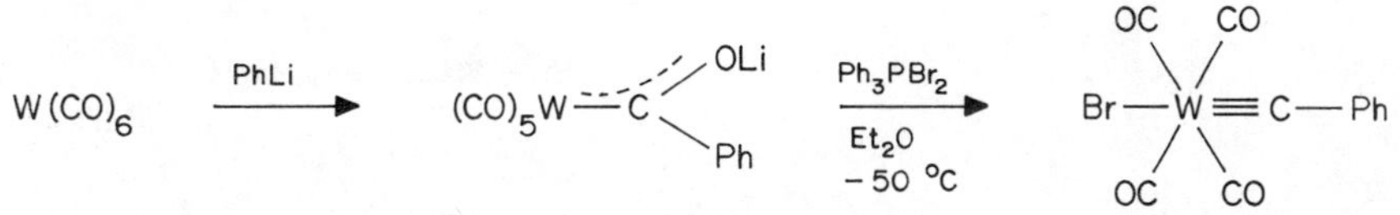

A plausible mechanism involves displacement of Br^- from Ph_3PBr_2 by the 'alkoxide' to form a C_{carb}—O—P linkage, which then eliminates triphenylphosphine oxide with concomitant attack of Br^- to give the carbyne complex.

Finally, alkylidyne complexes can also be obtained from Schrock's alkylidene–tantalum complexes by treatment with trialkylphosphine:

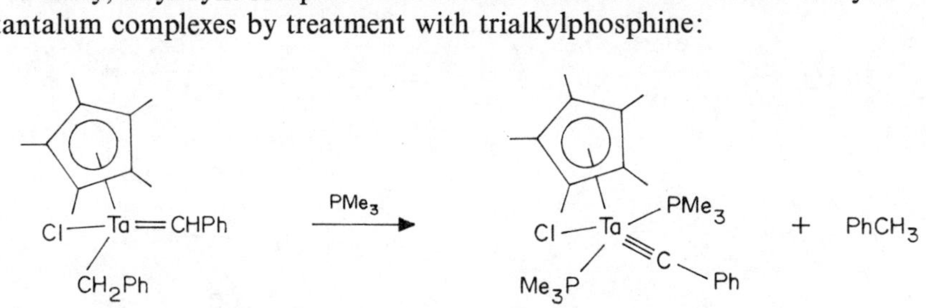

Reactions of Carbyne Complexes

It was mentioned earlier that most of the carbyne complexes are thermally labile. Indeed, at temperatures of 30–50 °C in nonpolar solvents the ligand appears to be liberated and undergoes dimerization:

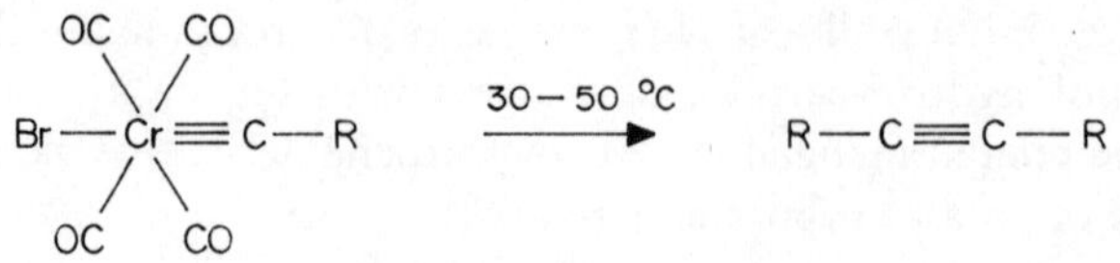

$$Br-\underset{\underset{OC}{|}}{\overset{\overset{OC}{|}}{Cr}}\equiv C-R \xrightarrow{30-50\ °C} R-C\equiv C-R$$

An interesting reaction of carbyne complexes containing an alkyne-substituted R group may be compared with Michael reactions of analogous acetylenic esters. Thus, reaction of the complex **25** with dimethylamine leads to the addition product **26**.

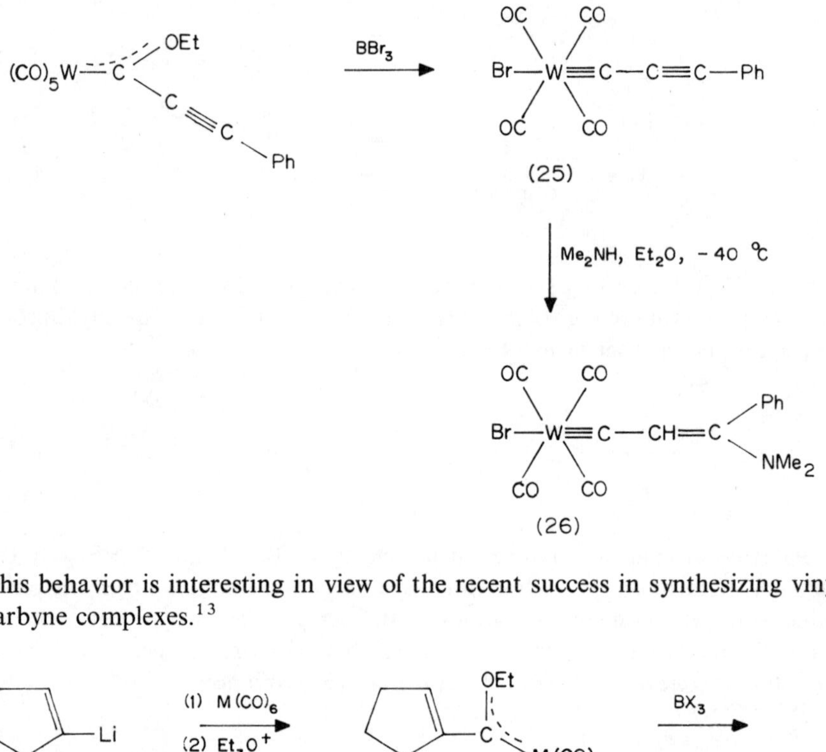

$$(CO)_5W=\!\!=C \cdots \!\!\overset{OEt}{\underset{\underset{C\equiv C}{|}}{}} \xrightarrow{BBr_3} Br-\underset{\underset{OC}{|}}{\overset{\overset{OC}{|}}{W}}\equiv C-C\equiv C-Ph$$

(25)

$$\downarrow \ Me_2NH,\ Et_2O,\ -40\ °C$$

$$Br-\underset{\underset{CO}{|}}{\overset{\overset{OC}{|}}{W}}\equiv C-CH=C\overset{Ph}{\underset{NMe_2}{}}$$

(26)

This behavior is interesting in view of the recent success in synthesizing vinyl carbyne complexes.[13]

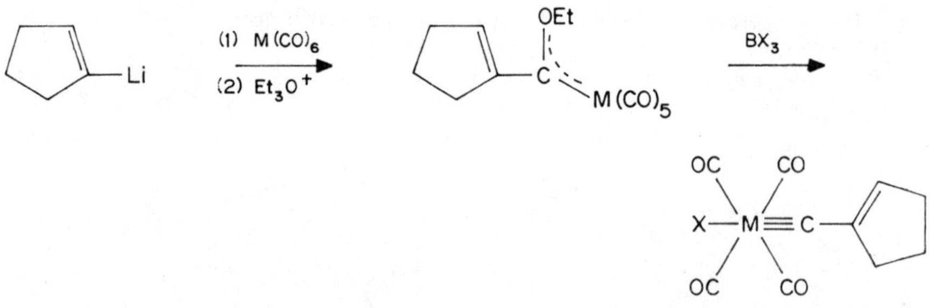

Treatment of the cationic carbyne complex **27** with azide ion as a nucleophile results in a nitrile complex, the formation of which can be rationalized by nucleophilic addition of azide to the carbyne carbon atom, followed by rearrangement:[14]

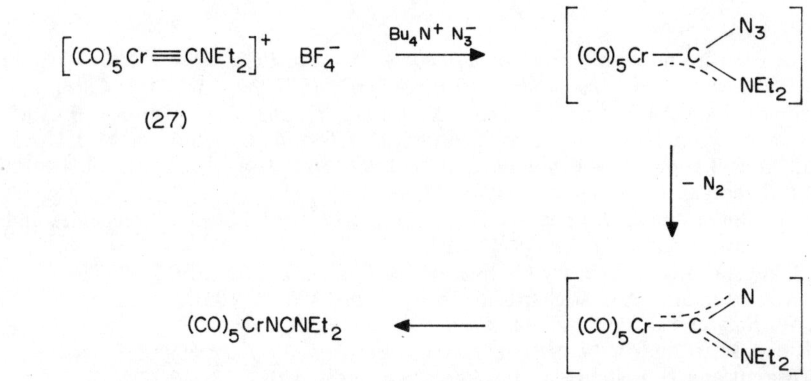

An interesting conversion of certain tungsten carbyne complexes, e.g. **28**, into ketene derivatives occurs on treatment with either carbon monoxide or trimethylphosphine. The products may be converted into σ-alkyne derivatives as shown below.[15]

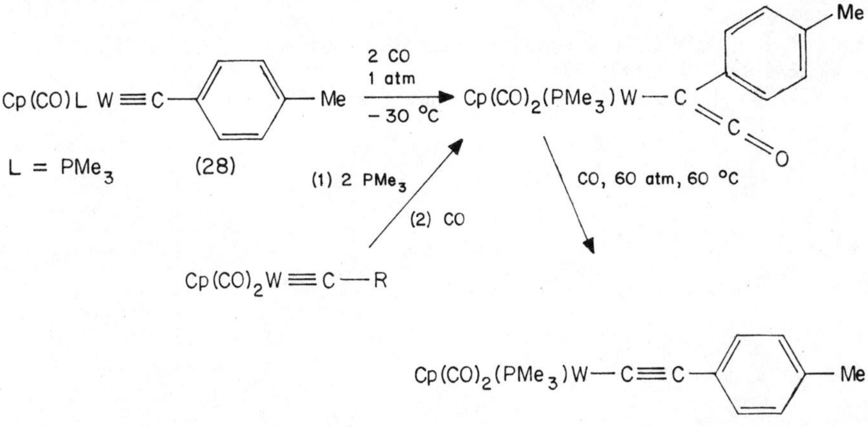

Halide ion may be displaced with formation of a metal—metal bond when carbyne–W(CO)$_4$Br complexes are treated with NaRe(CO)$_5$:[16]

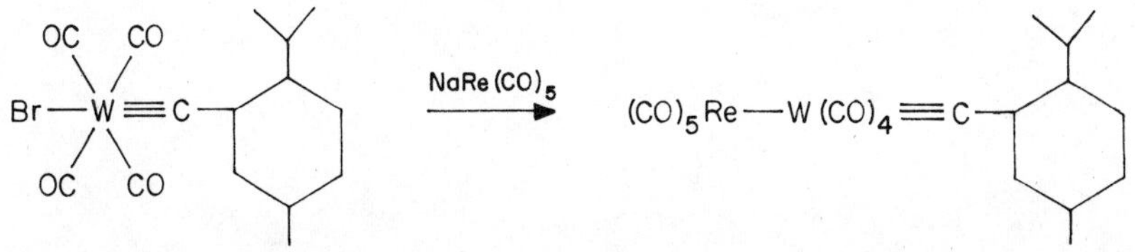

Although real synthetic applications of metal–carbyne complexes have not yet emerged, the availability of these substances would now make possible the development of novel reactions which might one day receive the attention of synthetic organic chemists.

162

REFERENCES

1. E. O. Fischer and A. Maasbol, *Angew. Chem., Int. Ed. Engl.*, 1964, **3**, 580.
2. E. O. Fischer, *Pure Appl. Chem.*, 1970, **24**, 407; 1972, **30**, 353; D. J. Cardin, B. Cetinkaya and M. F. Lappert, *Chem. Rev.*, 1972, **72**, 545; F. A. Cotton and C. M. Lukehart, *Prog. Inorg. Chem.*, 1972, **16**, 487; D. J. Cardin, B. Cetinkaya, M. J. Doyle and M. F. Lappert, *Chem. Soc. Rev.*, 1973, **2**, 99.
3. R. R. Schrock, *Acc. Chem. Res.*, 1979, **12**, 98.
4. W. M. Butler and J. H. Enemark, *Inorg. Chem.*, 1971, **10**, 2146; G. Rowochias and B. L. Shaw, *J. Chem. Soc. A*, 1971, 2097.
5. L. Chugaev and M. Skanary-Grigorizeva, *J. Russ. Chem. Soc.*, 1915, **47**, 776.
6. F. A. Cotton and D. C. Richardson, *Inorg. Chem.*, 1966, **5**, 1851.
7. P. W. Jolly and R. Pettit, *J. Am. Chem. Soc.*, 1966, **88**, 5044.
8. M. Brookhart and W. O. Nelson, *J. Am. Chem. Soc.*, 1977, **99**, 6099.
9. S. Brandt and P. Helquist, *J. Am. Chem. Soc.*, 1979, **101**, 6473.
10. K. Takai, Y. Hotta, K. Oshima and H. Nozaki, *Tetrahedron Lett.*, 1978, 2417.
11. F. N. Tebbe, G. W. Parshall and G. S. Reddy, *J. Am. Chem. Soc.*, 1978, **100**, 3611.
12. M. A. McGuire and L. S. Hegedus, *J. Am. Chem. Soc.*, 1982, **104**, 5538.
13. E. O. Fischer, W. R. Wanger, F. R. Kreissl and D. Neugebauer, *Chem. Ber.*, 1979, **112**, 1320.
14. E. O. Fischer, W. Kleine, U. Schubert and B. Neugebauer, *J. Organomet. Chem.*, 1978, **149**, C40.
15. F. R. Kreissl, W. Uedelhoven and K. Eberl, *Angew. Chem., Int. Ed. Engl.*, 1978, **17**, 859; *Chem. Rev.*, 1979, **112**, 3376.
16. R. E. Wright and A. Vogler, *J. Organomet. Chem.*, 1978, **160**, 197.

CHAPTER 5

η^2-Alkene and η^2-Alkyne Complexes

Historically, η^2-olefin complexes are very important, as illustrated by the well known Zeise's salts (ethylene–Pt complexes), first prepared in *ca.* 1830.[1] Olefin–metal complexes also are formed as transient intermediates in a wide range of catalytic processes, some examples of which are given in Chapter 2. This chapter deals with some of the basic aspects of stable isolable η^2-complexes of transition metals. We begin with a discussion of the chemistry of olefin–iron complexes and follow with a presentation of alkene complexes of other metals. The second section is concerned with side-bound alkyne complexes and their reactions.

η^2-ALKENE COMPLEXES

Iron, Ruthenium and Osmium Group

Olefin–Fe(CO)₄ Complexes

These complexes are intermediates in the formation of diene–Fe(CO)$_3$ complexes during reactions of a diene with an iron carbonyl. Although butadiene–Fe(CO)$_3$ and a number of other diene–Fe(CO)$_3$ complexes have been known since 1930, it was not until *ca.* 1961 that the first stable monoolefin–Fe complexes were prepared. Monoalkene–Fe(CO)$_4$ complexes are rather unstable. Ethene–Fe(CO)$_4$ may be prepared by reaction of Fe$_2$(CO)$_9$ or Fe$_2$(CO)$_{12}$ with ethene under pressure, to give the complex as a yellow oil, b.p. 34 °C/12 mm Hg, m.p. -21.8 °C, which decomposes slowly at room temperature to give Fe$_3$(CO)$_{12}$. It can, however, be stored for prolonged periods under nitrogen at -80 °C.

$$\text{Fe}_2(\text{CO})_9 \ + \ \text{C}_2\text{H}_4 \ \xrightarrow[\text{2 days}]{\text{50 atm}} \ \begin{array}{c}\text{CH}_2\\||\\\text{CH}_2\end{array}\!\!\!-\text{Fe(CO)}_4 \ + \ \text{Fe(CO)}_5$$

Other monoolefinic hydrocarbons, such as propene, cyclohexene and styrene, have been used as π-ligands in olefin–Fe(CO)$_4$ complexes, prepared by either

163

irradiation of a solution of $Fe(CO)_5$ in the presence of the olefin or mild reaction between $Fe_2(CO)_9$ and the olefin in solution. The products are very sensitive to air oxidation.

Olefin–$Fe(CO)_4$ complexes are characterized by three bands in the infrared spectrum due to the coordinated CO stretching vibrations, at *ca.* 2085, 1990 and 1970 cm^{-1}, while the complexed C=C bond shows an infrared band at *ca.* 1500 cm^{-1}. The mass spectra of the complexes show successive loss of the four CO ligands and simultaneous loss of the alkene. X-ray crystal structure determinations indicate that the olefin in general replaces an equatorial CO ligand of the trigonal bipyramidal complex $Fe(CO)_5$.

Olefin–$Fe(CO)_4$ complexes of alkenes having the C=C bond conjugated with an electron-withdrawing group (e.g. —C≡N, —C=O) are much more stable than the simple alkene complexes. This is due to the fact that conjugation lowers the energy of the olefin LUMO, bringing it closer to the filled iron d orbitals. As a consequence, a better bonding interaction occurs. The modified olefin is said to be a better π-acceptor, e.g.

Many of these complexes are obtained as crystalline solids and are relatively stable to heat and, in the absence of light, stable in air.

Olefin–$Fe(CO)_4$ complexes which contain an adjacent small ring undergo some interesting and sometimes useful ring-opening reactions. For example, treatment of vinylcyclopropanes with pentacarbonyliron at elevated temperature results in intermediate formation of an olefin–$Fe(CO)_4$ complex which undergoes spontaneous ring opening and hydride shift to give a diene–$Fe(CO)_3$ complex:

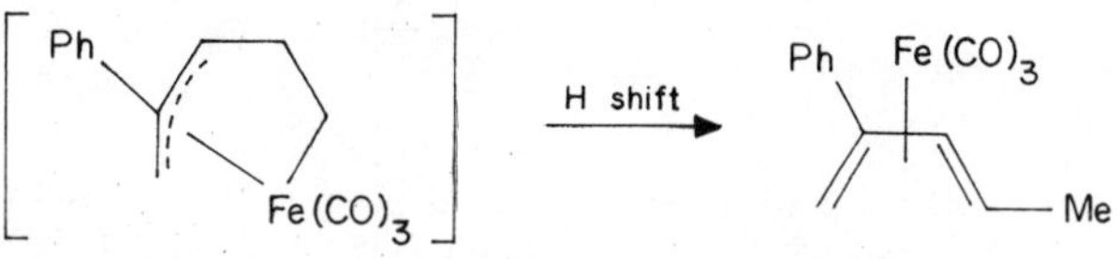

When the adjacent small ring is an epoxide, some interesting and potentially very useful results are obtained. A few years ago Heck and Boss[2] had observed that treatment of certain chloroallylic alcohols with $Fe(CO)_5$ results in the formation of π-allyl ferrelactone complexes, e.g.

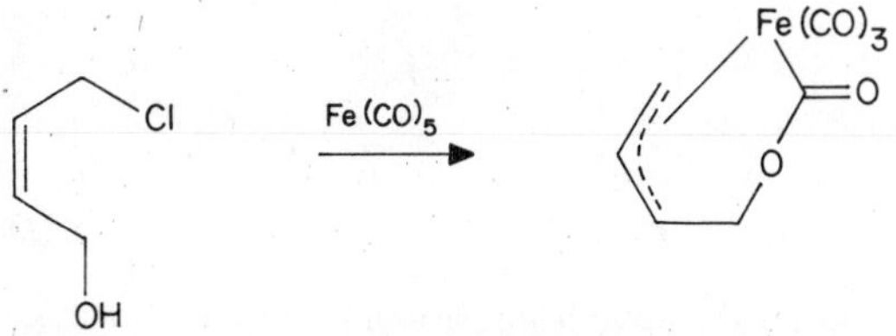

At the same time, similar complexes were characterized by Murdoch[3] and later could be obtained by reaction of a vinyl epoxide with iron carbonyl.[4] This type of behavior has been turned into a useful synthetic method for the construction of β-lactone and β-lactam synthesis, since treatment of the ferralactone complexes with $Ce^{(IV)}$ results in decomplexation and concomitant lactone formation. Some examples are given below.[5]

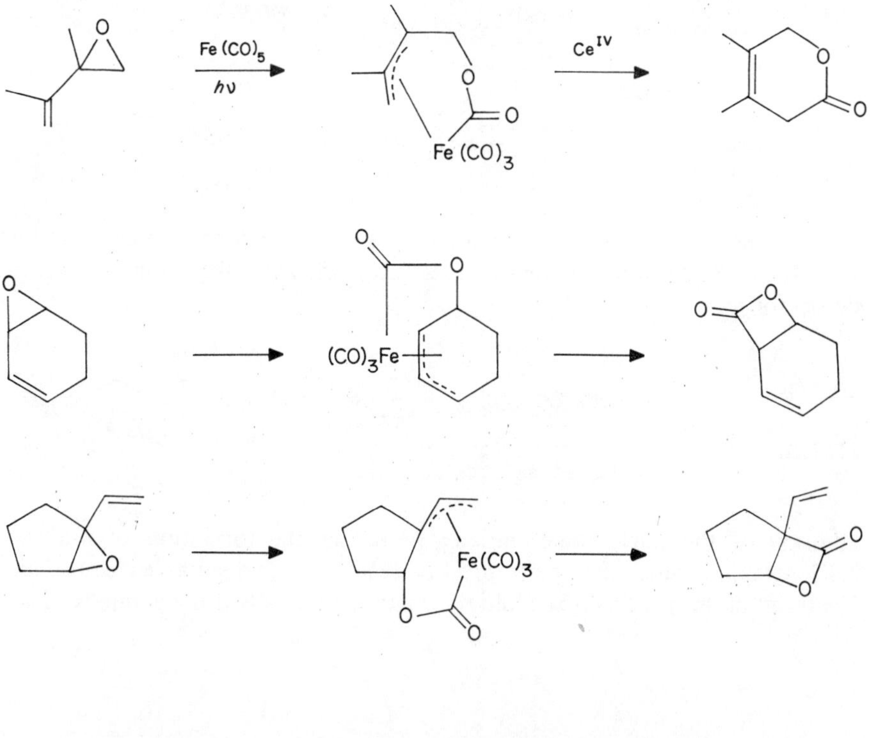

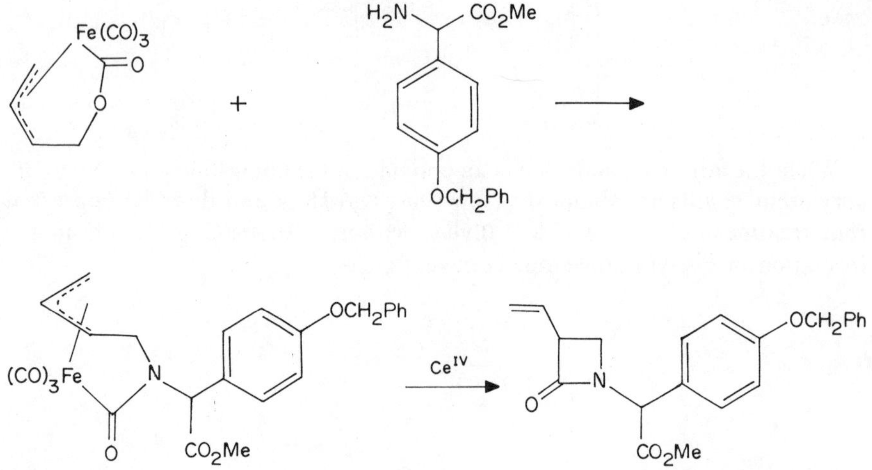
Fe(CO)3
O
O
+
H2N
CO2Me
OCH2Ph
(CO)3Fe
N
O
CO2Me
OCH2Ph
Ce^IV
O
N
CO2Me
OCH2Ph

Olefin–$Fe(CO)_4$ complexes have been shown to undergo attack by stable enolate nucleophiles.[6] Whilst most complexes studied were the more stable derivatives of, e.g., methyl acrylate, it was also found that ethene–$Fe(CO)_4$ undergoes addition of, e.g., dimethyl malonate anion in moderate yield. This reflects that the metal serves to activate the olefin toward nucleophilic attack by some means, e.g.

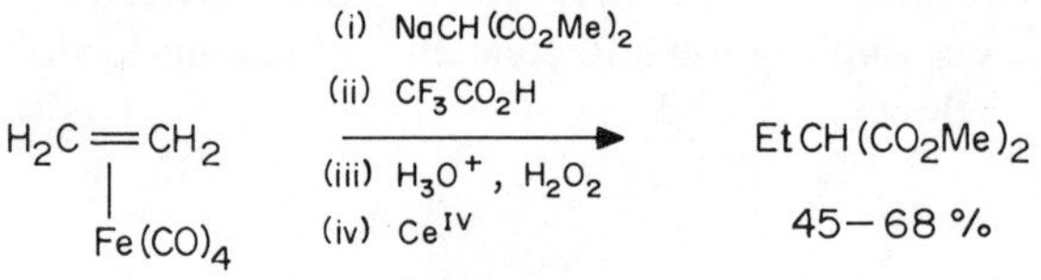
H2C=CH2
Fe(CO)4
(i) NaCH(CO2Me)2
(ii) CF3CO2H
(iii) H3O+, H2O2
(iv) Ce^IV
EtCH(CO2Me)2
45–68 %

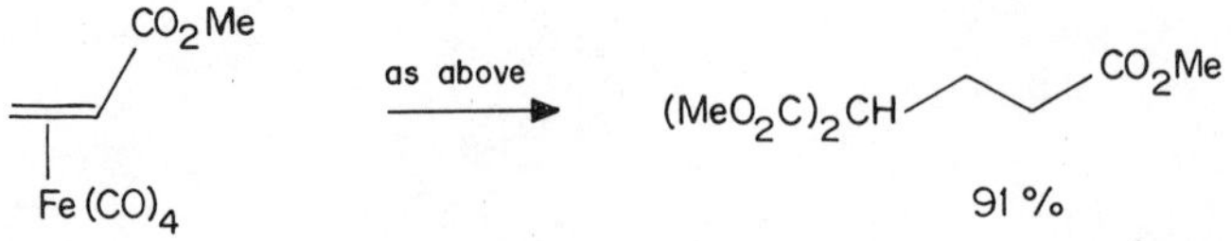
CO2Me
Fe(CO)4
as above
(MeO2C)2CH
CO2Me
91 %

It may be noted that under similar conditions methyl malonate anions undergo no Michael addition to methyl acrylate, whereas the complex reacts satisfactorily:

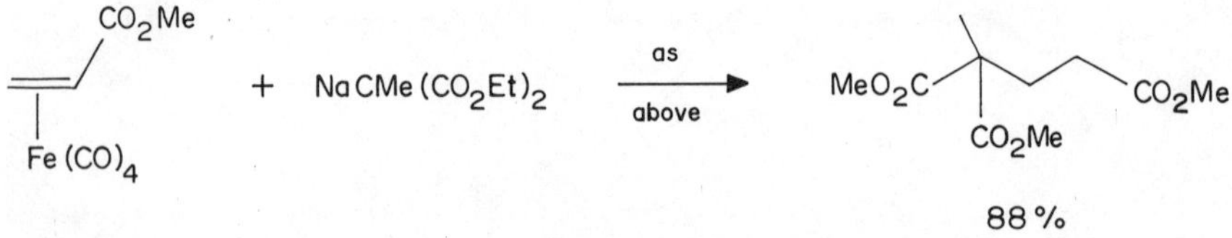
CO2Me
Fe(CO)4
+
NaCMe(CO2Et)2
as
above
MeO2C
CO2Me
CO2Me
88 %

Addition of the nucleophile initially results in the formation of σ-alkyl–$Fe(CO)_4$ anion, which may be trapped by reaction with a suitable electrophile. The result of this process is the initial formation of a σ-alkyl–Fe complex which undergoes carbonyl insertion and a reductive elimination of the organic ligands, e.g.

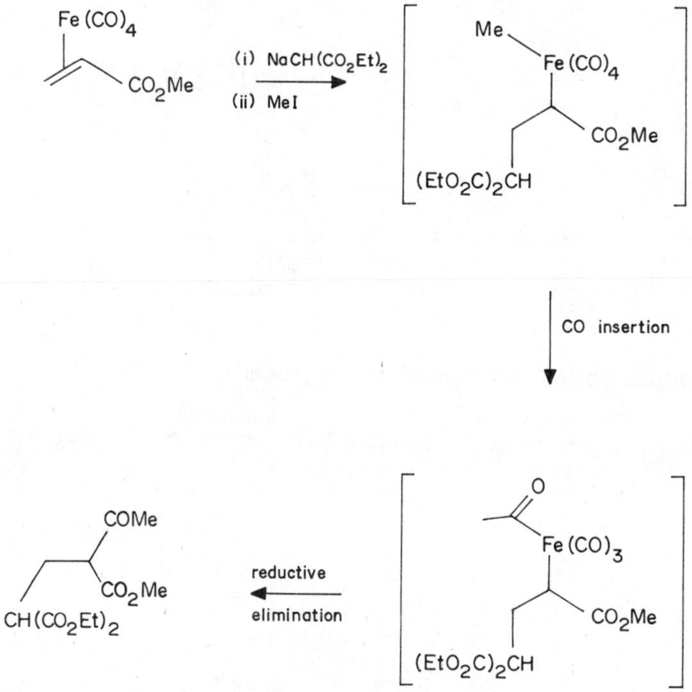

The latter behavior is analogous to the reaction of Collman's reagent [NaHFe(CO)$_4$] with electrophiles (see Chapter 3).

Cyclopentadienyl–M(CO)$_2$–olefin Complexes

Complexes of the type $(\eta^5\text{-Cp})M(CO)_2(\eta^2\text{-olefin})^+$ are known for iron, ruthenium and osmium. In general, reactions of CpM(CO)$_2$Br with olefin in the presence of Lewis acid results in the formation of the cationic complex. The iron complex $(\eta^5\text{-Cp})Fe(CO)_2(C_2H_4)^+PF_6^-$ is obtained as a pale yellow solid showing CO frequencies in the IR spectrum at 2083 and 2049 cm^{-1}, and the C=C band at 1527 cm^{-1}. The ruthenium analogue is a white air-stable solid (m.p. 215 °C) showing v_{CO} 2090, 2048 and $v_{C=C}$ 1523 cm^{-1}, while the osmium complex is also a white solid, showing v_{CO}, 2080, 2033 and $v_{C=C}$ 1546 cm^{-1}. All three complexes are almost insoluble in water and soluble in polar organic solvents such as acetone. A number of alternative methods have been worked out for the preparation of the CpFe(CO)$_2$–olefin complexes, commonly known as Fp–olefin complexes, and these are summarized below.[7]

(a) Direct reaction of olefin with $CpFe(CO)_2Br$.

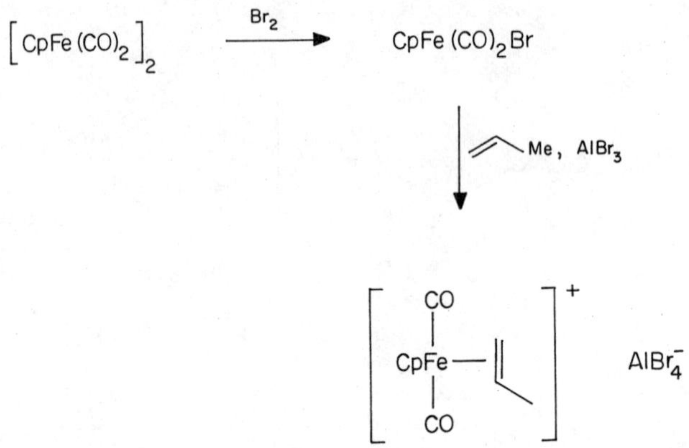

$$\left[CpFe\,(CO)_2 \right]_2 \xrightarrow{\ Br_2\ } CpFe\,(CO)_2\,Br$$

(b) Hydride abstraction from alkyl complexes.

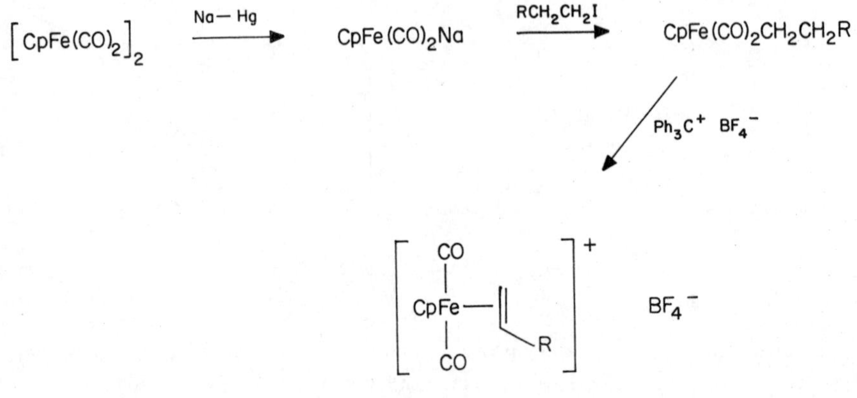

$$\left[CpFe(CO)_2 \right]_2 \xrightarrow{\ Na-Hg\ } CpFe\,(CO)_2\,Na \xrightarrow{\ RCH_2CH_2I\ } CpFe\,(CO)_2\,CH_2CH_2R$$

(c) Protonation of σ-allyl complexes.

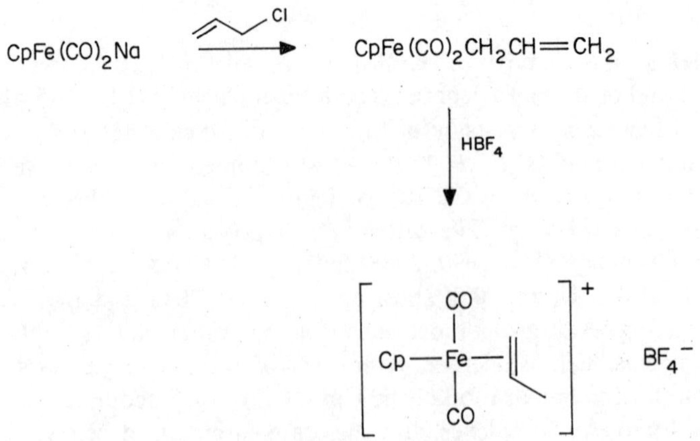

$$CpFe\,(CO)_2\,Na \xrightarrow{\ \ } CpFe\,(CO)_2\,CH_2\,CH=CH_2$$

(d) From epoxides.

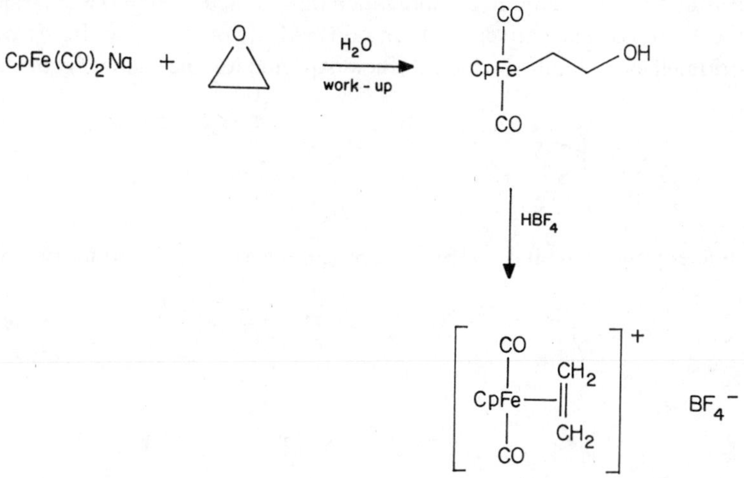

(e) By olefin exchange.

Hence there exists a wide range of methods available and the one most appropriate to the task at hand can usually be chosen. The last method above, olefin exchange, appears to be potentially useful for the preparation of complexes of functionalized or substituted olefins. Some degree of selectivity is observed in this reaction, for example dienes generally give products resulting from complexation of the less substituted double bond, while enynes lead to Fp(alkene) complexes, probably indicative of the greater stability of η^2-alkene over η^2-alkyne complexes.[8]

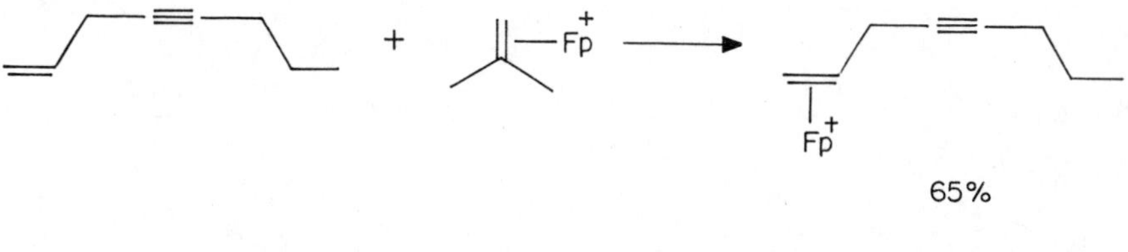

65%

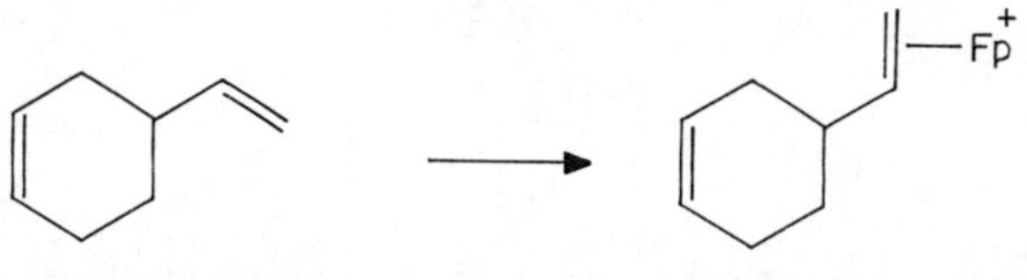

76 %

Olefin–Fp complexes, because of their positive charge, are excellent substrates for the addition of nucleophiles, being much more reactive than the corresponding olefin–Fe(CO)$_4$ derivatives. Sodium borohydride or, better, sodium cyanoborohydride effect reduction which is chemospecific for the olefin ligand.

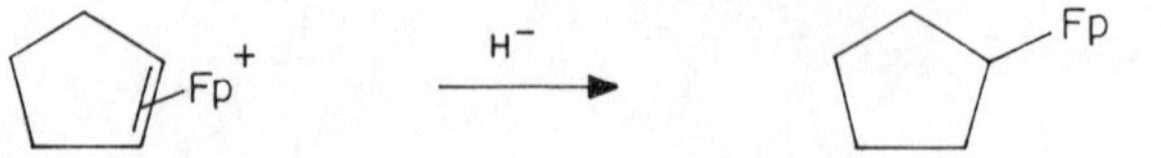

77%

Other nucleophiles which are effective are summarized in the equations below.

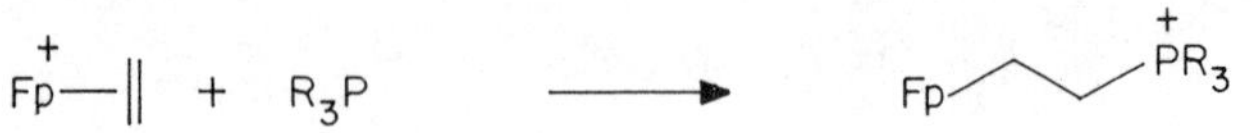

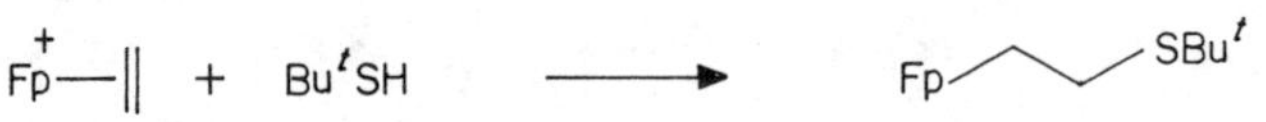

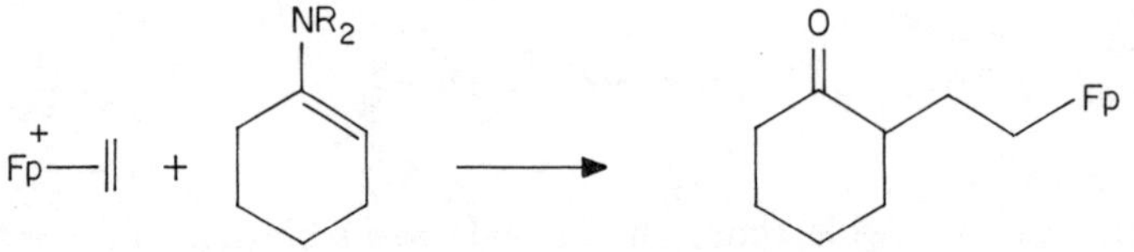

48%

60%

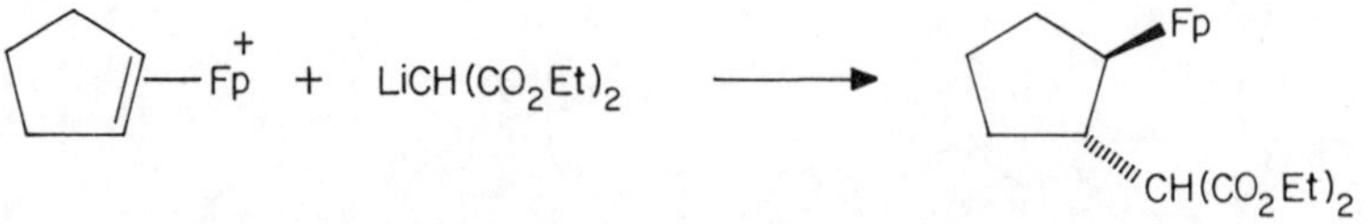

82%

In systems such as the last example, where stereochemistry of the reaction can be determined, the nucleophile is found to add *trans* to the Fp group (see later).

Of particular usefulness appears to be the metal assisted Michael addition shown by Fp derivative of α,β-unsaturated ketones, although the scope of this reaction has not been fully explored. Some examples are given below, in which the initially formed σ-bonded Fp derivative is treated with alumina, effecting annelation of the diketone and removal of the Fp group.[9]

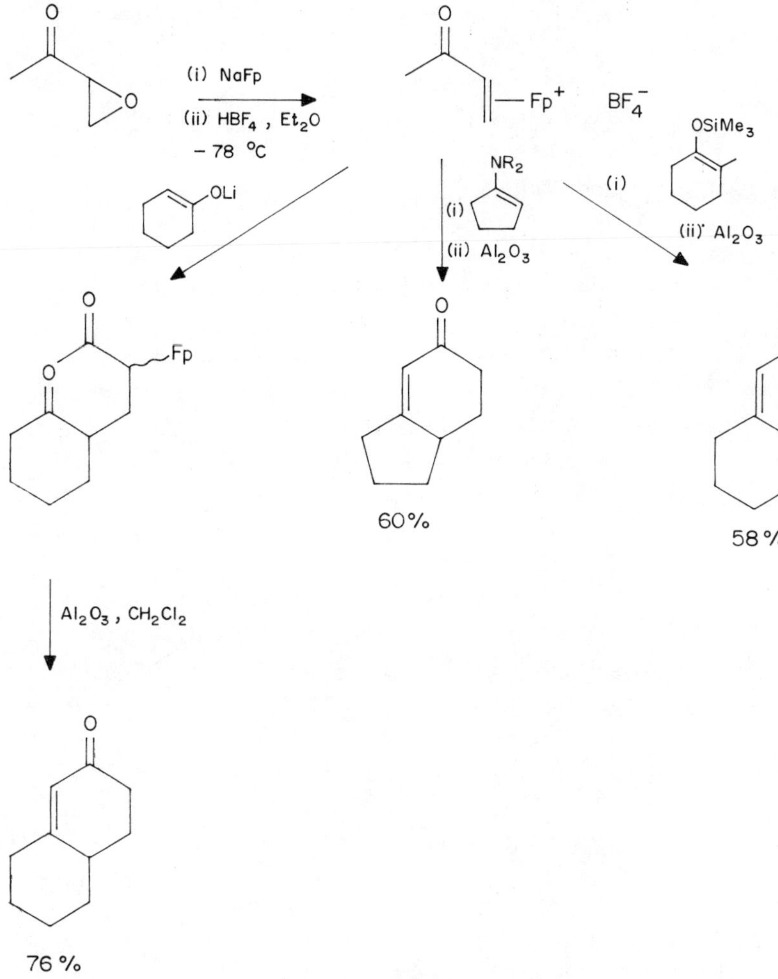

(i) NaFp
(ii) HBF$_4$, Et$_2$O
− 78 °C

Fp$^+$ BF$_4^-$

OSiMe$_3$

OLi

NR$_2$

(i)

(i)
(ii) Al$_2$O$_3$

(ii) Al$_2$O$_3$

60%

58%

Al$_2$O$_3$, CH$_2$Cl$_2$

76 %

The clean reaction with silyl enol ethers noted above is interesting because, although silyl enol ethers are known to undergo conjugate addition to α,β-unsaturated ketones under the influence of titanium tetrachloride, regiochemical scrambling of substituted enol ethers often occurs.

As might be expected, the attachment of the Fp group causes deactivation of an alkene toward electrophilic reagents. In fact, this offers an excellent means of protecting carbon—carbon double bonds during manipulation of other unsaturated moieties in the molecule. Some examples are given below.[10]

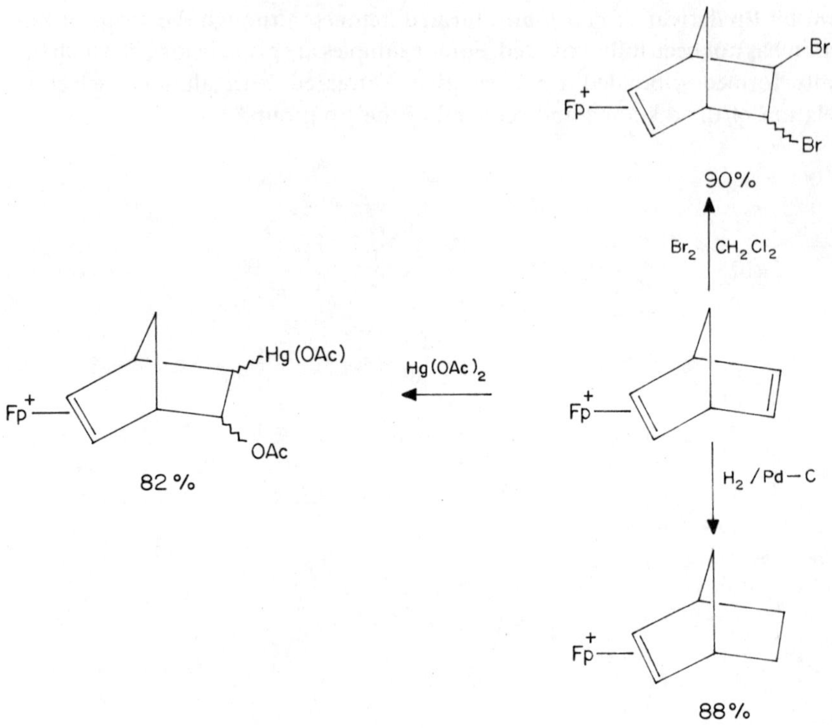
Br
Fp+
Br
90%
Br2
CH2Cl2
Hg(OAc)
Hg(OAc)2
Fp+
Fp+
OAc
82 %
H2 / Pd—C
Fp+
88%

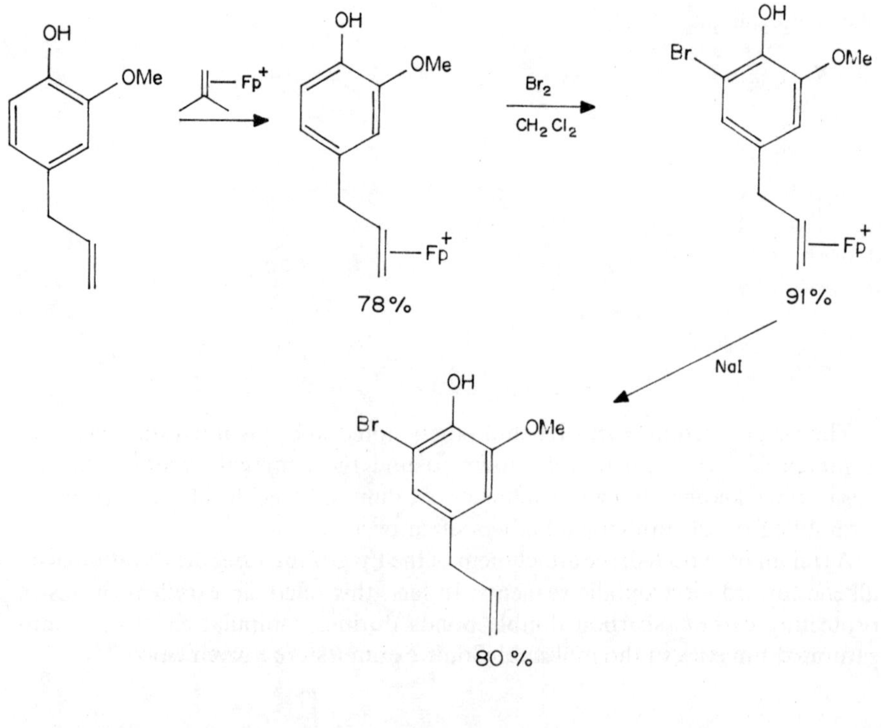
OH
OMe
Fp+
OH
OMe
Br2
CH2Cl2
OH
Br
OMe
Fp+
78 %
Fp+
91%
NaI
OH
Br
OMe
80 %

The Fp group is readily removed from the resultant complex in good yield by treatment with sodium iodide in acetone.

Molecular orbital basis for nucleophile additions

While the olefin–Fp complexes, which carry a full positive charge, might be expected to undergo facile nucleophile addition, the activation of olefins by uncharged Fe(0) is less readily explained.

The Fe(CO)$_4$ group is known to be an acceptor of electrons (cf. Coleman's reagent), but the reactivity of olefin–Fe(CO)$_4$ complexes to relatively unreactive nucleophiles is nevertheless puzzling. This behavior led Eisenstein and Hoffman[11] to examine the reaction using molecular orbital calculations. While the metal undoubtedly removes electron density from the olefin by interaction between empty metal d orbitals and the olefin HOMO, there is also a synergistic back-donation into the olefin LUMO, which would deactivate it towards nucleophilic attack. The net effect is difficult to assess. Eisenstein and Hoffman attribute the olefin activation to the potentiality of a geometrical deformation. They calculated overlap populations between H$^-$ (as the approaching nucleophile) and the entire unactivated molecule as a function of approach angle and found that if they allowed a slipping of the metal along the olefin double bond to occur as the nucleophile approached, then the overlap population increased significantly. If no slipping was allowed then the complexed olefin was found to be *less* prone to nucleophilic attack than was the uncomplexed olefin. In some ways this corresponds to what might be expected during a reaction where the complex changes from π-olefin to σ-alkyl:

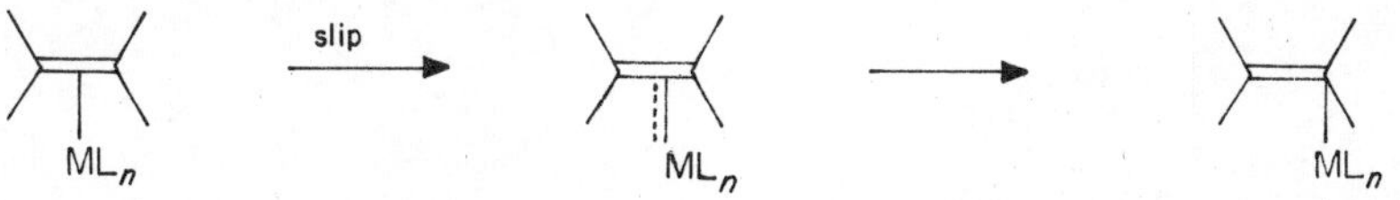

Also of interest is the observation that the LUMO of the complex corresponds approximately to an antibonding combination of the olefin π^* orbital and metal d orbitals (although some mixing of a π/metal antibonding combination is important):

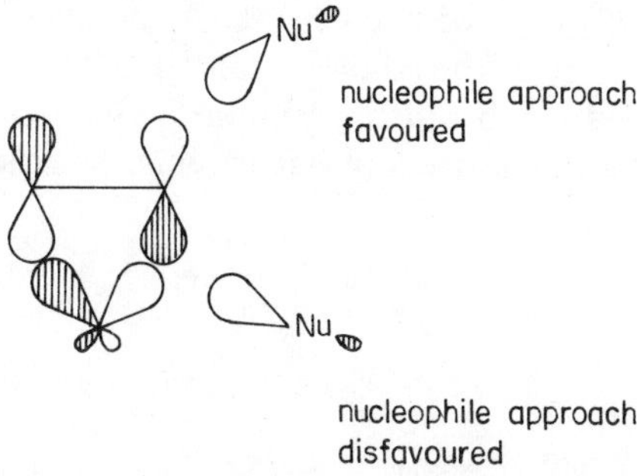

nucleophile approach
favoured

nucleophile approach
disfavoured

This would also explain the stereospecificity of nucleophile addition since attack from the metal side would lead to some antibonding interactions between the nucleophile HOMO and complex LUMO. On the other hand, attack trans to the metal gives a good bonding interaction, leading to a preference for this mode of addition.

Manganese, Technetium and Rhenium

Mono olefin complexes of this group of metals can be roughly divided into two groups: (a) cationic complexes of general structure $M(CO)_5(olefin)^+$ or cis-$M(CO)_4(olefin)_2{}^+$, being formally substitution products of the hexacarbonyl cation $M(CO)_6{}^+$; (b) uncharged $C_5H_5M(CO)_2(olefin)$ complexes formally derived from $C_5H_5M(CO)_3$ and $RMn(CO)_4(olefin)$.

Cationic Complexes

Treatment of $EtMn(CO)_5$ with Ph_3C^+ $BF_4{}^-$ gives only a small yield of $Mn(CO)_5(C_2H_4)^+$ $BF_4{}^-$ resulting from hydride abstraction from the σ-bonded ethyl group. Fair yields of the ethylene complex can be obtained by treatment of $Mn(CO)_5Cl$ with $AlCl_3$ and ethylene under pressure, and substituted olefin complexes are conveniently prepared by treatment of an appropriate σ-allyl complex with acid, e.g. anhydrous $HClO_4$ in benzene. Acids whose anion can function as a ligand cause the formation of $Mn(CO)_5X$ and release of the olefin:

$$Mn(CO)_5CH_2CH\!\!=\!\!CH_2 \quad \xrightarrow{\text{HClO}_4} \quad Mn(CO)_5(CH_2\!\!=\!\!CHCH_3)^+ \quad ClO_4{}^-$$

$$\downarrow \text{HCl}$$

$$Mn(CO)_5\,Cl$$

The cationic complexes are usually colorless or lightly colored, showing infrared bands at $ca.$ 2165, 2075 (CO ligand) and $ca.$ $1540\,cm^{-1}$ ($C\!\!=\!\!C$).

Reaction of $(Tc(CO)_4Cl)_2$ with ethene/$AlCl_3$ at 60–$65\,°C$ and $250\,atm$ in cyclohexane as solvent gives the colorless salt $[Tc(CO)_4(C_2H_4)_2]AlCl_4$ in 60% yield. Both mono- and disubstituted rhenium derivatives can be obtained in about 80% yield using this procedure but varying the ethene pressure:

$$Re(CO)_5Cl \ + \ C_2H_4 \ + \ AlCl_3 \quad \xrightarrow[\substack{\text{then} \\ NH_4PF_6}]{60\ atm} \quad \left[Re(CO)_5(C_2H_4)\right]PF_6$$

$$\xrightarrow[\text{then}\ \ NH_4PF_6]{250\ atm,} \quad \left[Re(CO)_4(C_2H_4)_2\right]PF_6$$

$[Re(CO)_5C_2H_4]PF_6$ shows IR signals at 2178, 2075 and 1582 cm^{-1}, while $[Re(CO)_4(C_2H_4)_2]PF_6$ shows peaks at 2146, 2053, 2016 and 1539 cm^{-1}.

All of the cationic complexes are colorless and unstable. The manganese compound decomposes immediately in water, while $[Re(CO)_4(C_2H_4)_2]^+$ is sufficiently stable to allow NMR measurement in D_2O solution.

Uncharged Complexes

These complexes are usually prepared using photochemical methods. For example, irradiation of $Me_3Sn–Mn(CO)_5$ in pentane solution, at elevated temperature and pressure of ethene, causes replacement of one CO ligand by ethene:

$$Me_3Sn–Mn(CO)_5 \xrightarrow[h\nu]{C_2H_4} Me_3Sn–Mn(CO)_4C_2H_4$$

Perhaps the most popular complexes are derivatives of type $(\eta^5\text{-}Cp)Mn(CO)_2(olefin)$, prepared by photochemical methods, e.g.

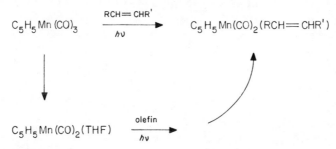

The product olefin complexes are generally crystalline, air-stable, deep yellow solids which can usually be purified by sublimation *in vacuo*. The cyclopentene complex melts at 64–65 °C and shows IR bands at 1969 and 1905 cm^{-1} (in cyclohexane solution). The parent ethene complex melts at 125–127 °C, has IR peaks at 1976 and 1916 cm^{-1} and the NMR spectrum shows two singlets at δ 3.37 (5H, cyclopentadienyl) and 1.50 p.p.m. (4H, ethene), the high field of these absorptions indicating considerable shielding by the manganese. Olefin–$Mn(CO)_2Cp$ complexes of cyclic olefins tend to be formed most readily by relatively strained double bonds, e.g. cyclopentene, cycloheptene, *cis*-cyclooctene and norbornene, but not by the strain-free cyclohexene.

Treatment of the olefin complexes with stronger ligands usually results in displacement of the olefin:

$$CpMn(CO)_2(olefin) + PR_3 \longrightarrow CpMn(CO)_2PR_3 + olefin$$

Complexes of α,β-unsaturated ketones, in which only the C=C bond is coordinated to manganese, may be obtained by photochemical substitution of $CpMn(CO)_3$.[12]

Interestingly, treatment of the acetylenic compound $Me_2C{=}CHC{\equiv}$ CCO_2Me with $CpMn(CO)_2(thf)$ leads to rearrangement giving vinylallene complexes:[13]

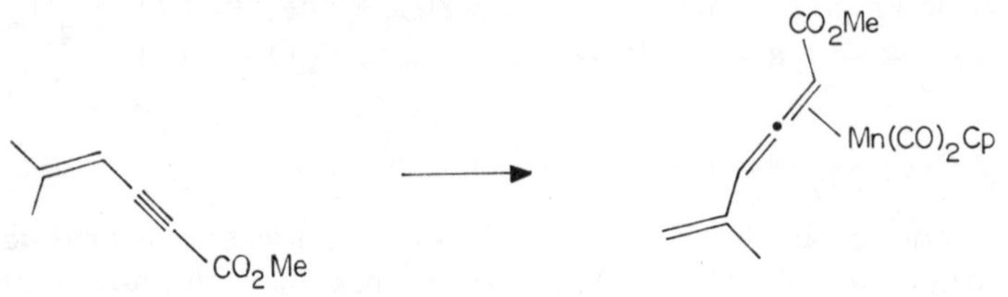

Cobalt, Rhodium and Iridium

Despite the implications of cobalt–olefin complexes in various catalyzed reactions, such as hydroformylation (Chapter 2), only recently have examples of cobalt complexes of monoolefinic hydrocarbons been described. Two examples of ethene complex preparations are given below. [14,15]

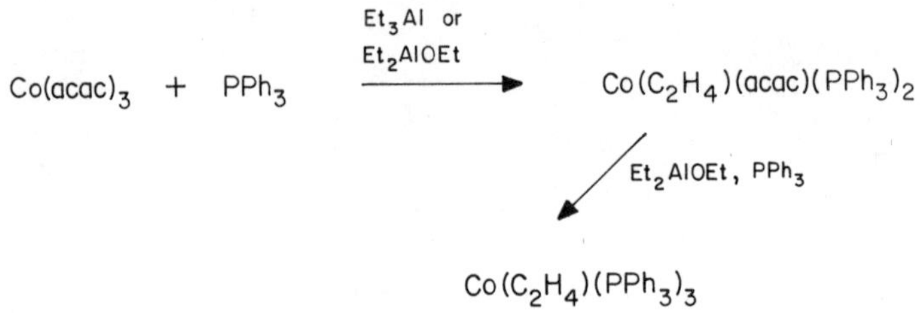

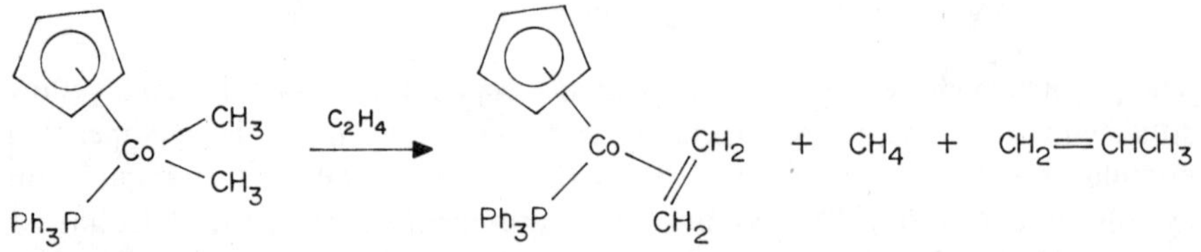

Complexes of olefins conjugated with electron-withdrawing groups have attracted more attention. Fumaronitrile (fmn) displaces one Ph_3P ligand from η^5- $CpCo(PPh_3)_2$ to give the dark red crystalline complex η^5-$CpCo(fmn)(PPh_3)$.[16] Similarly, maleic anhydride complexes can be prepared:[17]

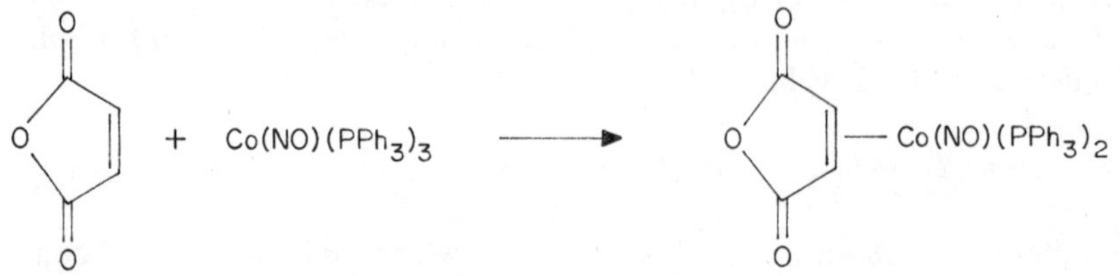

A number of interesting complexes are formed when unsaturated acid chlorides are treated with sodium tetracarbonylcobaltate, $NaCo(CO)_4$, as shown below.

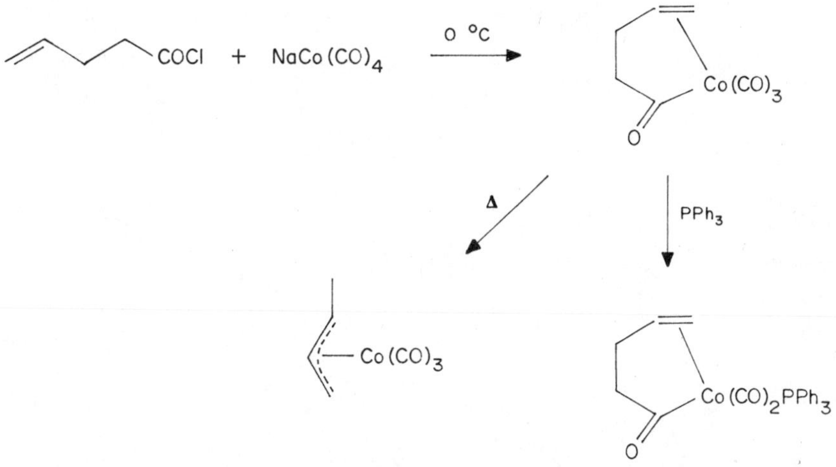

The above acyl olefin cobalt tricarbonyl derivatives are unstable, undergoing decarbonylation and hydride shift to give π-allyl complexes at temperatures above 25 °C. Treatment with triphenylphosphine displaces a CO ligand to give a more stable phosphine complex. As might be expected, the range of ring sizes obtainable by this method is very limited.

Owing to the ability of rhodium complexes to catalyze a wide variety of conversions of alkenes, there has been considerable interest in the preparation and characterization of stable η^2-alkene–rhodium complexes. Given the transient existence of such species in catalytic cycles, the facility with which the first ethene complex was prepared and its stability is remarkable. Thus, the compound $[(C_2H_4)_2RhCl]_2$ is readily prepared by simply bubbling ethene through a solution of $RhCl_3 \cdot 3H_2O$ in aqueous methanol:[18]

$$2\ RhCl_3\ +\ 2\ H_2O\ +\ 6\ C_2H_4\ \longrightarrow$$
$$\left[(C_2H_4)_2RhCl\right]_2\ +\ 2\ CH_3CHO\ +\ 4\ HCl$$

The complex is obtained as orange–red crystals which are almost insoluble in most organic solvents, indicating a considerable lattice energy in the solid state. In the solid state the complex is stable in air 5 °C, and does not decompose until heated to 115 °C. Its structure is shown below, there being a dihedral angle of 116° between the two $(C_2H_4)_2RhCl_2$ planes.

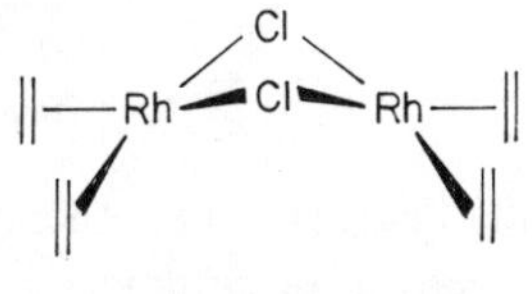

The coordinated ethylenes may be displaced by a variety of ligands without causing cleavage of the Cl bridges. Some reactions of the complex are shown below.

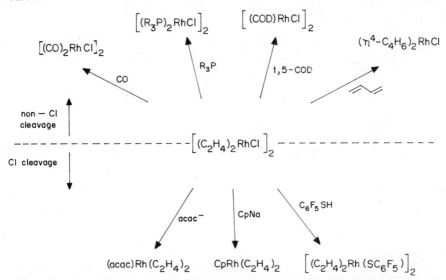

A limited number of cycloalkenes form analogous complexes: cycloheptene, cyclooctene and norbornene do form π-complexes, whereas cyclopentene, cyclohexene and camphene do not. The cycloalkene complexes decompose fairly readily in solution and are highly reactive. This reactivity makes the cyclooctene complex a very useful precursor for the preparation of a wide range of other olefin complexes, some of which are summarized below.

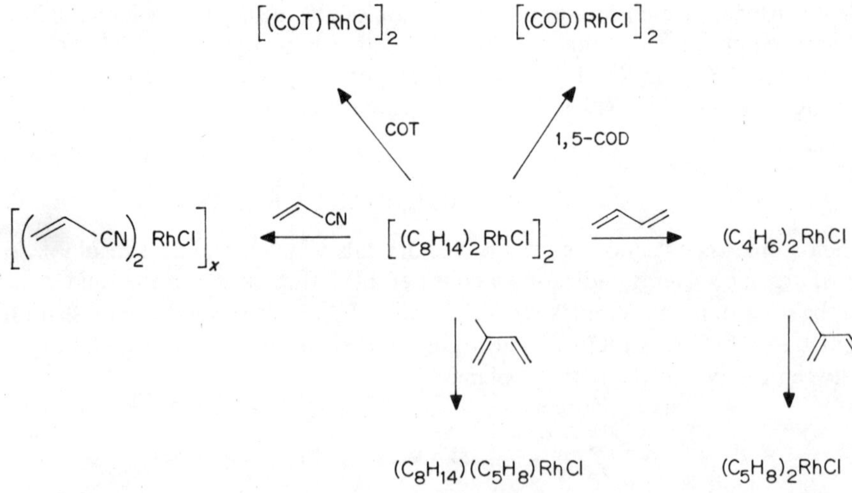

The splitting of the chloride bridges in the ethene complex by acac⁻ or Cp⁻ (see above) results in the formation of complexes having different reactivities. Thus, whereas the complex $(C_2H_4)_2Rh(acac)$ is very labile to ligand exchange with CO, PR_3 and other olefins, the cyclopentadienyl derivative $(C_2H_4)_2RhCp$ is kinetically inert:

$$(acac)Rh(C_2H_4)_2 \xrightarrow{\quad \wedge\text{Me}\quad} (acac)Rh(C_3H_6)_2$$

$$\downarrow CO$$

$$\searrow PR_3$$

$$(acac)Rh(PR_3)_2$$

$$(acac)Rh(CO)_2$$

$$CpRh(C_2H_4)_2 \xrightarrow{\quad L \quad} \text{no reaction}$$

The cyclopentadienyl complex is, however, very reactive towards electrophilic reagents, evolving ethene very rapidly when treated with, e.g., $AgNO_3$, I_2 or SO_2:

$$CpRh(C_2H_4)_2 + SO_2 \xrightarrow{\; -80\ ^\circ C \;} CpRh(C_2H_4)(SO_2) + C_2H_4$$

These reactivity patterns are most easily explained as follows. The complex $CpRh(C_2H_4)_2$ is formally an 18-electron species (Rh surrounded by 54 electrons → xenon configuration), so that there are no vacant sites which will allow addition of another ligand. On the other hand, $(acac)Rh(C_2H_4)_2$ does possess such a vacant site, being a 16-electron complex, so that the ligand-exchange reaction may proceed by the following mechanism, via an 18-electron species:

$$(acac)Rh(C_2H_4)_2 + L \; \rightleftharpoons \; (acac)Rh(C_2H_4)_2L$$

$$\updownarrow -C_2H_4$$

$$(acac)Rh(C_2H_4)L \quad \text{etc.}$$

The reactivity of $CpRh(C_2H_4)_2$ towards electrophiles is probably due to the presence of a pair of unshared electrons localized in a Rh orbital (lone pair) which can therefore interact with electron-deficient species. This mechanism is supported by kinetic measurements.

Reaction of $CpRh(C_2H_4)_2$ with HCl in chloroform solution at $-60\ ^\circ C$ gives a complex identified as $\{CpRh(C_2H_5)(C_2H_4)Cl\}$ on the basis of its NMR spectrum. The compound is stable at $-20\ ^\circ C$ for a few hours.

Reaction of chlorotris(triphenylphosphine)rhodium(I) with ethene results in the establishment of an equilibrium between phosphine and olefin complexes:

$$(Ph_3P)_3RhCl + C_2H_4 \; \rightleftharpoons \; (Ph_3P)_2Rh(C_2H_4)Cl + PPh_3$$

If the solution is concentrated under an atmosphere of ethene, the olefin complex can be isolated as a yellow crystalline air-stable solid. Dissociation occurs readily in solution. As might be expected, use of a better π-acceptor olefin, e.g. C_2F_4, results in the formation of the much more stable complex $(Ph_3P)_2RhCl(C_2F_4)$. The carbonyl analogue of this complex, i.e. $(Ph_3P)_2RhCl(CO)$, is only able to add olefins which are very good π-acceptors, e.g. tetracyanoethene (tcne), giving $(Ph_3P)RhCl(CO)(tcne)$,[19] in contrast to Vaska's iridium complex, trans-$\{(Ph_3P)_2IrCl(CO)\}$, which reacts with a variety of unactivated molecules.

Anionic olefin–rhodium complexes are formed when solutions of $\{(C_2H_4)_2RhCl\}_2$ or $(acac)Rh(C_2H_4)_2$ are treated with HCl, as a result of chlorobridge splitting:

$$\{(C_2H_4)RhCl\}_2 \; + \; 2\;Cl^- \; \rightleftharpoons \; 2\;\{(C_2H_4)_2RhCl_2\}^-$$

Dicationic rhodium–alkene complexes can be prepared by hydride abstraction from the corresponding σ-alkyl complexes. These, and also monocationic olefin complexes, have been found to react with simple nucleophiles such as phosphines, tertiary amines and thiocyanate:[20]

$$CpRh(PMe_3)_2 \; + \; EtI \; \longrightarrow \; \{CpRh(PMe_3)_2Et\}^+$$

$$\Big\downarrow Ph_3C^+$$

Reaction between chloroiridic acid $(H_2IrCl_6 \cdot 6H_2O)$ and various cyclic monoolefins results in the formation of alkene–iridium complexes, all of which are air-sensitive. Thus, treatment of cycloheptene in aqueous ethanol at 55 °C with $H_2IrCl_6 \cdot 6H_2O$ produces a complex of composition $(C_7H_{12})_3IrCl(CO)$. The carbonyl ligand is derived from the ethanol solvent. Similarly, the cyclooctene complex $(C_8H_{14})_3IrCl(CO)$ may be prepared. The complexes are obtained as colorless solids, showing the characteristic $\nu(C{\equiv}O)$ peak at ca. 1990 cm^{-1}. The monomeric complexes are convertible to dimers by treatment with chloroform or even washing with ether, owing to shifting of the equilibrium

$$2 \ (C_8H_{14})_3 IrCl (CO) \ \rightleftharpoons \ \{(C_8H_{14})_2 IrCl(CO)\}_2 \ + \ 2 \ C_8H_{14}$$

Analogous to the rhodium complex, the olefin ligands of the cyclooctene–iridium dimer can be displaced by other ligands:

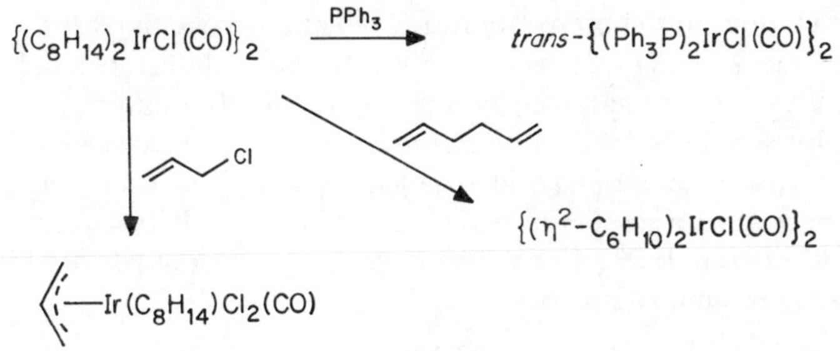

Indeed, the lability of the dimeric cyclooctene complex makes it a useful precursor for the preparation of other olefin complexes which are not readily accessible by direct methods. For example, treatment with ethene in heptane affords a complex containing four η^2-ethene ligands, $(C_2H_4)_4 IrCl$, which is stable below $-50\,°C$ under nitrogen. It decomposes at $30\,°C$ in an atmosphere of ethene:[21]

$$(C_2H_4)_4 IrCl \ \xrightarrow{30\ °C} \ \{(C_2H_4)_2 IrCl\}_2$$

Treatment of the pentamethylcyclopentadienyl complex $\{(C_5Me_5)IrCl_2\}_2$ with ethene in ethanol containing Na_2CO_3 at $70\,°C$ results in the formation of a white crystalline bis-ethene complex:[22]

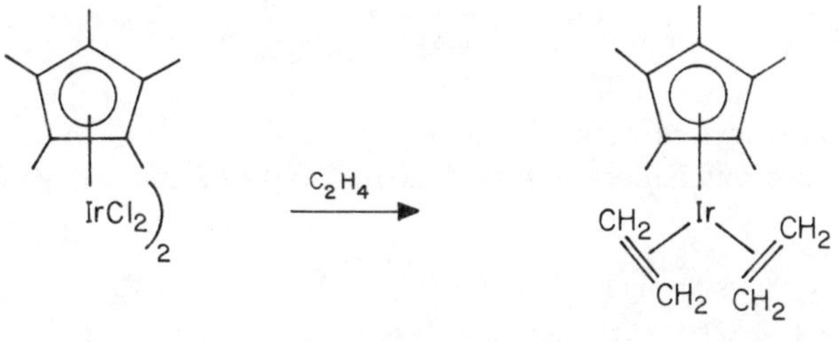

Vaska's complex, trans-$\{(Ph_3P)_2 IrCl(CO)\}$, and the related iodo derivative $\{(Ph_3P)_2 IrI(CO)\}$ have been used to study olefin complex formation. While the chloro compound does not react with ethene, it does give η^2-olefin complexes with, e.g., maleic anhydride, tetracyanoethene and tetrafluoroethene, and forms a very unstable adduct with cyclohexene. The iodo compound is more reactive and undergoes reaction with ethene in toluene solution at $26\,°C$ and $700\,mm$ Hg ethene.

$$(Ph_3P)_2 IrI (CO) \ \xrightarrow{R_2C=CR_2} \ (Ph_3P) IrI (CO) (C_2R_4)$$

Nickel, Palladium and Platinum

Much of the reactivity of π-olefin–Pd complexes has been met in Chapter 2 in connection with Pd-catalyzed olefin conversions. This section will be devoted to the preparation and properties of complexes of Ni, Pd and Pt and aspects of their reactivity not already covered.

The propensity of nickel, like cobalt, to form π-allyl complexes presents a severe limitation on the preparation of η^2-olefin nickel complexes.

Displacement of CO ligands from $Ni(CO)_4$ can only be achieved by olefins which bear electron-withdrawing groups, e.g. acrylonitrile and maleic anhydride, and not by simple olefins. Monoolefin complexes produced in this way are usually polymeric, containing bridging olefin ligands. When $Ni(CO)_4$ is heated at reflux temperature in acrylonitrile solvent, $Ni(CH_2\!=\!CHCN)_2$ is formed in good yield. Related complexes of unactivated carbonyl compounds are formed in a similar way, as summarized below:

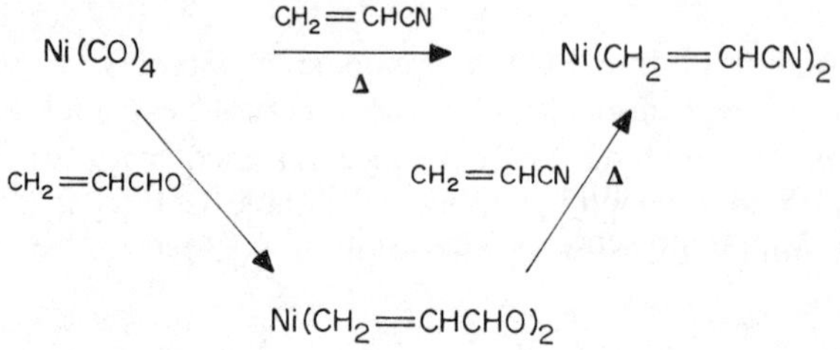

The acrylonitrile complex is obtained as bright red, pyrophoric crystals, and is diamagnetic. The complex is formally coordinatively unsaturated and a polymeric structure has been proposed. A fairly stable monomeric bipyridyl adduct has been obtained:

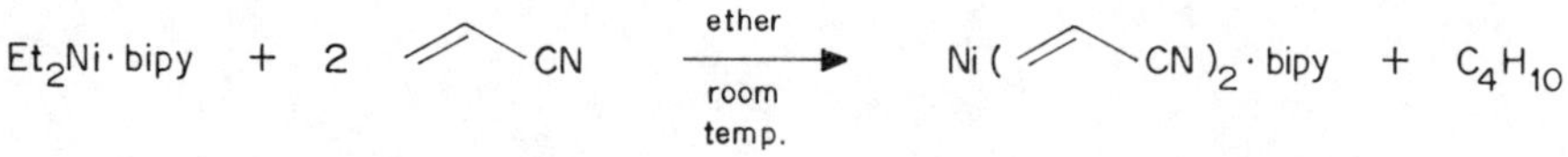

Reaction of bis(acrylonitrile)nickel with allylic halides results in the formation of polymeric organonitrile complexes, which may be decomposed by treatment with acid.

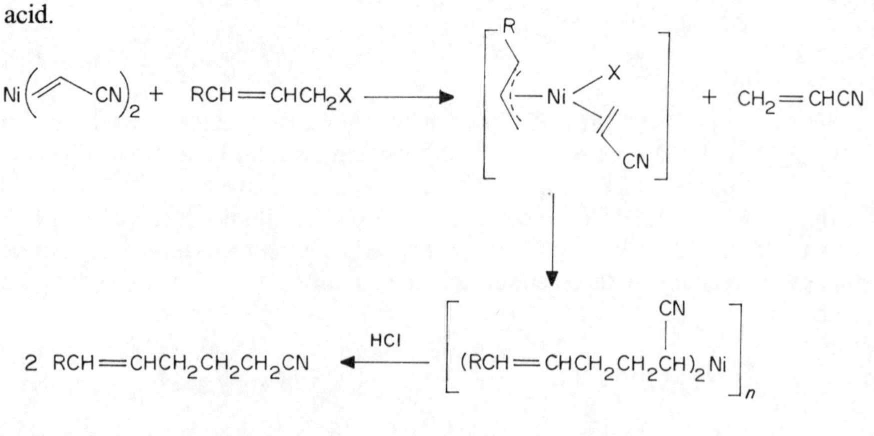

The overall reaction corresponds to the Michael addition of the allylic halide moiety to acrylonitrile. It appears to be regio-controlled, there being no allylic inversion as so often occurs in the reaction of, e.g., allyl Grignard reagents with electrophiles.

This behavior contrasts with that of the monotriphenylphosphine adduct:

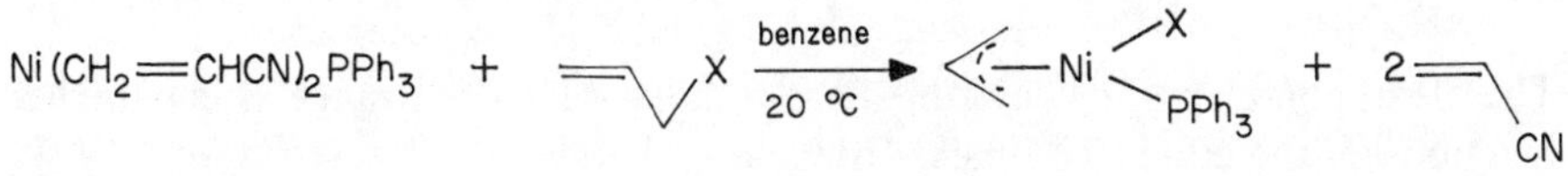

Reaction of $Ni(CH_2{=}CHCN)_2$ with a bulky phosphite results in formation of a monoacrylonitrile bisphosphite complex, which is stable up to 135 °C:

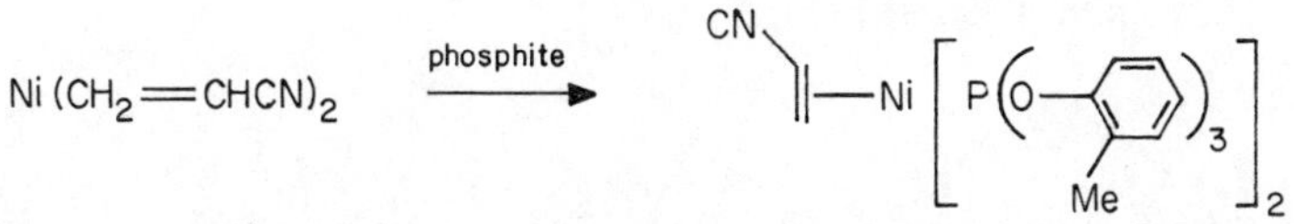

Complexes in which nickel(0) is bound to a monoolefin hydrocarbon are most often produced in the presence of a phosphine ligand, e.g.

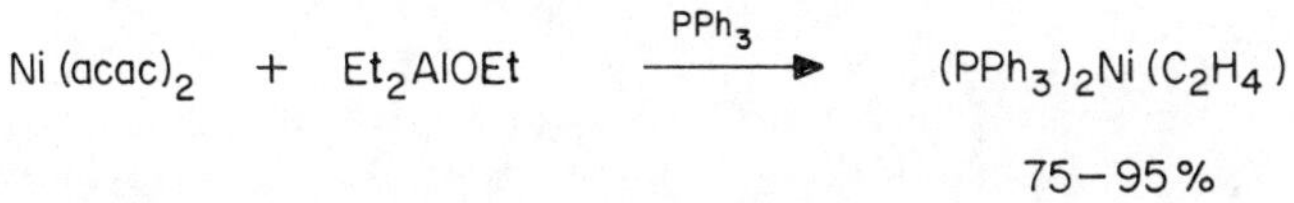

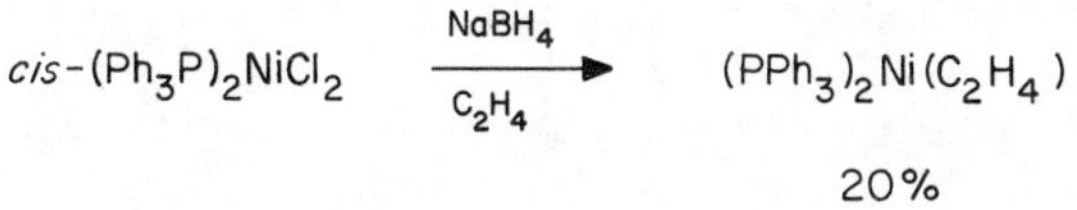

$$Ni(acac)_2 \;+\; Et_2AlOEt \xrightarrow{\;PPh_3\;} (PPh_3)_2Ni(C_2H_4)$$

$$75{-}95\%$$

$$cis\text{-}(Ph_3P)_2NiCl_2 \xrightarrow[C_2H_4]{NaBH_4} (PPh_3)_2Ni(C_2H_4)$$

$$20\%$$

The ethene complexes thus obtained are air-sensitive and undergo thermal decomposition in solution at room temperature in the absence of excess ethene. An X-ray structure determination reveals that the above compound is approximately planar with the ethene ligand slightly twisted out of the $Ni(PPh_3)_3$ plane.

Other π-olefin complexes are readily obtained by displacing the ethene ligand of $(R_3P)_2NI(C_2H_4)$ with an aryl-substituted olefin and certain alkynes, e.g.

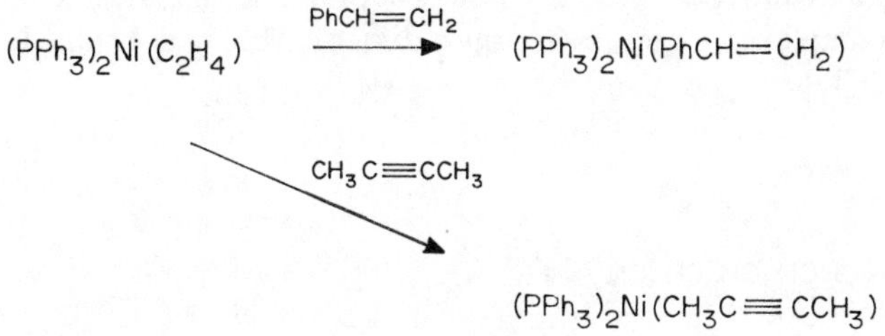

$$(PPh_3)_2Ni(C_2H_4) \xrightarrow{PhCH{=}CH_2} (PPh_3)_2Ni(PhCH{=}CH_2)$$

$$\xrightarrow{CH_3C{\equiv}CCH_3}$$

$$(PPh_3)_2Ni(CH_3C{\equiv}CCH_3)$$

η^2-Alkene complexes containing aliphatic monoolefins cannot be prepared in this way, and are best obtained directly from Ni(acac)$_2$, e.g.

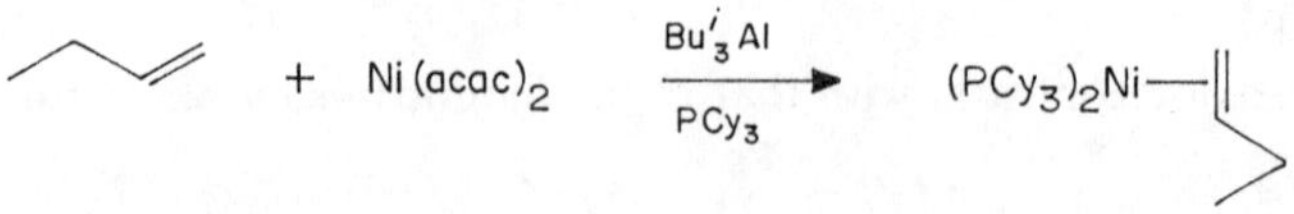

The olefin complexes thus obtained are usually yellow to orange. Some further reactions of the ethene complexes are shown below.

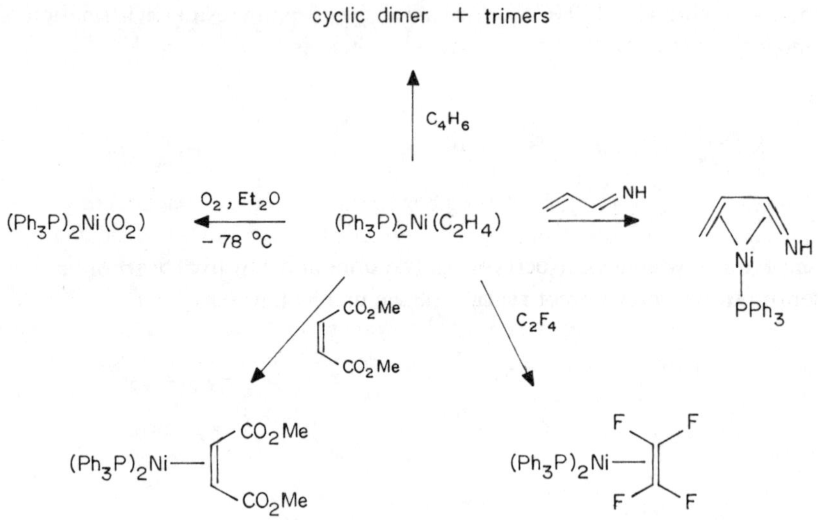

Palladium

As is observed with nickel, palladium shows a pronounced tendency to form π-allyl complexes, which again limits the types of π-olefin complex which can be prepared. This is in marked contrast to platinum (see later), which forms a wide range of π-olefin complexes. While the instability associated with alkene complexes of palladium is decidedly unhelpful if the complexes are to be isolated, it is in fact responsible for these functioning as reactive intermediates in a number of important catalytic processes (see Chapter 2). The most common stable isolable complexes are invariably dimeric halogen-bridged derivatives [(olefin)PdCl$_2$]$_2$:

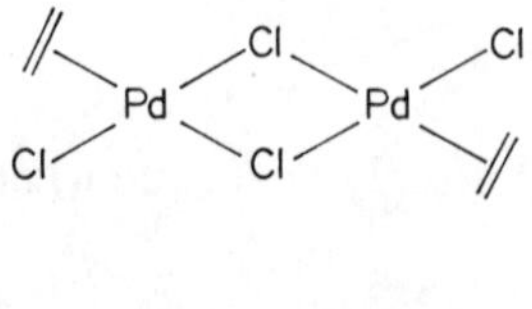

Some examples of the preparation of these types of complex are given below.

$$PdCl_2 \ + \ C_2H_4 \ \xrightarrow[\substack{10 \ atm \\ 20 \ °C}]{EtCl} \ \left[(C_2H_4)PdCl_2\right]_2$$

$$(PhCN)_2PdCl_2 \ + \ \text{olefin} \ \xrightarrow{\text{benzene}} \ \left[(\text{olefin})PdCl_2\right]_2$$

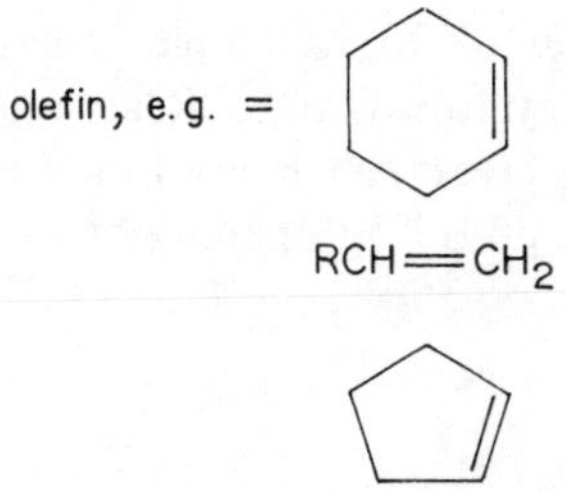

olefin, e.g. =

$RCH{=\!=}CH_2$

Obviously, there are limitations on the preparation of these types of complex; for example, the complexes of pinene or camphene are very unstable and complexes of, e.g., stilbene and dichloroethene are not obtainable by the above method. The coordinated olefin does not react with the usual reagents which attack double bonds. For example, diazoacetonitrile attacks the metal, and the unchanged olefin can be liberated from the resulting complex by treatment with a phosphine ligand:

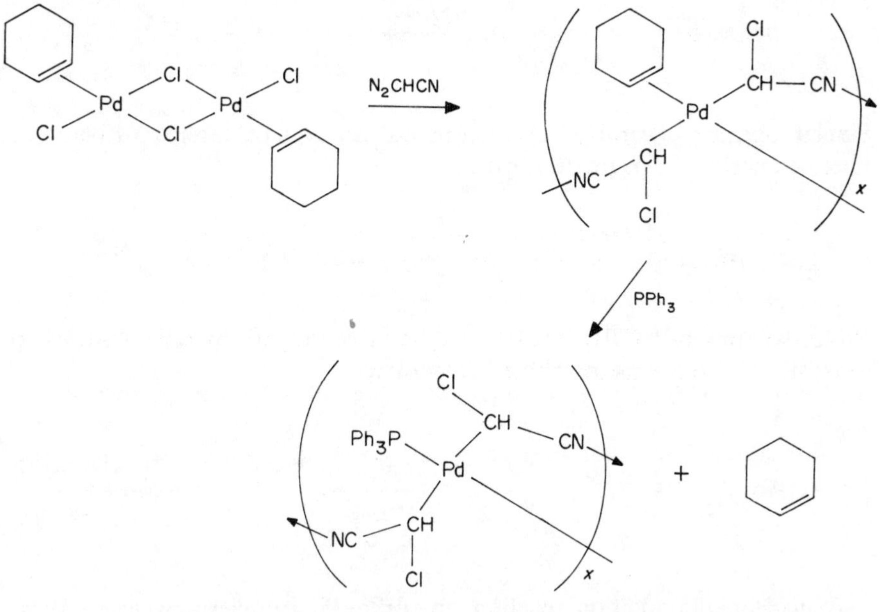

(Olefin–PdCl$_2$)$_2$ complexes partially dissociate when dissolved in either chlorinated or aromatic solvents. This leads to easy exchange between free and complexed monoolefins, providing a useful method for the preparation of new complexes, e.g.

$$\left[(R_2C\!=\!CR_2)\,Pd\,Cl_2\right]_2 \ + \ R'_2C\!=\!CR'_2 \ \rightleftharpoons$$

$$\left[(R'_2C\!=\!CR'_2)Pd\,Cl_2\right]_2 \ + \ R_2C\!=\!CR_2$$

The olefin exchange reaction may be used to effect transformations of vinylcyclopropanes. Reaction of $[(C_2H_4)PdCl_2]_2$ with vinylcyclopropane at 25 °C gives the vinylcyclopropane complex, which on heating in benzene at 40 °C undergoes ring opening of the three-membered ring to give a π-allyl complex:

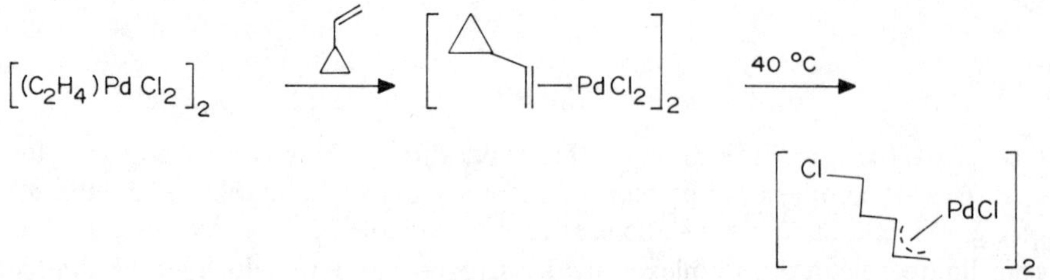

and

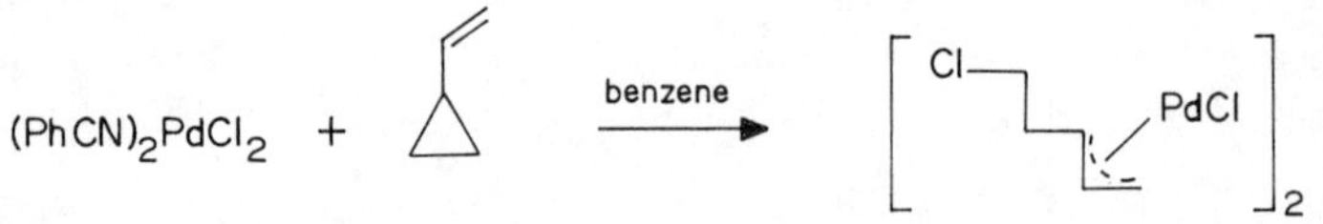

Similar opening of strained cyclopropanes can also be effected without the intermediacy of a π-olefin derivative:

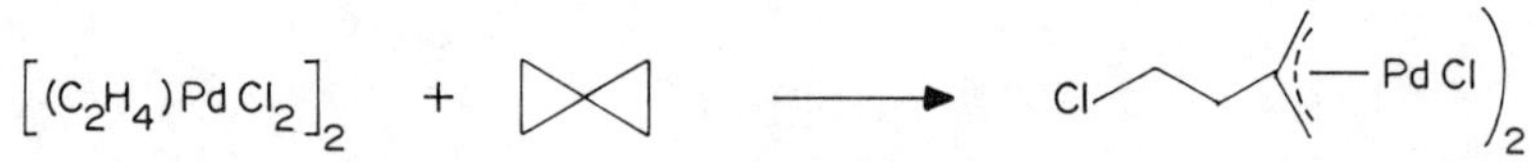

Bridge-splitting of $\{(C_2H_4)PdCl_2\}_2$ can be achieved with pyridine N-oxide to give unstable crystalline monomeric complexes:

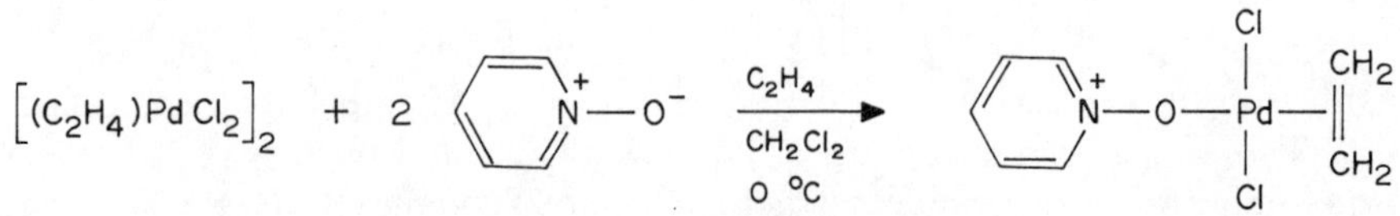

A nucleophilic addition reaction on olefin–Pd complexes which is complementary to the catalytic carbonylation, and carbonylation of olefins discussed in Chapter 2, is the reaction with carbomethoxy mercuric chloride, e.g

Platinum

Historically, the η^2-alkene complexes of platinum are extremely important. The well known complexes of stoichiometry, e.g., $PtCl_2 \cdot C_2H_4$, now known to be a dimer with Cl bridges, and characterized by the Danish pharmacist Zeise around 1830, were probably the first organometallic derivatives of the transition metals to be prepared. There is not space in this section to review olefin–Pt complexes completely, and the interested reader is referred to the excellent book by Herberhold.[1] Olefin complexes are formed by platinum in the $+2$ and 0 oxidation states, e.g. $\{(olefin)PtX_2\}$ and $(Ph_3P)_2Pt(olefin)$, respectively.

Zeise's salt is readily prepared:

$$K_2PtCl_4 \ + \ C_2H_4 \ \xrightarrow{H_2O} \ K\{C_2H_4PtCl_3\} \cdot H_2O$$

This, and other olefin complexes, have a square-planar arrangement of ligands about the metal, although coordination numbers higher than four are often assumed for transition states or reactive intermediates during reaction of such complexes.

Binuclear halogen-bridged complexes, $[(olefin)PtX_2]_2$, analogous to the palladium complexes discussed above, may be prepared by the following methods:

$$PtX_4 \ + \ C_2H_4 \ \xrightarrow[\substack{or\ AcOH \\ (reduction)}]{benzene} \ (C_2H_4PtCl_2)_2$$

$$K(C_2H_4PtCl_3) \cdot H_2O \ \xrightarrow[EtOH]{HCl} \ {}'H(C_2H_4PtCl_3)' \ \xrightarrow{60\ {}^\circ C} \ (C_2H_4PtCl_2)_2$$

Olefin exchange:

$$(C_2H_4PtCl_2)_2 \ \underset{\longleftarrow}{\overset{R_2C=CR_2}{\longrightarrow}} \ (C_2R_4PtCl_2)_2$$

There are many limitations to all of these methods and certain olefins, e.g. highly substituted derivative, do not behave well. Attempts to use the last method were unsuccessful when the less volatile olefins have a poor ability to form complexes, e.g. 1, 1-diphenylethene. The parent complex $(C_2H_4PtCl_2)_2$ is an orange, light-sensitive, crystalline compound, stable up to 100°C.

The chloro bridges are broken, to varying degrees, in solvents which have donor properties, and by certain donor ligands, e.g. amines and pyridine *N*-oxide, although phosphines and certain other ligands cause displacement of the olefin:

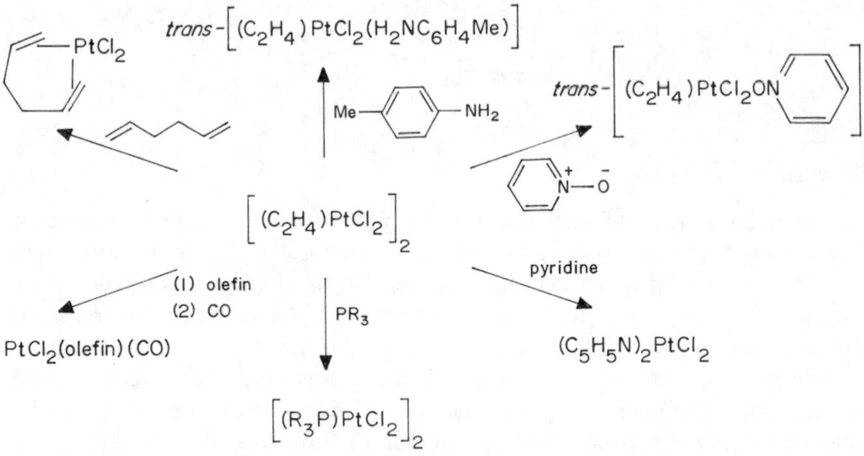

Reaction with hydrogen occurs rapidly with concomitant reduction of the olefin and liberation of the corresponding alkane, but scrambling of label is observed when deuterium is used.

$$(C_2H_4PtCl_2)_2 \xrightarrow{\;\;H_2\;\;} 2\ Pt\ +\ 4\ HCl\ +\ 2\ C_2H_6$$

The dimeric complexes, on treatment with excess of halide ions, undergo bridge-splitting reactions, to yield anionic monomeric olefin complexes analogous to Zeise's salt. These anionic complexes may also be prepared by the methods outlined below:

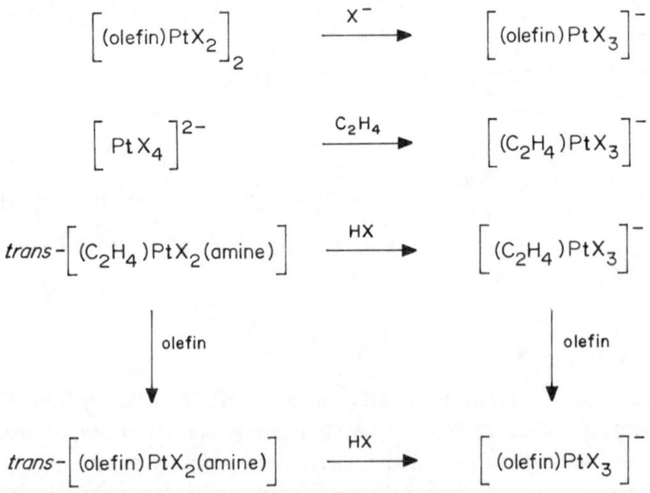

X-ray crystal structure determination of Zeise's salt, $K(C_2H_4PtCl_3) \cdot H_2O$, reveals a square-planer arrangement with the alkene occupying one coordination site, and with the $C{=}C$ bond axis perpendicular to the coordination plane:

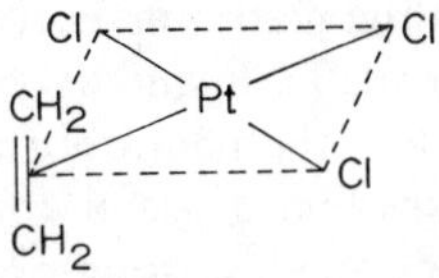

The chloride ligand *trans* to ethene in Zeise's salt is very labile and is substituted rapidly even by water in aqueous solutions:

$$(C_2H_4PtCl_3)^- \xrightarrow{H_2O} trans\text{-}\left[(C_2H_4)PtCl_2(H_2O)\right]$$

In aqueous systems at elevated pH the complex undergoes hydrolysis to produce acetaldehyde, but in acidic solution it is stable. The formation of acetaldehyde may be compared with the palladium-catalyzed hydration of ethene (Chapter 2).

$$K(C_2H_4PtCl_3)^- \longrightarrow \left[(C_2H_4)PtCl_2(H_2O)\right] \xrightarrow{OH^-}$$

$$KCl + 2\ HCl + \overset{.}{Pt} + CH_3CHO$$

Some other reactions of Zeise's salt are summarized below.

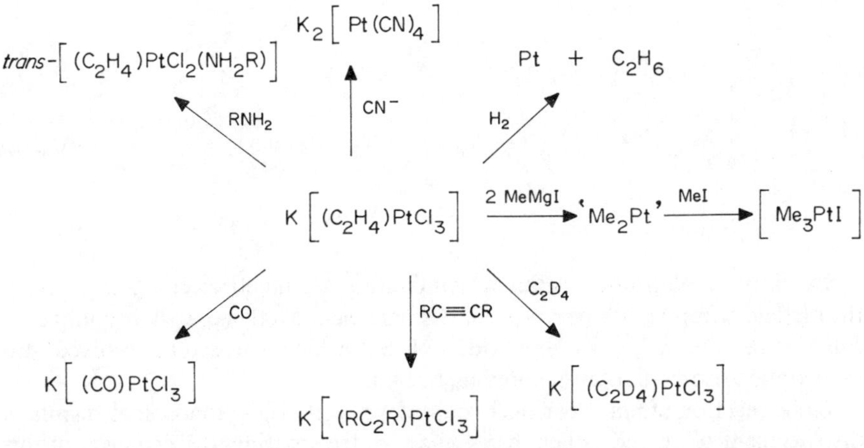

Treatment of Zeise's salt with amines or ammonia results in the formation of a neutral *trans*-[(olefin)PtCl$_2$(amine)]] complex. If the corresponding *cis*-NH$_3$ complex is treated with silver nitrate, a cationic olefin–Pt complex is produced.

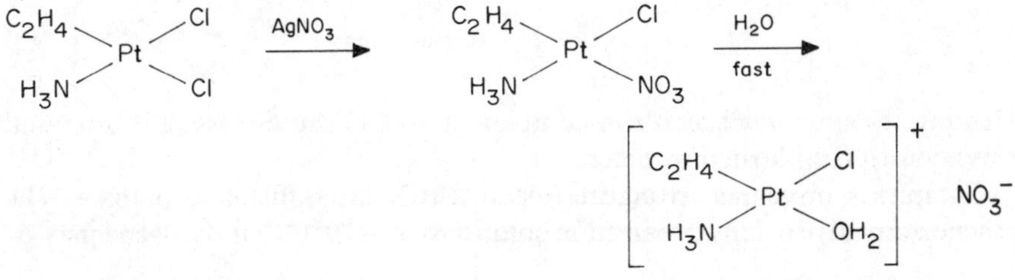

Owing to the contraction of the platinum d orbitals as a result of the positive charge, back-bonding to the olefin ligand is considerably weaker in these complexes, so they are rarer than the neutral or anionic species.

The neutral amine complexes find a useful application in the separation of racemic mixtures of asymmetric olefins, using optically active amines to effect diastereoisomer formation and then separation. As an example, *trans*-cyclooctene may exist in two enantiomeric forms, since the ring is so twisted that the olefin C=C group cannot rotate relative to the remainder of the molecule. Resolution may be effected as outlined in the following scheme (Am = α-phenylethylamine):

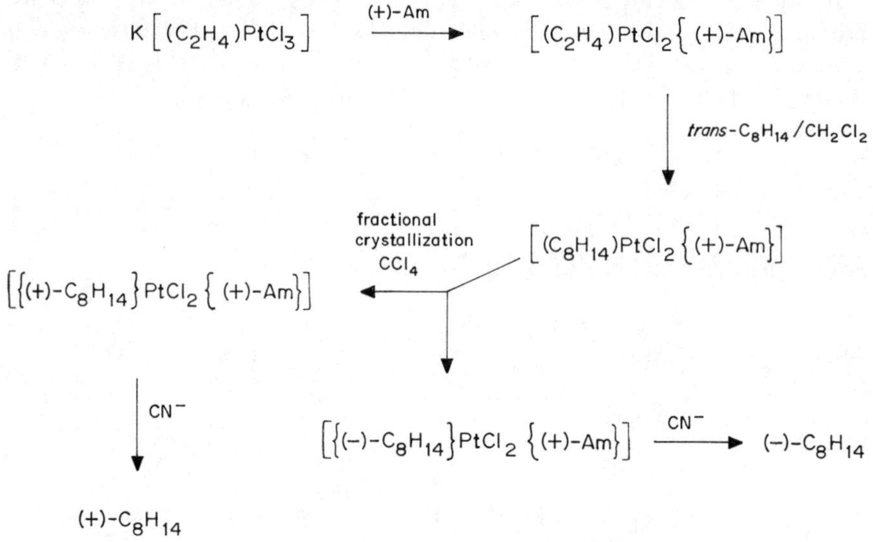

K[(C2H4)PtCl3]
(+)-Am
[(C2H4)PtCl2{(+)-Am}]
trans-C8H14/CH2Cl2
fractional
crystallization
CCl4
[(C8H14)PtCl2{(+)-Am}]
[{(+)-C8H14}PtCl2{(+)-Am}]
CN⁻
[{(−)-C8H14}PtCl2{(+)-Am}]
CN⁻
(−)-C8H14
(+)-C8H14

Similarly, resolution of olefins which contain asymmetric centers α- or β- to the olefinic grouping can be resolved. Displacement of ethene in these complexes can also be achieved with sulphoxides RR′SO, which can also be resolved into their optical isomers in an analogous fashion.

While reaction of binuclear olefin complexes with trialkylphospines results in displacement of alkene rather than halide-bridge splitting, the products of this reaction can be split by olefins:

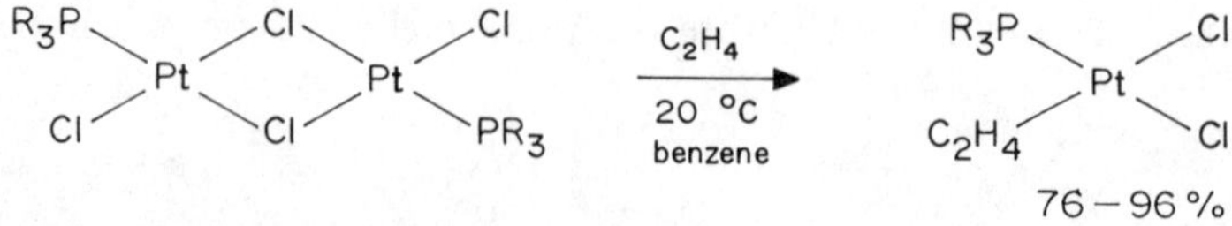

R3P
Cl
Pt
Cl
Cl
Pt
Cl
PR3
C2H4
20 °C
benzene
R3P
Cl
Pt
C2H4
Cl
76−96 %

Heating the mononuclear olefin complex at 100 °C causes loss of ethene and reversion to the phosphine dimer.

Complexes of general structure $(olefin)_2PtCl_2$ are difficult to prepare. The ethene derivative is only obtained in solutions at −70 °C, but oct-1-ene may be used successfully as a ligand in these complexes:

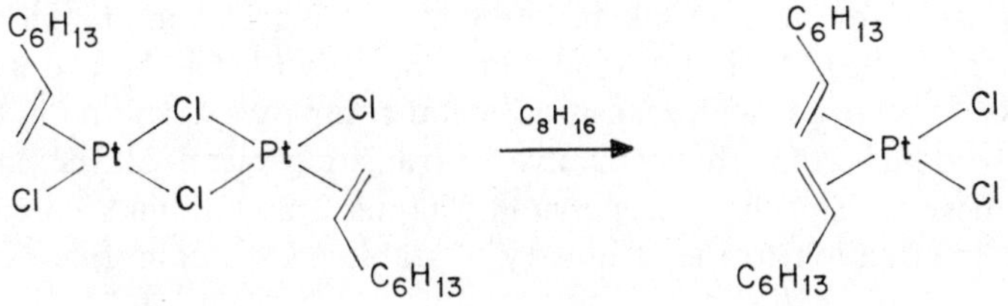

The product bis(octene) complex is a colorless, crystalline material which can be recrystallized from organic solvents. The number of olefins which form such complexes is very limited.

Olefin complexes which contain platinum in the formally zero oxidation state are readily prepared by treatment of an appropriate triphenylphosphine–Pt(II) complex with olefin in the presence of a suitable reducing agent, e.g.

$$cis-\left[(PPh_3)_2PtCl_2\right] \xrightarrow[N_2H_4 \cdot H_2O]{\text{olefin}} (PPh_3)_2Pt(\text{olefin})$$

This method is very limited, and is unsuccessful for a large number of olefins. A better method employs $(PPh_3)_2Pt(O_2)$:

$$(PPh_3)_2Pt(O_2) \xrightarrow[\substack{EtOH, \\ N_2H_4 \text{ or } NaBH_4}]{C_2H_4} (PPh_3)_2Pt(C_2H_4)$$

The ethene complex is obtained as off-white crystals, m.p. 122–125 °C, which are stable in air. The ^1H NMR spectrum shows a signal at $\delta\,2.15$ p.p.m. due to four equivalent ethene protons, showing a $^{195}Pt–H$ coupling constant of 62 Hz. Some reactions of this complex, most of which proceed via pre-dissociation to give $(Ph_3P)_2Pt$, are summarized below.

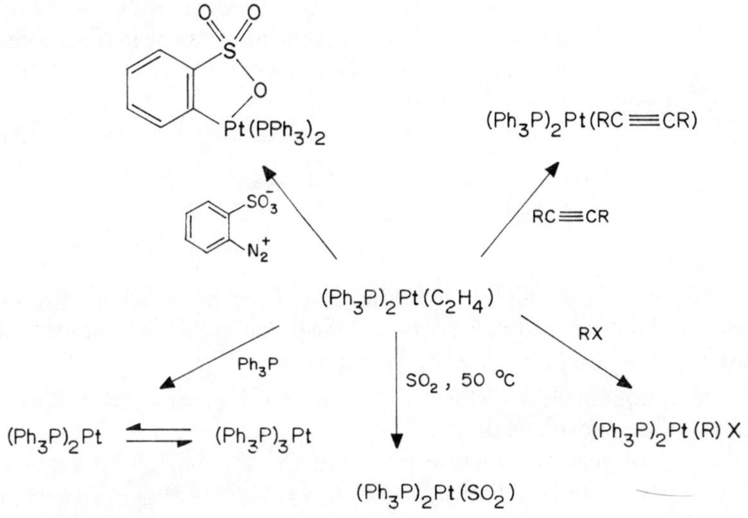

η^2-ALKYNE COMPLEXES

A very useful review of η^2-alkyne complexes as reagents in organic synthesis has been published.[23] An enormous amount of work has been done on the reactions of acetylenes with transition metal complexes, resulting in a diverse range of products, thought to involve η^2-transition metal–alkyne complexes. Some of these concepts have been met in Chapter 2, and in this section we shall confine our attention to the chemistry of stable alkyne complexes, mainly of cobalt.

The synergistic interaction so often cited for alkene–metal complexes also occurs with η^2-alkyne derivatives, and results in considerable distortion from linearity of the alkyne ligand, possibly owing to some degree rehybridization of the carbon atoms. Some examples are shown below.

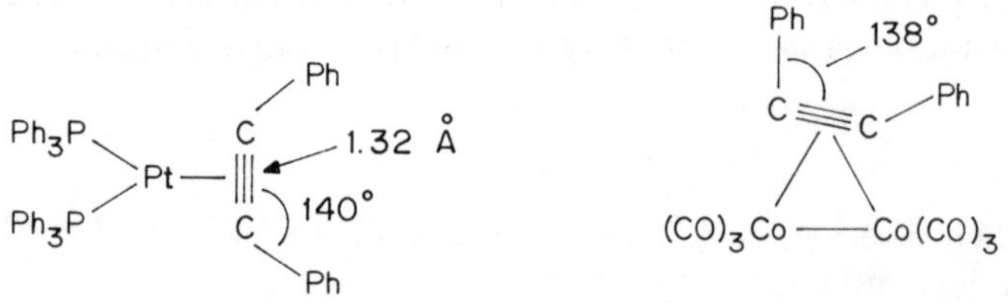

Evidence for a tendency toward a $C=C$ double bond in these complexes is also found from IR and ^1H and ^{13}C NMR spectral data.

In the main there are two methods for the preparation of metal–alkyne complexes, ligand substitution and reductive complexation, the former being the most common and the ligand displaced usually being CO.

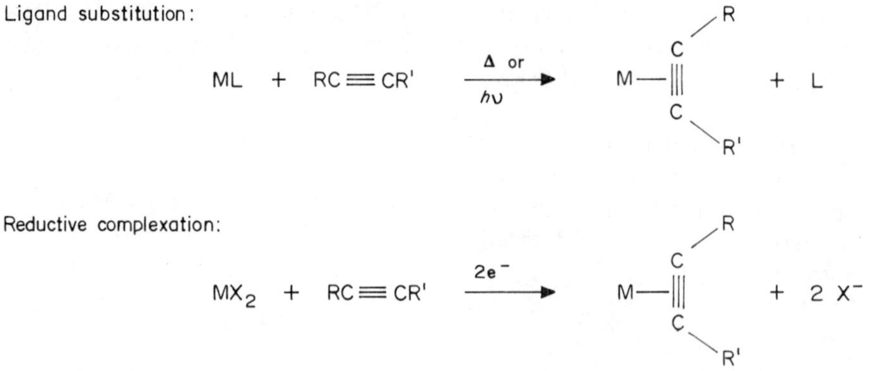

Ligand substitution:

Reductive complexation:

A consequence of the slight rehybridization of acetylene carbon atoms is that cycloalkyne complexes are stable compared with the free ligands. Some examples of these types of complexes are shown below.

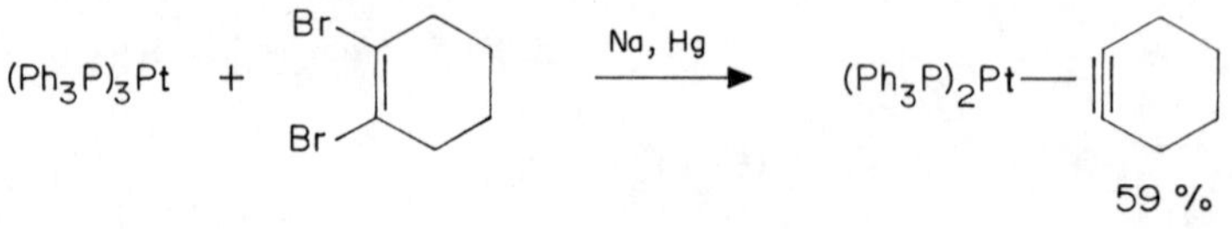

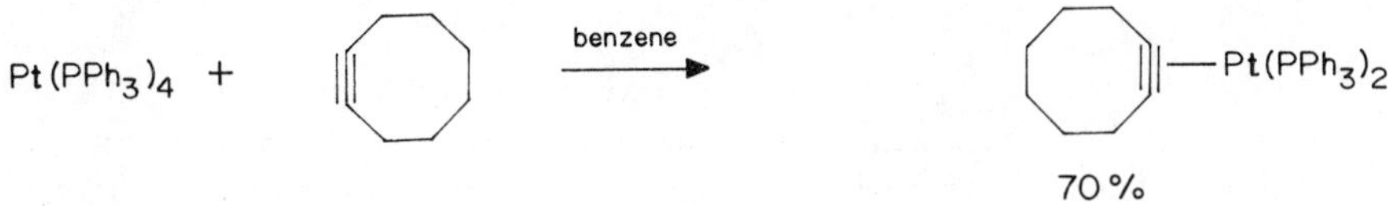

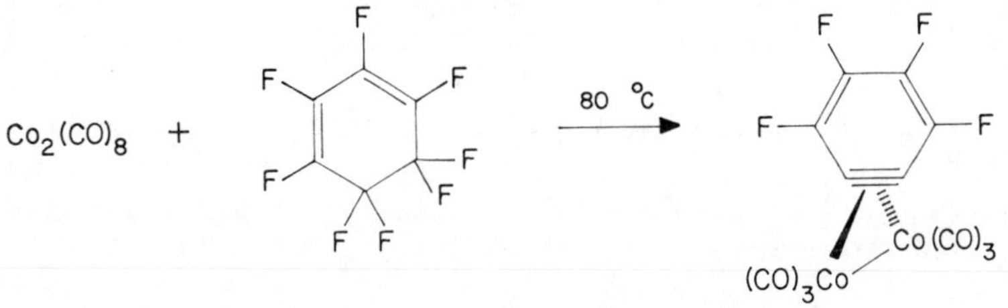

70 %

Highly reactive acetylenes may also be trapped at low temperatures by treatment with dicobalt octacarbonyl, e.g.

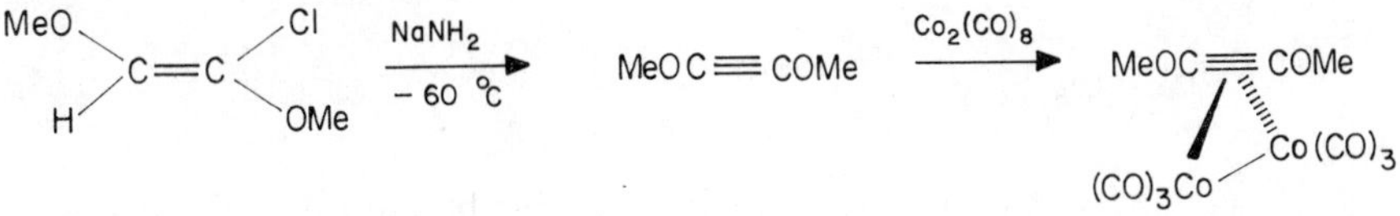

While both π-bonds of the acetylene are involved in bonding to the two cobalt atoms of alkyne–$Co_2(CO)_6$ complexes, we shall use the following simple representation (left) for our further discussion:

Attachment of the $Co_2(CO)_6$ unit to an acetylenic grouping considerably reduces the reactivity of the C≡C triple bond. This offers a good means of protecting the acetylenic group while synthetic operations are carried out on, e.g., olefinic or aromatic groups in the same molecule. The protecting group is readily removed by treatment with cerium(IV) ammonium nitrate or iron(III) nitrate. Some examples which show this utility and the stability of the acetylene–$Co_2(CO)_6$ entity to a range of reaction conditions are given in the following equations.

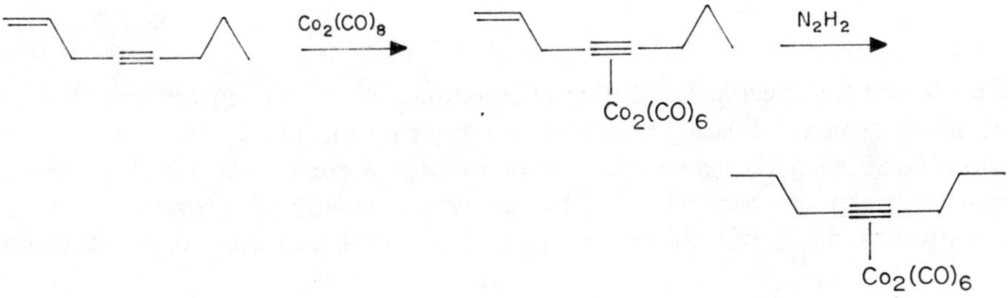

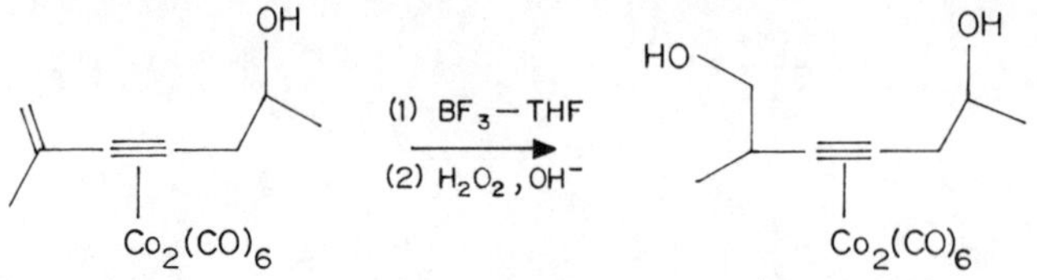

OH

(1) BF$_3$—THF
(2) H$_2$O$_2$, OH$^-$

HO
OH
Co$_2$(CO)$_6$

Co$_2$(CO)$_6$

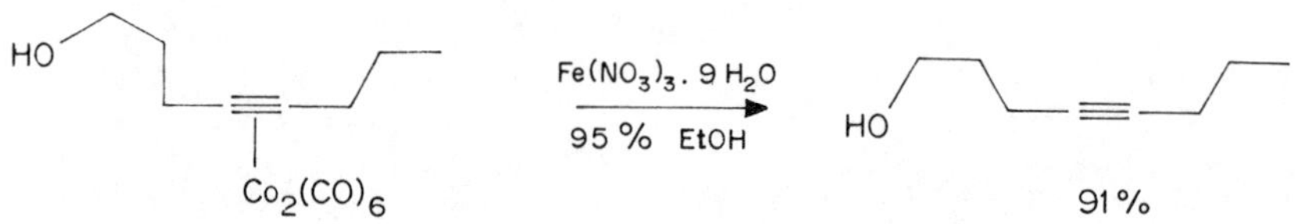

HO

Co$_2$(CO)$_6$

Fe(NO$_3$)$_3$ · 9 H$_2$O
95 % EtOH

HO

91 %

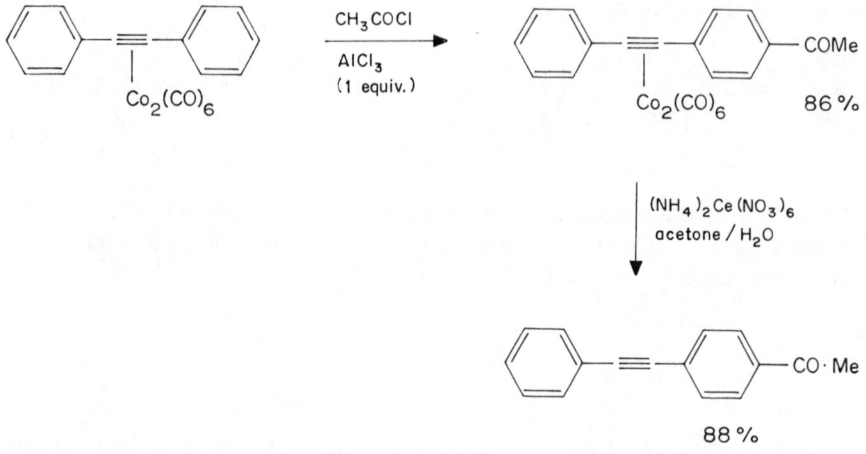

Co$_2$(CO)$_6$

CH$_3$COCl
AlCl$_3$
(1 equiv.)

COMe
Co$_2$(CO)$_6$ 86 %

(NH$_4$)$_2$Ce(NO$_3$)$_6$
acetone / H$_2$O

CO·Me

88 %

The acetylene–Co$_2$(CO)$_8$ unit effects stabilization of carbocation centers α- to itself, which probably accounts for the *para*-directing effect of this group in the acylation of attached benzene rings:

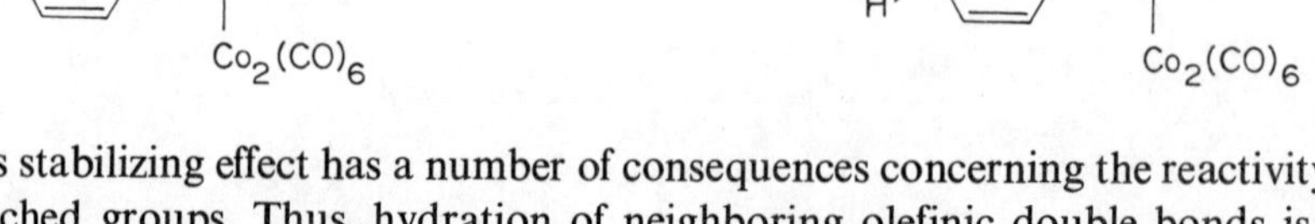

R

$\overset{+}{CH_3CO}$

CH$_3$CO

$+$
R

H

Co$_2$(CO)$_6$

Co$_2$(CO)$_6$

This stabilizing effect has a number of consequences concerning the reactivity of attached groups. Thus, hydration of neighboring olefinic double bonds is far more facile than the same reaction of isolated systems, and occurs without rearrangement or conversion of the acetylenic group. The reverse reaction, dehydration, is also facilitated, as is the stereospecific opening of neighboring cyclopropylcarbinol groups. These reactions are collected in the following equations.

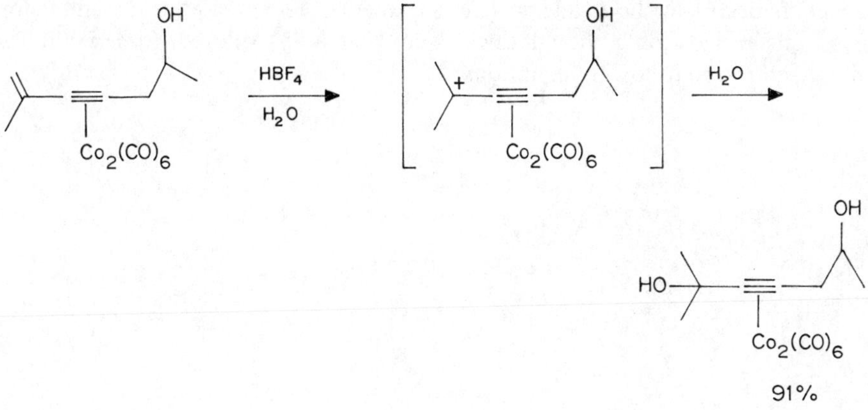

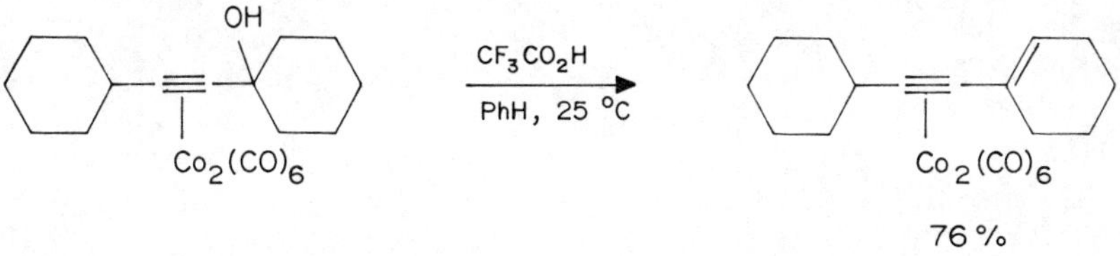

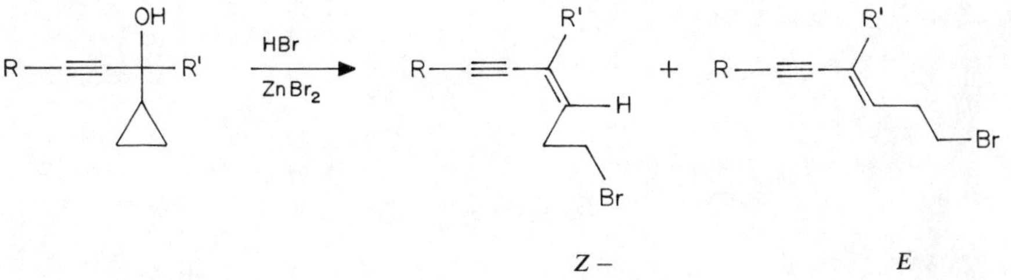

91%

76%

Compare these with

90–99%   E –

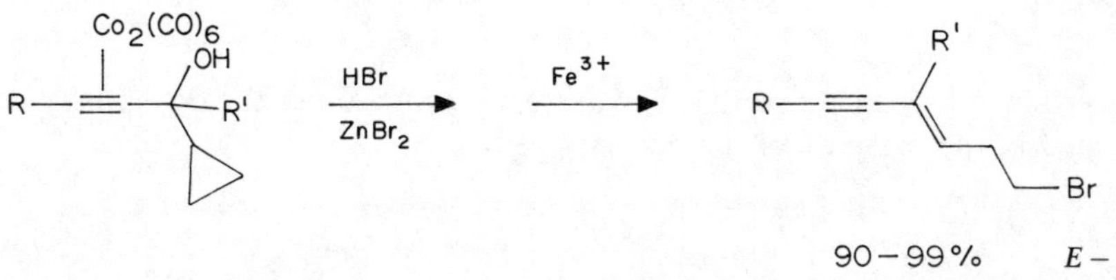

In fact, a number of the stabilized carbocations have been isolated as crystalline solids, and NMR and IR spectral studies indicate delocalization of charge over the acetylene–$Co_2(CO)_6$ moiety. The cations have similar stability to that of the triphenylmethyl cation. The salts are highly reactive towards a range of nucleophilic reagents, including trimethylsilyl enol ethers and allylsilanes, and reaction occurs without allenic rearrangements. There is some potential for synthetic application, although this has not been fully explored. Some examples are given in the following equations.[24]

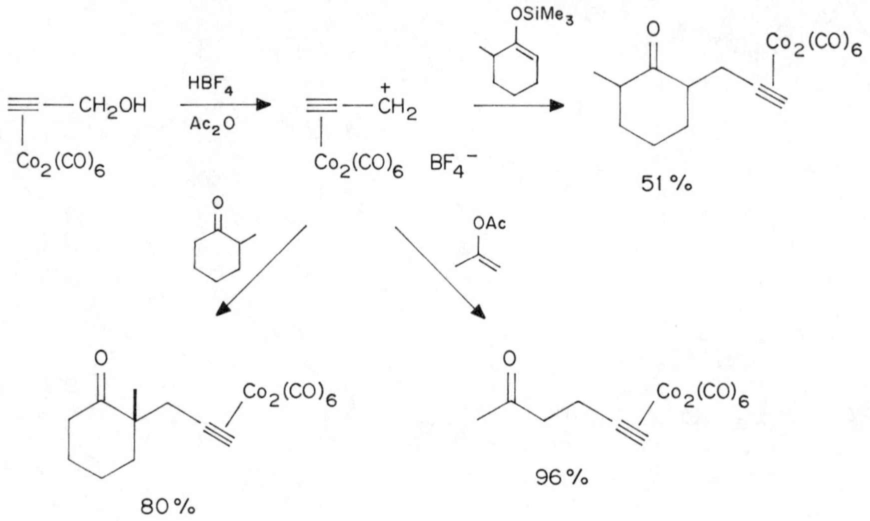

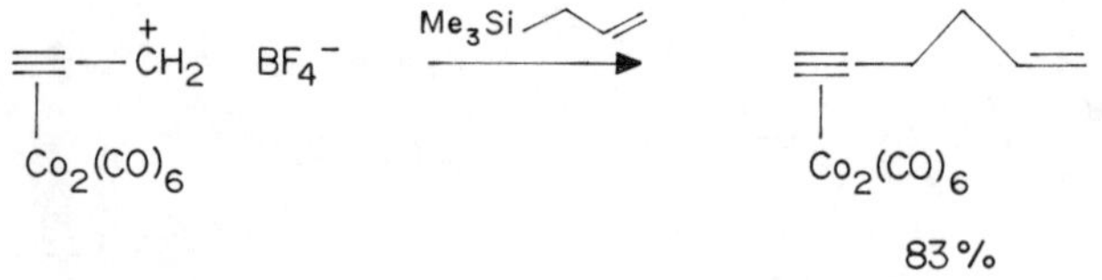

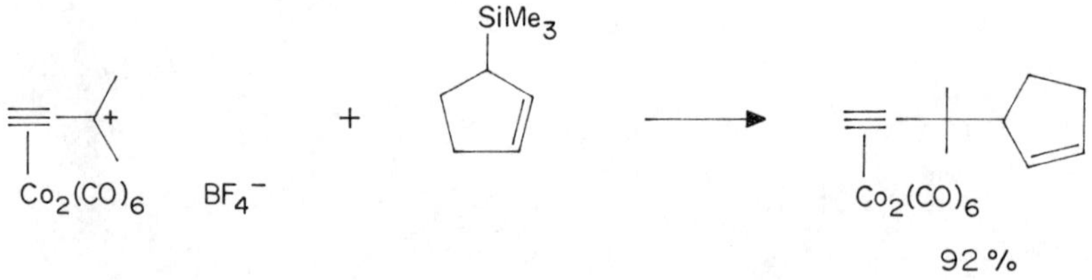

In certain reactions, e.g. with active aromatic compounds, the cation may be generated in the presence of the nucleophile.[25]

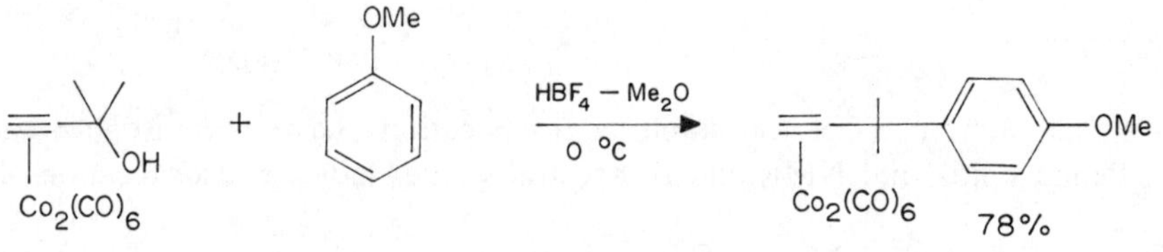

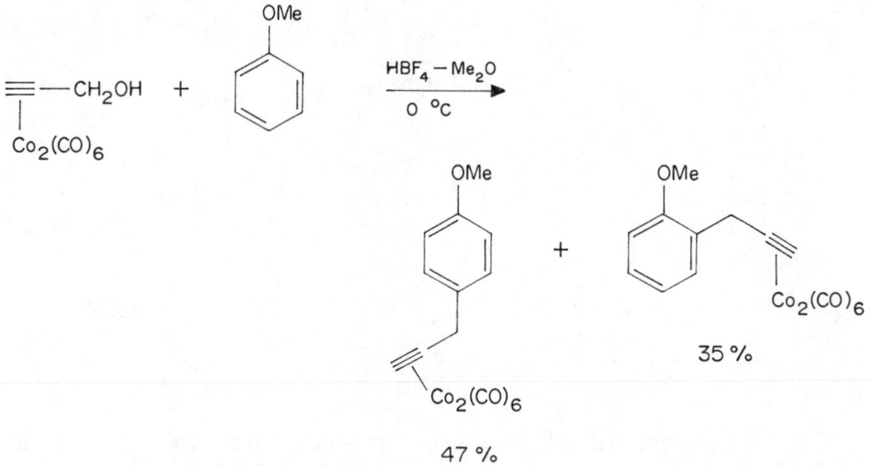

Of course, the reactivity of the coordinated triple bond will to some extent be governed by the nature of the metal involved. Alkyne complexes of metals of low oxidation state are invariably found to undergo reactions with electrophiles, while those complexes of transition metals of higher oxidation state, or cationic complexes, are usually reactive towards nucleophilic species. Examples of these types of behavior are summarized below.

(a) *Reactions with electrophiles:*

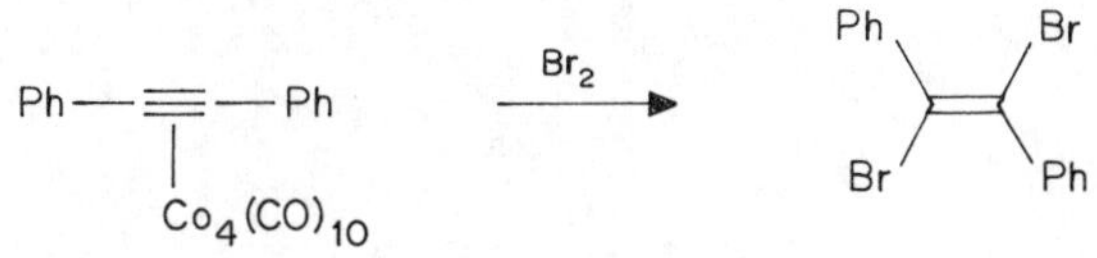

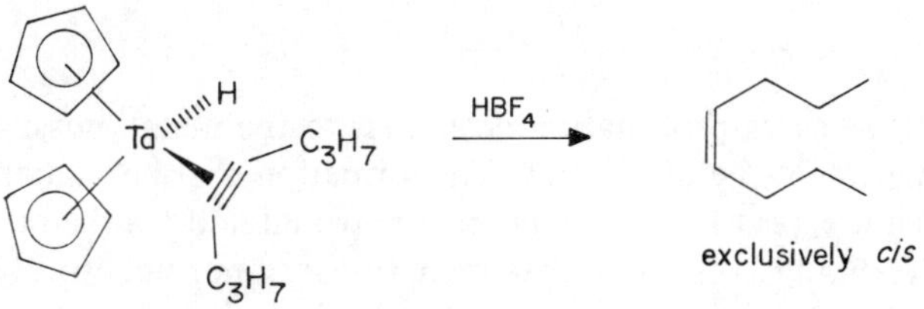

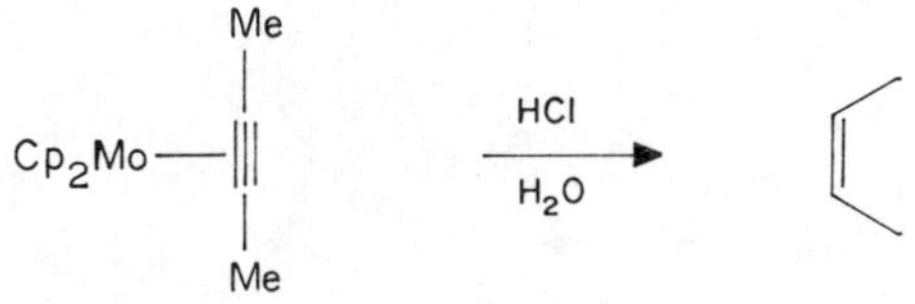

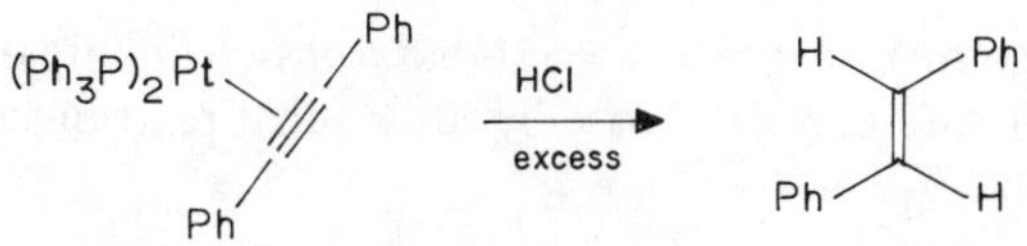

(v)

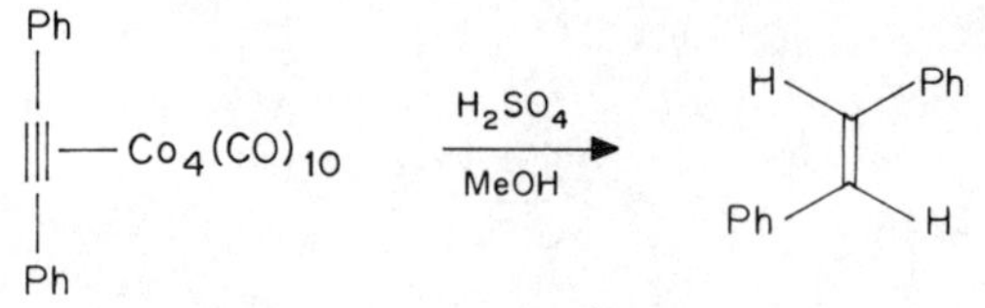

(vi)

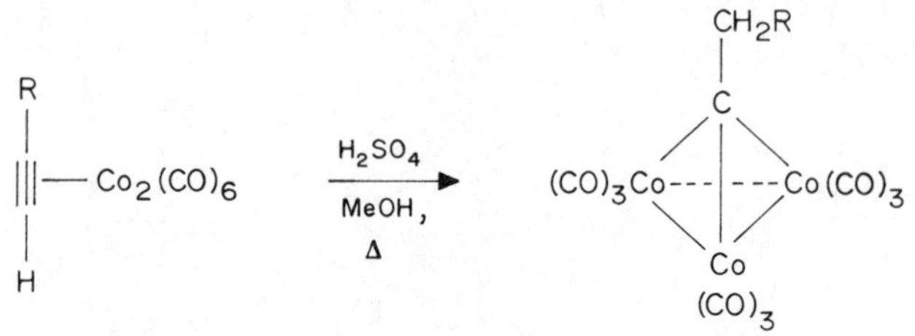

It is readily observed that a range of products from acid treatment are available, depending on the metal and conditions used. Most strikingly, some metal complexes produce *cis* and others *trans*-olefins. The mechanism appears to involve the intermediacy of a σ-vinyl–metal complex, and these species have been isolated in a number of instances:

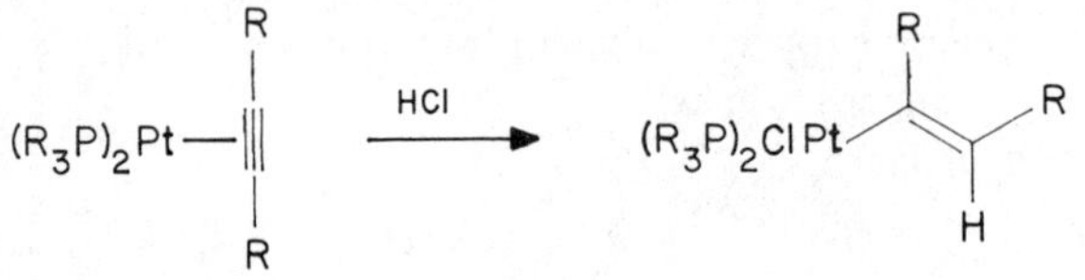

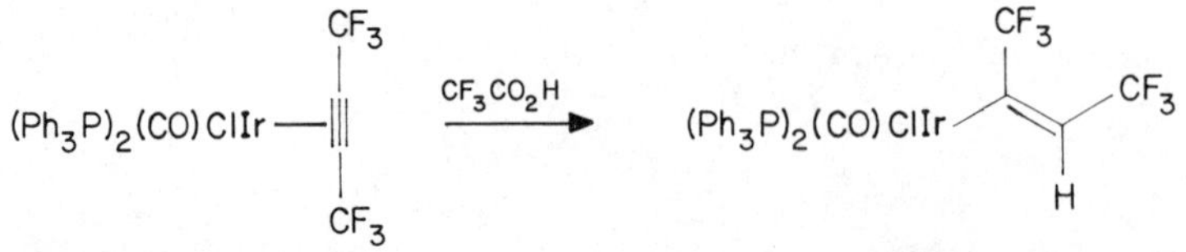

In these cases protonation occurs *cis* to the metal, possibly reflecting prior protonation at the metal itself. The formation of *trans*-olefins is puzzling, since one might expect that a second proton would add via the metal to give the *cis*-olefin resulting from reductive elimination, or a normal protolytic cleavage of the M—C bond in which the stereochemistry is retained:

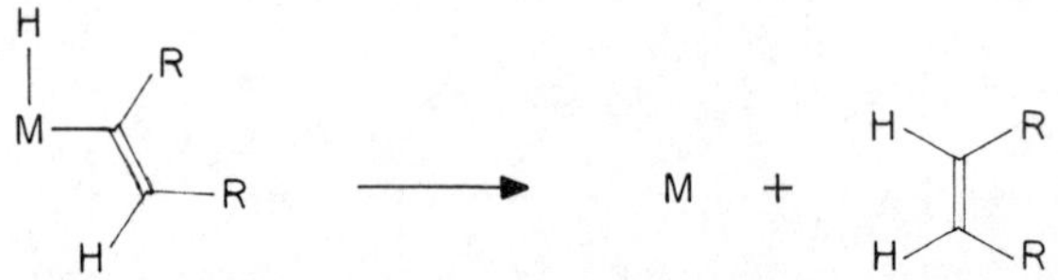

However, there is no evidence to suggest that *trans*-olefins are the products of a concerted reaction and they may arise by subsequent rearrangement under the reaction conditions.

(b) Reactions with nucleophiles:[23,26]

Nucleophilic attack on the coordinated triple bond may occur either *cis* or *trans* to the metal, depending on whether prior attachment of nucleophile to the metal ($\rightarrow cis$) or direct addition to the alkyne occurs ($\rightarrow trans$). Usually the *cis* mode of addition will only occur for coordinatively unsaturated metal derivatives. Some examples are given below.

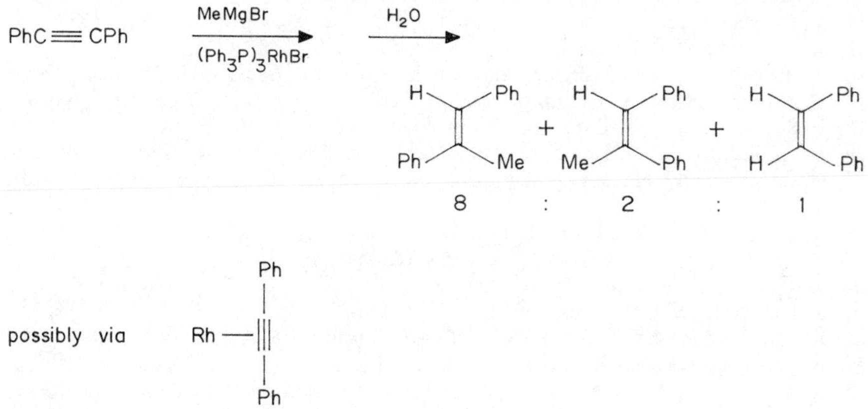

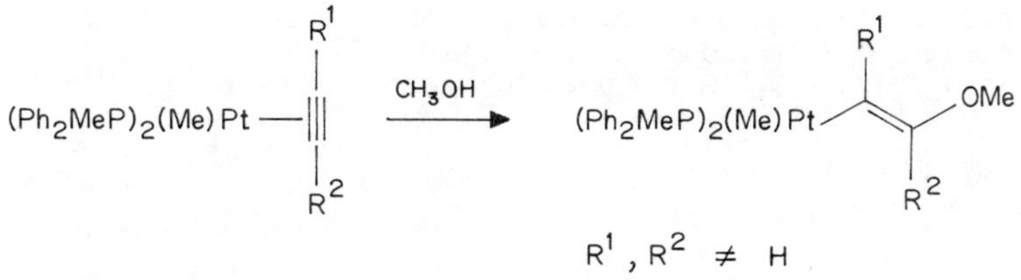

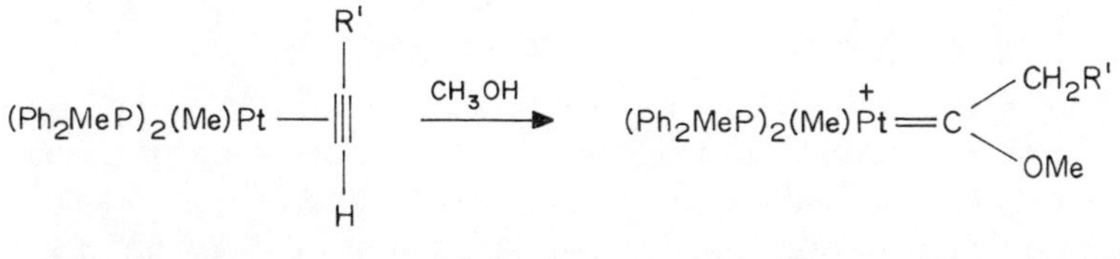

$$PhC \equiv CPh \xrightarrow[\text{(Ph}_3\text{P)}_3\text{RhBr}]{\text{MeMgBr}} \xrightarrow{\text{H}_2\text{O}}$$

8 : 2 : 1

possibly via

$$(Ph_2MeP)_2(Me)Pt \xrightarrow{CH_3OH}$$

$R^1, R^2 \neq H$

$$(Ph_2MeP)_2(Me)Pt \xrightarrow{CH_3OH}$$

Acetylene complexes also undergo a number of interesting insertion reactions, resulting in, e.g., oligomerization and hydrometallation. These have been discussed in Chapter 2.

REFERENCES

1. Review to 1968: M. Herberhold, *Metal π-Complexes*, Elsevier, Amsterdam, 1972, Vol. II, Part 1.
2. R. F. Heck and C. R. Boss, *J. Am. Chem. Soc.*, 1964, **86**, 2580.
3. H. D. Murdoch, *Helv. Chim. Acta*, 1964, **47**, 936.
4. R. Aumann, K. Frohlich and H. Ring, *Angew. Chem., Int. Ed. Engl.*, 1974, **13**, 275.

200

5. G. D. Annis and S. V. Ley, *J. Chem. Soc., Chem. Commun.*, 1977, 581; G. D. Annis, E. M. Hebblethwaite and S. V. Ley, *J. Chem. Soc., Chem. Commun.*, 1980, 297.
6. B. W. Roberts and J. Wong, *J. Chem. Soc., Chem. Commun.*, 1977, 20; B. W. Roberts, M. Ross and J. Wong, *J. Chem. Soc., Chem. Commun.*, 1980, 428.
7. M. Rosenblum, *Acc. Chem. Res.*, 1974, **7**, 122.
8. P. F. Boyle and K. M. Nicholas, *J. Org. Chem.*, 1975, **40**, 2682.
9. A Rosan and M. Rosenblum, *J. Org. Chem.*, 1975, **40**, 3621.
10. K. M. Nicholas, *J. Am. Chem. Soc.*, 1975, **97**, 3254.
11. O. Eisenstein and R. Hoffman, *J. Am. Chem. Soc.*, 1980, **102**, 6148; 1981, **103**, 4308.
12. M. Gifford and P. Dixneuf, *J. Organomet. Chem.*, 1975, **85**, C26.
13. M. F. Neumann and F. Brion, *Angew, Chem., Int. Ed. Engl.*, 1979, **18**, 688.
14. Y. Kubo, L. S. Pu, A. Yamamoto and S. Ikada, *J. Organomet. Chem.*, 1975, **84**, 369.
15. P. R. Evitt and R. G. Bergman, *J. Am. Chem. Soc.*, 1979, **101**, 3973.
16. H. Yamazaki and N. Hagihara, *Bull. Chem. Soc., Jpn.*, 1971, **44**, 2260.
17. G. LaMonica, G. Navazio, P. Sandrini and S. Cenini, *J. Organomet. Chem.*, 1971, **31**, 89.
18. R. Cramer, *Inorg. Chem.*, 1962, **1**, 722.
19. W. H. Baddley, *J. Am. Chem. Soc.*, 1968, **88**, 4545.
20. H. Werner, R. Feser and W. Buckner, *Chem. Ber.*, 1979, **112**, 834; H. Werner, R. Feser and R. Werner, *J. Organomet. Chem.*, 1979, **181**, 27.
21. A. van der Ent and T. C. van Soest, *Chem. Commun.*, 1970, 225.
22. K. Moseley, J. W. Kang and P. M. Maitlis, *J. Chem. Soc. A*, 1970, 2875.
23. K. M. Nicholas, M. O. Nestle and D. Seyferth, in *Transition Metal Organometallics in Organic Synthesis* (Ed. H. Alper), Academic Press, New York, 1978, Vol. 2, p.1.
24. K. M. Nicholas, M. Mulvaney and M. Bayer, *J. Am. Chem. Soc.*, 1980, **102**, 2508; J. O'Boyle and K. M. Nicholas, *Tetrahedron Lett.*, 1980, **21**, 1595.
25. R. Lockwood and K. M. Nicholas, *Tetrahedron Lett.*, 1977, 4163.
26. Review: M. H. Chisholm and H. C. Clark, *Acc. Chem. Res.*, 1973, **6**, 202.

CHAPTER 6

η^3-Allyl Complexes

Of the large number of π-allyl complexes known for many transition metals, those of palladium have come to the forefront in recent years as intermediates for organic synthesis. A very large contribution in this respect has been made by Trost's group. We shall deal first with π-allyl complexes of the nickel group of metals (Ni, Pt, Pd), followed by consideration of the chemistry of other metal complexes. Structure and bonding and NMR spectra have already been discussed. An early review on π-allyl complexes is a useful source of preparation methods.[1]

η^3-ALLYL COMPLEXES OF NICKEL AND PALLADIUM

Two reviews are available describing in detail the applications of π-allyl nickel intermediates in organic synthesis,[2] and synthetic applications of π-allyl-palladium complexes have been reviewed.[3]

Allylnickel Complexes

There are basically two types of π-allylnickel complex. The first, bis-π-allylnickel(0), which is a 16-electron complex of nickel in its (formally) zero oxidation state, is prepared by reaction of nickel chloride with allylmagnesium chloride:

$$NiCl_2 \;+\; C_3H_5MgCl \quad \xrightarrow[-10\ ^{\circ}C]{Et_2O} $$

This compound is very air-sensitive.

The other type of complex, which is more stable, is an 18-electron species composed of the π-allyl ligand, nickel, an anion (after bridging to give a dimer) and sometimes a second ligand (e.g. phosphines or CO). A common example of this type of complex is π-allylnickel(I)-μ-dibromo-π-allylnickel(I), **1**, in which the

201

metal is formally considered to be in the $+1$ oxidation state. There are a number of methods available for the preparation of these complexes, illustrated below.

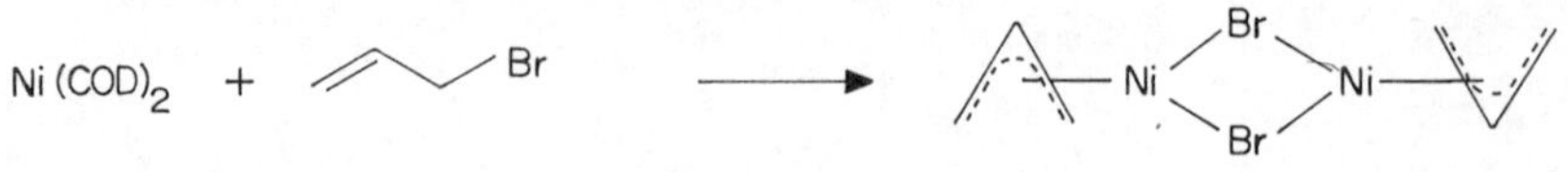

(1) 100%

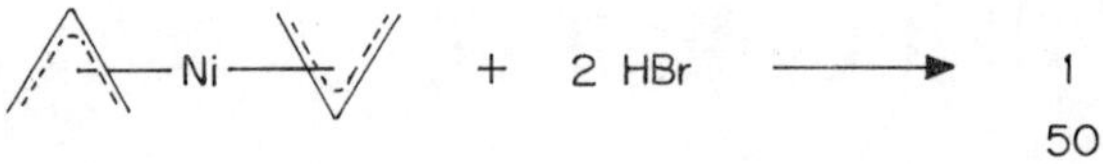

50%

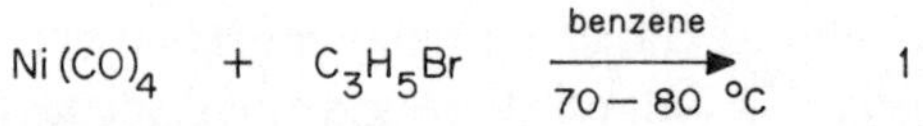

The last method involves the use of nickel carbonyl, and it must be remembered that this compound is exceedingly toxic by inhalation. It is very volatile (b.p. 43 °C) and must be handled in an efficient fume hood.

The allylnickel halide complexes disproportionate in strongly coordinating solvents, e.g. DMF and HMPA, to give the bis-π-allylnickel species, e.g.

Stability of the bis-π-allylnickel complexes is dependent on the nature of the ligand, the 2-methylallyl complex being more stable than the unsubstituted derivative, while π-cyclohexenyl and π-cycloheptenyl complexes are very unstable. They have found very little use as stoichiometric reagents for organic synthesis, possibly owing to their low stability, whilst this lability makes them useful as catalysts for diene oligomerization. The latter aspect has been outlined in Chapter 2. As stoichiometric reagents, two reactions are promising, but rather limited in scope: coupling of the allyl ligands to give hexa-1,5-dienes, and formation of ketones by carbon monoxide insertion during this coupling. Some examples are given below.

100%

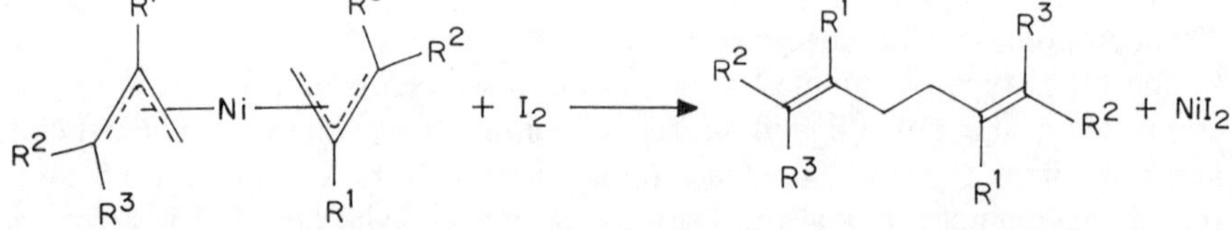

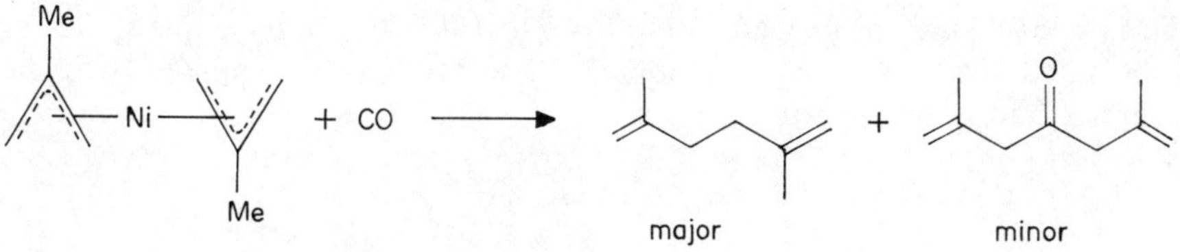

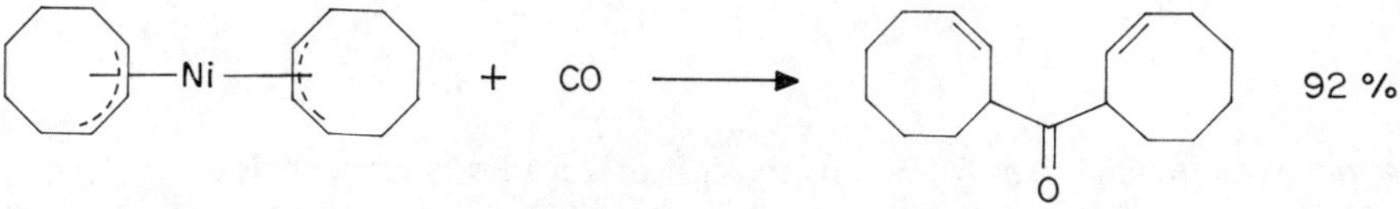

Of course, the ease with which the coupling reaction takes place limits the synthetic utility of these reagents, since it is more difficult to bring the coordinated allyl group into reaction with other substrates. For example, a mixture of products is obtained during the reaction of a π-allylnickel complex with allyl halides:

$(C_3H_5)_2Ni + 2 CH_2{=}CMe CH_2Br \longrightarrow$

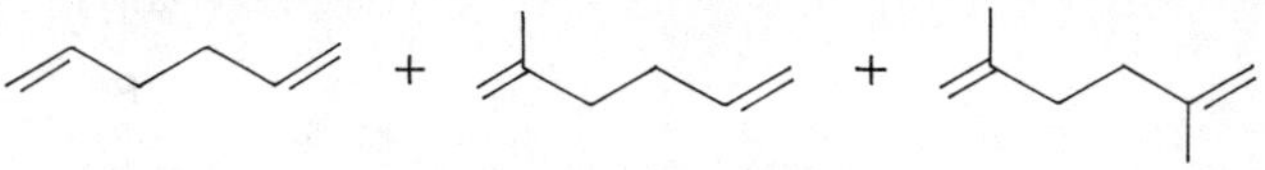

As outlined in Chapter 2, the complex serves as a source of active nickel during the cyclotrimerization of butadiene to cyclododeca-1,5,9-triene.

The mechanism of allyl coupling is probably as shown in the following scheme, involving a sequence of 16- and 18-electron complexes, and either a radical-like coupling of the σ-allyl groups or their reductive elimination to give diallyl.

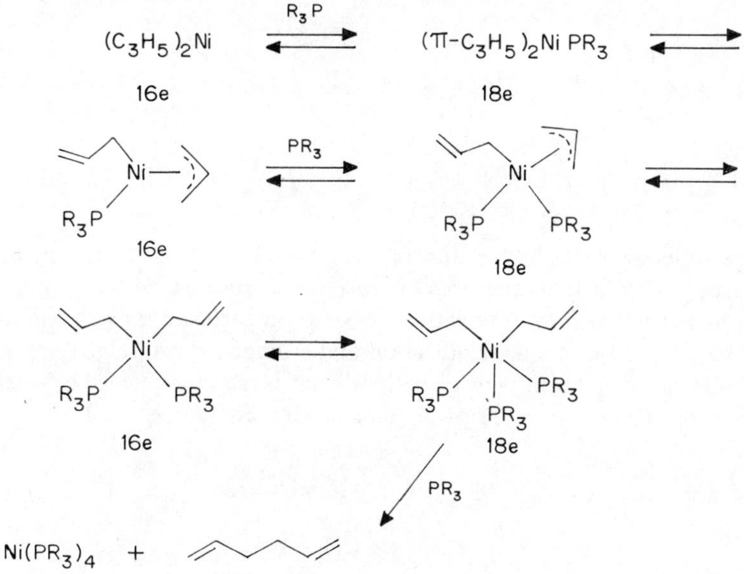

This would explain the propensity of substituted allyl groups to couple at the less substituted terminus, since, e.g. complex **2** is disfavored compared with the complex **3** for steric reasons.

(2)   (3)

A plausible mechanism for the formation of ketones is given below.

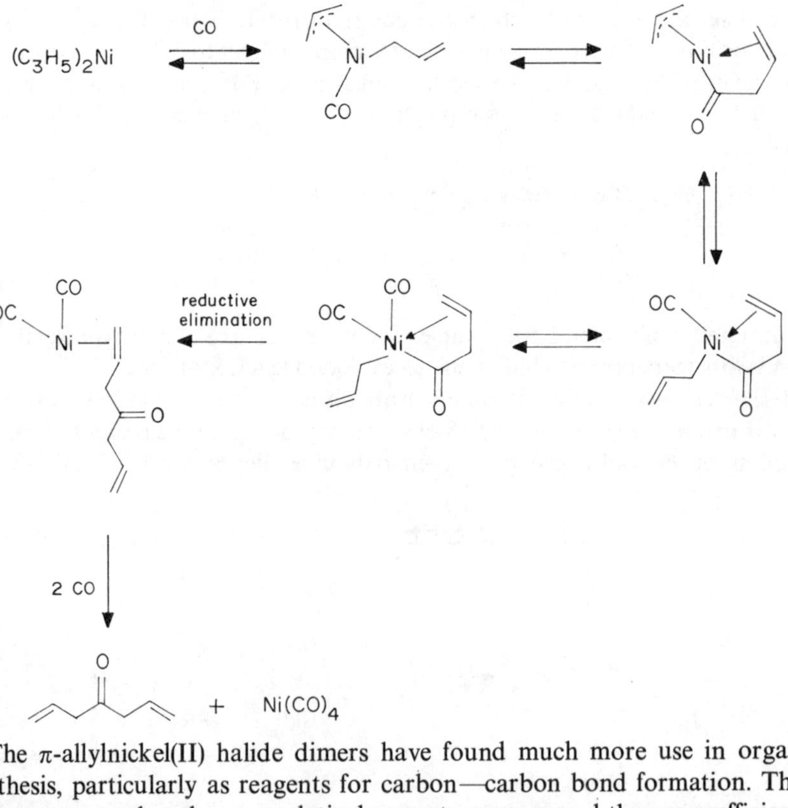

The π-allylnickel(II) halide dimers have found much more use in organic synthesis, particularly as reagents for carbon—carbon bond formation. Their advantages are that they are relatively easy to prepare and they are sufficiently stable to allow isolation, crystallization and storage for prolonged periods.

Carboxylation of the allylnickel halide dimers is readily achieved by treatment with carbon monoxide to afford β,γ-unsaturated esters:

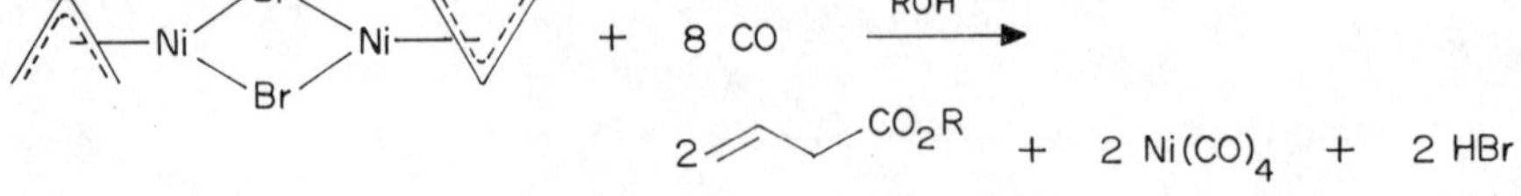

In the presence of an acetylene and the absence of an alcohol solvent, a *cis*-hexa-2,5-dienoyl bromide is formed, which may be converted into the carboxylic acid or ester:

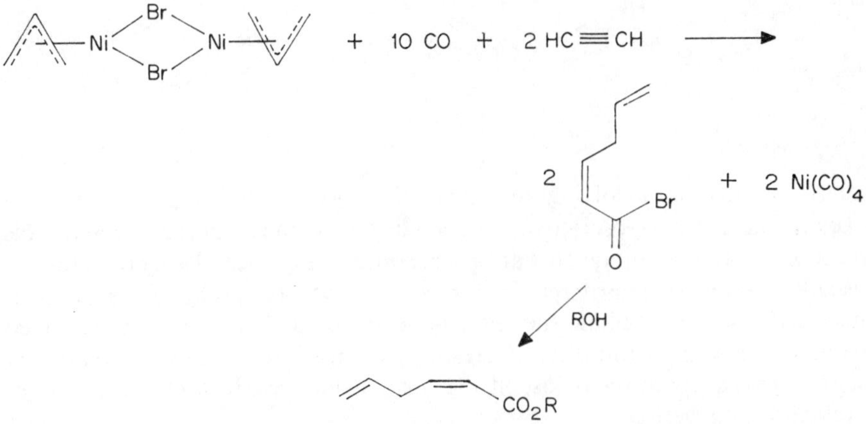

The same reaction may be achieved catalytically, using $Ni(CO)_4$:

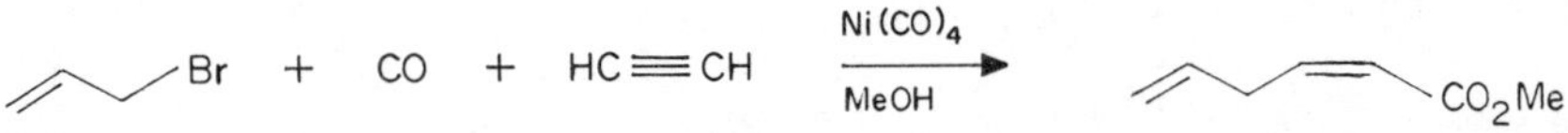

Carbon—carbon bond formation is readily achieved by reaction with organic halides in a polar solvent. This reaction is particularly useful for the introduction of substituted allyl groups, e.g. the isoprenyl unit, since the reaction tends to be regiospecific, alkylation occurring at the unsubstituted allyl terminus. This contrasts with Grignard reagents, which tend to give mixtures, and is illustrated below for Corey's synthesis of α-santalene.[4]

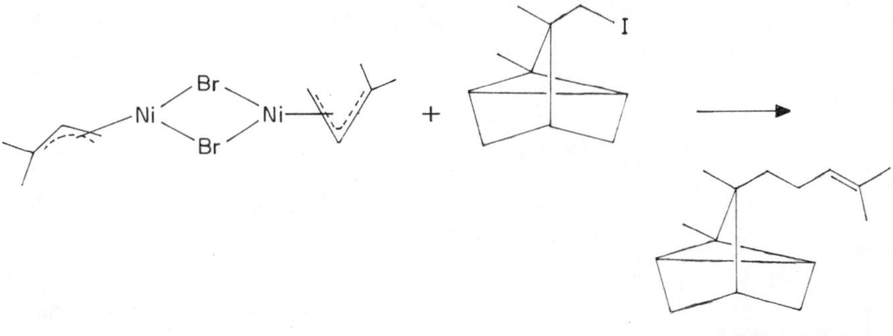

95%

α-santalene

These reagents have another advantage over allyl-Grignard reagents in that they are very selective, being relatively unreactive towards common organic functional groups. This undoubtedly is a mechanistic feature, since the coupling probably involves an oxidative addition of the alkyl halide at some stage, followed by elimination of the coupled product:

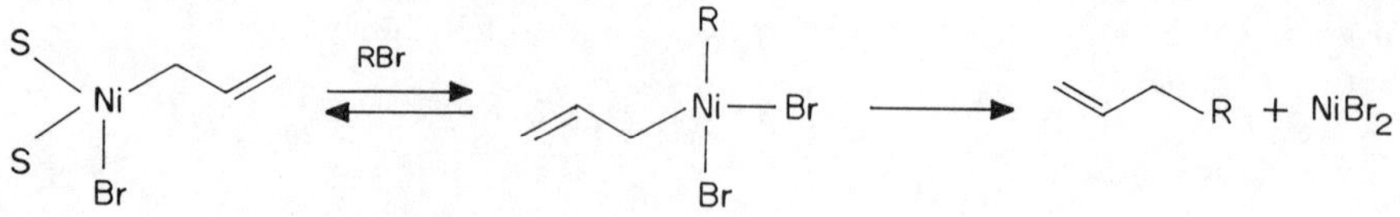

(S = solvent)

Such a mechanism, involving the intermediacy of a σ-allyl complex, would also explain the high regioselectivity observed, for reasons outlined above. This reaction has the advantage that an *aryl* bromide can be used. Also, problems can usually be overcome fairly readily, for example, whereas cyclohexyl bromide is unreactive in the coupling, the corresponding iodide behaves in the expected manner. These and some other interesting examples are listed below. Note that stereochemical integrity is lost in the alkyl halide, as shown by the 4-iodo-cyclohexanol reaction:

98 %

70 %

62 : 38 epimer mixture

96 %

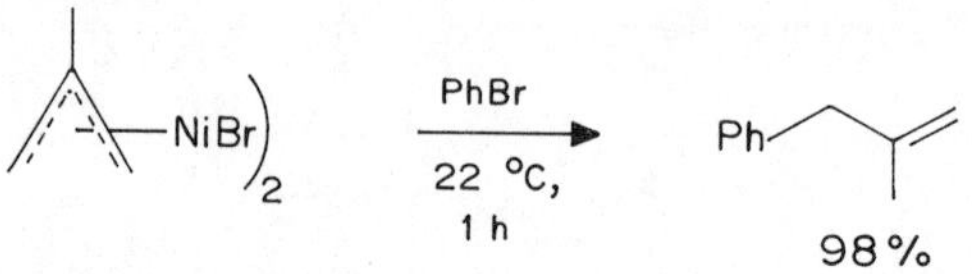

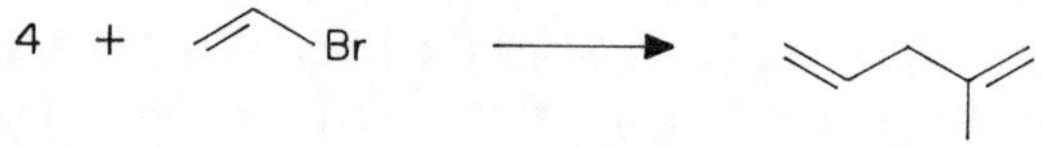

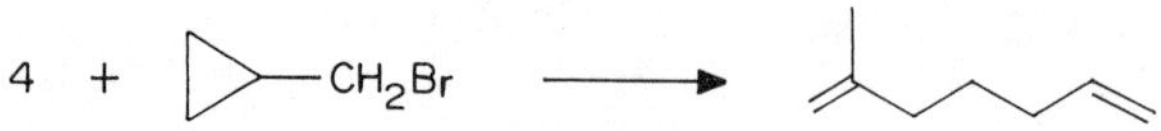

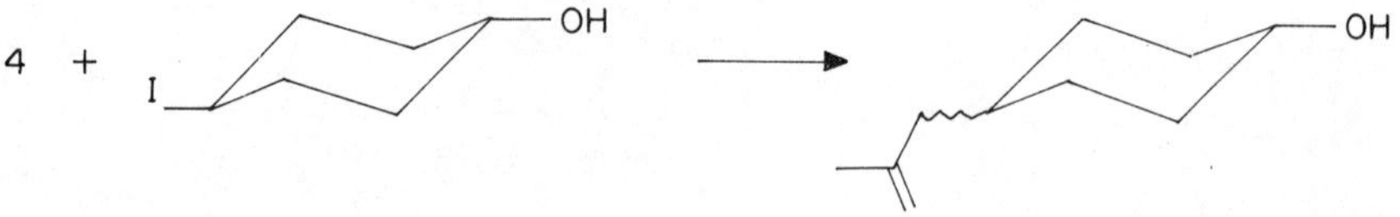

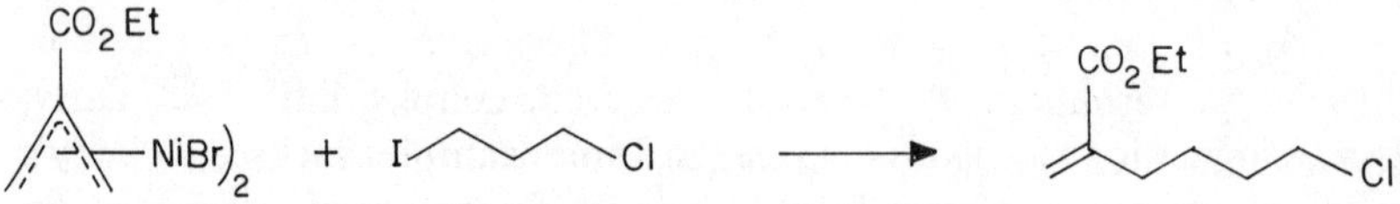

While the reagents are fairly selective for alkyl halides, it has been found that under more forcing conditions they will react with other functional groups, behaving rather like other organometallic species, e.g. Grignard reagents. Thus, in polar solvents at 50–60 °C some ketones and aldehydes will react.

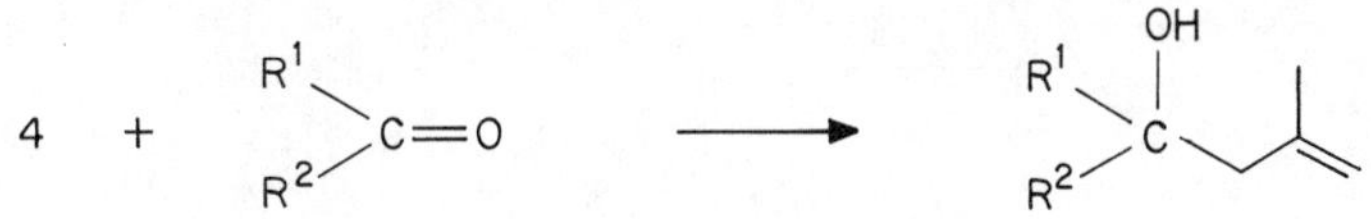

Esters and deactivated ketones are not reactive under these conditions, but α,β-unsaturated esters undergo conjugate addition. In contrast, α,β-unsaturated aldehydes undergo 1,2-addition:

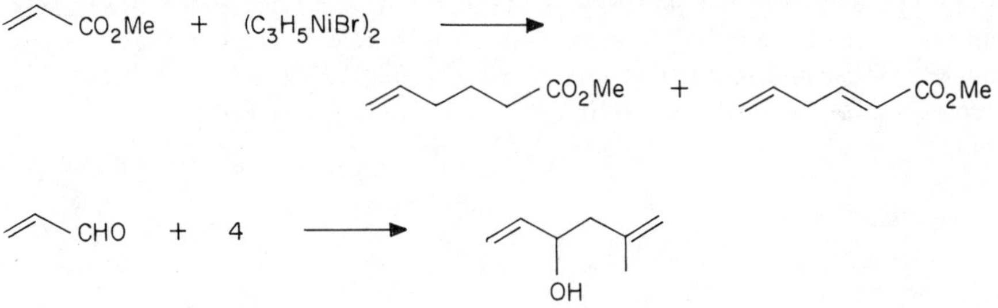

Epoxides are opened, but in the manner expected for acid-catalyzed reaction, rather than direct nucleophilic attack, suggesting participation by the nickel:

The early observations by Webb and Borcherdt[5] on the nickel carbonyl-catalyzed coupling of allylic halides has had significant consequences for organic synthesis:

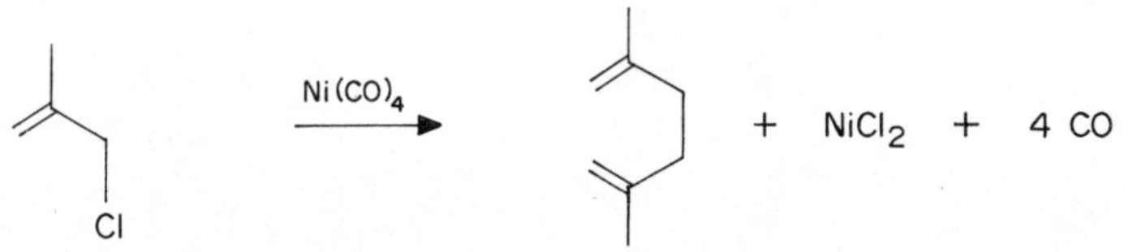

The reaction undoubtedly proceeds via the formation of allylnickel bromide dimers, followed by their disproportionation to bis-π-allylnickel complexes, and coupling of the allyl ligands, as discussed above. The usefulness of this procedure lies in the fact that terminal bis-allyl bromides can be coupled intramolecularly, leading to the formation of various-sized rings. Some examples, including Corey's synthesis of humulene, are shown below.[6] For the formation of large rings, an internal double bond which will coordinate to the nickel and force the acyclic precursor to adopt a favorable arrangement appears to be advantageous in some cases (see examples below).

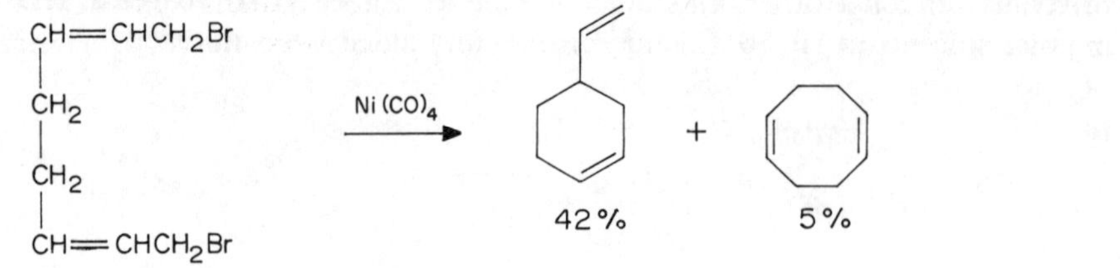
CH=CHCH2Br
CH2
CH2
CH=CHCH2Br
Ni(CO)4
+
42%
5%

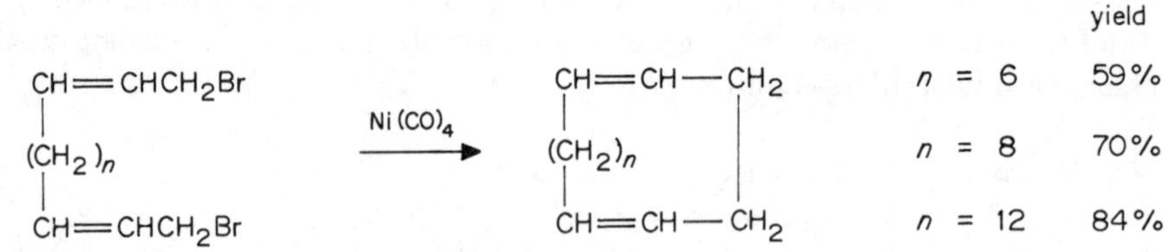
CH=CHCH2Br
(CH2)n
CH=CHCH2Br
Ni(CO)4
CH=CH—CH2
(CH2)n
CH=CH—CH2
yield
n = 6 59%
n = 8 70%
n = 12 84%

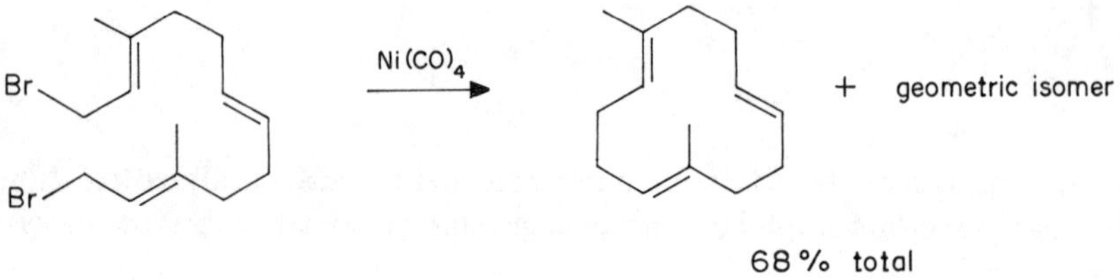
Br
Br
Ni(CO)4
+ geometric isomer
68% total

Humulene Synthesis:

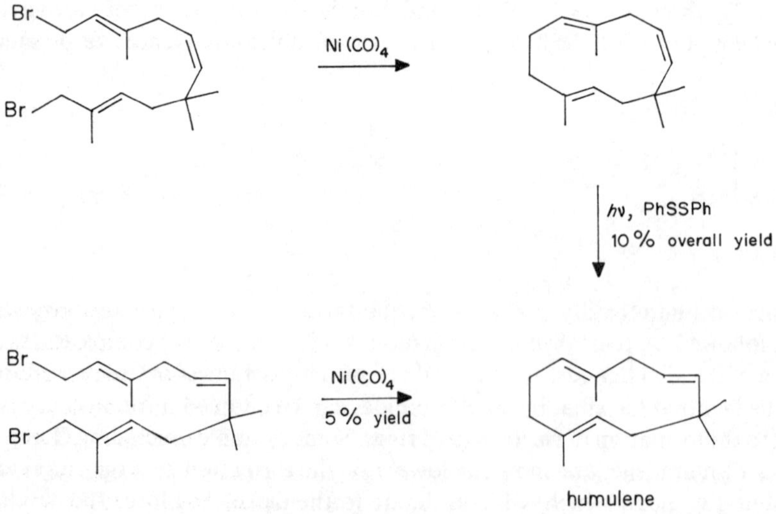
Br
Br
Ni(CO)4
hν, PhSSPh
10% overall yield
Br
Br
Ni(CO)4
5% yield
humulene

It can be seen from the above discussion that under normal circumstances allylnickel halide dimers act as a source of nucleophilic allyl group. However, under certain conditions π-allylnickel complexes behave as electrophiles, reacting with amines to give allylamines. For example, the nickel-catalyzed amination of butadiene to give butenyl- and octadienylamines has been suggested[7] to proceed via the bis-π-allylnickel complexes 5 and 6.

(5)　　　　　　　　　　　　(6)

Protonation of these complexes prior to attack by the amine is suggested by its lack of reactivity in the absence of acid.[8]

Treatment of cyclopentadiene with $[(EtO)_3P]_4Ni$ and acid gives the cationic π-allyl complex 7, which undergoes reaction with morpholine as shown. That the M-*endo* deuteriated complex gives the amine product in which the morpholine group is *trans* to deuterium suggests that nucleophile attack proceeds direct on the allyl ligand rather than by initial addition to the metal.[9]

(7)

The behavior of these cationic π-allylnickel complexes has not been exploited for a range of synthetic applications, in contrast to the similar complexes of palladium (see the next section).

Allylpalladium Complexes

Organopalladium chemistry and its application to organic synthesis have undergone very rapid growth during the past 20 years, and there are many reviews of this important area, the most recent being those of Trost.[3] In particular, much of the early chemistry, which was performed using π-allylpalladium complexes as (expensive!) stoichiometric reagents, has been developed to such a level that many of the useful bond-forming reactions can be performed with catalytic amounts of palladium.

Preparation

1. Reaction of dienes, allenes or vinylcyclopropanes with palladium chloride in the presence of various nucleophiles leads to π-allylpalladium chloride dimeric

complexes, which are usually crystalline, air-stable materials that can be isolated and handled with ease. Some examples are summarized below. The reaction is equivalent to nucleophilic addition to an olefin–Pd complex, but the presence of the extra conjugated or cumulated double bond leads to the formation of the allyl complex:

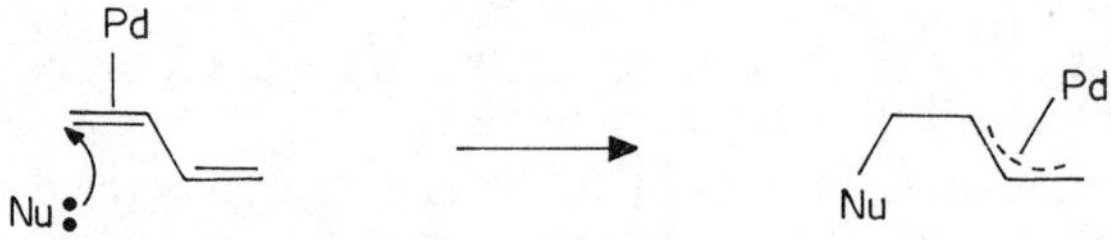

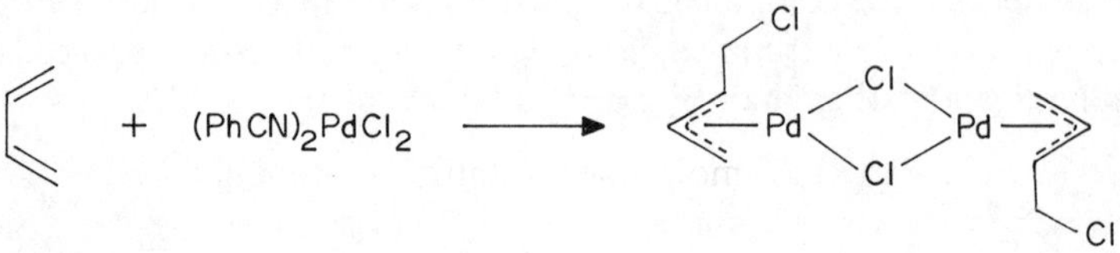

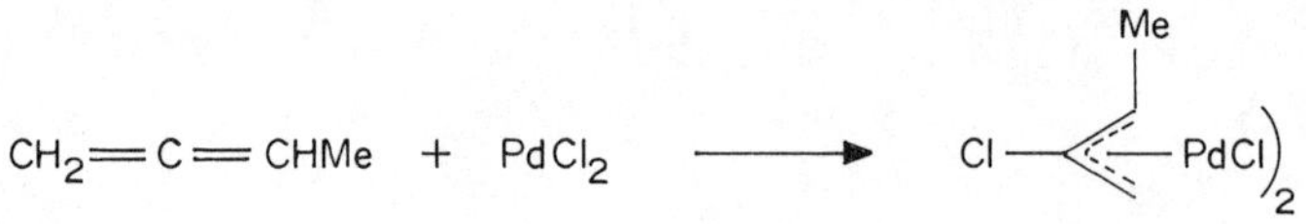

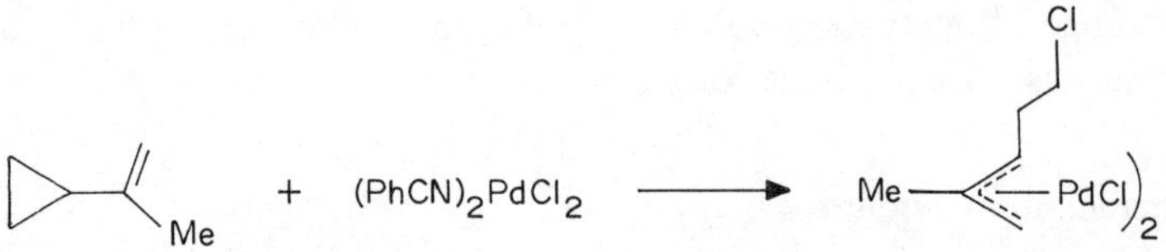

When the reaction is conducted in aqueous acetic acid, using an excess of diene which acts as a reducing agent, a π-allyl complex is formed without nucleophile incorporation, e.g.

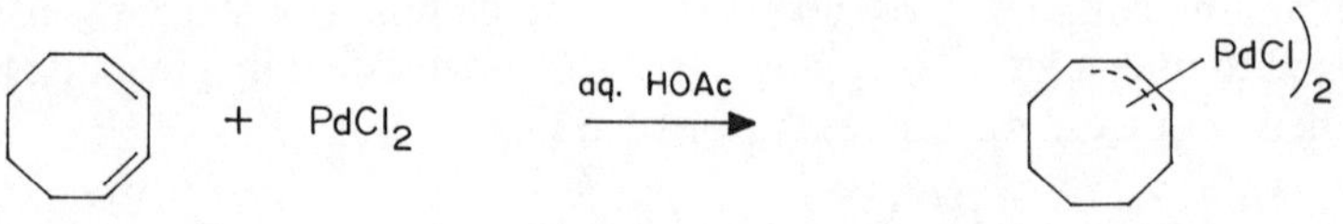

81 %

If there is a nucleophile present in the diene, e.g. a remote double bond as in myrcene, then cyclization may occur:

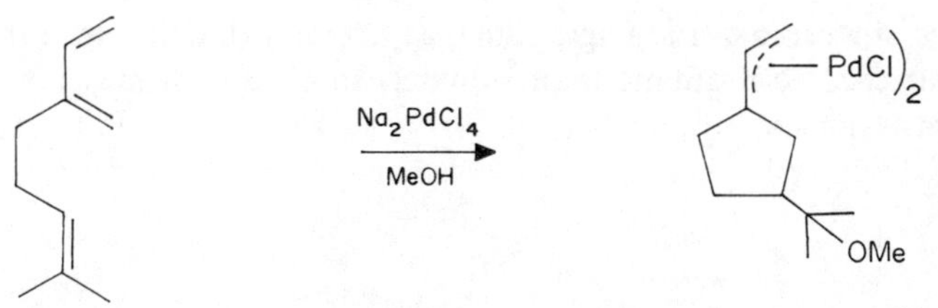

2. A second method of forming π-allyl–PdCl dimers is by reaction of olefins, having allylic hydrogen, with palladium chloride. Mechanistically, this reaction is simple:

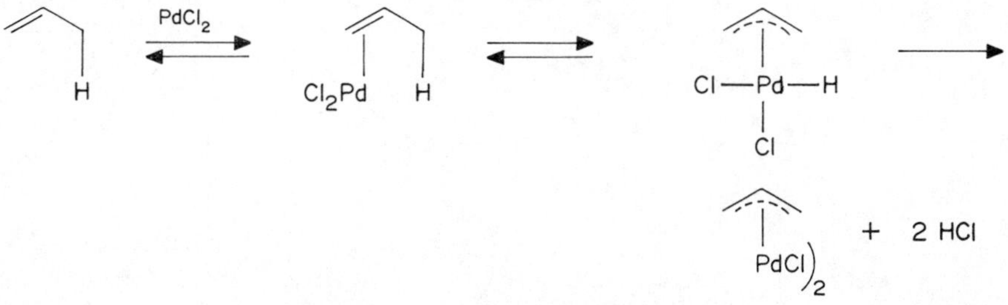

Thus there occurs an intramolecular oxidative addition of the allylic C—H bond, followed by reductive elimination of HCl. The conversion may be achieved using a variety of palladium salts, e.g. $PdCl_2$, Na_2PdCl_4 or $(PhCN)_2 PdCl_2$, in alcohol or aqueous acetic acid as solvent. Some examples are as follows:

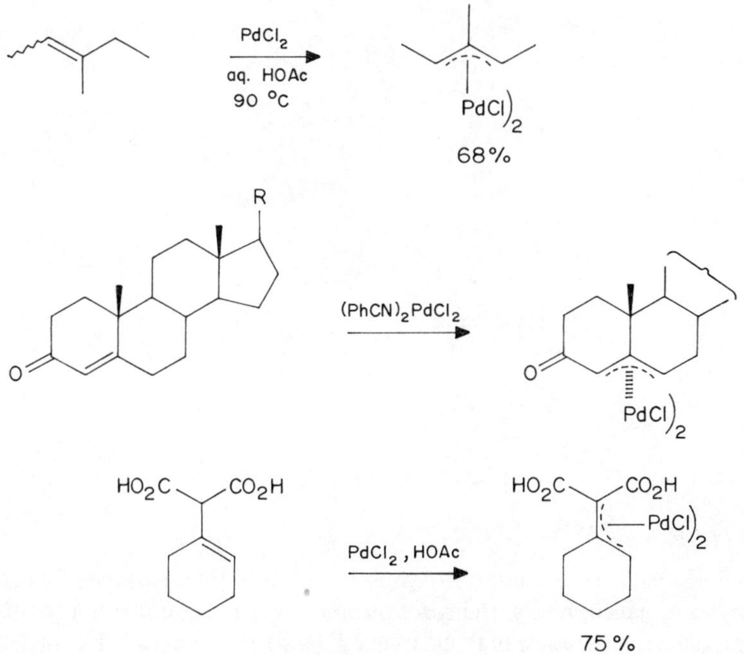

Inclusion of a weak oxidizing agent such as Copper (II) chloride in the reaction mixture can lead to dramatic improvements in yield, and may also affect the regiochemistry, e.g.

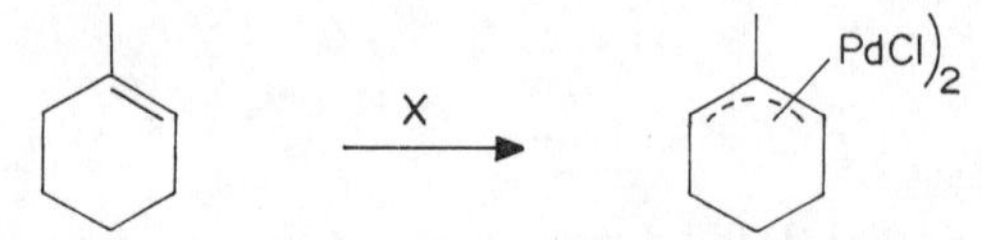

X = PdCl$_2$, HOAc, H$_2$O : yield 2.5%

X = NaCl, PdCl$_2$, CuCl$_2$, NaOAc, HOAc: yield 86%

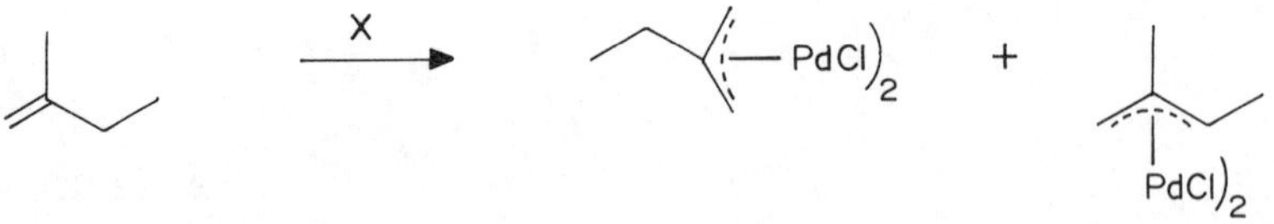

X = NaCl, PdCl$_2$, NaOAc, HOAc : 29 : 71 (49%)

X = as above + CuCl$_2$: 74 : 26 (89%)

A large number of complexes have been prepared and studied using this procedure. Where there is a choice, complex formation appears to occur at a non-conjugated double bond, e.g.

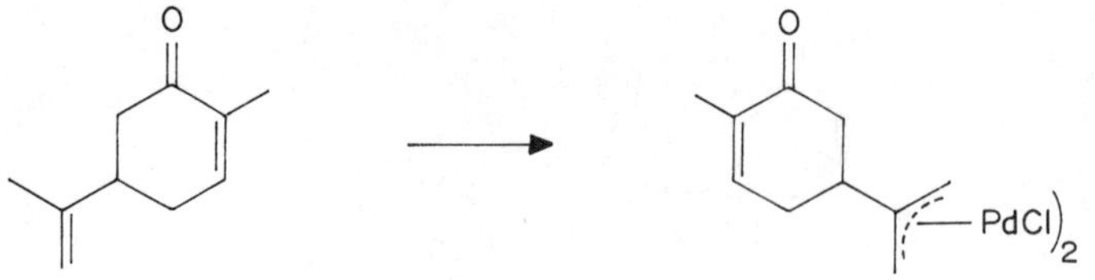

52%

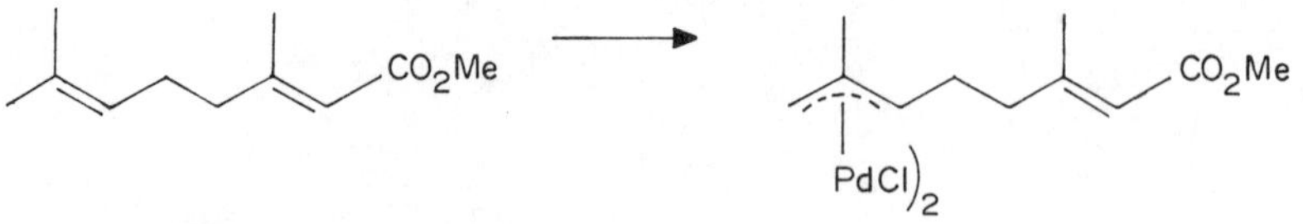

68%

Reactions of π-allyl–PdCl dimers

The most usual reaction is one of nucleophile addition. However, in order for this to occur satisfactorily, the reaction must be carried out in a coordinating solvent, or in the presence of a phosphine ligand. This probably has the effect of splitting off the chloride ligands to produce a cationic allylpalladium complex, which is then reactive towards nucleophiles:

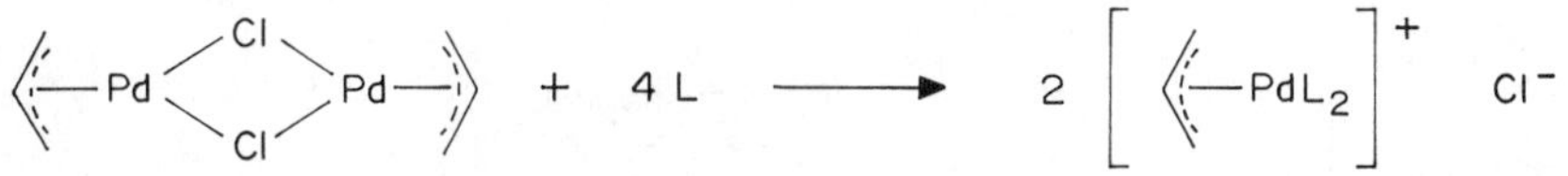

$$\left(\begin{array}{c}\text{—Pd} \end{array}\begin{array}{c}\text{Cl} \\ \text{Cl}\end{array}\text{Pd—}\right) + 4\,L \longrightarrow 2\left[\begin{array}{c}\text{—PdL}_2\end{array}\right]^{+} \text{Cl}^{-}$$

Reaction occurs readily with a variety of nucleophiles, including amines, enamines and stabilized enolates, and usually occurs by attack *trans* to the metal. While silyl enol ethers (ketone enolate equivalents) and allylsilanes do not react satisfactorily, it has recently been found that stannyl enol ethers and allyltin reagents are readily allylated.[10] A range of examples are given below.

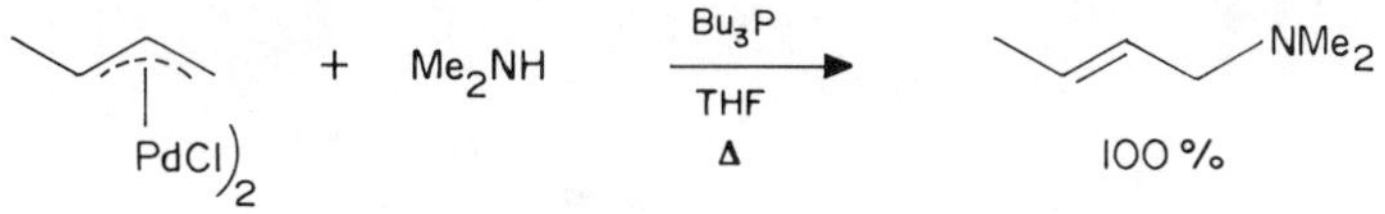

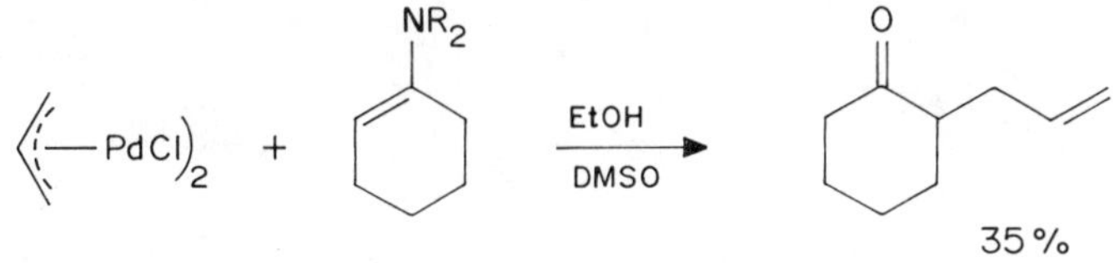

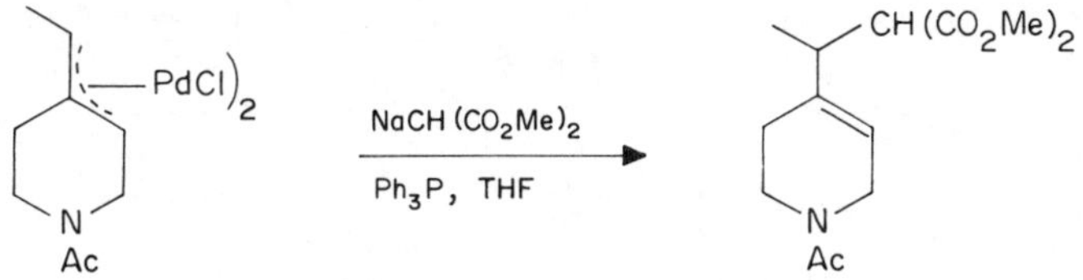

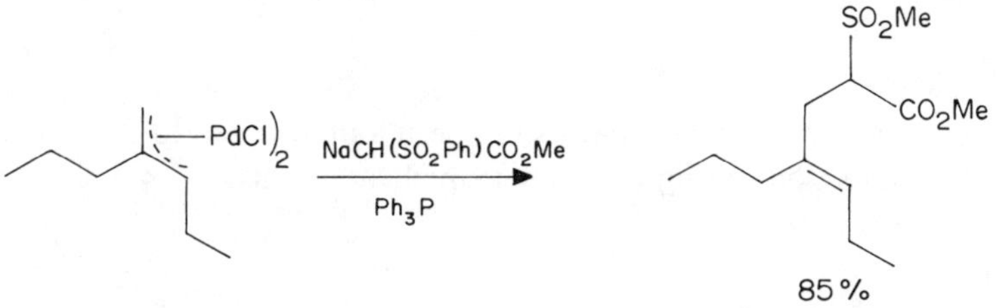

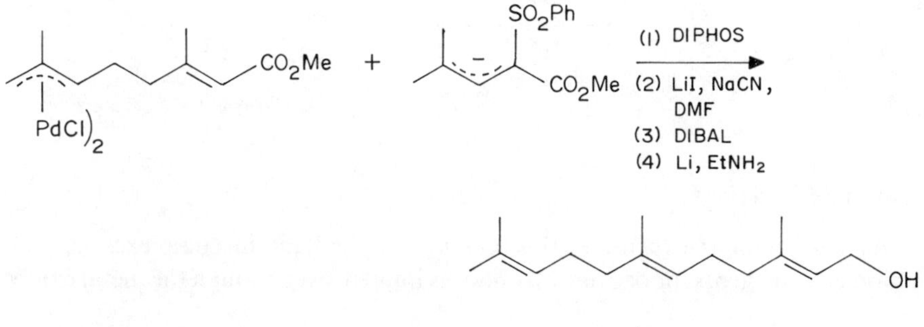

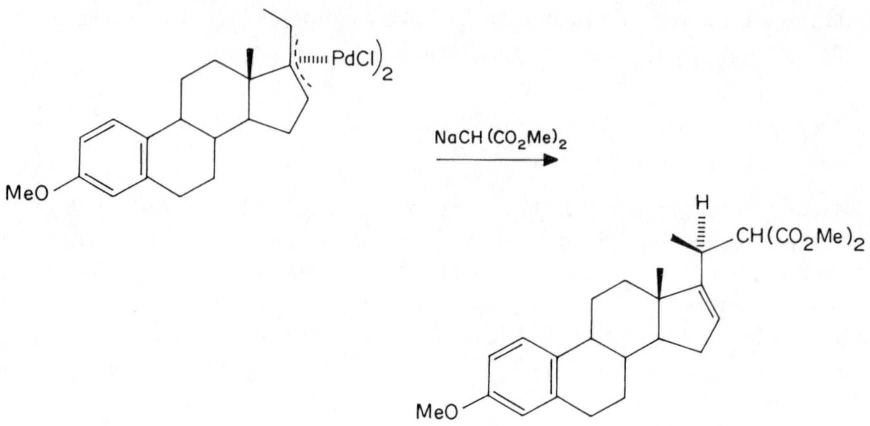

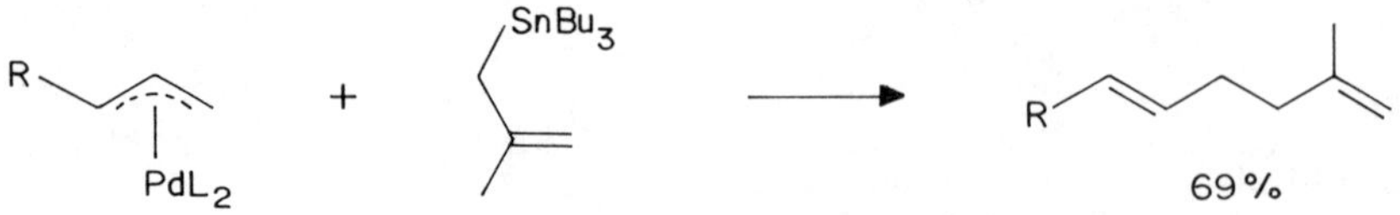

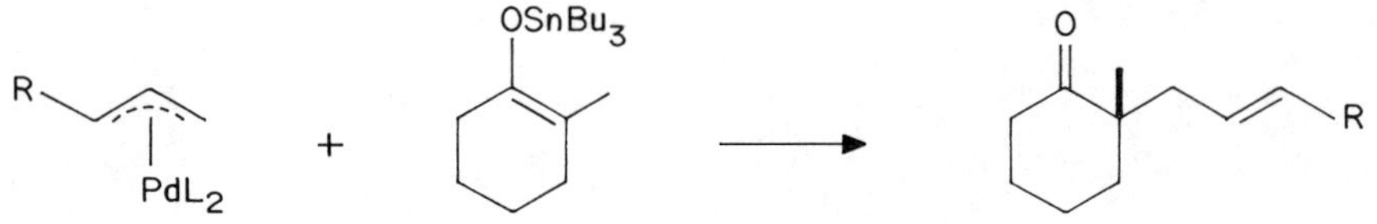

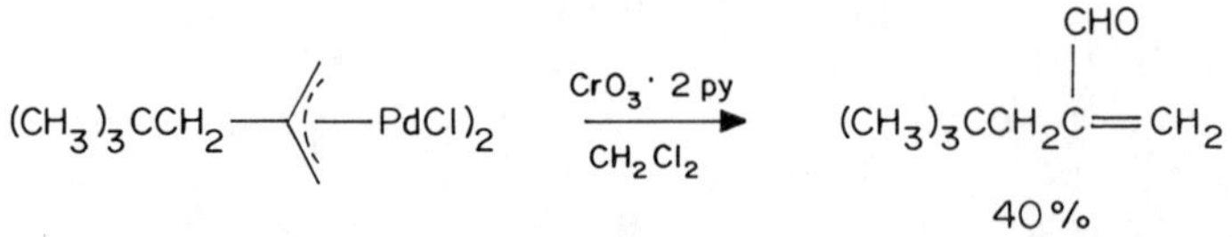

The use of optically active phosphines has been found to lead to some degree of asymmetric induction during these nucleophile additions:

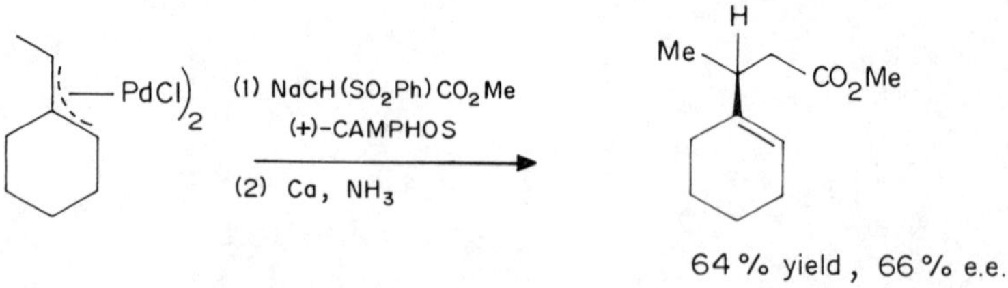

64 % yield , 66 % e.e.

Catalytic Reactions

As palladium is expensive, the use of π-allylpalladium complexes as stoichiometric reagents for organic synthesis is limited, even though the metal can be recovered. Consequently, the development of methods for generating the π-allyl complex using catalytic amounts of palladium was in some ways inevitable. Some of the first examples of catalytic reactions involving nucleophilic attack on π-allylpalladium complexes were the oligomerization of dienes in the presence of nucleophiles. Some examples are given in the following equations:

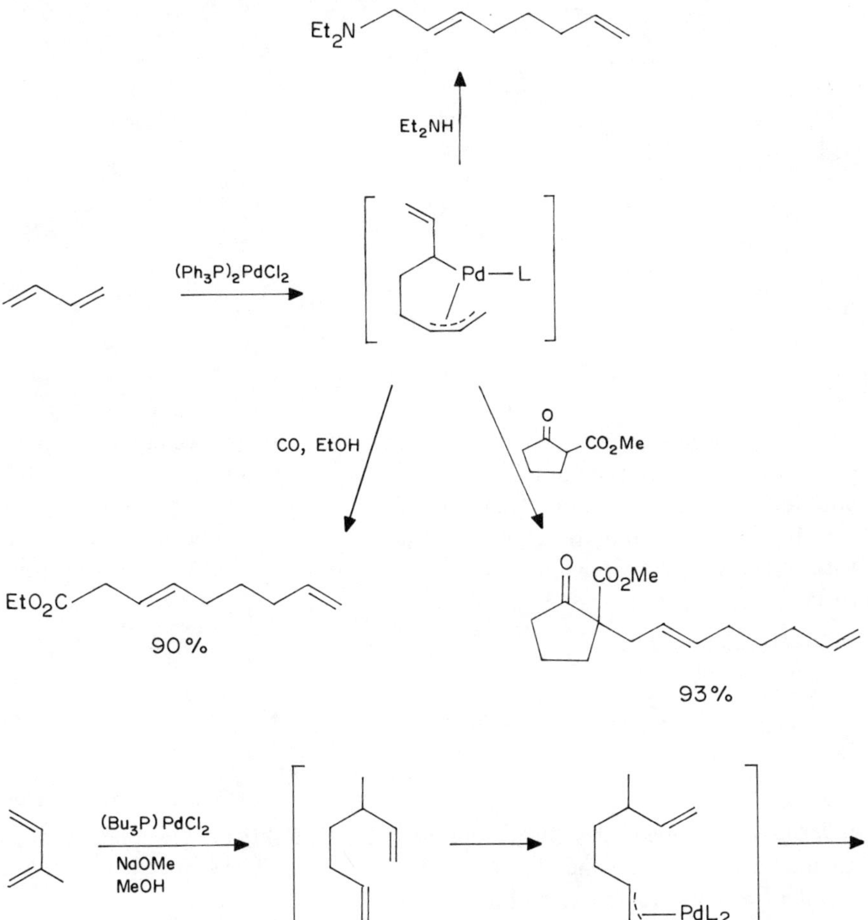

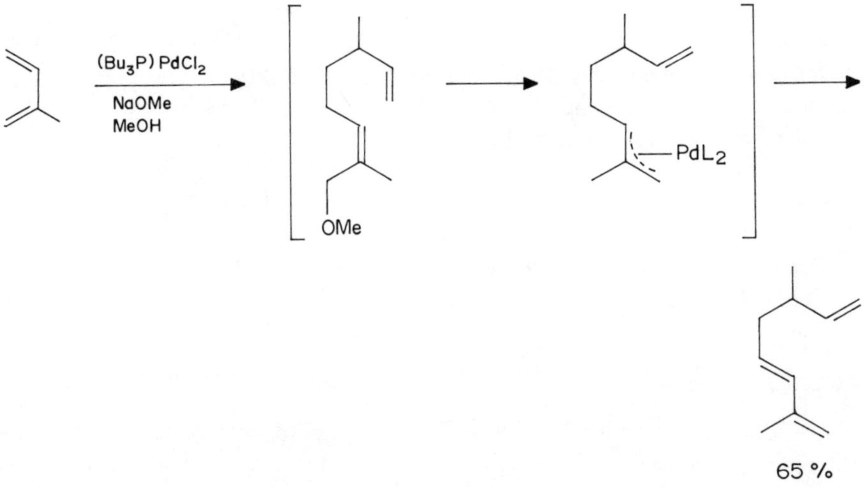

Oligomerization of the diene may be suppressed by using a stronger phosphine ligand, e.g. diphos. Under these conditions the nucleophilic reagent [e.g. $CH_2(CO_2Me_2)$] acts as a source of proton during the conversion of the diene to the allylpalladium complex, and the resultant nucleophilic counter anion enters into reaction with the complex:

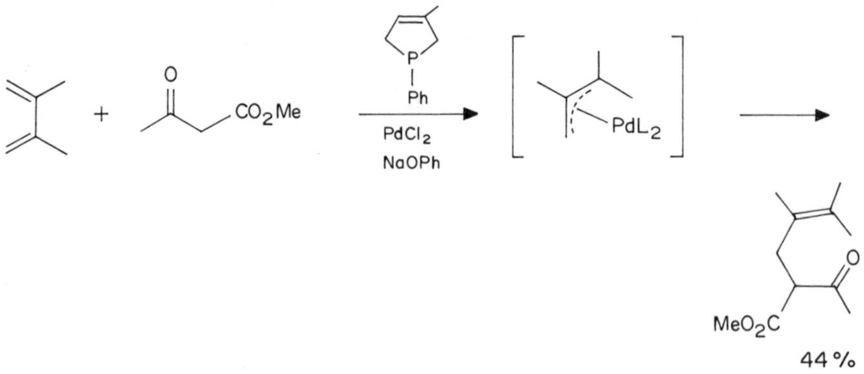

44%

This behavior has been extended to the use of allylic chlorides, ethers, acetates, alcohols, etc., in conjunction with a suitable palladium catalyst. Oxidative addition of the allylic acetate, etc., to the Pd(0) species occurs to give the complex. The most useful combination for these reactions appears to be an allylic acetate with tetrakis(triphenylphosphine)palladium, expulsion of OAc occurring *trans* to Pd.

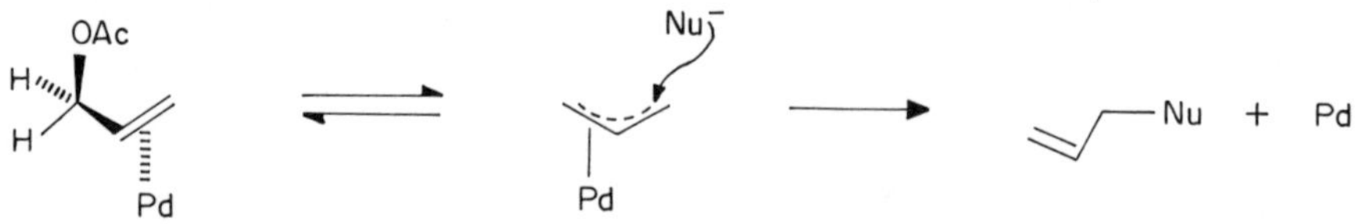

A few selected examples illustrating the synthetic utility are given below, but a detailed account is beyond the scope of this chapter. The reader is referred to excellent reviews by Trost[3] and a book by Tsuji.[11]

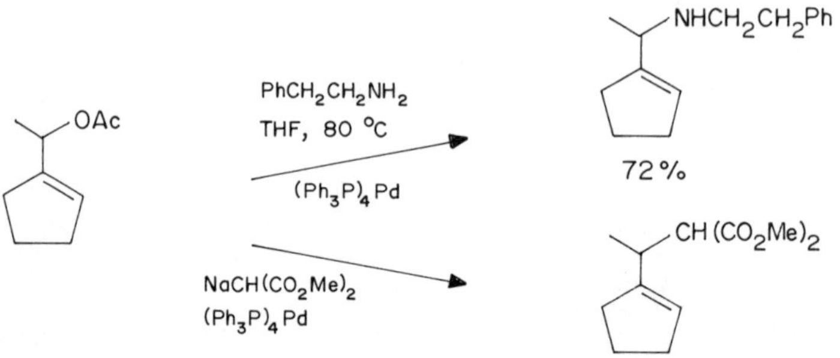

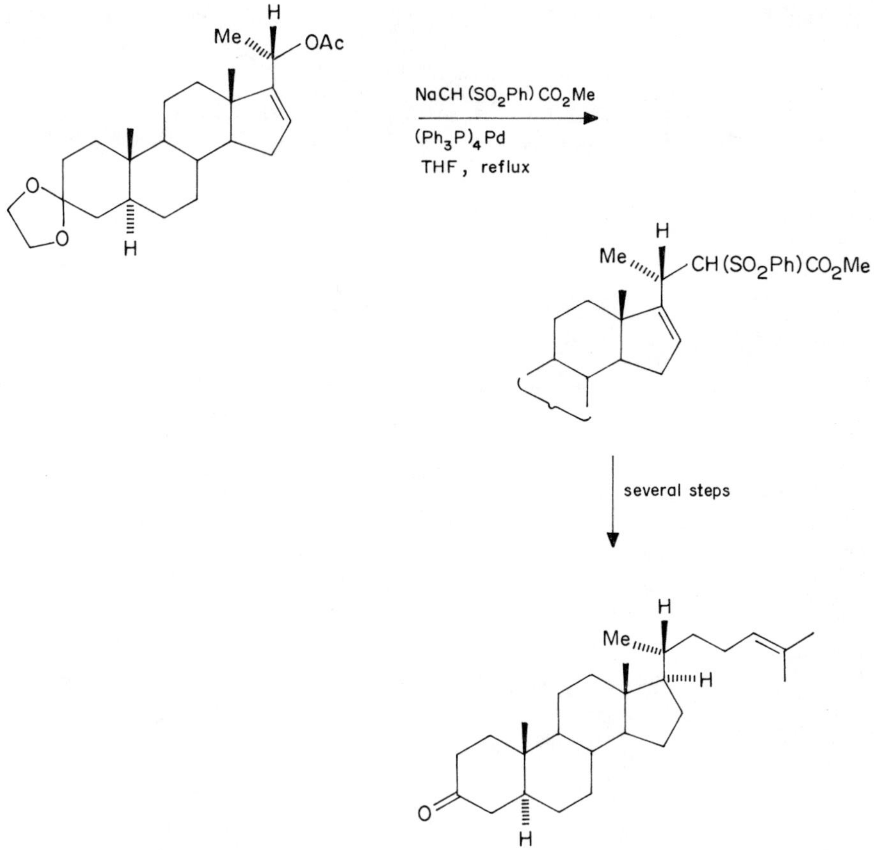
H
Me
OAc
NaCH (SO2Ph) CO2Me
(Ph3P)4 Pd
THF , reflux
O
O
H
H
Me
CH (SO2Ph) CO2Me
several steps
H
Me
H
O
H

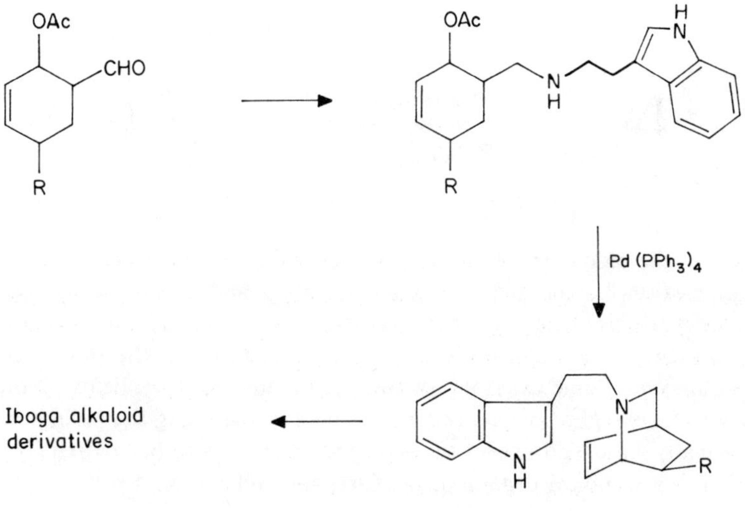
OAc
CHO
R
OAc
H
N
N
H
R
Pd (PPh3)4
Iboga alkaloid
derivatives
N
N
H
R

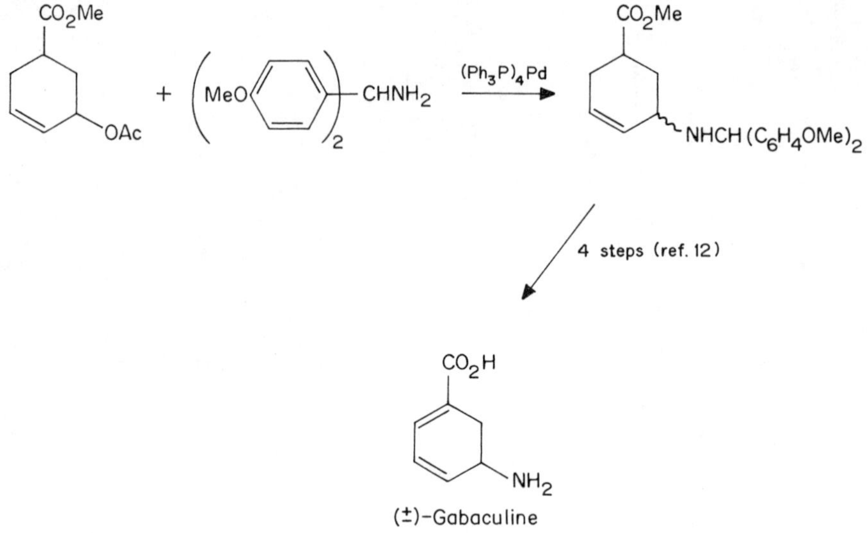

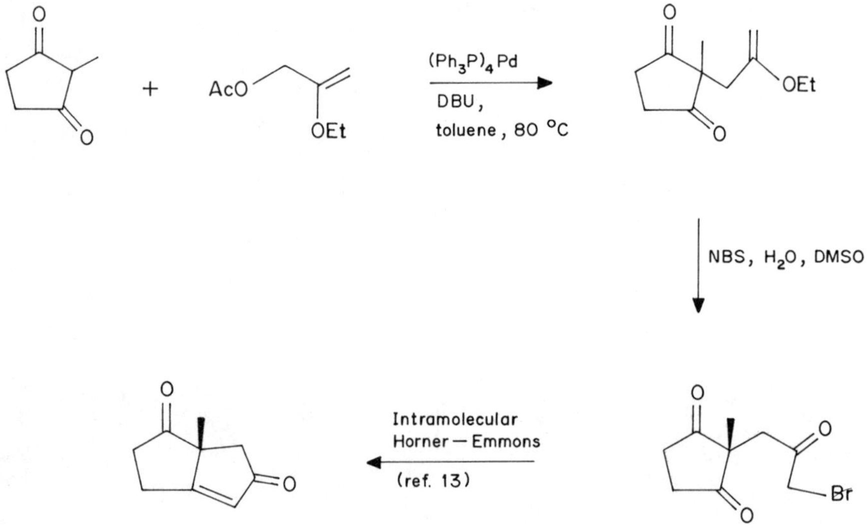

(±)-Gabaculine

Extension of the addition of carbon nucleophiles in these reactions to intramolecular examples has led to some interesting and useful cyclizations resulting in the synthesis of medium- and large-ring compounds. This is especially interesting in view of the occurrence of large-ring lactones in the important macrolide antibiotics. Some examples of these reactions are given below. Note the tendency to form eight- instead of six-membered rings using the palladium method. Normally we would expect the six-membered rings to be favored by a factor of 10^4 using standard organic procedures for cyclization.[14]

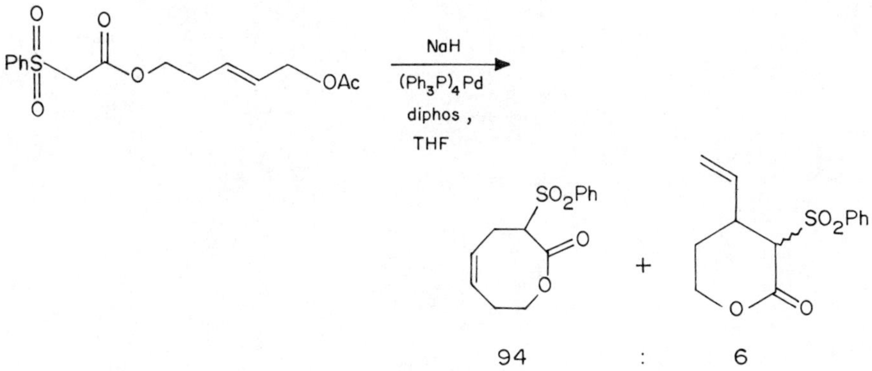

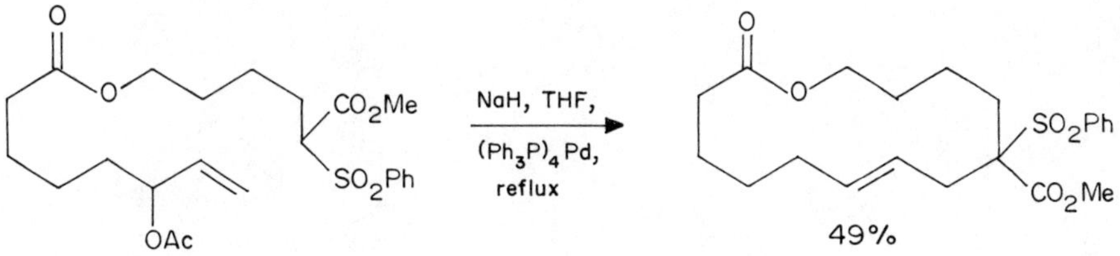

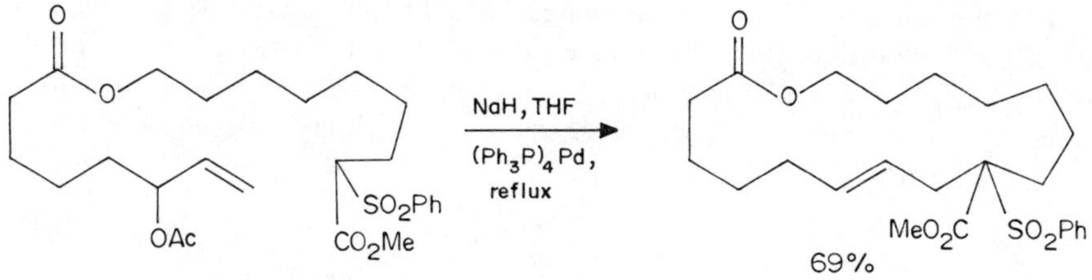

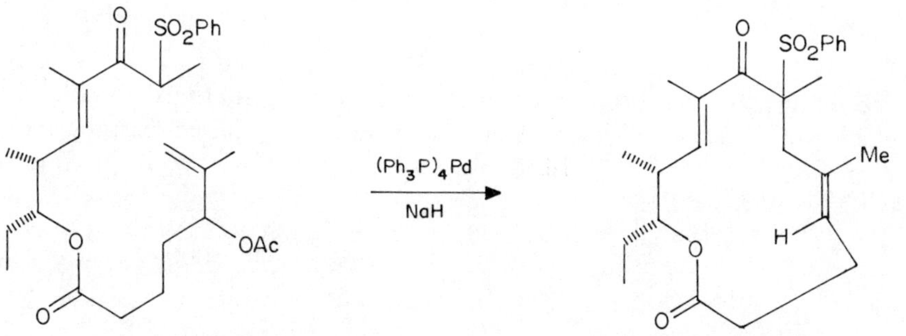

94 : 6

49%

69%

Model for erythronolide
synthesis

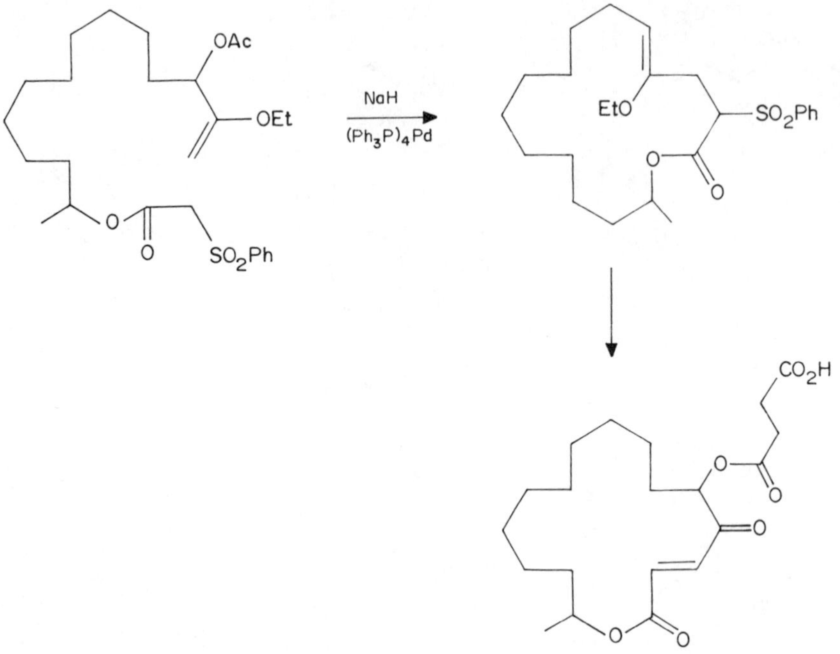

Antibiotic A 26771 B

Turning our attention to smaller rings, both Trost's and Tsuji's groups have described methodology for obtaining cyclopentane derivatives using cyclization of allylpalladium complexes. Normally the cyclization of a 4-haloketone or keto ester, by displacement of halide, proceeds on the carbonyl oxygen atom rather than the α-carbon atom, i.e.

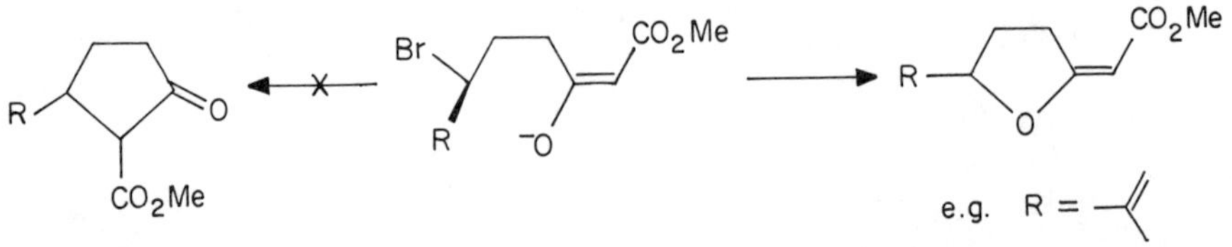

The tetrahydrofuran product when R = vinyl can be subjected to conditions which will induce a [3.3] sigmatropic rearrangement (Claisen rearrangement):

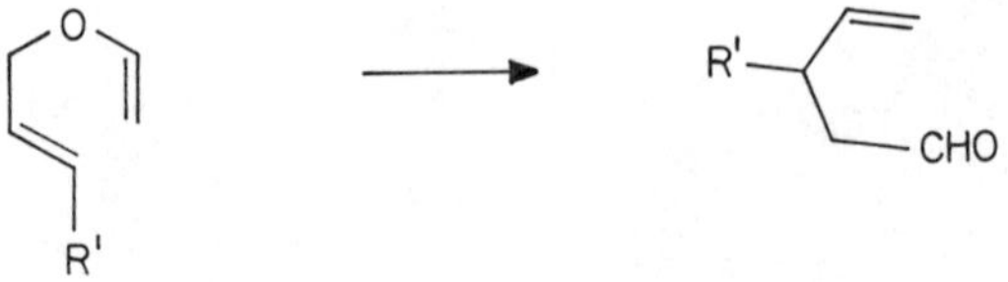

However, the observed product is a cycloheptenone derivative, not a cyclopentanone:

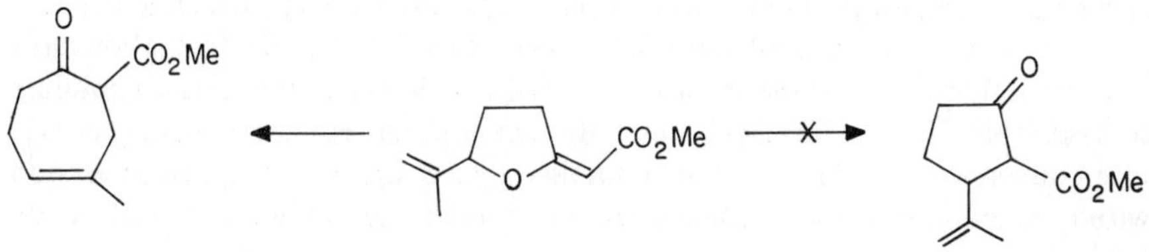

A reordering of this reactivity pattern is achieved by treatment of the oxygen-cyclized product with a palladium catalyst:

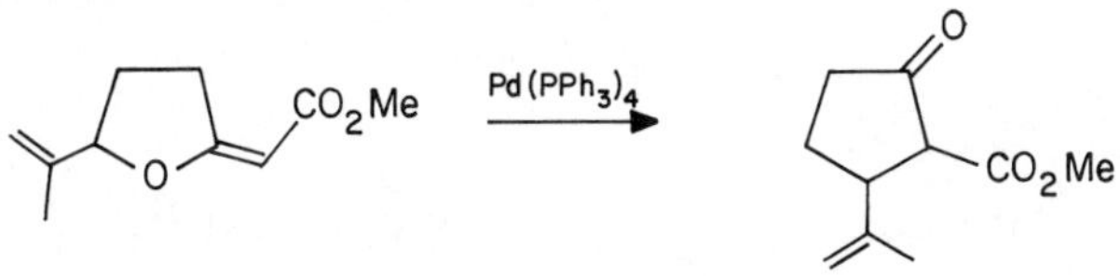

This may be used in an approach to prostaglandin analogues, shown below. Note the tendency of the acyclic precursor to cyclize on oxygen.

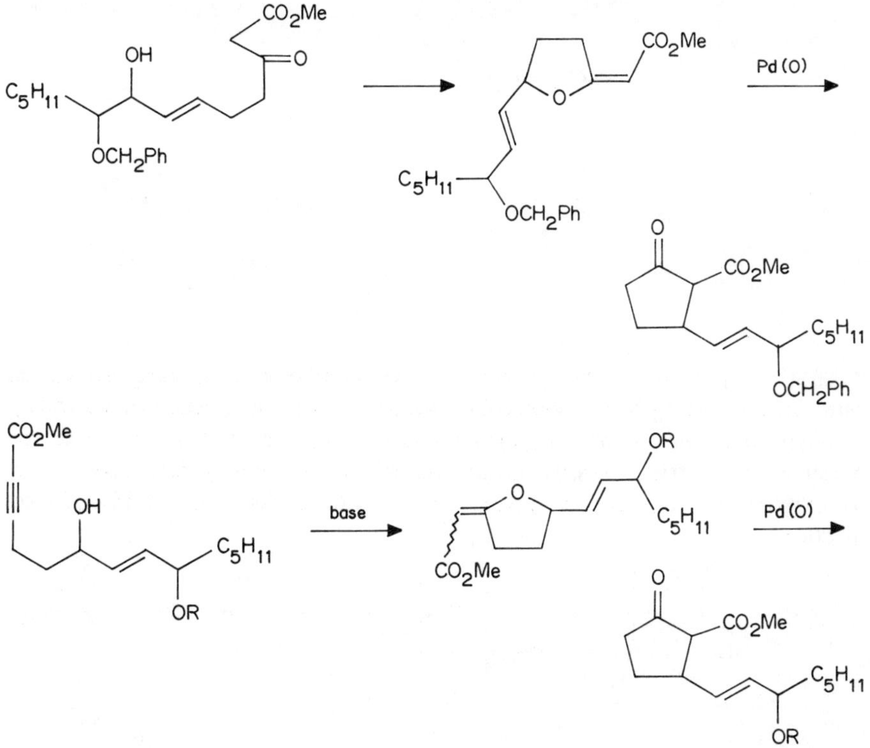

The reaction probably proceeds by palladium-promoted opening of the alkylidenevinyltetrahydrofuran to give an allylpalladium complex and enolate, which undergo intramolecular reaction. In principle, reaction could give either five- or seven-membered rings, and in fact this has been found to be dependent on the steric bulk of the ligands attached to the palladium catalyst, which affects the *syn–anti* interconversion of the π-allyl complex (allowing the formation of a *cis*-double bond to give the seven-membered ring which will not accommodate *trans*-double bonds). With a sterically demanding polymeric catalyst this interconversion is inhibited, but a smaller ligand allows fast interconversion with concomitant formation of a seven-membered ring, shown in the following example.

Another approach to the synthesis of five-membered rings using palladium catalysts has already been discussed in Chapter 2. This also depends on the ability of allylic acetates to form π-allylpalladium complexes, but in this case the presence of a trimethylsilyl group results in the ready formation of a trimethylenemethane-like complex, which then enters into cycloaddition reactions:

As can be seen from the above brief discussion, the greater potential of allylpalladium complexes for synthetic organic chemistry of a very diverse and complex nature has been demonstrated. No doubt this methodology will become commonplace, and the interested reader is referred to the cited reviews.

TITANIUM, ZIRCONIUM AND HAFNIUM

The first allyl complexes of titanium to be reported were the allylbiscyclopenta-dienyltitanium(III) derivatives, prepared by reaction of the appropriate metal chloride with allyl Grignard reagent.[15]

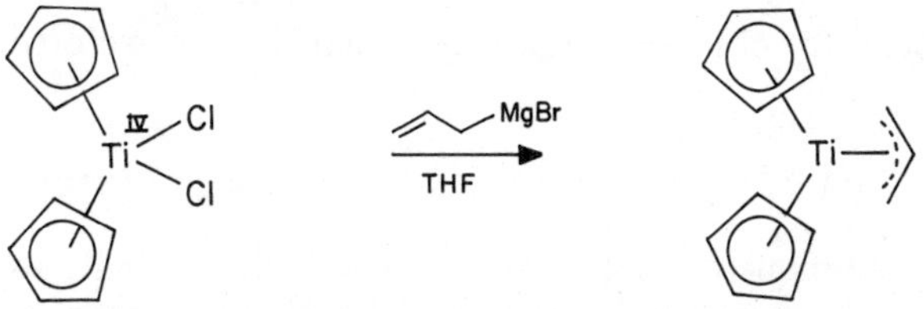

In this reaction, Ti(IV) is reduced to Ti(III) by the Grignard reagent. The complexes were formulated as η^3-allyl as opposed to σ-allyl derivatives, on the basis of infrared spectra, which showed no absorption in the region expected for a terminal $C=C$ bond. The titanium complex is a paramagnetic purple crystalline compound, m.p. 118 °C, which was reported to be remarkably stable thermally but susceptible to air oxidation. When air is passed through a solution in toluene/HCl immediate formation of Cp_2TiCl_2 occurs.

π-Allyltitanium complexes may also be prepared by reaction of a diene with Cp_2TiCl_2 in the presence of isopropylmagnesium bromide,[16] e.g.

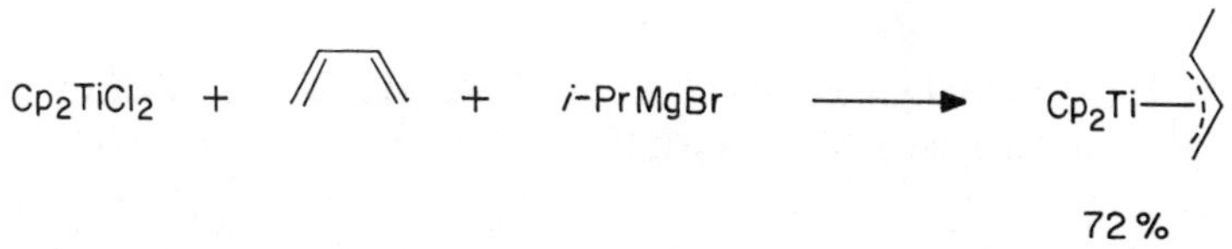

72%

Cyclooctatetraene dianion has also been used as a counter ligand for the formation of allyltitanium complexes. The green paramagnetic products are sensitive to air and water and decompose slowly at room temperature:[17]

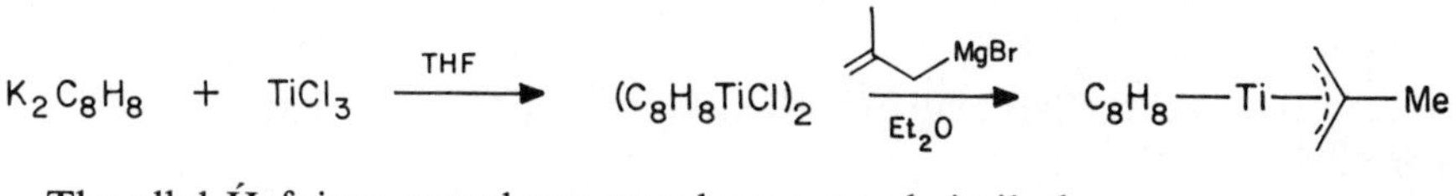

The allyl Hafnium complexes may be prepared similarly:

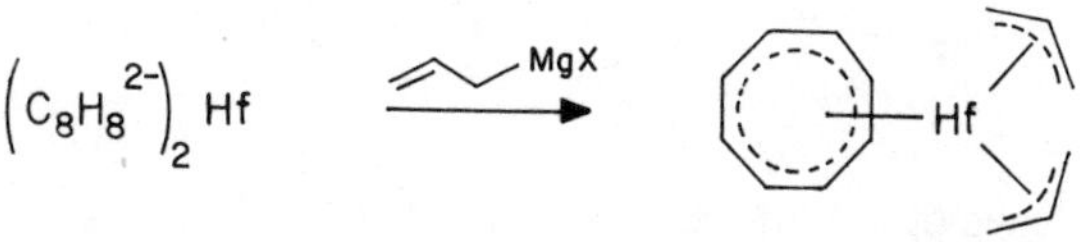

It is perhaps noteworthy that Hafnium maintains a $+4$ oxidation state in the allyl complex.[18]

Attempts to prepare complexes of type $(\eta^5\text{-Cp})\text{Ti(allyl)}_2$ have led only to the isolation of complexes $(\eta^5\text{-Cp})\text{Ti(allyl)(diene)}$, e.g.[19]

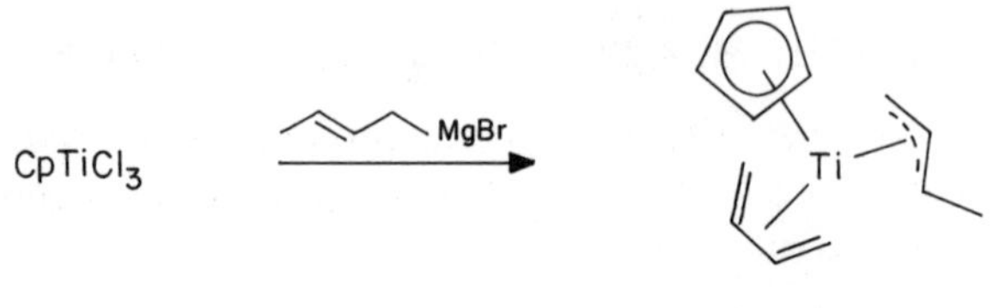

Further chemistry of these compounds remains to be explored.

VANADIUM, NIOBIUM AND TANTALUM

Vanadium is unique among transition metals in its ability to form a large number of monomeric neutral paramagnetic carbonyls which show little or no tendency to dimerize. Reaction of Cp_2VCl with allylmagnesium bromide gives a compound presumed to be $\text{Cp}_2\text{V}(\eta^3\text{-C}_3\text{H}_5)$ by analogy with the aforementioned titanium derivative, but which was too unstable to allow isolation.[20]

A later development in this group was the preparation of π-allylpentakis(trifluorophosphine)tantalum by treatment of the tetrakis(π-allyl) complex with PF_3 under pressure. The product can be isolated as a sensitive red crystalline compound:[21]

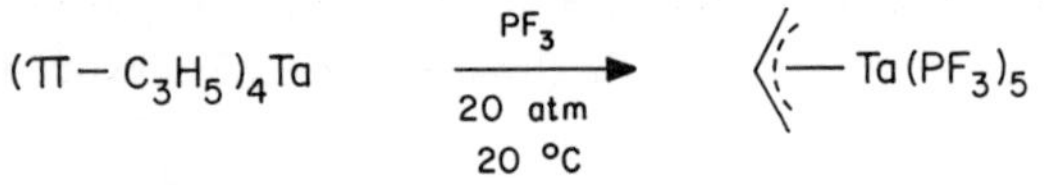

While the Cp_2V(allyl) complex is unstable, the corresponding niobium compounds may be isolated as crystalline solids.[22]

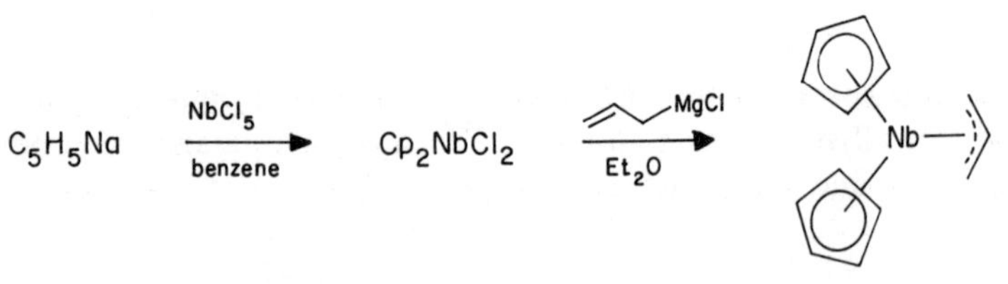

Treatment of the π-allyl complex with carbon disulfide results in the formation of a σ-allyl derivative:[23]

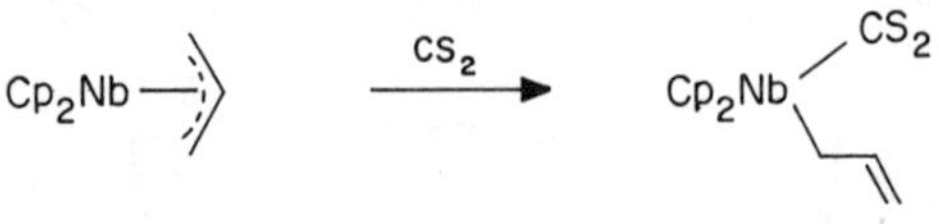

π-Allyl complexes of Nb and Ta have also been isolated from the reactions of

Cp$_2$NbH$_3$ and Cp$_2$TaH$_3$ with butadiene and allene:[24]

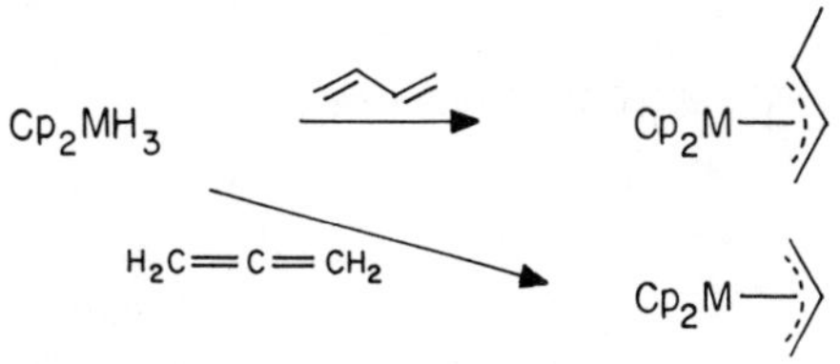

Reaction of sodium hexacarbonylvanadate and allyl chloride under irradiation yields $(\eta^3\text{-}C_3H_5)V(CO)_5$, and treatment of butadiene with HV(CO)$_5$ also gives the appropriate π-allyl complex:

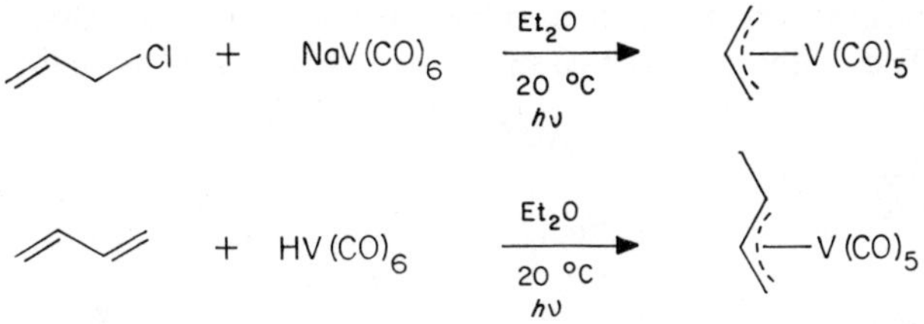

The parent allyl complex has a typical IR spectrum (2045, 1955, 1945, 1920 cm^{-1}) lacking a terminal C=C double bond, and the NMR spectrum is consistent with a π-allyl formulation.[25] Similar complexes containing the diars ligand have also been prepared:[26]

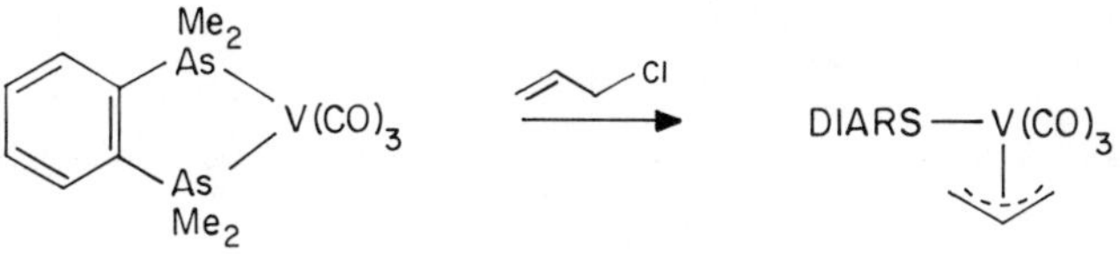

CHROMIUM, MOLYBDENUM AND TUNGSTEN

Work on these complexes prior to 1964 has been reviewed.[1] Some of the early work gave incorrect structure formulations owing to the unavailability of NMR spectra. Reaction of [(C$_5$H$_5$)Cr(CO$_3$)]$_2$ with butadiene under UV irradiation gives a product which was originally formulated as a diene complex, C$_5$H$_6$CrC$_4$H$_6$(CO)$_2$. Likewise, the product of reaction of (C$_5$H$_5$)$_2$ Cr with CO/H$_2$ was originally formulated as (C$_5$H$_6$)$_2$Cr(CO)$_2$, but revised to the π-allyl complex (C$_5$H$_5$)(C$_5$H$_7$)Cr(CO)$_2$. Thus, the complexes are prepared as follows:

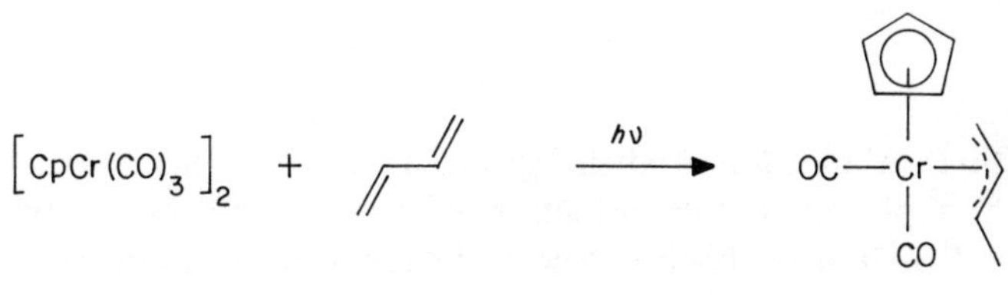

Cp_2Cr + H_2 + CO $\longrightarrow$

yellow crystals;
M.P. 74–75 °C

Reaction of $Na^+[C_5H_5Mo(CO)_3]^-$ with an excess of allyl chloride gives the π-allyl complex as a yellow oil in 40% yield. Irradiation of this compound causes expulsion of a CO ligand with concomitant formation of the π-allyl complex, obtained as lemon-yellow crystals, m.p. *ca.* 136 °C, (IR 1961, 1886, 1861 cm^{-1}) in 50% yield. The π-allyl tungsten complex may be prepared as yellow crystals in a similar manner.

Alternatively, reaction of $M(CO)_3(MeCN)_3$ with allyl halides gives the π-allyl complex:[27]

The formation of a π-allyl complex from a σ-allyl derivative in this way is very highly favored, as illustrated by the following example, where the driving force is sufficient to destroy the aromaticity of a benzene ring.[28]

In addition to the cyclopentadienyl group, a number of other auxiliary ligands have been employed in the formation of π-allylmolybdenum and -tungsten complexes. Reaction of dibenzenemolybdenum with allyl chloride and other allylic chlorides gives π-allyl dimeric complexes, obtained as purple crystalline solids.[29] The preparation and some reactions are shown below.

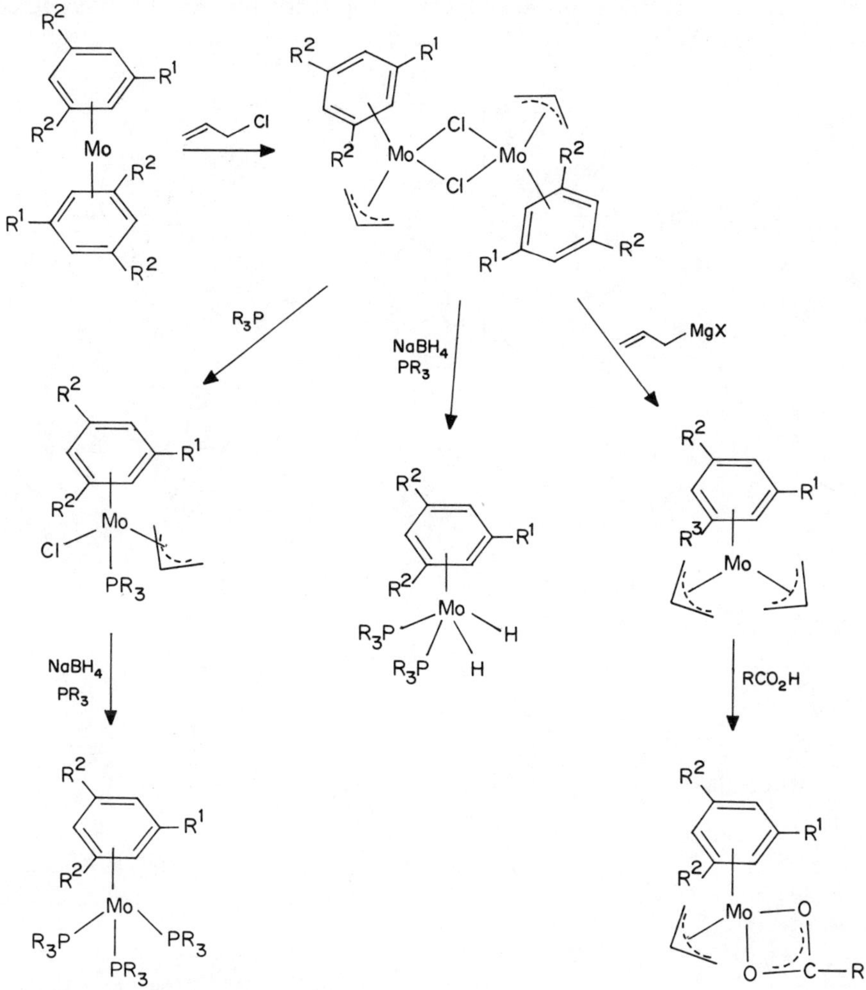

Conversion of the dimeric species to monomeric π-allyl cationic complexes, obtained as purple solids, has also been achieved by treatment with the mild Lewis acid EtAlCl$_2$, followed by addition of butadiene.[30]

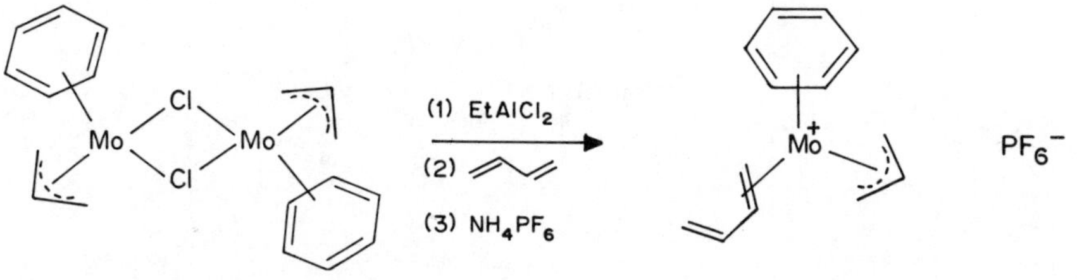

Bipyridyl has been found to be a useful ligand for the preparation of π-allyl complexes, including those of chromium which are not available by the above methods. The method is illustrated below,[31] together with uses of various other ancillary ligands.[32]

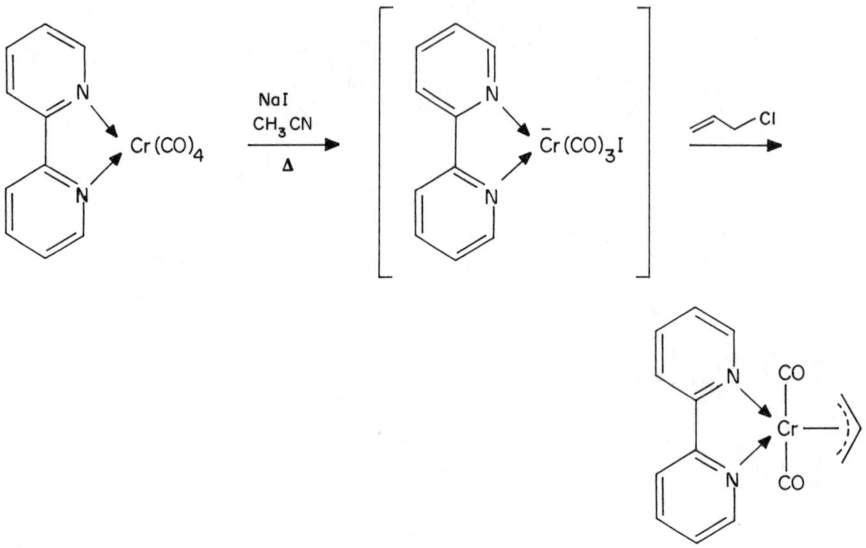

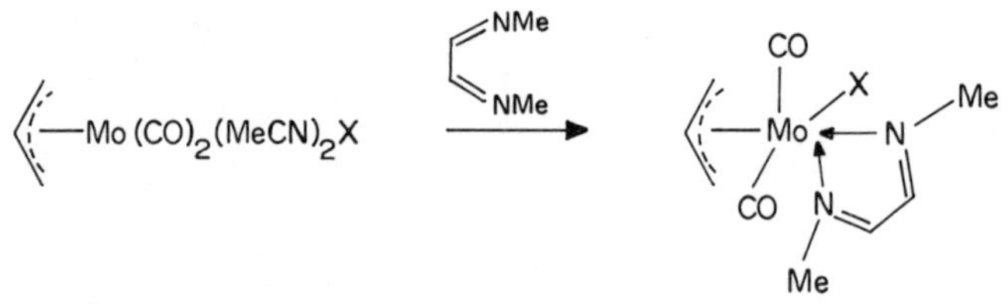

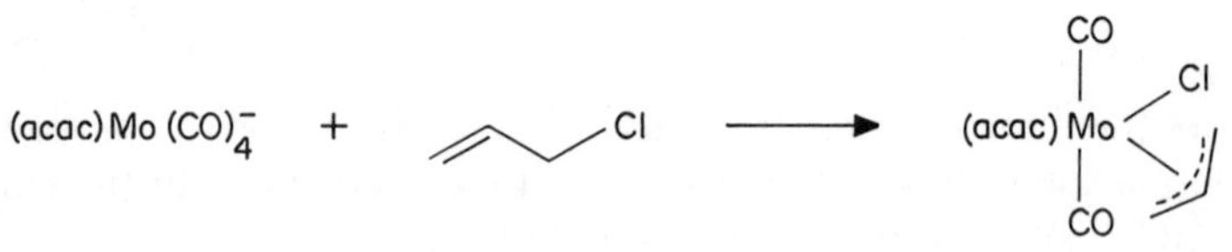

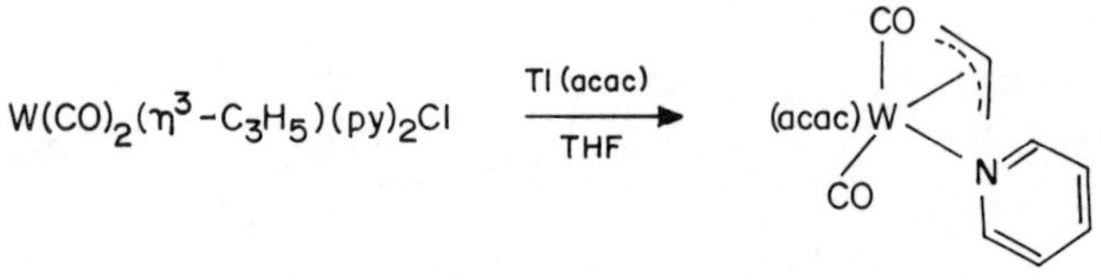

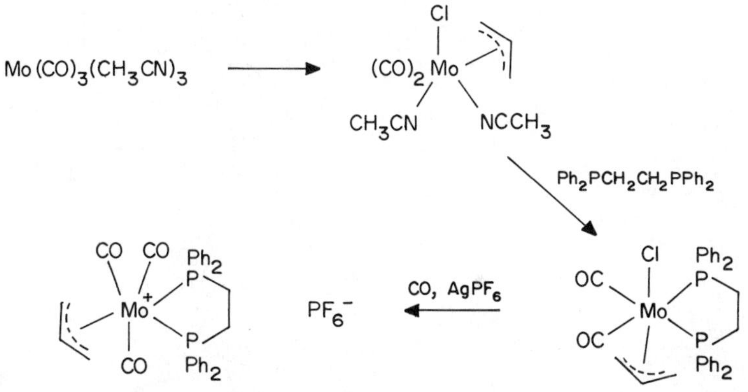

Cationic π-allyl complexes may be obtained using the nitrosyl complex in place of carbonyl complexes, as shown below. Reaction of these complexes with nucleophiles is reported to occur by addition at the allyl terminus to give an η^2-olefin complex.[33]

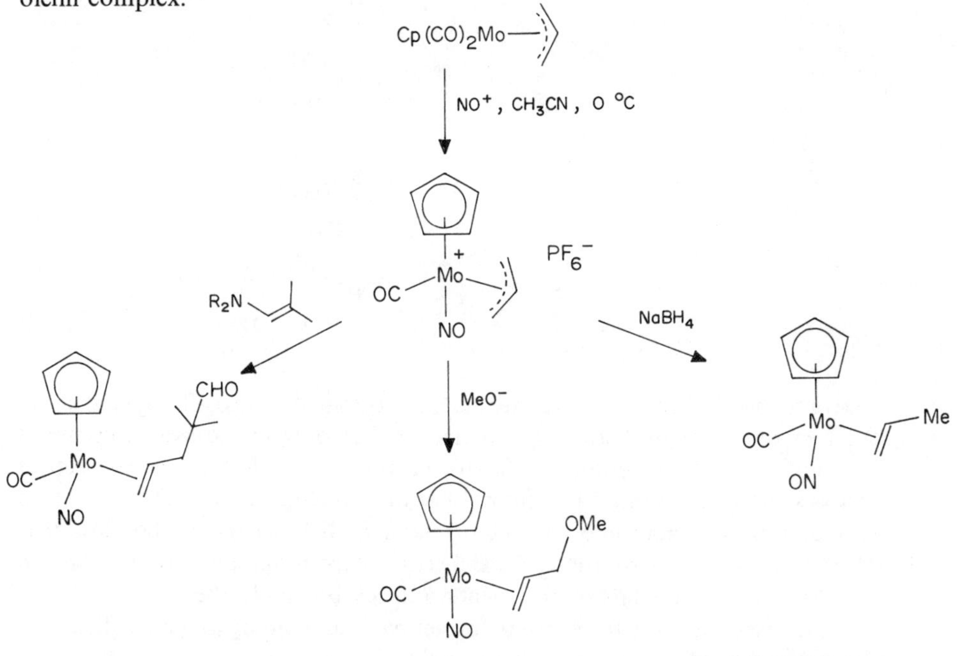

In fact, it has been observed that during these nucleophile additions there is a marked selectivity for one of the two non-equivalent ends of the allyl ligand,[34] invariably leading to production of a single diastereoisomer. The situation is complicated by the existence of the allyl complexes in two forms designated *exo* and *endo* (see below), but the addition may be summarized as follows: attack occurs *cis* to NO in the *exo* conformer and *trans* to NO in the *endo* conformer, giving an olefin in its most stable orientation. Reverse attacks give the olefin in a less stable orientation. The results are summarized below:

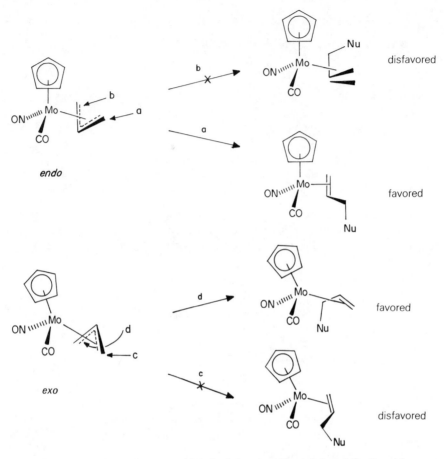

The two observed products are identical, by rotation of the olefin ligand, as are the two non-observed products, so that in fact only one diastereoisomer is obtained from the reaction. Hoffman's group have undertaken a theoretical analysis of this reaction.[35] It is found that the metal–allyl LUMO, which is an antibonding combination of metal d orbital and allyl ψ_2 (the non-bonding allyl MO), has uneven distribution of coefficients at the terminal carbons. The *exo* conformation has the larger coefficient on the carbon *cis* to the NO ligand, and this is marked, whereas the *endo* conformer has a marginally larger coefficient at the *trans* carbon:

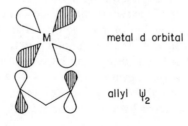

metal d orbital

allyl ψ_2

LUMO Coefficients for *exo/endo* **Conformers**

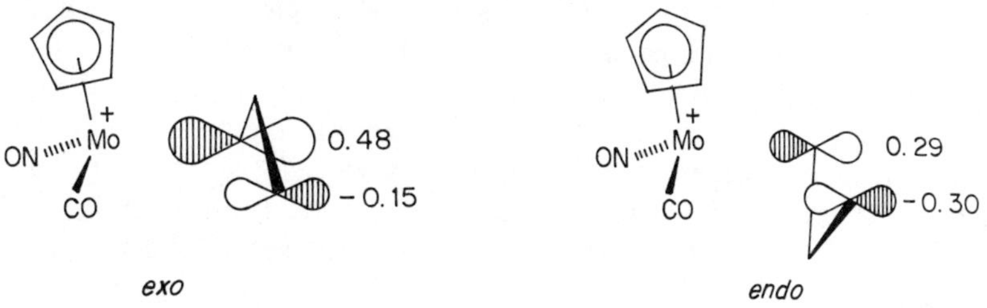

exo   *endo*

Calculations of overlap populations for the approach of H^- nucleophile to either terminus indicate a marginal, but significant, preference for attack *trans* to NO for the *endo* complex, and a very marked preference for attack *cis* to NO in the *exo* situation, reflecting the importance of the interaction between nucleophile HOMO and allyl–metal LUMO during the bond formation. This is also the important interaction proposed for a wide range of organic reactions and comparable, in this case, to a simple S_N2 substitution.[35]

The interaction also indicates that nucleophilic attack occurs from the side of the allyl ligand opposite to the metal, since this gives a favored bonding interaction. Approach of the nucleophile *cis* to the metal leads to a net non-bonding or anti-bonding interaction, as shown below.

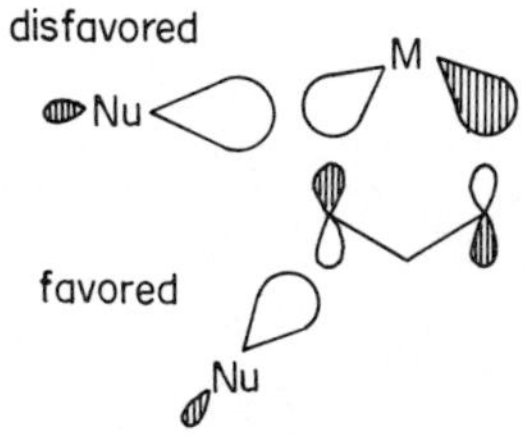

It seems likely that this is the main reason why nucleophile addition to a range of 18e π-allyl complexes occurs *trans* to the metal, and that the steric bulk of the metal grouping is of secondary importance. On the other hand, in 16e complexes the LUMO is probably almost pure metal orbital, and attack occurs by prior addition to the metal, followed by transfer to the ligand to give an overall result of addition *cis* to the metal.

In the case of the present complexes, it also turns out that there is an uneven distribution of charge on the terminal carbon atoms of the allyl ligand. In the *exo* complex the carbon *cis* to NO has the greater positive charge, whereas in the *endo* complex the higher positive charge is on the *trans* carbon. That the reaction is not under *overall* charge control is evidenced by the observation that in most cationic allyl complexes the *central* allyl carbon atom bears the highest positive charge (see Fe complexes later). If charge control were the major factor, we should expect attack at the central C atom:

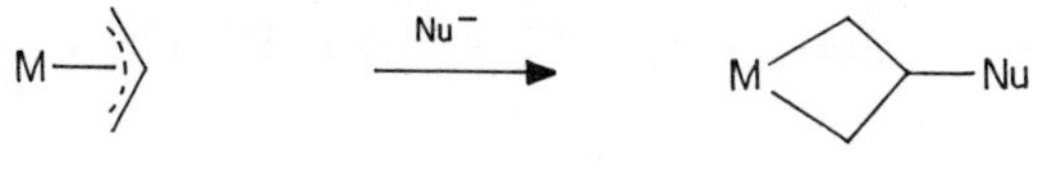

Consequently, in the nitrosyl complexes there is good reason to suppose that the addition involves HOMO/LUMO interaction, while selectivity between the two termini is due to (a) differences in LUMO coefficients, giving better interaction at one terminus than the other, and (b) differences in positive charge, giving a greater coulombic interaction at one of the termini. Since in the present case the two effects reinforce each other, it is difficult to assess which is the more important. As it turns out, there are a few reported cases of allylmolybdenum complexes which undergo nucleophile addition to the central allyl carbon atom. Consequently, the above reasoning is restricted only to a particular type of complex. When the ancillary ligands are changed, there is a complete change of behavior, illustrated by the following example.[37]

Now, this behavior is extremely interesting, since it reflects how the regiochemistry of nucleophile addition to coordinated polyenes is sensitive to the ligand environment, and therefore the electronic environment of the metal. It may be noted in this connection that η^5-C_5H_5 can be regarded as a poor π-acceptor compared with CO, NO, etc., so that this ligand will give relatively electron-rich metal centers compared with the isoelectronic $(CO)_3$ or $(\overset{+}{N}O)(CO)_2$ environments. Green's group have offered an explanation of the above behavior based on the electron-richness of the metal center, and have offered the following general rules for the addition of nucleophiles to unsaturated hydrocarbon ligands in 18-electron cationic complexes, for reactions which are kinetically rather than thermodynamically controlled.[38]

Rule 1:   Nucleophilic attack occurs preferentially at *even* coordinated polyenes which have no unpaired electrons in the HOMOs.

Rule 2:   Nucleophilic addition to open coordinated polyenes is preferred to addition to closed (cyclically conjugated) polyenes.

Rule 3:   For *even open* polyenes nucleophilic attack at the terminal carbon atom is always preferred, for *odd open* polyenyls attack at the terminal carbon atom occurs only if ML_n^+ is a strong electron-withdrawing group.

Thus, the results above illustrate rules 2 and 3, i.e. attack occurs at the allyl (open) rather than cyclopentadienyl (closed) ligands (rule 2), and when ML_n^+ is a strong electron-withdrawing group [CpMo(NO)(CO)] attack occurs at the allyl terminus, but when it is not (Cp$_2$Mo) attack occurs at the allyl C-2 (rule 3). No rationalization of this behavior on a theoretical basis has been offered. It seems likely, however, that there are two possibilities.

(1) The effect of increasing electron richness of the metal is to raise its d orbital energies, thereby changing the energy level of the allyl–M LUMO in such a way that the reaction becomes charge-controlled. As noted in Chapter 1, the central carbon atom, C-2, of a coordinated π-allyl bears the greatest positive charge (see above).

(2) The alteration of orbital energy levels of the metal changes the pattern of allyl–M orbital energy levels in the LUMO region in such a way that the LUMO is now a different orbital. It is not unreasonable to suppose that an orbital composed of allyl ψ_3 and metal d orbitals, which would be unoccupied, now becomes the complex LUMO. This has a large coefficient at C-2, thereby leading to favorable overlap with nucleophile HOMO at this position, illustrated below:

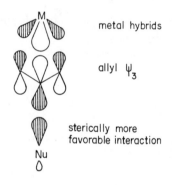

The proposition is not unreasonable since Hoffman's calculations show that there is probably a mixing of ψ_3 into the LUMO presented previously (i.e. that derived from ψ_2), so the extent of mixing might well change appreciably on changing M orbital energies. While this proposition is speculative and only tentative at this stage, there is a growing school of thought that these types of reactions are sensitive to such orbital-controlling effects. The ideas are presented here in anticipation that they will serve as an introduction to the concepts involved and will stimulate further thought on the matter.

MANGANESE, TECHNETIUM AND RHENIUM

Reactions of allyl chloride and sodium (or lithium) manganese pentacarbonyl at room temperature gives σ-allyl–Mn(CO)$_5$, obtained as a yellow distillable liquid in 71% yield. This compound shows the expected IR signal at 1620 cm^{-1} due to the terminal C=C bond of the allyl group.[39] When this complex is heated at 86 °C, expulsion of CO and conversion to the π-allyl complex in 88% yield occurs. This compound is obtained as a pale yellow solid, m.p. 55–57 °C, and no longer shows the allyl C=C stretch in the infrared, but shows a band at $1505\,cm^{-1}$, probably the C····C stretch vibration.

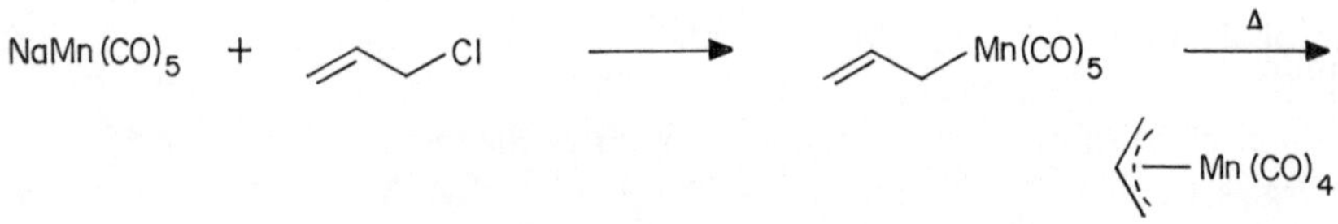

The addition of manganese carbonyl hydride to 1,3-dienes provides another route to σ-allyl- and π-allylmanganese complexes, e.g.

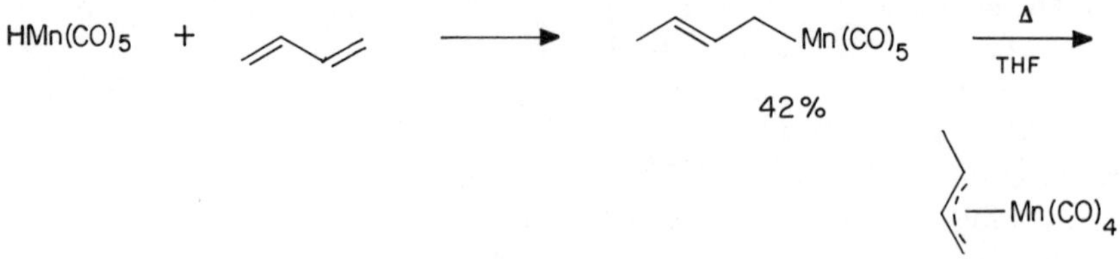

The geometry of the substituted allyl unit has been found to be retained during the conversion of σ- to π-allyl complexes.[40]

Carbonyl ligands can be displaced by phosphite ligands, under photochemical or thermal conditions.[41]

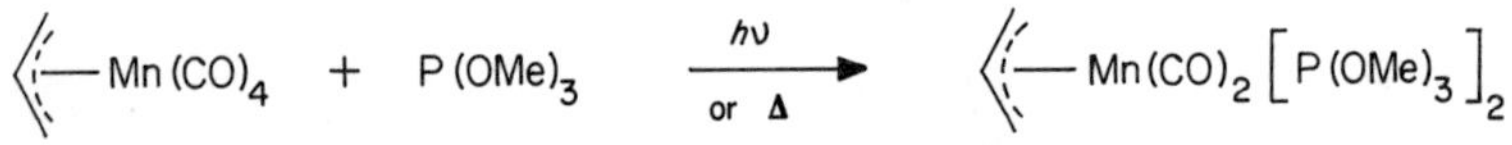

π-Allylrhenium complexes have been prepared by a method similar to that used for the manganese derivatives.[42]

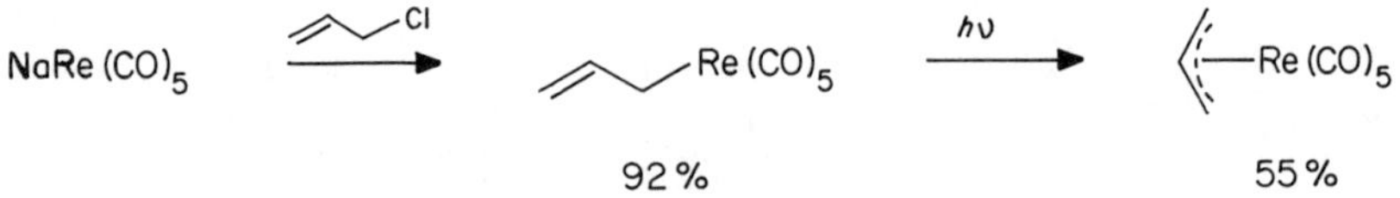

Carbon-13 NMR spectra have been determined for a few allyl–$Mn(CO)_4$ complexes, the data for the parent complex being shown below.[43]

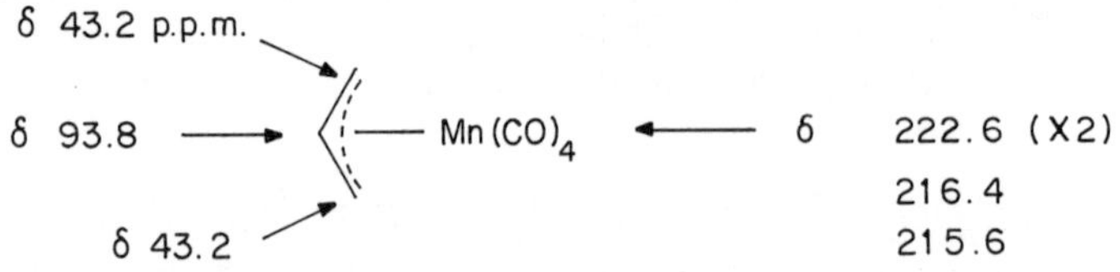

As with the allyl–$Fe(CO)_4$ derivatives discussed later, the central allyl carbon atom is the lowest field signal for this ligand, consistent with a bonding scheme which populates the allyl ψ_2 orbital (see Chapter 1). Extensive studies of the chemistry of these complexes have not been reported.

IRON, RUTHENIUM AND OSMIUM

(η^3-allyl)tricarbonyliron iodide is readily prepared as a dark brown crystalline solid, m.p. 79–80 °C, by heating allyl iodide with either iron pentacarbonyl or diiron nonacarbonyl. It is fairly stable and can be purified by vacuum sublimation.[44] Reaction of this complex with silver tetrafluoroborate under a carbon monoxide atmosphere gives (η^3-allyl)tetracarbonyliron tetrafluoroborate, a cationic complex.[45] While the former (neutral) complex is unreactive towards nucleophiles, the latter complex reacts readily with a range of nucleophilic reagents.

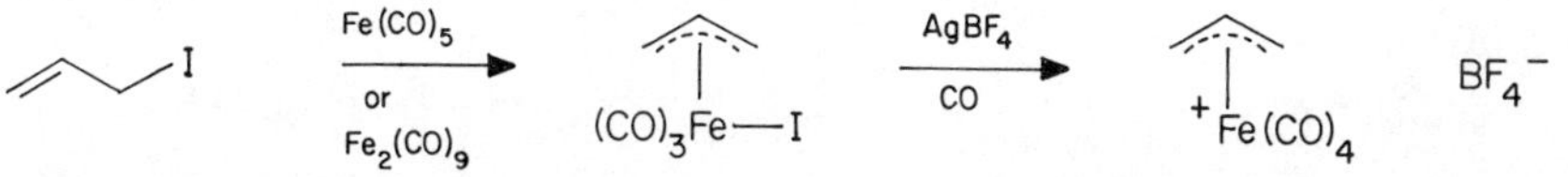

Treatment of diene–Fe(CO)₃ complexes with anhydrous HBF₄, preferably under a CO atmosphere for good yields, also gives π-allyl complexes, e.g.

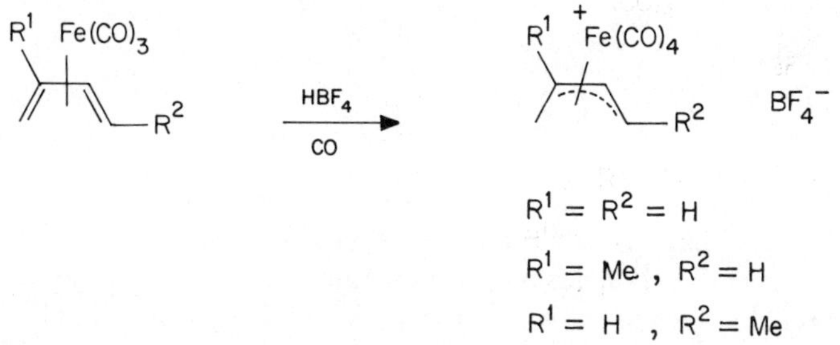

$$R^1 = R^2 = H$$
$$R^1 = Me, \quad R^2 = H$$
$$R^1 = H, \quad R^2 = Me$$

While this procedure normally occurs with retention of the diene geometry, prolonged acid treatment at a slightly elevated temperature results in isomerization to the more stable allyl complex.

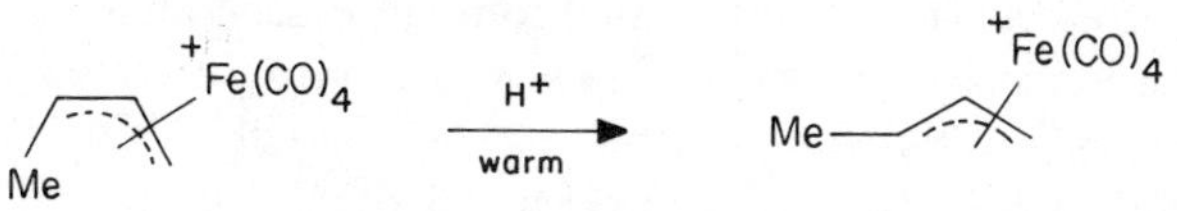

When there is a trisubstituted double bond appended to the diene–Fe(CO)₃ unit, this may protonate in preference and lead to cyclization (Chapter 7), e.g.

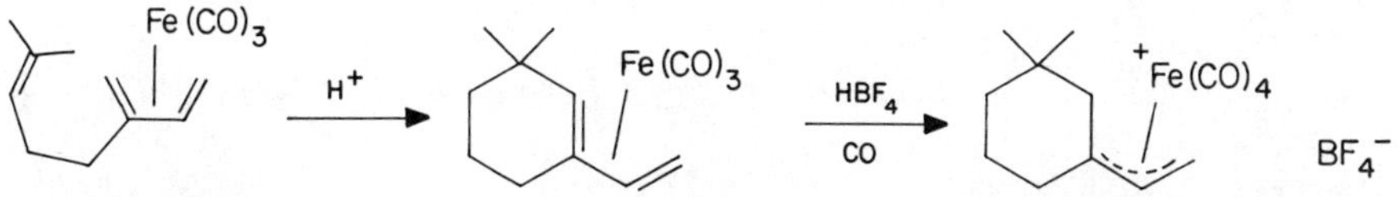

The cationic complexes are reactive towards nucleophiles, as might be expected, and undergo attack at the terminal allyl carbon atom. The resulting olefin–Fe(CO)₄ complex is often very unstable and either decomposes spontaneously or on exposure to air to give the substituted olefin. Some examples are given below.

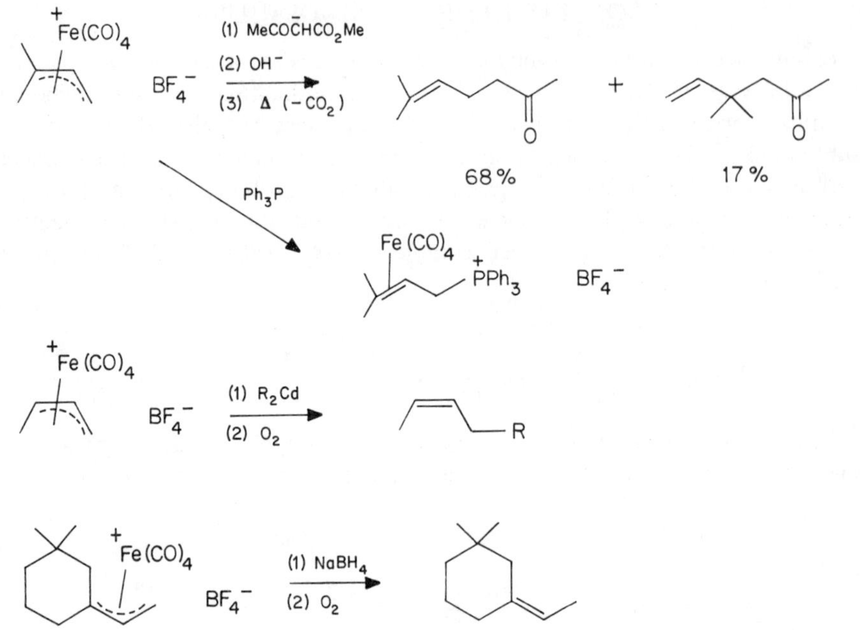

Recalling the results of nucleophilic addition to Mo and W π-allyl complexes, and also the fact (Chapter 1) that the greatest positive charge on the allyl ligand is found at the central carbon atom (C-2), it would appear that a major controlling factor in this reaction is the interaction of nucleophile HOMO and allyl-complex LUMO, and that the latter orbital is (approximately) the antibonding combination of ψ_2 and iron d orbital. Then we should expect maximum bonding interaction at C-1 or C-3 (see above discussion), as is indeed observed.

A very interesting variation on the nucleophile addition reaction has been developed by Roustan et al..[46] The π-allyl complex is generated in the presence of a carbon nucleophile, by treatment of an allylic halide or acetate with a catalytic amount of NaFe(CO)$_3$NO (0.1–0.3 equiv.) or (η^3-crotyl)Fe(CO)$_2$NO (0.1 equiv.). The nucleophile undergoes reaction with the allylic compound under these conditions, and this procedure may be compared with the analogous palladium-catalyzed reaction, although it has not been developed to the same extent. Some examples are given below.

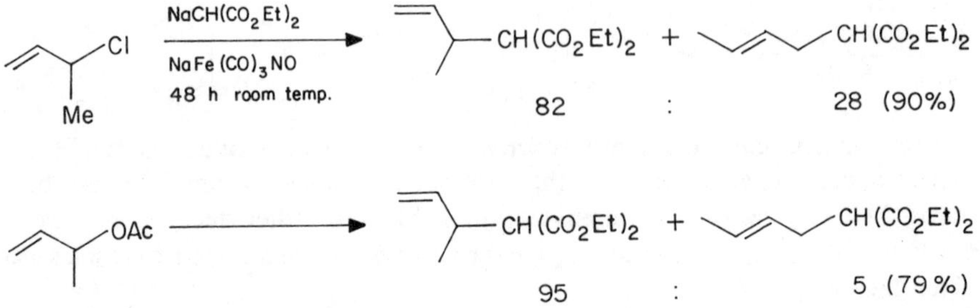

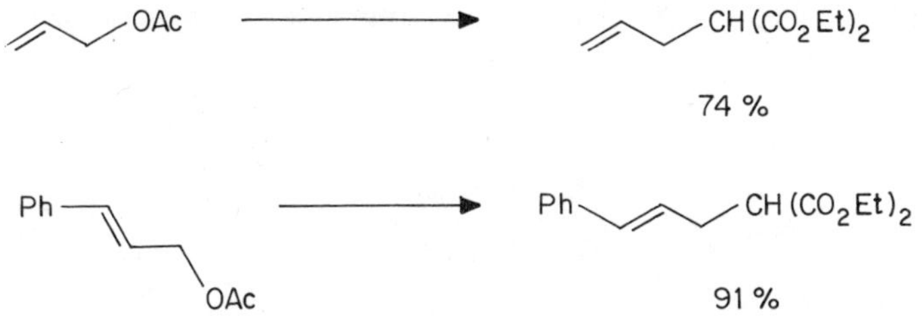

$\diagdown\diagup\diagdown$OAc $\longrightarrow$ $\diagdown\diagup\diagdown$CH$(CO_2Et)_2$

74 %

Ph$\diagdown\diagup\diagdown$OAc $\longrightarrow$ Ph$\diagdown\diagup\diagdown$CH$(CO_2Et)_2$

91 %

It may be noted that η^3-allyl complexes have also been characterized as intermediates during Friedel–Crafts acylation of diene–Fe(CO)$_3$ complexes (Chapter 7).

COBALT, RHODIUM AND IRIDIUM

Reaction of butadiene with cobalt hydridocarbonyl was first investigated in the late 1950s and the product, a red–brown liquid, was later formulated as a π-allyl complex.[47]

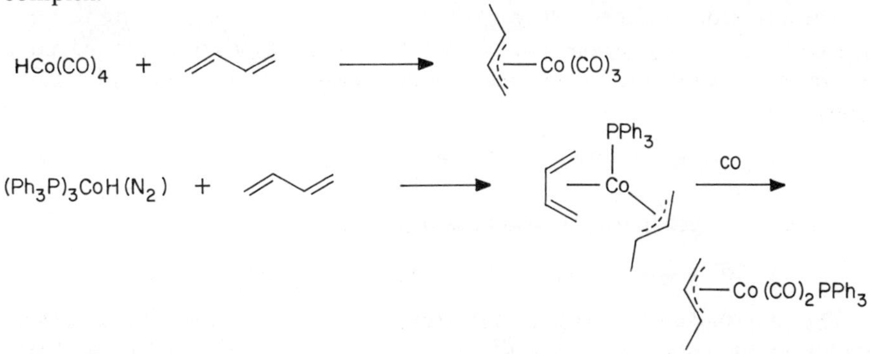

HCo(CO)$_4$ + $\diagup\diagdown\diagup$ $\longrightarrow$ $\diagup\!\!\!<$ —Co(CO)$_3$

(Ph$_3$P)$_3$CoH(N$_2$) + $\diagup\diagdown\diagup$ $\longrightarrow$ $\diagdown$—Co—PPh$_3$ $\xrightarrow{\text{CO}}$ $\diagup\!\!\!<$—Co(CO)$_2$PPh$_3$

A brief review[48] on allyl cobalt complexes appeared in 1973, although some of the early work on π-allylcobalt carbonyl complexes was not covered.

The preparation of the parent π-allyl–Co(CO)$_3$ follows a similar procedure to that for the Cr group complexes, viz.

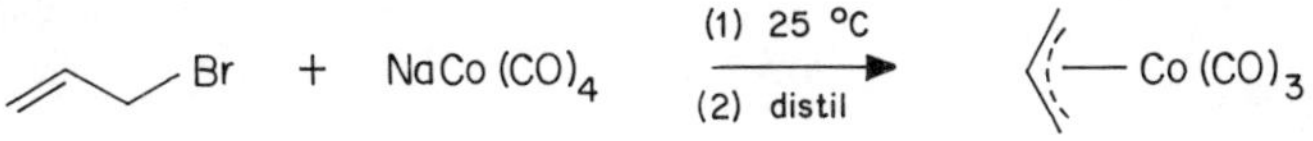

$\diagup\diagdown$Br + NaCo(CO)$_4$ $\xrightarrow[\text{(2) distil}]{\text{(1) 25 °C}}$ $\diagup\!\!\!<$—Co(CO)$_3$

The intermediate σ-allyl complex, while undoubtedly formed, is not usually isolated.[49] One of the CO ligands can be displaced with triphenylphosphine to give the allyl–Co(CO)$_2$PPh$_3$ complex. Interestingly, this is not obtained by direct reaction of allyl bromide with NaCo(CO)$_4$ in the presence of CO and Ph$_3$P. Instead, the following reaction occurs:[50]

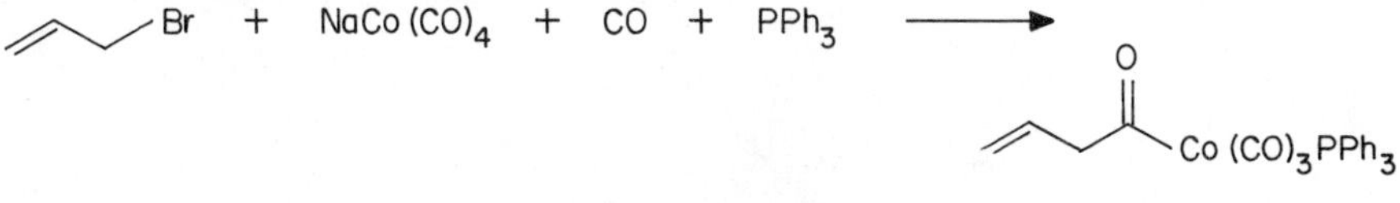

$\diagup\diagdown$Br + NaCo(CO)$_4$ + CO + PPh$_3$ $\longrightarrow$ $\diagup\diagdown\diagdown$C(=O)—Co(CO)$_3PPh_3$

238

The cationic complex CpĊo(allyl) X⁻, has also been prepared in low yield by the following method:[51]

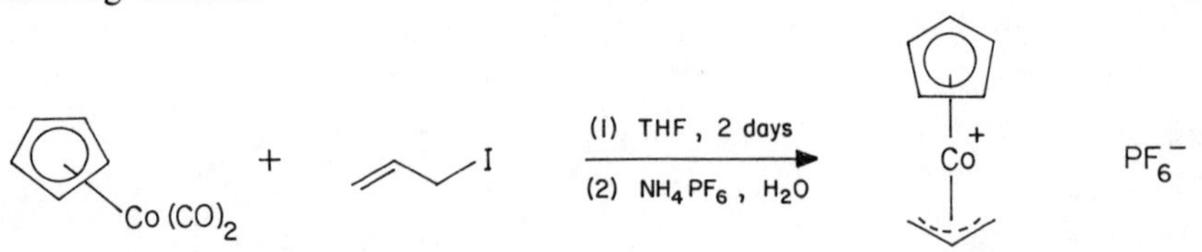

Tris(π-allyl)cobalt, a thermally very unstable complex containing no ancillary ligands, is conveniently prepared by reaction of, e.g., cobalt(II) chloride with allylmagnesium chloride:

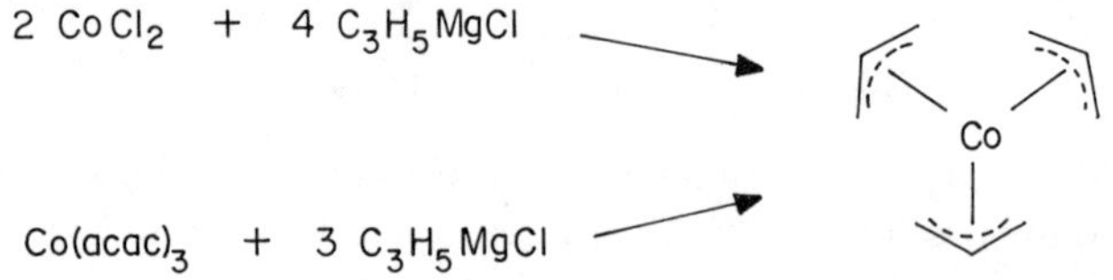

The allyl groups in tris(π-allyl)cobalt are cleaved at − 40 °C under a hydrogen atmosphere to give propane. Treatment with neutral ligands results in reductive coupling of two of the π-allyl ligands and formation of a mono-π-allyl–Co(I) complex:

The axial allyl group can be displaced by anions:

$$Co(C_3H_5)_3 \ + \ HX \longrightarrow (C_3H_5)_2CoX \ + \ C_3H_6$$

The intermediacy of π-allyl–Co complexes in polymerization reactions catalyzed by cobalt has been discussed in Chapter 2. Reactions of tris(π-allyl)cobalt with butadiene proceeds according to the following equation:

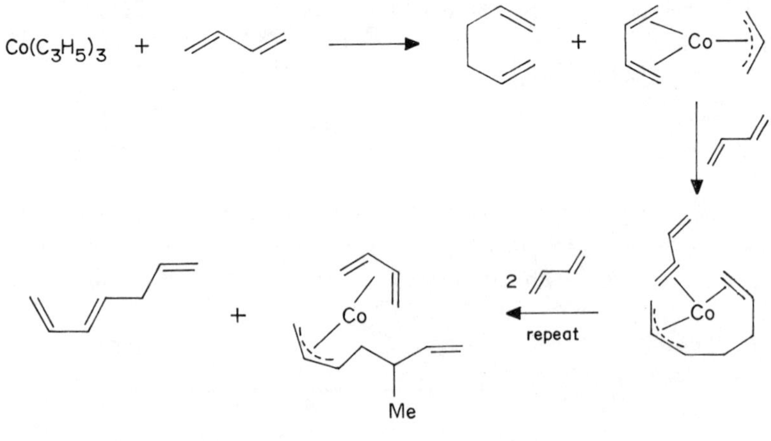

(A)

Complex **A** is the key intermediate in the dimerization of butadiene initiated by $(C_3H_5)_3Co$.

Treatment of the $(C_3H_5)_2CoX$ complexes with cyclopentadienyl anion results in displacement of halide and rearrangement of one π-allyl to σ-allyl ligand, to give an 18e complex:

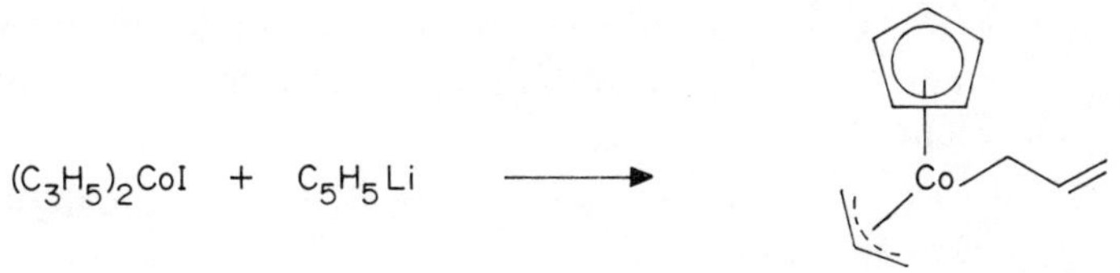

$$(C_3H_5)_2CoI + C_5H_5Li \longrightarrow$$

This latter complex, on reaction with a diene undergoes displacement of the allyl groups by reductive coupling:

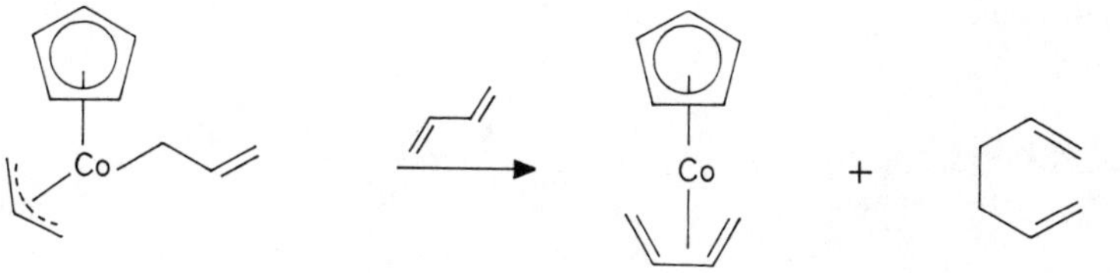

Reaction of $Co(acac)_3$ with organoaluminum derivatives in the presence of cycloocta-1,5-diene occurs smoothly to give a bis-cyclooctadiene complex:

$$Co(acac)_3 + 3\ AlR_3 + 2\ COD \longrightarrow (COD)_2CoR$$

When a dialkylaluminum hydride is used as the reducing agent, a complex is formed which in solution is an equilibrium mixture of **B** and **C**.

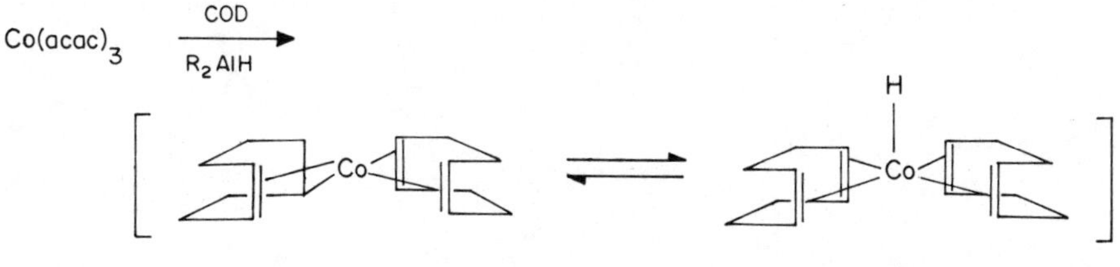

$$Co(acac)_3 \xrightarrow[\text{R}_2\text{AlH}]{\text{COD}}$$

(B)   (C)

At 40 °C the σ-cyclooctenyl group of this complex undergoes transannular ring contraction and dehydrogenation eventually giving the product below:

$$B + C \xrightarrow{40\ °C}$$

When Co(acac) is treated with an excess of dialkylaluminum hydride in the presence of cod, the (π-cyclooctenyl)(cod)cobalt complex is obtained as a by-product:

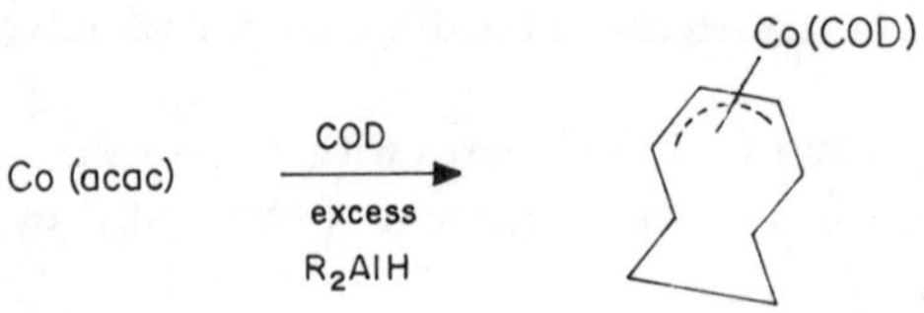

The π-cyclooctenyl complexes can also be obtained by treatment of the σ-cyclooctenyl complex **B** with neutral ligands, representing an interesting isomerization reaction:

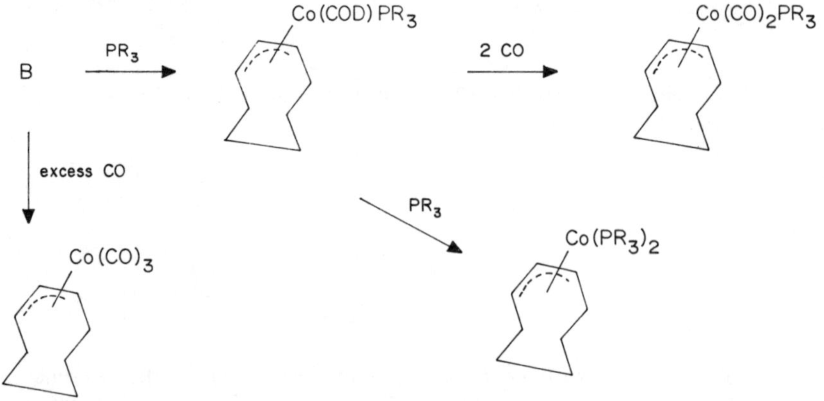

Reaction of $CoCl_2$ with an excess of cyclooctatetraene and sodium borohydride in ethanol at $-20\,°C$ gives a black, crystalline, air-sensitive complex, π-cyclooctatrienyl-η^4-cyclooctatetraenecobalt, treatment of which with carbon monoxide gives the η^3-cyclooctatrienyl complex:[52]

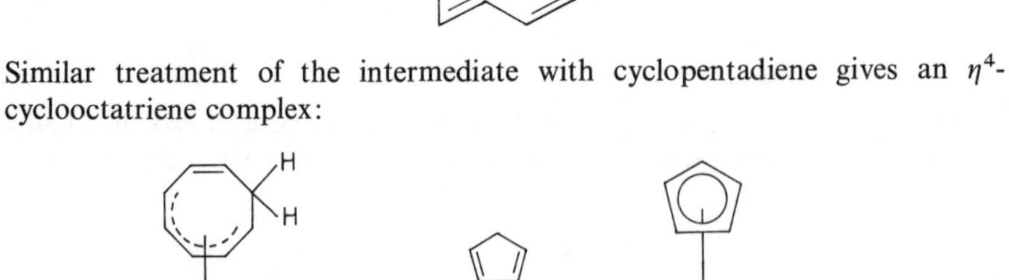

Similar treatment of the intermediate with cyclopentadiene gives an η^4-cyclooctatriene complex:

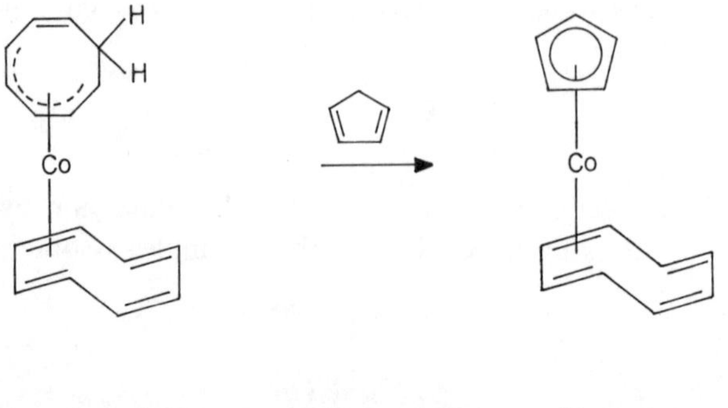

Displacement of a diene ligand from $(\eta^3\text{-allyl})(\eta^4\text{-}C_4H_6)Co(PPh_3)$ with PF_3 provides an entry into trifluorophosphine-containing π-allyl cobalt complexes.[53]

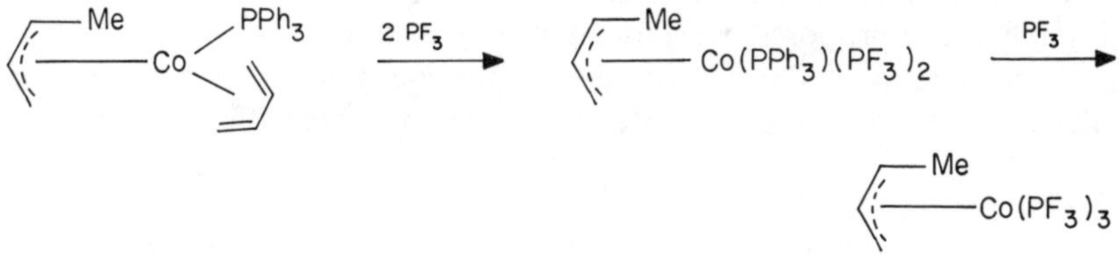

Similarly :

$$Co(\eta\text{-}C_3H_5)_3 \; + \; PF_3 \; \longrightarrow \; \text{Co}(PF_3)_3$$

In contrast to $(\pi\text{-}C_3H_5)Rh(PF_3)_4$, which is readily converted into $HRh(PF_3)_4$, on treatment with PF_3 and hydrogen at room temperature and 1 atm the above cobalt compounds are unreactive. In connection with the preparation of these types of phosphine complexes, it may be noted that $[(MeO)_3P]_3Co(C_3H_5)$ is an excellent catalyst for the hydrogenation of aromatic to saturated hydrocarbons under mild conditions.[54]

Reaction of π-allyl–$Co(CO)_3$ with potassium cyanide in the presence of iodine leads to cyanocobaltate anions containing a π-allyl ligands.[55]

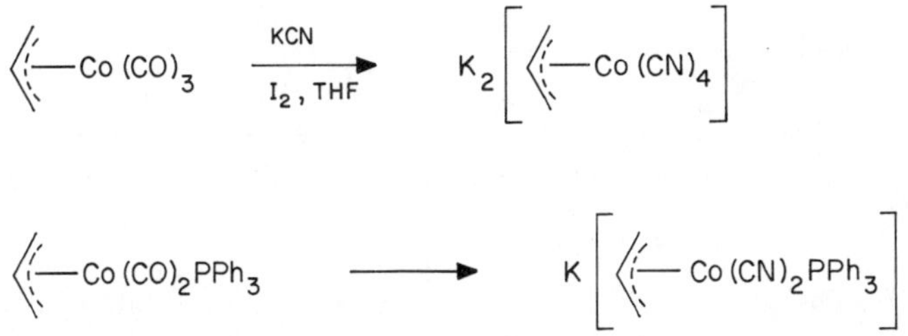

In an attempt to obtain further information regarding the regioselectivity of nucleophile addition to electrophile π-allyl complexes, Aviles and Green prepared a series of cobalt complexes $[Co(\eta\text{-}C_5H_5)(\pi\text{-allyl})L]PF_6$.[56] Similar complexes (L $=$ CO) had previously been prepared by two groups,[57] e.g.

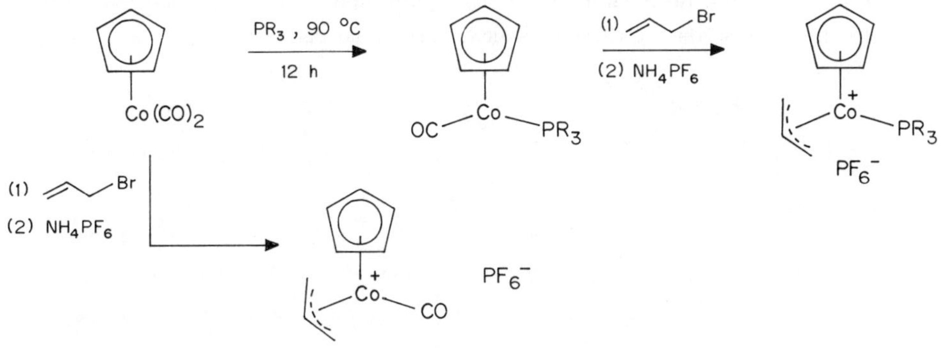

Unfortunately, reactions of these complexes with nucleophiles did not give stable isolable complexes, except in one case where reaction with methyllithium gave $Me_2Co(\eta\text{-}C_5H_5)(PMe_3)$. Consequently, no information was obtained concerning the above proposed rules for nucleophile addition.

The reaction of fluorinated olefin with π-allylcobalt complexes has been studied and found to give products of insertion, exemplified below.[58]

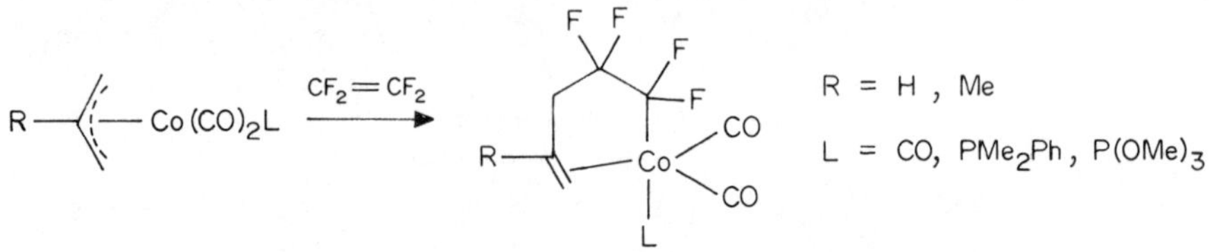

$R = H$, Me

$L = CO$, PMe_2Ph, $P(OMe)_3$

This type of reaction is of importance in the study of allylcobalt catalyzed polymerization reactions. The mechanism has not been delineated.

Allyl complexes of rhodium and iridium can be prepared in similar ways to those of cobalt, although often these are not completely general for preparation of a range of substituted complexes, e.g.[59]

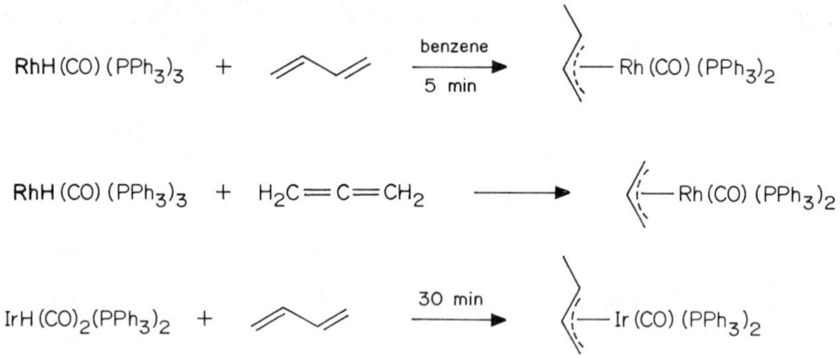

The iridium complexes so produced are pale yellow crystalline solids which are stable towards air oxidation. The rhodium complexes are much less stable, slowly decomposing even under nitrogen or *in vacuo*, and decomposing much faster in solution. The interaction of hydride metal complexes with dienes is very limited, failing to give pure characterizable compounds even with isoprene. Other methods of preparing these complexes are given below.[60]

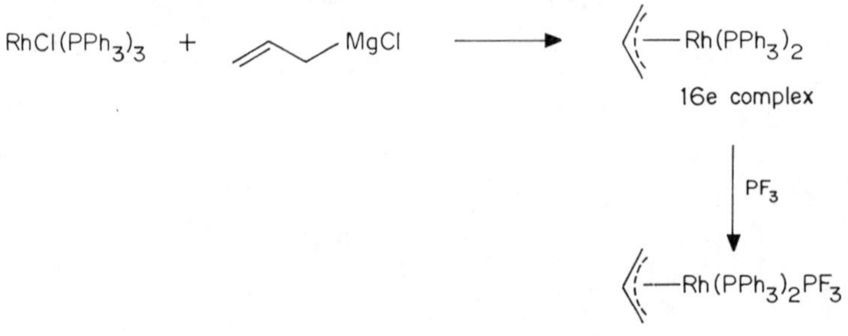

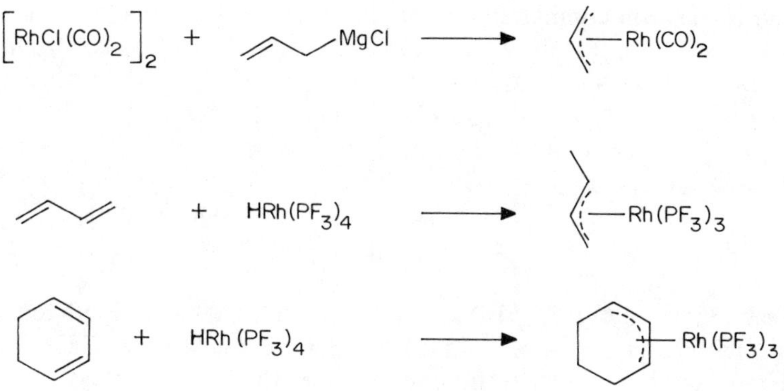

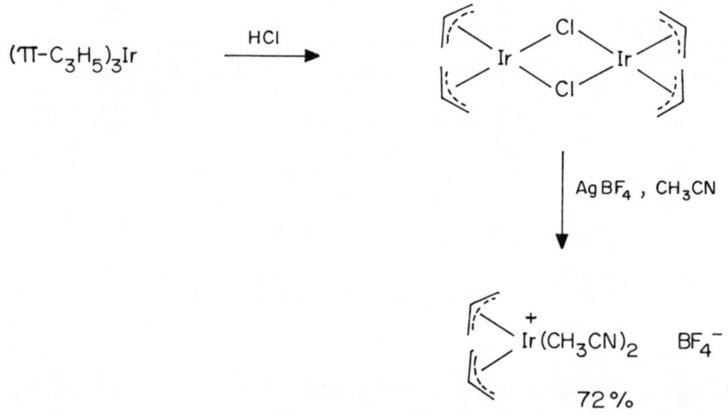

$$\left[RhCl(CO)_2 \right]_2 \;+\; \underset{}{\diagup\!\!\!\diagup\!\!\!\diagdown} MgCl \;\longrightarrow\; \diagup\!\!\!\diagdown\!\!\!-Rh(CO)_2$$

$$\diagup\!\!\!\diagdown\!\!\!\diagup \;+\; HRh(PF_3)_4 \;\longrightarrow\; \diagup\!\!\!\diagdown\!\!-Rh(PF_3)_3$$

$$\bigcirc\!\!\!\diagdown \;+\; HRh(PF_3)_4 \;\longrightarrow\; \bigcirc\!\!-Rh(PF_3)_3$$

$$(\pi\text{-}C_3H_5)_3Ir \;\xrightarrow{\;HCl\;}\; \left[\; Ir\diagdown_{Cl}^{Cl}Ir \; \right]$$

$$\downarrow \; AgBF_4 \,,\; CH_3CN$$

$$Ir(CH_3CN)_2^{+} \quad BF_4^{-}$$

72%

The 16-electron complexes readily coordinate a molecule of fluorobutyne followed by insertions to the Rh—allyl bond with concomitant coordination of solvent (ether),[61] e.g.

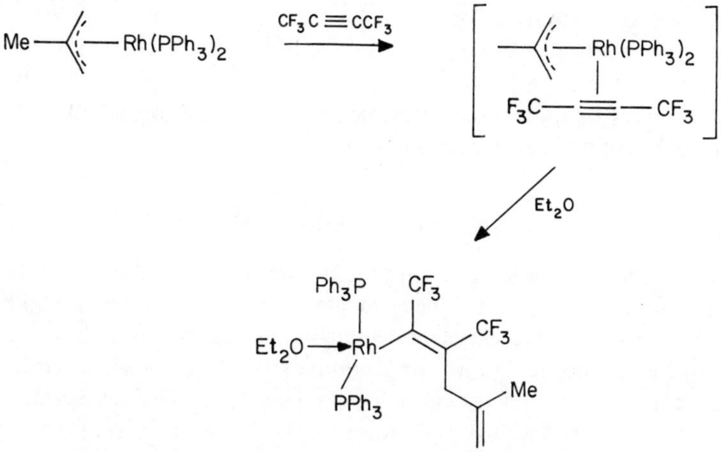

$$Me-\!\!\!\diagup\!\!\!\diagdown\!\!-Rh(PPh_3)_2 \;\xrightarrow{\;CF_3C\equiv CCF_3\;}\; \left[\;\diagup\!\!\!\diagdown\!\!-Rh(PPh_3)_2 \atop F_3C-\!\!\equiv\!\!-CF_3 \; \right]$$

$$\swarrow Et_2O$$

The 18-electron complexes give similar insertion products:[62]

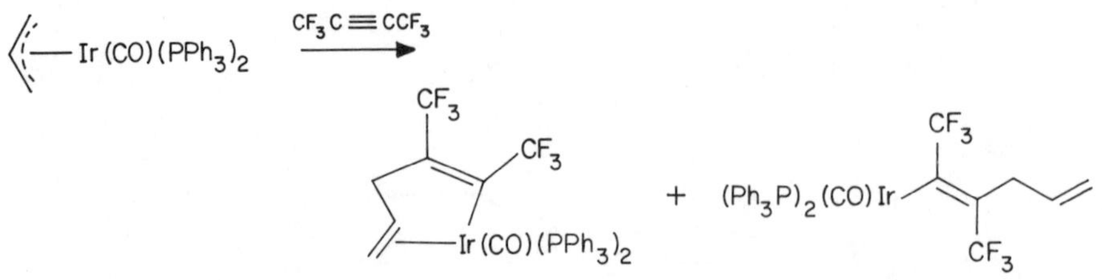

Reaction of (allyl)Ir(CO)(PPh$_3$)$_2$ with tetrafluoroethene at room temperature occurs with replacement of a phosphine ligand by the alkene. These complexes undergo insertion of the tetrafluoroethene into the Ir—allyl bond on treatment with CO or Ph$_3$P. The same insertion products may be obtained directly by reaction of the π-allyl complex with C$_2$F$_4$ at 80 °C:[63]

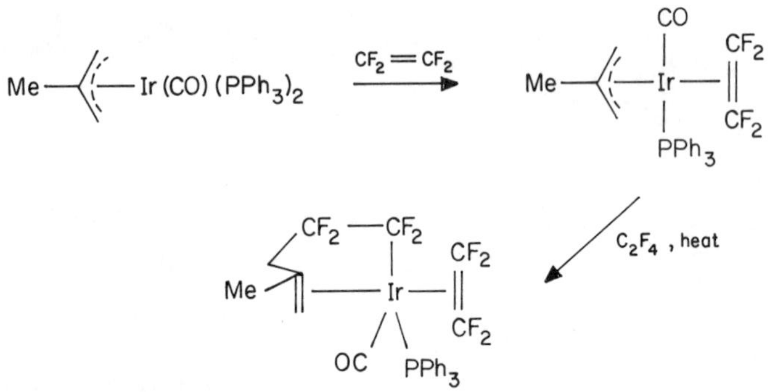

Addition of amines to butadiene has been shown to be catalyzed by rhodium trichloride, suggesting the intermediacy of π-allyl complexes which react with the amine, somewhat analogous to the similar palladium-catalyzed reaction, e.g.[64]

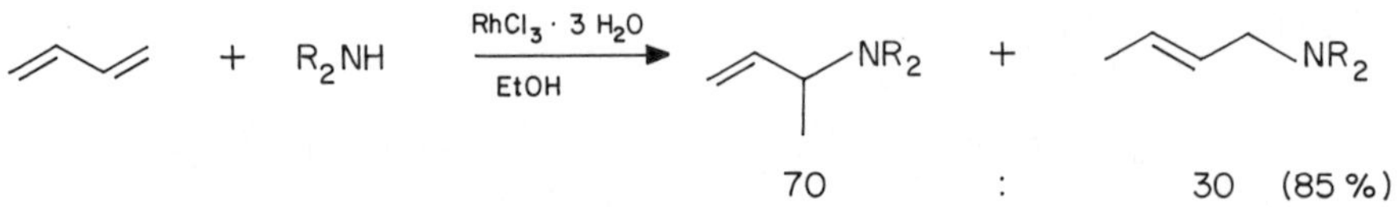

$$70 \quad : \quad 30 \quad (85\%)$$

The ability of cationic π-allyl complexes of this type to undergo addition of nucleophiles has not been extensively studied.

GENERAL COMMENTS

From the above discussion, it is apparent that π-allyl complexes are chemically interesting for several reasons. They provide useful, or potentially useful, reagents for the formation of strategic carbon—carbon bonds, enabling the construction of a range of molecules of varying complexity. This aspect is perhaps best illustrated by the π-allylpalladium derivatives. They are accessible to both theoretical and empirical investigations into the factors which control reactivity, such as nucleophile addition (as ambient electrophiles). They often function as catalysts in polymerization reactions and preparations of complexes with modified reactivity might allow the isolation of compounds which are analogous to proposed intermediates in the catalytic cycle. It is hoped both that this chapter has provided a glossary of complexes which illustrate these aspects, and that an insight has been given into the current trends in interpreting reactivity phenomena using frontier molecular orbital concepts.

REFERENCES

1. M. L. H. Green and P. L. I. Nagy, *Adv. Organomet. Chem.*, 1964, **2**, 325.
2. P. Heimbach, P. W. Jolly and G. Wilke, *Adv. Organomet. Chem.*, 1970, **8**, 29; M. F. Semmelhack, *Org. React.*, 1972, **19**, 115.
3. B. M. Trost, Tetrahedron, 1977, **33**, 2615; *Acc. Chem. Res.*, 1980, **18**, 385; *Pure Appl. Chem.*, 1981, **53**, 2357.
4. E. J. Corey and M. F. Semmelhack, *J. Am. Chem. Soc.*, 1967, **89**, 2755.
5. I. D. Webb and G. T. Borcherdt, *J. Am. Chem. Soc.*, 1951, **73**, 2654.
6. E. J. Corey and E. Hamanaka, *J. Am. Chem. Soc.*, 1967, **89**, 2758.
7. P. W. Jolly and G. Wilke, *The Organic Chemistry of Nickel*, Academic Press, New York, 1965, Vol. 2, and references cited therein.
8. B. Akermark, G. Akermark, C. Moberg, C. Bjorklund and K. Siirala-Hansén, *J. Organomet. Chem.*, 1979, **164**, 97; J. Kiji, K. Yamamoto, E. Sasakawa and J. Furakawa, *J. Chem. Soc., Chem. Commun.*, 1973, 770.
9. C. Moberg, *Tetrahedron Lett.*, 1980, **21**, 4539.
10. B. M. Trost and E. Keinan, *Tetrahedron Lett.*, 1980, **21**, 2591, 2595; J. Godschalx and J. K. Stille, *Tetrahedron Lett.*, 1980, **21**, 2599.
11. J. Tsuji, *Organic Synthesis with Palladium Compounds*, Springer Verlag, Berlin, 1980.
12. B. M. Trost and E. Keinan, *J. Org. Chem.*, 1979, **44**, 3451.
13. B. M. Trost and D. P. Curran, *J. Am. Chem. Soc.*, 1980, **102**, 5699.
14. B. M. Trost and T. R. Verhoeven, *J. Am. Chem. Soc.*, 1980, **102**, 4743.
15. H. A. Martin and F. Jellinek, *Angew. Chem., Int. Ed. Engl.*, 1964, **3**, 311.
16. H. A. Martin and F. Jellinek, *J. Organomet. Chem.*, 1966, **6**, 293; 1968, **12**, 149.
17. H. K. Hofstee, H. O. van Oven and H. J. de Liefde Meijer, *J. Organomet. Chem.*, 1972, **42**, 405; H. K. Hofstee, C. J. Groenenboom, H. O. van Oven and H. J. de Liefde Meijer, *J. Organomet. Chem.*, 1975, **85**, 193.
18. H.-J. Kablitz, R. Kallweit and G. Wilke, *J. Organomet. Chem.*, 1972, **44**, C49.
19. A. Zwijnenberg, H. O. van Oven, C. J. Groenenboom and H. J. de Liefde Meijer, *J. Organomet. Chem.*, 1975, **94**, 23.
20. H. A. Martin and F. Jellinek, *Angew. Chem., Int. Ed. Engl.*, 1964, **3**, 311.
21. T. Kruck and H.-U. Hempel, *Angew. Chem., Int. Ed. Engl.*, 1971, **10**, 408.
22. F. W. Siegert and H. J. de Liefde Meijer, *J. Organomet. Chem.*, 1970, **23**, 177.
23. G. W. A. Fowles, L. S. Pu and D. A. Rice, *J. Organomet. Chem.*, 1973, **54**, C17.
24. E. K. Barefield, G. W. Parshall and F. N. Tebbe, *J. Am. Chem. Soc.*, 1970, **92**, 5234; F. N. Tebbe and G. W. Parshall, *J. Am. Chem. Soc.*, 1971, **93**, 3793.
25. M. Schneider and E. Weiss, *J. Organomet. Chem.*, 1974, **73**, C7.
26. J. E. Ellis and R. A. Faltynek, *J. Organomet. Chem.*, 1975, **93**, 205.
27. R. G. Hayter, *J. Organomet. Chem.*, 1963, **13**, P1.
28. R. B. King and A. Fronzaglia, *J. Am. Chem. Soc.*, 1966, **88**, 709.
29. M. L. H. Green, J. Knight, L. C. Mitchard, G. C. Roberts and W. E. Silverthorn, *J. Chem. Soc., Chem. Commun.*, 1972, 987; M. L. H. Green and W. E. Silverthorn, *J. Chem. Soc., Dalton Trans.*, 1973, 301; M. L. H. Green, L. C. Mitchard and W. E. Silverthorn, *J. Chem. Soc., Dalton Trans.*, 1973, 1403; 1973, 1952.
30. M. L. H. Green and J. Knight, *J. Chem. Soc., Dalton Trans.*, 1974, 311.

246

31. B. J. Brisdon and G. E. Griffin, *J. Chem. Soc., Dalton Trans.*, 1975, 1999.
32. G. Doyle, *J. Organomet. Chem.*, 1977, **132**, 243; 1978, **150**, 67; B. J. Brisdon, *J. Organomet. Chem.*, 1977, **125**, 225; J. W. Faller and D. A. Haitko, *J. Organomet. Chem.*, 1978, **149**, C19; B. J. Brisdon and K. E. Paddick, *J. Organomet. Chem.*, 1978, **149**, 113; M. G. B. Drew, B. J. Brisdon, D. A. Edwards and K. E. Paddick, *Inorg. Chim. Acta*, 1979, **35**, L381.
33. N. A. Bailey, W. G. Kita, J. A. McCleverty, A. J. Murray, B. E. Mann and N. W. J. Walker, *J. Chem. Soc., Chem. Commun.*, 1974, 592; J. W. Faller and A. M. Rosan, *J. Am. Chem. Soc.*, 1976, **98**, 3388.
34. R. D. Adams, D. F. Chodosh, J. W. Faller and A. M. Rosan, *J. Am. Chem. Soc.*, 1979, **101**, 2590.
35. B. E. R. Schilling, R. Hoffman and J. W. Faller, *J. Am. Chem. Soc.*, 1979, **101**, 5921.
36. I. Fleming, *Frontier Orbitals and Organic Chemical Reactions*, Wiley, New York, 1976.
37. M. Ephritikhine, M. L. H. Green, and R. E. MacKenzie, *J. Chem. Soc., Chem. Commun.*, 1976, 619; M. Ephtitikhine, B. R. Francis, M. L. H. Green, R. E. MacKenzie and M. J. Smith, *J. Chem. Soc., Dalton Trans.*, 1977, 1131.
38. S. G. Davies, M. L. H. Green and D. M. P. Mingos, *Tetrahedron*, 1978, **34**, 3047.
39. W. R. McClellan, H. H. Hoehn, H. N. Cripps, E. L. Muetterties and B. W. Howk, *J. Am. Chem. Soc.*, 1961, **83**, 1601.
40. N. N. Druz, V. I. Klepikova, M. I. Lobach and V. A. Kormer, *J. Organomet. Chem.*, 1978, **162**, 343.
41. L. S. Stuhl and E. L. Muetterties, *Inorg. Chem.*, 1978, **17**, 2148.
42. B. J. Brisdon, D. A. Edwards and J. W. White, *J. Organomet. Chem.*, 1979, **175**, 113.
43. A. Oudeman and T. S. Sorenson, *J. Organomet. Chem.*, 1978, **156**, 259.
44. R. A. Plowman and F. G. A. Stone, *Z. Naturforsch., Teil B*, 1962, **17**, 575; H. D. Murdoch and E. Weiss, *Helv. Chim. Acta*, 1962, **45**, 1927.
45. T. H. Whitesides, R. W. Arhart and R. W. Slaven, *J. Am. Chem. Soc.*, 1973, **95**, 5792.
46. J. L. Roustan, J. Y. Mérour and F. Houlihan, *Tetrahedron Lett.*, 1979, 3721.
47. C. L. Aldridge, H. B. Jonassen and E. Pulkkinen, *Chem. Ind. (London)*, 1960, 374; R. F. Heck and D. S. Breslow, *J. Am. Chem. soc.*, 1961, **83**, 1097; see also P. V. Rinze and H. Noth, *J. Organomet. Chem.*, 1971, **30**, 115.
48. H. Bönnemann, *Angew, Chem., Int. Ed. Engl.*, 1973, **12**, 964.
49. R. F. Heck and D. S. Breslow, *J. Am. Chem. Soc.*, 1960, **82**, 750; R. F. Heck, *J. Am. Chem. Soc.*, 1963, **85**, 655.
50. R. F. Heck and D. S. Breslow, *J. Am. Chem. Soc.*, 1960, **82**, 4438.
51. R. F. Heck, *J. Org. Chem.*, 1963, **28**, 6064.
52. A. Greco, M. Green and F. G. A. Stone, *J. Chem. Soc. A*, 1971, 285.
53. M. A. Cairns and J. F. Nixon, *J. Organomet. Chem.*, 1973, **51**, C27; *J. Chem. Soc., Dalton Trans.*, 1974, 2001.
54. E. L. Muetterties and E. J. Hirsekorn, *J. Am. Chem. Soc.*, 1974, **96**, 4063.
55. J. A. Dineen and P. L. Pauson, *J. Organomet. Chem.*, 1974, **71**, 87.
56. T. Aviles and M. L. H. Green, *J. Chem. Soc., Dalton Trans.*, 1979, 1116.
57. E. O. Fischer and R. D. Fischer, *Z. Naturforsch., Teil B*, 1961, **16**, 475; P. Powell and L. J. Russell, *J. Organomet. Chem.*, 1977, **129**, 415.
58. A. Greco, M. Green and F. G. A. Stone, *J. Chem. Soc. A*, 1971, 3476; M. Bottrill, R. Goddard, M. Green and P. Woodward, *J. Chem. Soc., Dalton Trans.*, 1979, 1671.
59. C. K. Brown, W. Mowat, G. Yagupsky and G. Wilkinson, *J. Chem. Soc. A*, 1971, 850.
60. J. F. Nixon, B. Wilkins and D. A. Clement, *J. Chem. Soc., Dalton Trans.*, 1974, 1993; M. Green and G. J. Parker, *J. Chem. Soc., Dalton Trans.*, 1974, 333.
61. D. A. Clement, J. F. Nixon and J. S. Poland, *J. Organomet. Chem.*, 1974, **76**, 117.
62. M. Green and S. H. Taylor, *J. Chem. Soc., Dalton Trans.*, 1975, 1142.
63. M. Green and S. H. Taylor, *J. Chem. Soc., Dalton Trans.*, 1975, 1128.
64. R. Baker and D. E. Halliday, *Tetrahedron Lett.*, 1972, 2773.

CHAPTER 7

η^4-Diene Complexes

INTRODUCTORY COMMENTS

There are basically two types of metal–diene complexes, from the point of view of the ligand—those containing an unconjugated (e.g. 1,4- or 1,5-) diene, which acts as a chelating ligand, and those containing a conjugated 1,3-diene. In terms of reactivity the complexes of unconjugated dienes closely resemble those of monoolefins, while 1,3-diene complexes have special properties due to their different bonding arrangement. We shall deal with these two classifications separately.

TRANSITION METAL COMPLEXES OF CHELATING, NON-CONJUGATED DIENES

These are mostly typified by metal complexes of cycloocta-1,5-diene (cod) and norbornadiene (nbd), and have the structures **1** and **2**.

(1) COD complex   (2) NBD complex

The non-conjugated cod complex is formed only with metals which will not readily isomerize the diene, or which have a radius too large to allow preferential formation of the 1,3-diene complex. The nbd complexes can even be formed with metals which prefer to coordinate to a 1,3-diene. For example, Fe(0) will normally only form 1,3-diene complexes, but readily forms the NBD–Fe(CO)$_3$ complex.

Nickel, Platinum and Palladium

Apart from the observation that unconjugated diene complexes of these metals are more reactive, their behavior is very similar to that of the monoolefin complexes.

247

The platinum and palladium complexes of cycloocta-1,5-diene and norbornadiene are readily prepared by reaction of the diene with an appropriate salt of the metal, e.g.

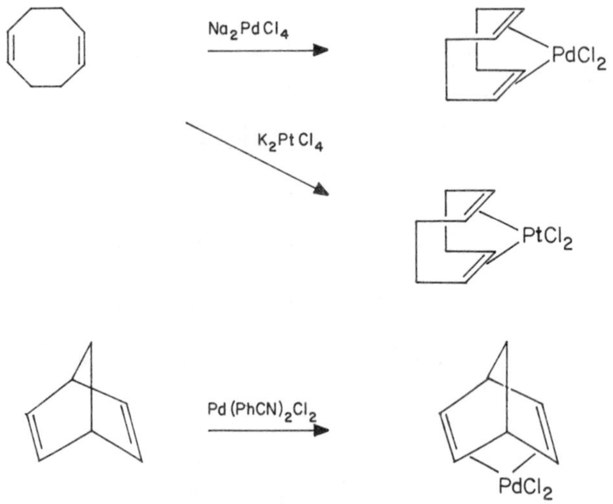

(cod)$PdCl_2$ is obtained as pale orange needles (decomposition point 205–210 °C), whilst the platinum derivative is a stable white, crystalline (needles) substance (decomposition range 220–278 °C). They are generally insoluble in most organic solvents except boiling chloroform, dichloromethane and acetic acid.

Interesting diolefin nickel complexes are obtained on reaction of nickelocene with the diene in the presence of tetrafluoroboric acid:

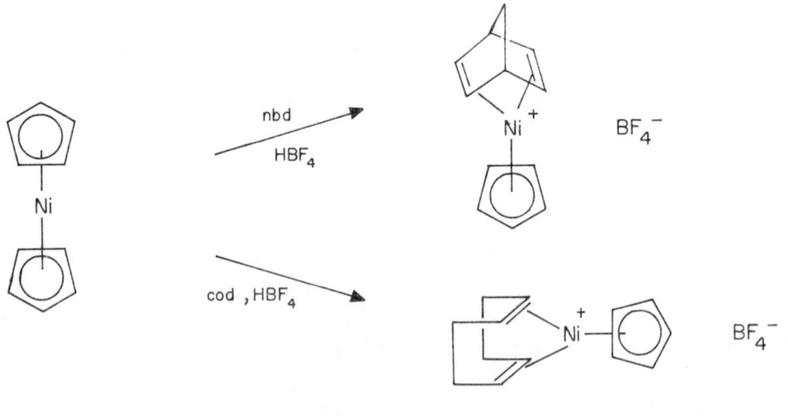

The diene is activated towards nucleophile addition in all of these complexes. The following experiment demonstrates that hydroxide anion adds to the olefin—Pd bond *trans* to the metal, assuming of course that carbonyl insertion occurs at the metal face (see Chapter 2).[1]

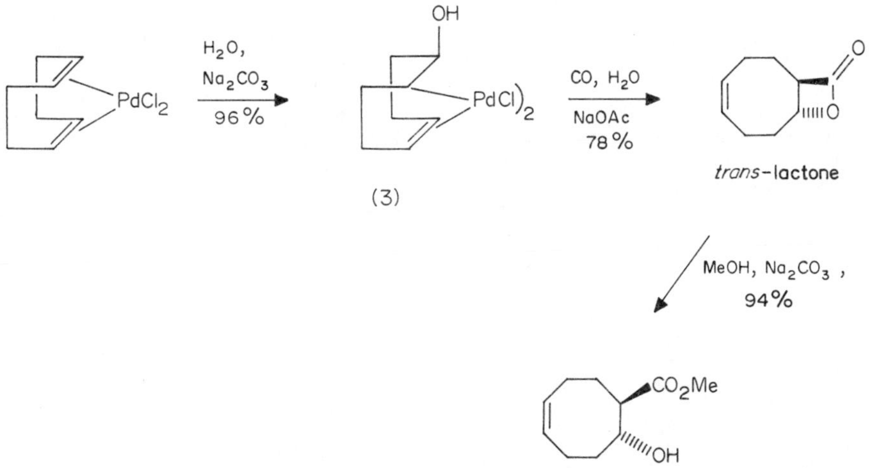

(3)

This result is fairly important in establishing the stereochemistry of nucleophilic attack on alkene—Pd complexes in more general terms. It therefore appears that direct attack occurs on the ligand rather than prior attack at the metal followed by some sort of migration. It may be noted at this point that the direct product of nucleophile addition 3 is a σ-, η^2-complex which is a dimeric species having Pd—Cl—Pd bridges. This again is a fairly general observation. Some other well known nucleophile addition reactions are summarized below.[2]

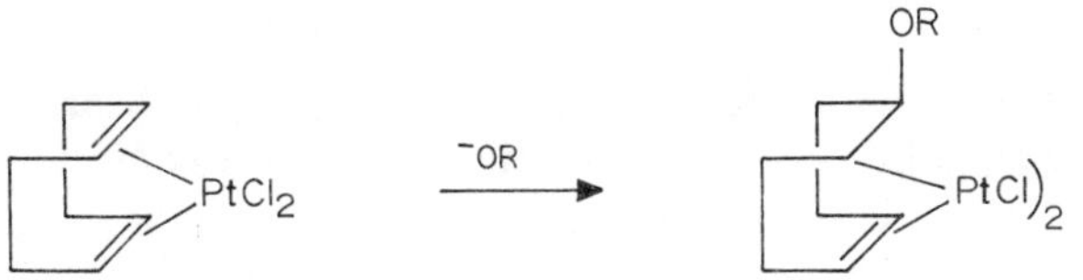

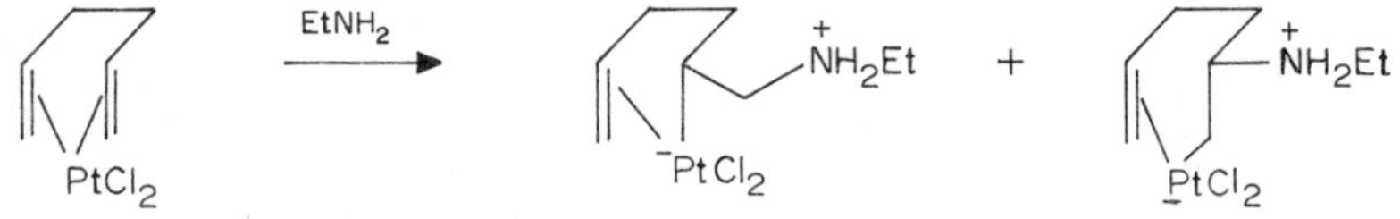

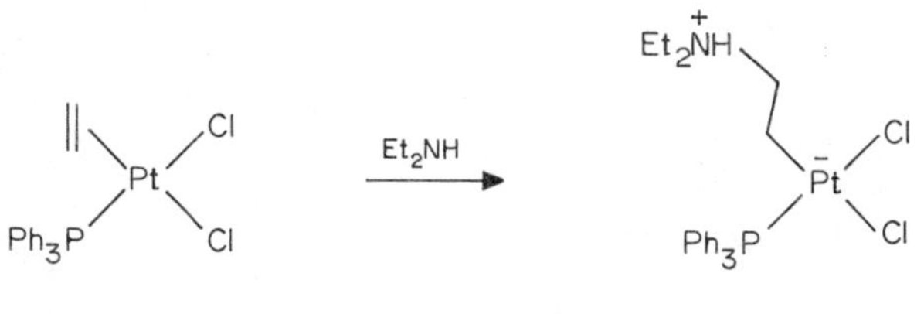

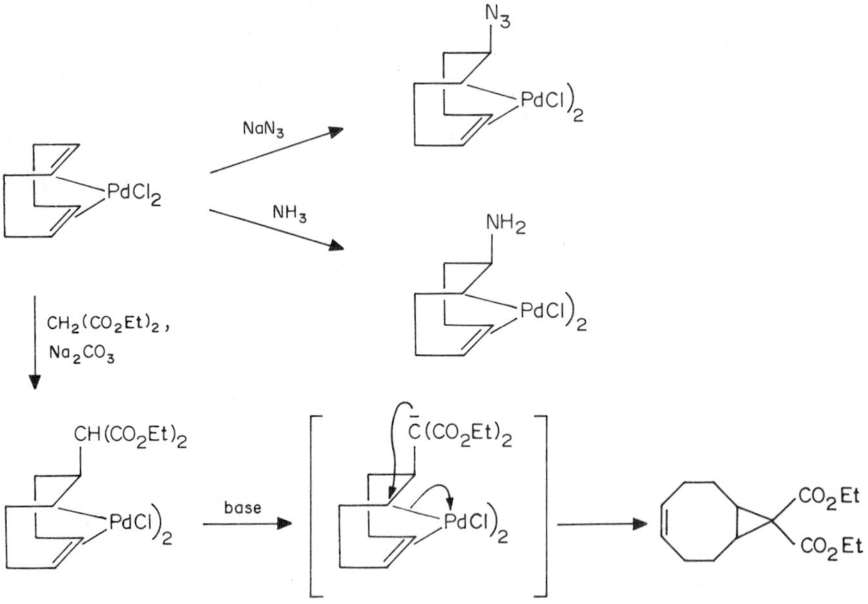

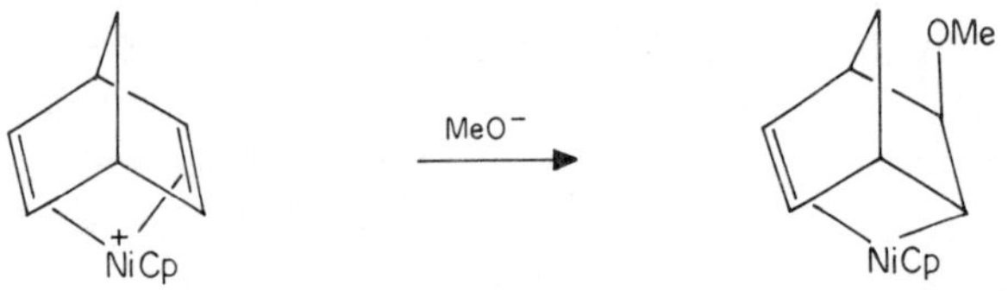

In most cases platinum and palladium complexes behave in a similar fashion. However, they do show contrasting reactivity towards diphenylmercury, palladium complexes behaving normally and platinum complexes undergoing displacement of halides. However, the stereochemistry of the product from the palladium complexes indicates that Ph_2Hg first attacks the metal, displacing one chloride, and the product undergoes migratory insertion:[3]

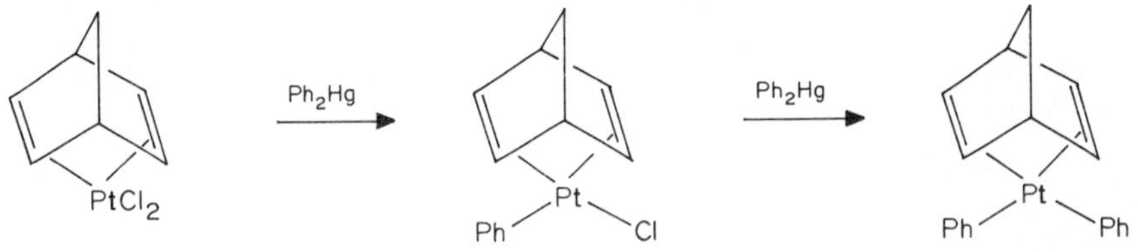

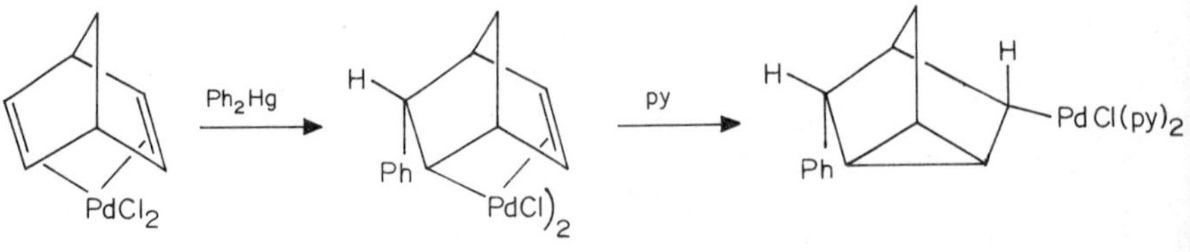

Reaction with certain sulphur nucleophiles results in displacement of the diene,[4] e.g.

The products of nucleophile addition may be converted into organic compounds by a number of methods. Sodium borohydride reduces the σ-, π-allyl system to an alkane, whilst treatment with base results in elimination to give a substituted diene, e.g.

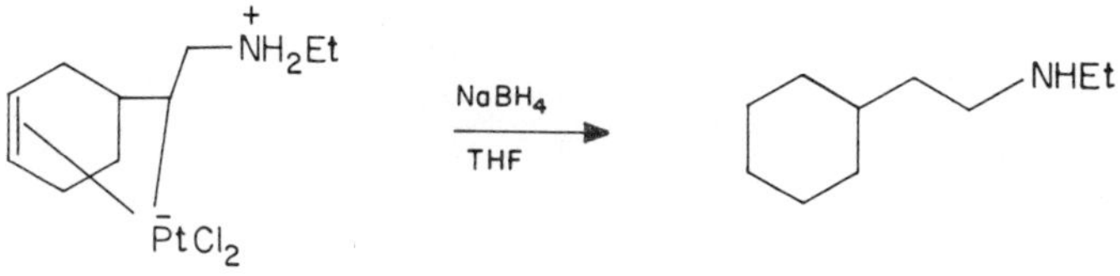

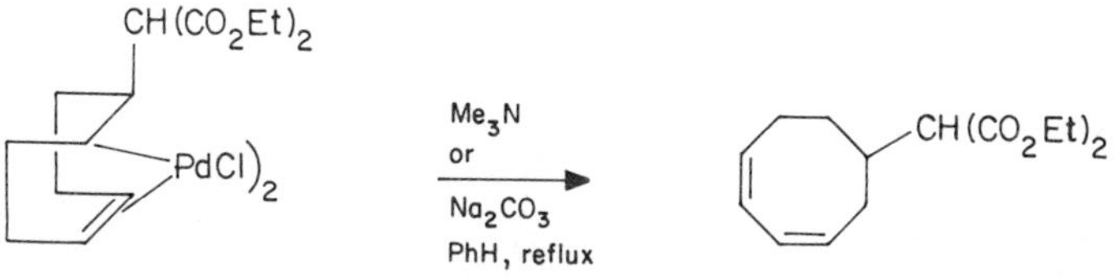

The chloride bridges to the dimeric species can be broken by treatment with amine, e.g.

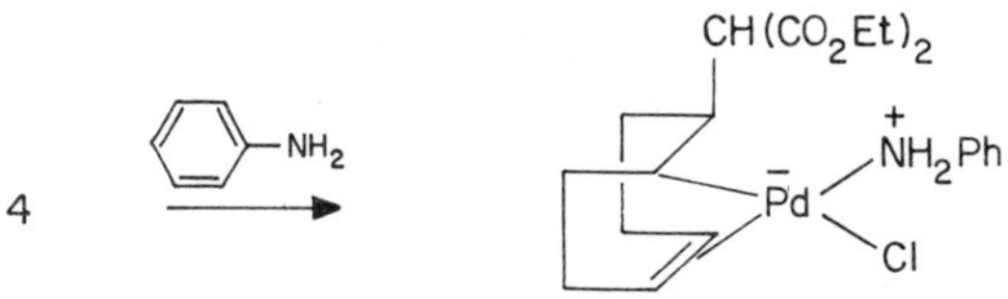

Bis(cod)platinum complexes have been prepared by reaction of (cod)PtCl$_2$ with cyclooctatetraene dianion in the presence of excess of cycloocta-1,5-diene at low temperature. Alternatively, cobalticene can be used as the reducing agent.[5] The corresponding palladium complex can only be prepared if propene is incorporated into the reaction mixture.[6] The complexes react with electrophiles,[7] occasionally giving dinuclear species as products:[8]

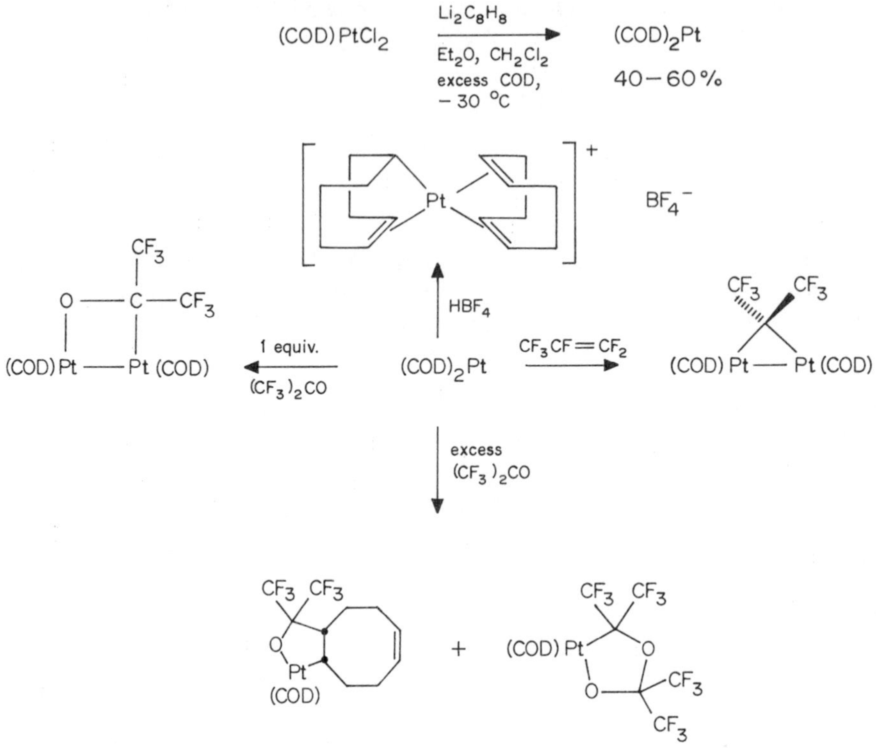

$(COD)PtCl_2$ $\xrightarrow[\substack{Et_2O,\ CH_2Cl_2 \\ excess\ COD, \\ -30\ °C}]{Li_2C_8H_8}$ $(COD)_2Pt$ 40–60%

$$\left[\text{Pt}(COD)_2 \right]^+ \quad BF_4^-$$

$$\begin{array}{c} CF_3 \\ | \\ O-C-CF_3 \\ | \\ (COD)Pt-Pt(COD) \end{array} \xleftarrow[(CF_3)_2CO]{1\ equiv.} (COD)_2Pt \xrightarrow{CF_3CF=CF_2} \begin{array}{c} CF_3\ \ CF_3 \\ \diagdown\diagup \\ (COD)Pt-Pt(COD) \end{array}$$

(COD)₂Pt ↑ HBF₄

excess $(CF_3)_2CO$ ↓

$$\begin{array}{c} CF_3\ CF_3 \\ \text{Pt} \\ (COD) \end{array} \quad + \quad (COD)Pt \begin{array}{c} CF_3\ CF_3 \\ O \\ O-CF_3 \\ CF_3 \end{array}$$

Cobalt, Rhodium and Iridium

These metals also form π-complexes with unconjugated diolefins, although rhodium and iridium have been most studied. The dimeric rhodium complex, having chloride bridges, was prepared during the 1950s, a time when much increased activity in the preparation of organometallic compounds was witnessed.[9] The chloride bridges can be broken by treatment with e.g., p-toluidine:

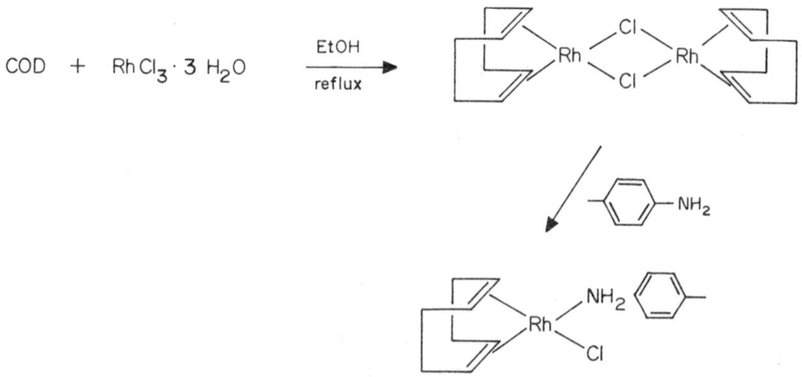

$$COD + RhCl_3 \cdot 3\ H_2O \xrightarrow[\text{reflux}]{\text{EtOH}} \begin{array}{c} Rh \diagup\diagdown Cl \diagdown Rh \\ Cl \end{array}$$

$$\xrightarrow{\text{—⬡—NH}_2} \begin{array}{c} Rh \diagdown NH_2-⬡- \\ Cl \end{array}$$

Medium-ring dienes are readily converted to the diolefin–rhodium complexes, isomerization of a double bond often occurring if the existing one does not have appropriate geometry to be accommodated:[10]

RhCl3
EtOH
RhCl)2

RhCl3 · 3 H2O
95 % EtOH
reflux, 1 h
RhCl)2

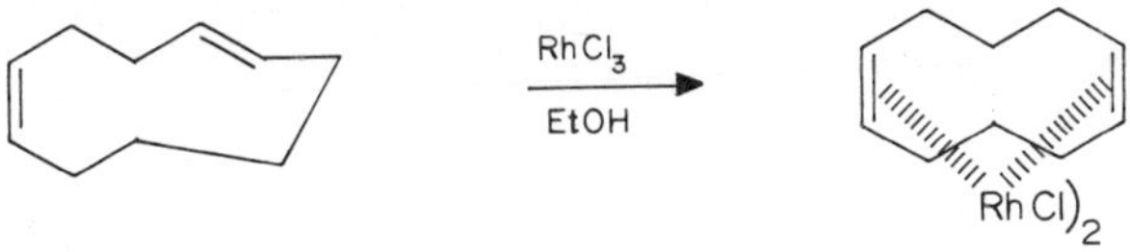

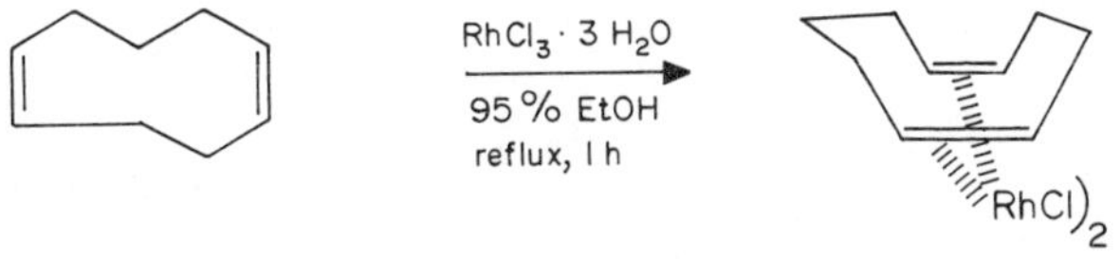

Preparation of the analogous cod–iridium and –cobalt complexes requires more manipulation. A high yield is obtained by removal of HCl from the initially formed HCl adduct using a mild base such as sodium acetate. The deep red dimeric product can then be converted into a yellow monomeric species by treatment with an appropriate phosphorus ligand in non-polar solvents:[11]

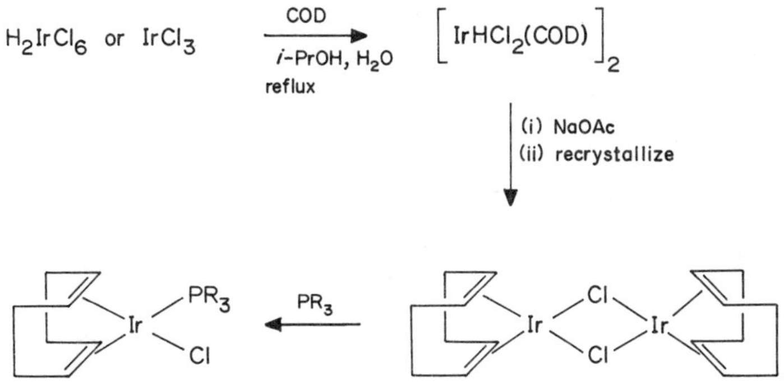

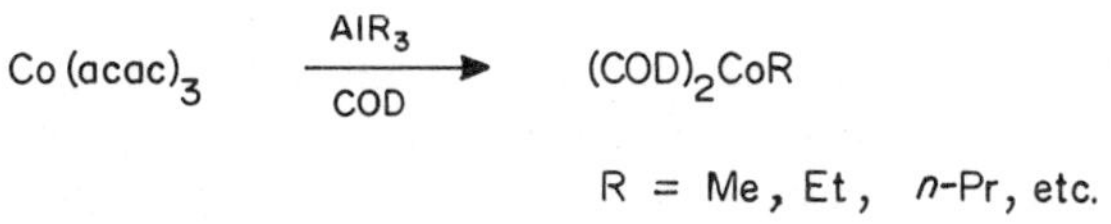

H2IrCl6 or IrCl3
COD
i-PrOH, H2O
reflux
[IrHCl2(COD)]2
(i) NaOAc
(ii) recrystallize
PR3
Ir
Cl
PR3
Cl
Ir
Cl
Ir

Co (acac)3
AlR3
COD
(COD)2CoR
R = Me, Et, n-Pr, etc.

The use of bulky phosphine ligands in the presence of ammonium hexafluorophosphate allows the preparation of cation bisphosphine derivatives, which have been found to show some catalytic activity for hydrogenation of relatively unsubstituted alkenes, especially terminal alkenes.[12]

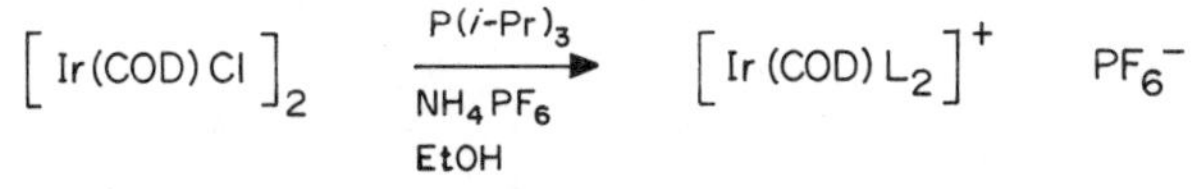

[Ir(COD)Cl]2
P(i-Pr)3
NH4PF6
EtOH
[Ir(COD)L2]+
PF6−

Chemically, these complexes are different to those of the nickel group discussed above. There appear to be no reports of nucleophile addition to the coordinated diolefin. Instead, reaction with electron-deficient entities seems to be the norm, although very limited information is available. An interesting cycloaddition occurs with hexafluorobut-2-yne,[13] whilst the (cod)M(η^5-Cp) derivatives have been shown to undergo protonation readily.[14]

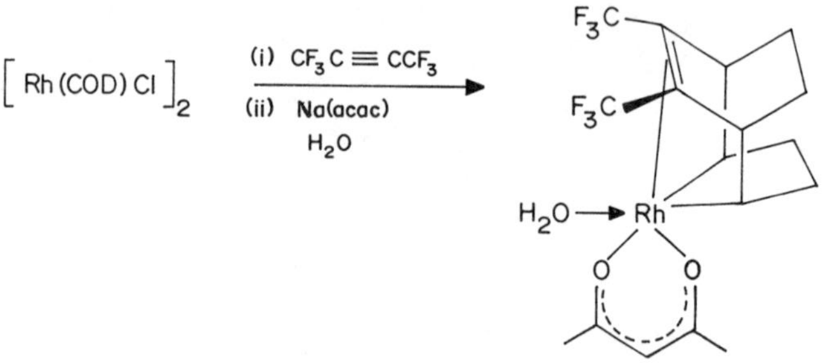

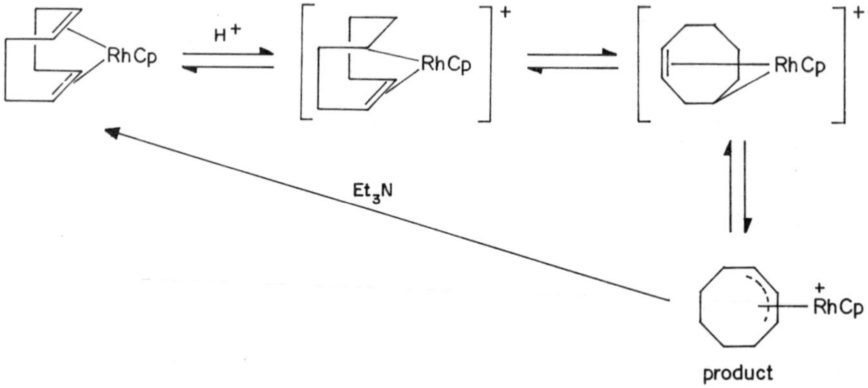

Iron and Ruthenium

Iron and ruthenium differ in their ability to form stable complexes with unconjugated dienes. In fact, iron carbonyl will isomerize such olefins to give a 1,3-diene complex, except in cases where this is a disallowed process, e.g. norbornadiene.[15] Mild conditions are necessary in order to form 1,5-cod complexes of iron.[16] In contrast, ruthenium forms stable complexes with cycloocta-1,5-diene. These appear to be polymeric species, the chloro bridges of which are readily split.[17]

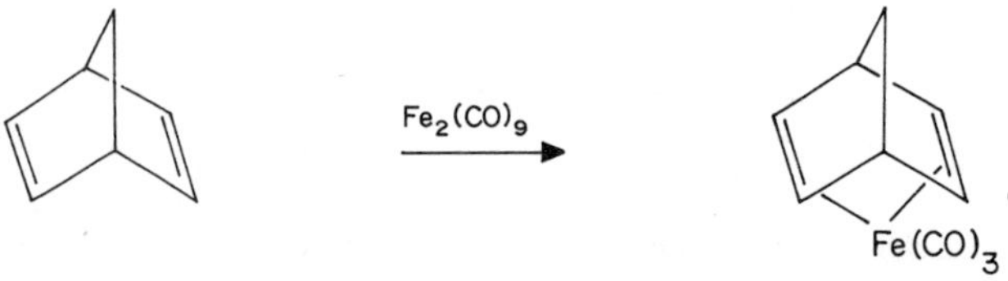

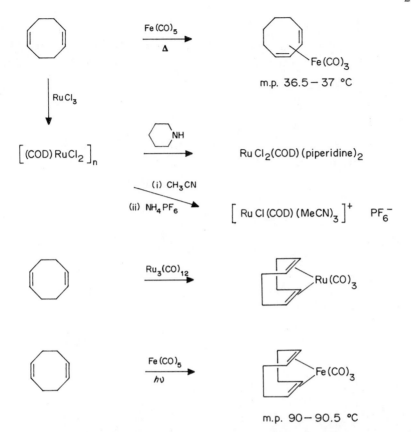

m.p. 36.5 – 37 °C

m.p. 90 – 90.5 °C

Interestingly, reaction of the 1,5-cod complexes with trityl tetrafluoroborate results in hydride abstraction to give η^3-allyl-η^2-alkene cationic complexes. These react with nucleophiles at the allyl or olefinic group, depending on the metal and the nucleophile, but very few details have been given in the literature:[17,18]

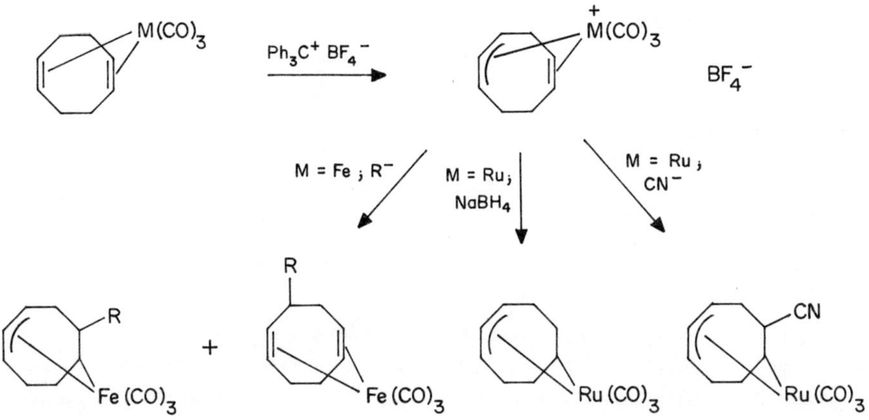

Other Metals

A variety of other metals have been found to form complexes with cycloocta-1,5-diene, although studies of subsequent chemistry are limited. Some examples are given below.[19]

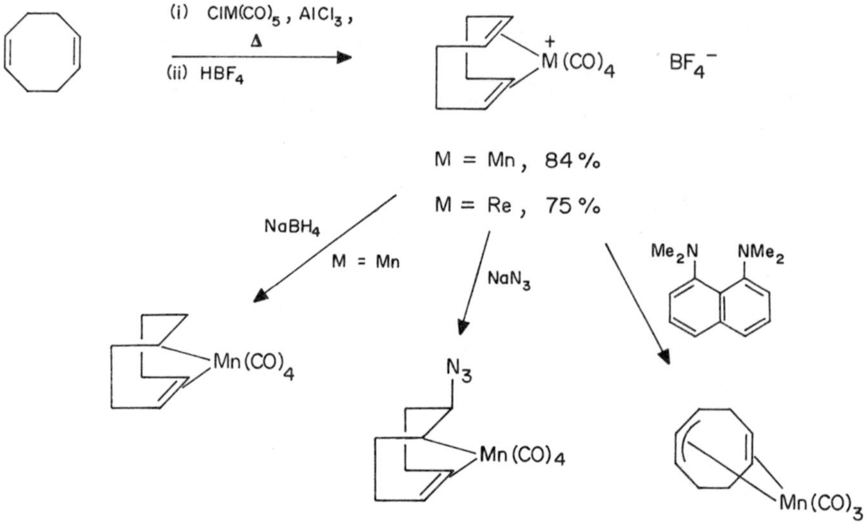
(i) ClM(CO)5 , AlCl3 ,
Δ
(ii) HBF4
+
M (CO)4
BF4−
M = Mn, 84 %
M = Re, 75 %
NaBH4
M = Mn
NaN3
Me2N
NMe2
Mn (CO)4
N3
Mn (CO)4
Mn (CO)3

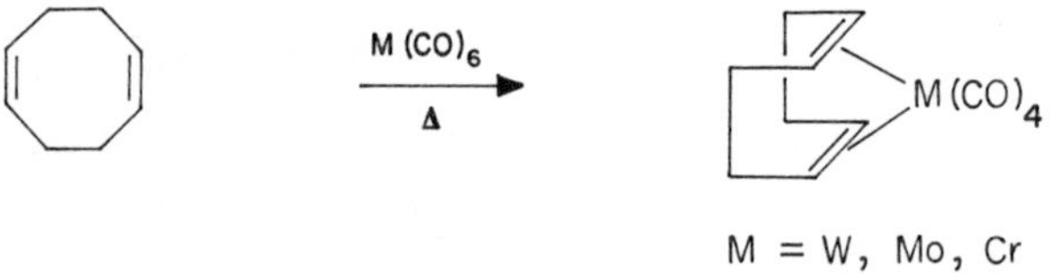
M (CO)6
Δ
M (CO)4
M = W, Mo, Cr

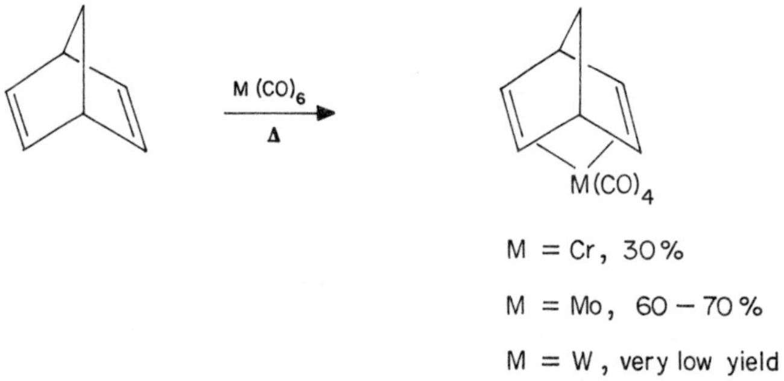
M (CO)6
Δ
M (CO)4
M = Cr, 30 %
M = Mo, 60 – 70 %
M = W, very low yield

Whilst interesting chemistry is observed for some of these cod complexes, there have been no attempts to exploit them for organic synthesis purposes, in some cases possibly owing to the expense of the metals as stoichiometric reagents.

COMPLEXES OF CONJUGATED DIENES

Iron[20]

Butadiene–$Fe(CO)_3$ was prepared as long ago as 1930 by Reihlen et al.,[21] but practically no development in the area took place until more recently, when a number of acyclic 1,3-diene complexes were studied by Pettit's group and others, and when cyclohexadiene complexes became available. The chemistry of these complexes has received careful attention in recent years, so that a full discussion of FMO effects is at least partly possible.

Preparation

The synthetic methods for these compounds already presented in Pettit and Emerson's review[22] are still widely employed, viz. direct reaction of dienes (conjugated or unconjugated) with $Fe(CO)_5$, $Fe_2(CO)_9$ or $Fe_3(CO)_{12}$ under the influence of either heat or ultraviolet irradiation. This induces loss of CO from the iron carbonyl and/or the breaking of an Fe —Fe bond to form the coordinatively unsaturated $Fe(CO)_4$ species, which will react with an olefinic double bond. The mechanism is as follows:

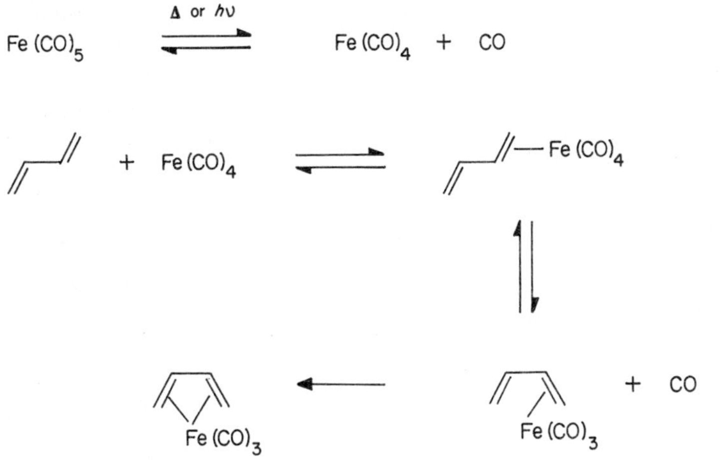

The structure of butadiene–$Fe(CO)_3$ is generally written as **6**, indicating that the iron is η^4-coordinated.

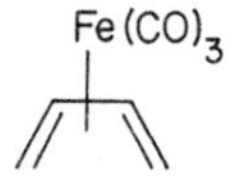

(6)

Since the product complexes and the iron carbonyl starting materials are oxygen-sensitive, all preparations are conducted under an atmosphere of dry nitrogen or argon, as is the case with most organometallics. The most commonly used method is to heat the appropriate diene with Fe(CO)$_5$ at reflux temperature in a high-boiling inert solvent, usually di-n-butyl ether. In order to avoid excessive decomposition of complexes due to peroxides present in the ether, this is generally filtered through a column of basic alumina immediately prior to use. The advantage of using an open system such as this is that the CO produced escapes to the atmosphere, displacing the equilibrium to the complexes. Using this method 1,4-dienes, particularly dihydrobenzene derivatives, are converted to the corresponding 1,3-diene–Fe(CO)$_3$ complexes. Rearrangement of substituted cyclohexadienes invariably occurs, so that for example 1-methoxycyclohexa-1,4-diene gives a mixture of tricarbonyl(1–4-η-1-methoxycyclohexa-1,3-diene)iron **7** and tricarbonyl(1–4-η-2-methoxycyclohexa-1,3-diene)iron **8**, which are readily separated by chromatography.

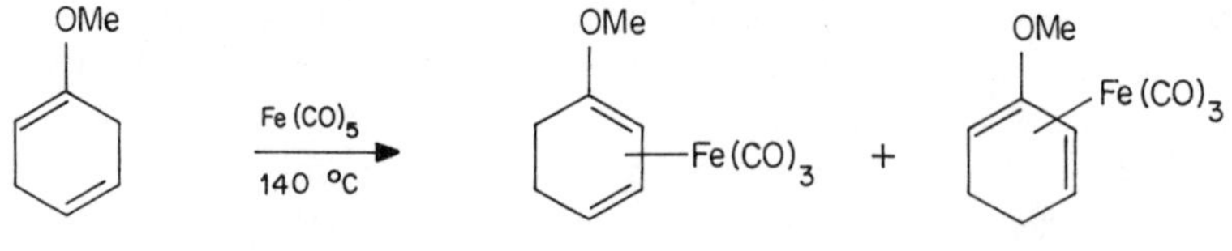

This can be avoided by preconjugation of the diene, as shown below.

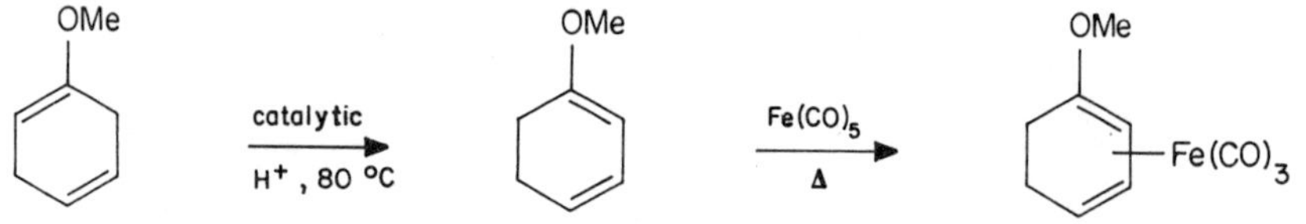

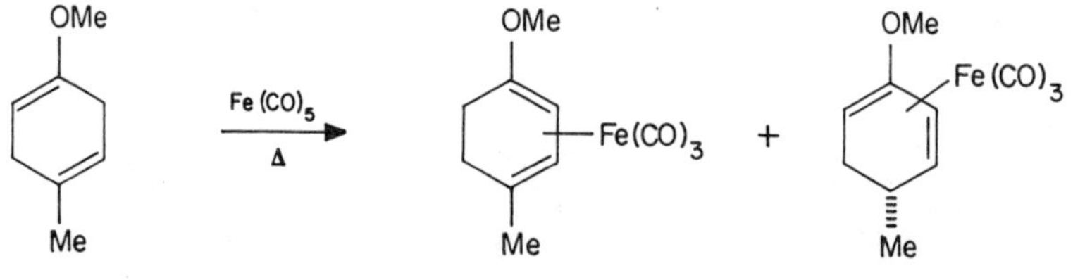

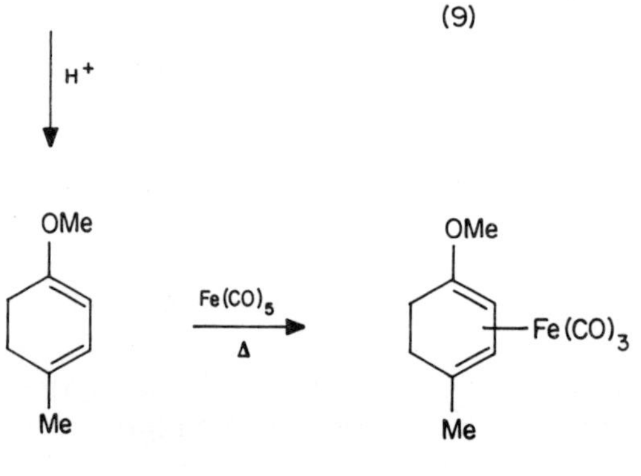

Even the unconjugated diene, 4-vinylcyclohexene, will produce a 1,3-diene–$Fe(CO)_3$ complex:

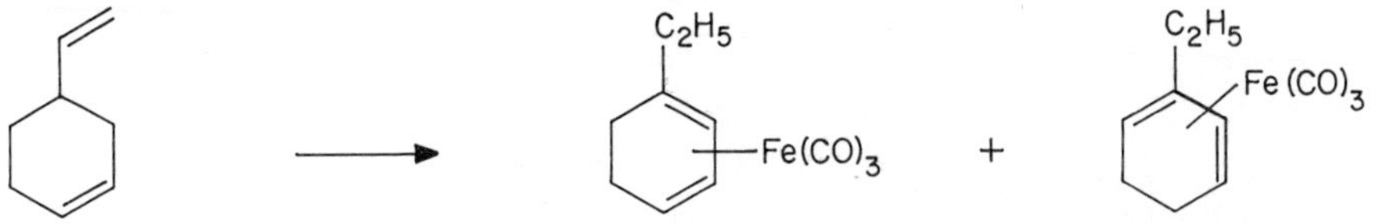

A number of investigations into the mechanism of this rearrangement have led to the general agreement that a transient π-allyliron hydride species is involved, formed by transfer of an *endo*-hydrogen to the iron (*endo* refers to substituents on the same side of the molecule as the $Fe(CO)_3$ group).

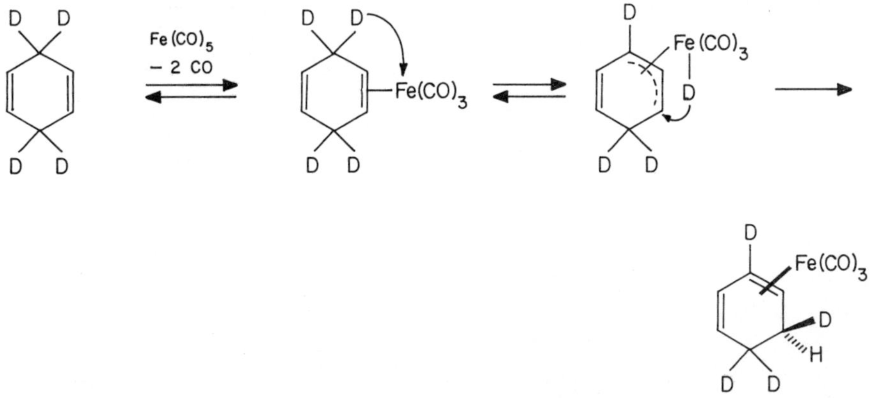

Nonacarbonyldiiron can be used under milder conditions than $Fe(CO)_5$ (e.g. 40–60 °C), but is generally more useful with conjugated dienes. It has been found particularly good for the formation of heterodiene complexes (C=C—C=X):

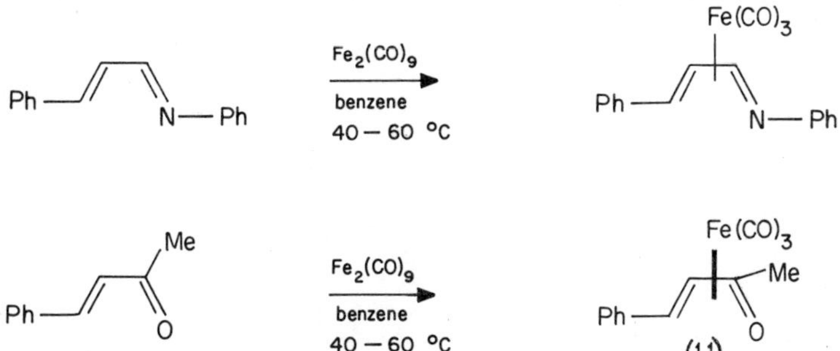

The benzylideneacetone complex **11** and related systems have proved to be very useful as $Fe(CO)_3$ transfer reagents for the synthesis of diene complexes that are otherwise difficult to obtain, or available in only low yields using the standard procedures. The high-yield preparation of the tricarbonyliron complex of ergosteryl acetate reported by Barton's group, using $Fe_2(CO)_9$ in the presence of *p*-methoxybenzylideneacetone, undoubtedly proceeds with the intermediacy of the heterodiene complex.[23]

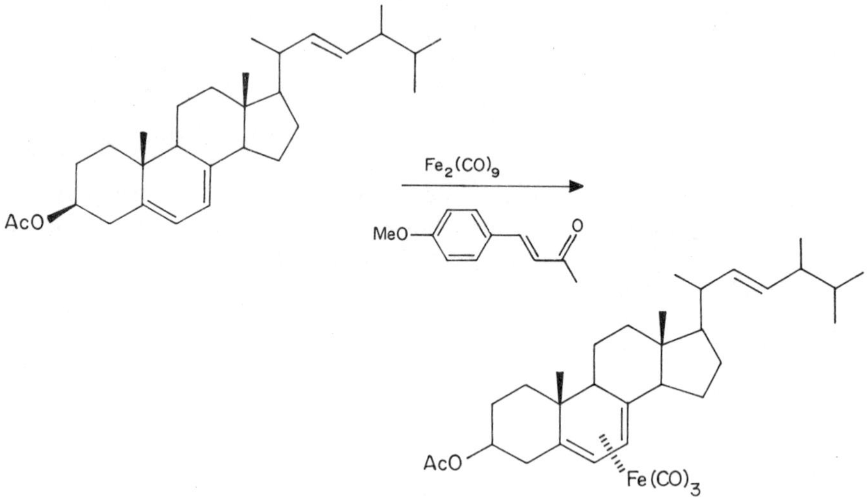

(12)

Perhaps the most promising application of heterodiene complexes is in the preparation of optically active diene complexes. Birch and co-workers have found that treatment of methoxycyclohexa-1,3-dienes with $Fe_2(CO)_9$ in the presence of optically active α,β-unsaturated ketone (e.g. pulegone or steroidal enones) leads to the formation of optically active methoxycyclohexadiene–$Fe(CO)_3$ complexes. The enantiomeric excess is low (< 40%), but this represents a significant breakthrough in view of the applicability of these and derived dienyl complexes to natural product synthesis e.g.

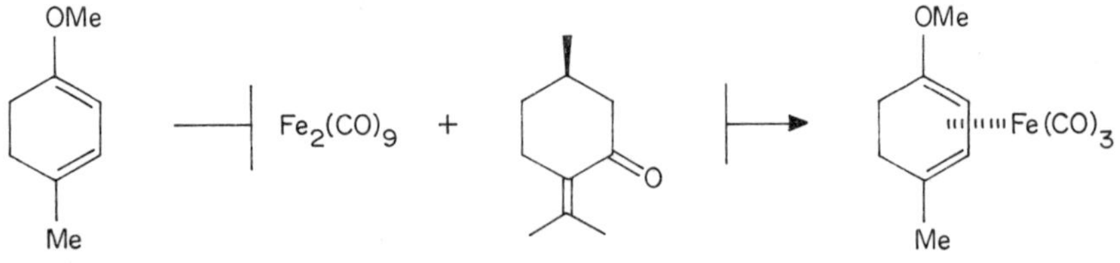

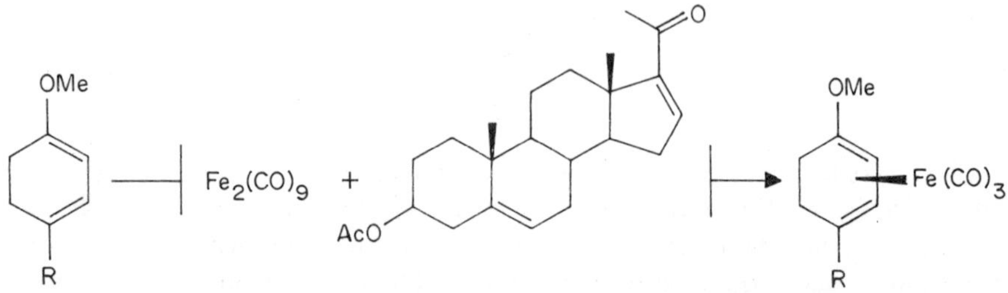

Shvo and Hazum[24] have reported that reaction of a diene with pentacarbonyl-iron (2 mol) and a controlled amount of trimethylamine N-oxide (4 mol) at $0\,^\circ$C in benzene, or at $-78\,^\circ$C in acetone, gives moderate yields of complexes. Reaction of the amine oxide and Fe(CO)$_5$ is thought to generate Fe(CO)$_4$.

Some interesting ring-opening reactions of vinylcyclopropanes and related compounds with pentacarbonyliron should be included here, since in some cases the products are diene–Fe(CO)$_3$ complexes. Irradiation of vinylcyclopropane and Fe(CO)$_5$ at $-50\,^\circ$C in diethyl ether solution leads to an olefin–Fe(CO)$_4$ complex 13, which undergoes ring opening in benzene at $50\,^\circ$C to give a mixture of complexes, together with the product of CO insertion 14. Similar reactions are included below.

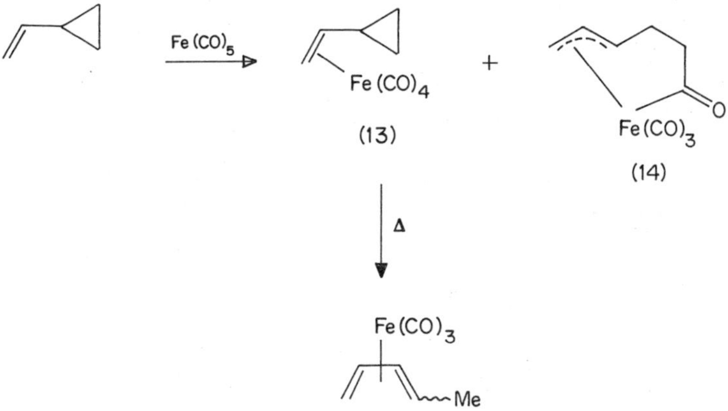

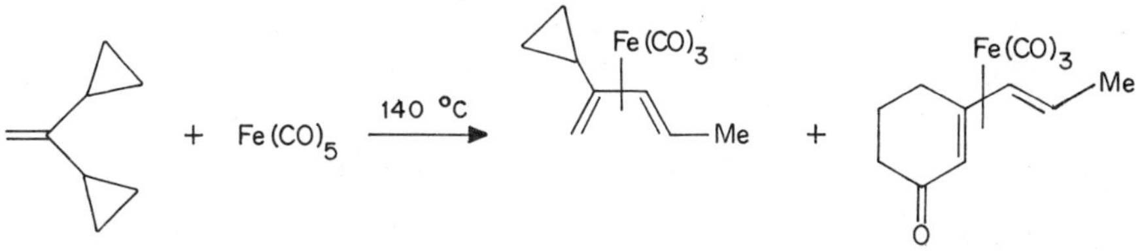

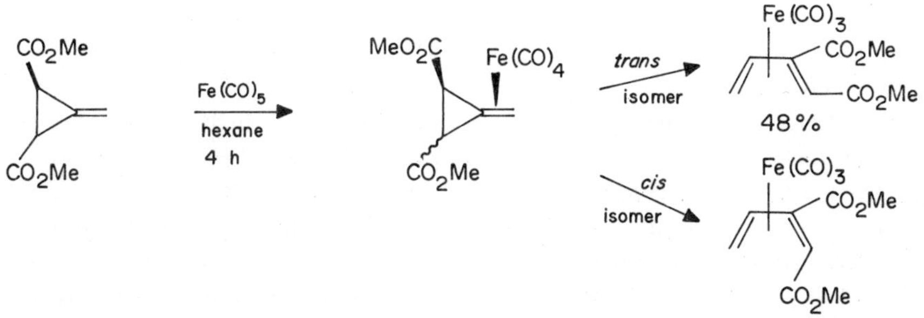

Predictably, the reaction of conjugated diacetylenes with Fe(CO)$_5$ under photochemical conditions does not lead to coordination of both acetylene groups to iron, owing to the spatial arrangement, which does not permit the proximity necessary for two π-bonds to form. The reaction proceeds with carbonyl insertion:

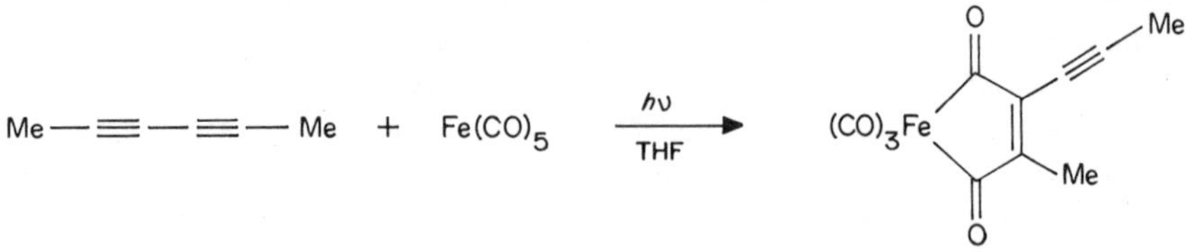

The synthesis of complexes derived from conjugated polyenes, in which one diene unit is complexed, has been described. These compounds undergo shift isomerization on heating:

Cycloheptatriene also forms a diene–Fe(CO)$_3$ complex **15**, in which one of the triene double bonds is free.

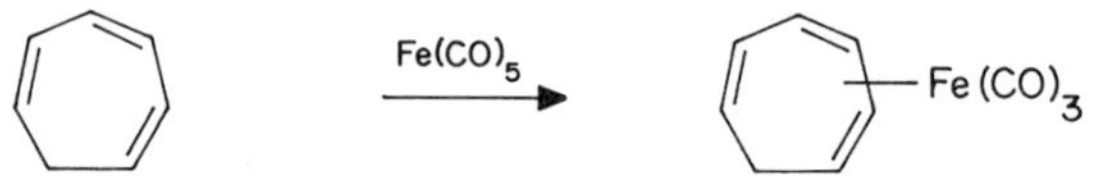

(15)

The difference in reactivity between complexed and uncomplexed double bonds may be used for synthetic purposes (see later).

Properties

The bonding in these complexes has already been discussed. It is of interest here to discuss some of the spectral and physical properties of diene–Fe(CO)$_3$ complexes as representative of the general trends observed for olefinic π-complexes, and the usefulness of, e.g., NMR for determining the structure of substituted derivatives.

The complexes are generally obtained as yellow oils which may or may not have a boiling point sufficiently low to permit distillation, or as yellow crystalline solids. The oils are generally sensitive to aerobic oxidation, whereas the crystalline compounds may usually be stored at 0 °C without exclusion of oxygen.

All tricarbonyldieneiron complexes exhibit two strong absorptions in the infrared at *ca.* 2050 and 1970 cm^{-1}, the latter sometimes being resolvable into two bands, characteristic of the carbonyl stretching frequencies for the Fe(CO)$_3$ group. Electron-withdrawing substituents on the diene moiety generally shift these bands to higher frequencies, whereas electron donors shift them to lower frequencies, a consequence of their effect on the diene–metal and hence metal–

CO $d\pi$–$p\pi$ interactions, and probably attributable to shifts in the energy levels of diene HOMO and LUMO. There are no absorptions above 1600 cm^{-1} due to diene C=C stretching, this being shifted to much lower frequency [1464 cm^{-1} in butadiene–Fe(CO)$_3$] owing to coordination with iron.

In most cases, the mass spectrum shows a molecular iron, followed by successive loss of the three CO ligands, giving $M - 28$, $M - 56$ and $M - 84$ peaks, with subsequent loss of iron to give $M - 140$. Sometimes the molecular ion is too weak to allow accurate mass determinations, and in a few cases the highest mass seen is $M–84$. In some cyclohexadiene complexes there is the added possibility of loss of hydrogen during fragmentation, the driving force being aromatization. Hydrogen loss usually occurs after loss of two of the CO ligands, so that the spectra show M, $M - CO$, $M - 2CO$, $M - (2CO + 2H)$ and $M - (3CO + 2H)$. The last ion is often the base peak.

For structural information NMR spectra, both proton and carbon-13, are indispensable. Tricarbonylcyclohexadieneiron 16 shows H-2 and H-3 in the proton NMR spectrum as the A$_2$ part of an A$_2$X$_2$ group, in the normal olefinic region at δ 5.23 p.p.m. ($J_{2,3} = 4.1$, $J_{1,2} = J_{3,4} = 6.6$, $J_{2,4} = J_{1,3} = 1.5$ Hz), H-1 and H-4 as a multiplet at δ 3.14 p.p.m. (near the *allylic* region) and the methylene protons as a multiplet at δ 1.63 p.p.m. The separation of these signals makes the task of determining substitution patterns of derived diene complexes simpler. Some examples are given in the structures below.

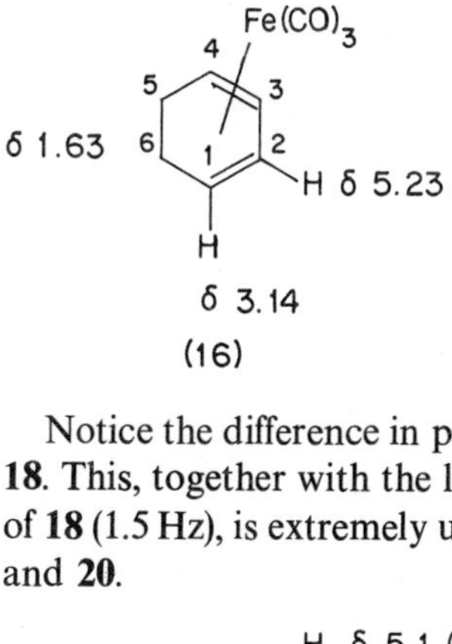

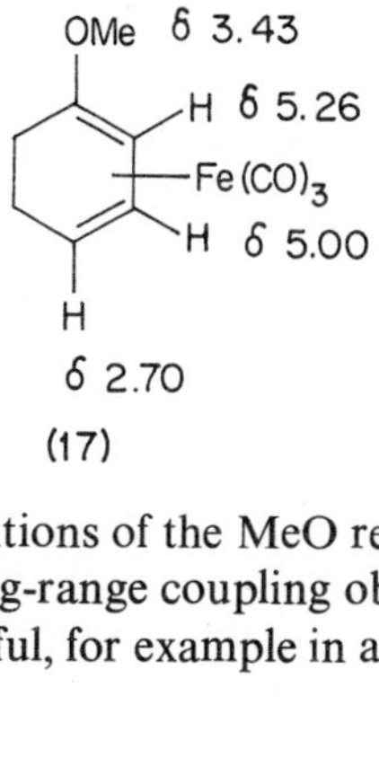

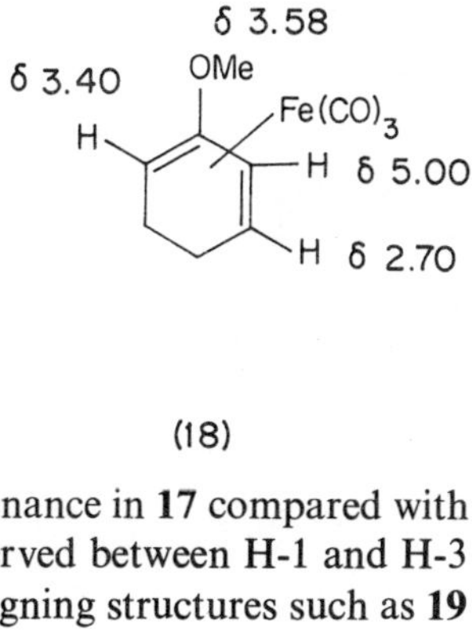

Notice the difference in positions of the MeO resonance in 17 compared with 18. This, together with the long-range coupling observed between H-1 and H-3 of 18 (1.5 Hz), is extremely useful, for example in assigning structures such as 19 and 20.

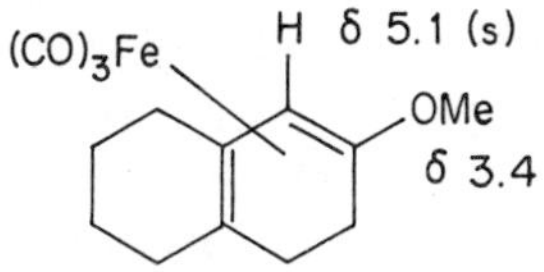

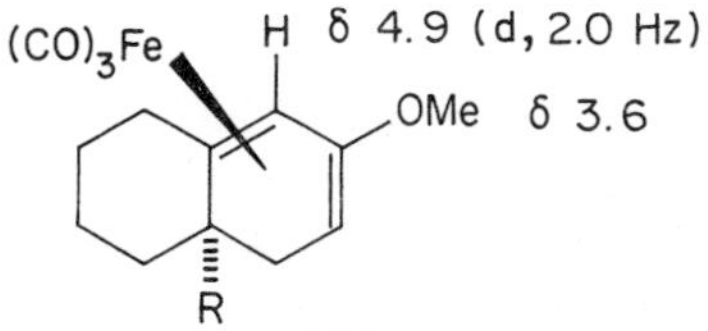

The tricarbonyliron complexes of acyclic dienes show analogous proton NMR spectra, H-2 and H-3 of the parent butadiene complexes occurring at δ 5.28 and H-1a and H-4a at δ 1.68 p.p.m. The most striking difference between this and the cyclic diene complexes is that we now have two additional protons, H-1b and H-4b, which resonate at very high field (δ 0.22 p.p.m.):

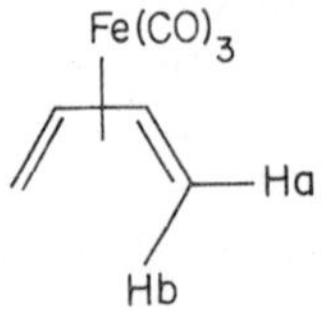

Some discussion of ^{13}C NMR spectra was presented in Chapter 1, mainly the effects of substituents. In diagrammatic form below are given NMR data (δ) for some simple complexes and related dienes, for comparison.

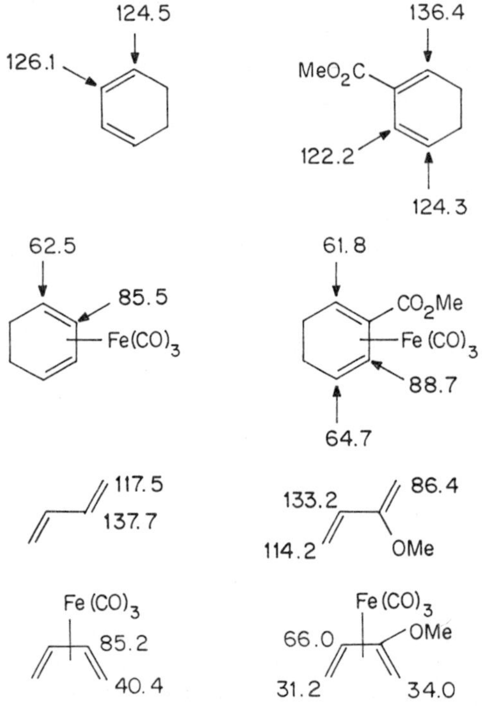

On first inspection, the relatively high-field resonances of the terminal diene carbon atoms of the complexes might be taken to imply that these have some sp^3 character. However, the observation that the C–H coupling constants $^1J_{CH}$ for the terminal carbons (172.8 Hz) and C-2/C-3 (172.8 Hz) are the same in the cyclohexadiene complex and in the range expected for sp^2 hybridized carbons (ca. 160–170 Hz) rather than sp^3 carbons (ca. 130 Hz) quickly dispels this notion.

It is interesting that at normal temperatures the Fe(CO)$_3$ group shows a single resonance at ca. 212 p.p.m for the CO carbons. When the temperature is lowered, two resonances are seen in an intensity ratio 2:1, which are expected for a complex of fixed geometry. This is due to the fluxionality phenomenon discussed in Chapter 2.

Removal of the Fe(CO)₃ Group

To be useful as a stoichiometric reagent in organic synthesis, a metal complex must be capable of having the metal removed in order to release, in high yield, the organic moiety which will presumably be involved in a synthetic scheme. Although the diene ligand can in a few limited cases be displaced by another ligand, e.g. triphenylphosphine, this method is not general and consequently more reliable means are sought. The most widely applicable procedures involve the use of a mild oxidizing agent that is capable of oxidizing the metal but not the liberated ligand. With acyclic dienes, over-oxidation is not usually a problem, but cyclohexadienes may aromatize very readily if oxidation is too severe. A number of transition metal oxidants have been used, cerium(IV) ammonium nitrate being particularly successful for a number of acyclic diene complexes, and iron(III) chloride or ethanolic copper(II) chloride for cyclohexadiene complexes. Shvo and Hazum[24] reported the use of trimethylamine *N*-oxide, and this is now the reagent of choice. It has the advantage that reaction conditions are neutral to basic (Me₃N is generated, so this depends to some extent on solvent and temperature), so that methoxy-substituted dienes can be liberated without hydrolysis of the acid-sensitive enol ether moiety.

Chemistry of Diene–Fe(CO)₃ complexes

Generally, the complexes do not undergo the same reactions as free dienes. Thus, they are resistant to Diels–Alder reactions, catalytic hydrogenations, hydroxylation, etc. This makes Fe(CO)₃ an attractive protecting group for conjugated dienes (see later). The complexes will react with electrophilic reagents. The butadiene complexes appear to react more readily than cyclohexadiene complexes, and the reactions proceed without polymerization of the diene. Butadiene–Fe(CO)₃ is acetylated under mild conditions to give mixtures of **21** and **22** in varying proportions, depending on the work-up used. Thus, quenching the reaction mixture with cold aqueous ammonia gives **21**, while use of sodium methoxide in methanol gives exclusively **22**, provided sufficient time is allowed for isomerization.

The stereochemistry of acylation, while not important for butadiene complexes, deserves comment. The reaction appears to take place by reaction of the acyl cation on the diene face *endo* to the metal, as evidenced by the isolation and X-ray structure determination of the π-allyl intermediate **23**.[25]

(23)

The most thorough study of acetylation of acyclic derivatives was carried out by Graf and Lillya,[26] who investigated the regiochemistry of the reaction with a number of substituted diene complexes. Of particular interest is that tricarbonyl(2-methoxybutadiene)iron gives exclusively the product **24** on acetylation, although in only moderate yield. For an uncomplexed diene, we might have expected electrophilic attack at the terminus α- to the MeO substituent. A plausible explanation is that acylation occurs first on the Fe 'lone pair' (see above) and the acyl cation transfers to the diene carbon atom having the largest coefficient in the complex HOMO. Since the complex HOMO corresponds to the diene LUMO, we expect that acylation will occur at C-4, as observed. The implication is that the reaction is under overall frontier orbital control:

(24)   complex HOMO

Acetylation of tricarbonyl(cyclohexadiene)iron occurs less readily and gives the M–*endo* acetylated product in low yield (30–40%), in accordance with the above stereochemical results. Apparently the π-allyl intermediate is extremely unstable (see later) and readily transfers the only available, but albeit less acidic, proton *endo* to Fe, to give the product **25** directly without the need for aqueous work-up. The triphenylphosphine dicarbonyl complex **26** reacts much more readily, even at $-78\,°C$, to give the product **27** in 85% yield. This undoubtedly reflects the enhanced availability of electron density on the Fe t_{2g} orbitals.[20]

(25)

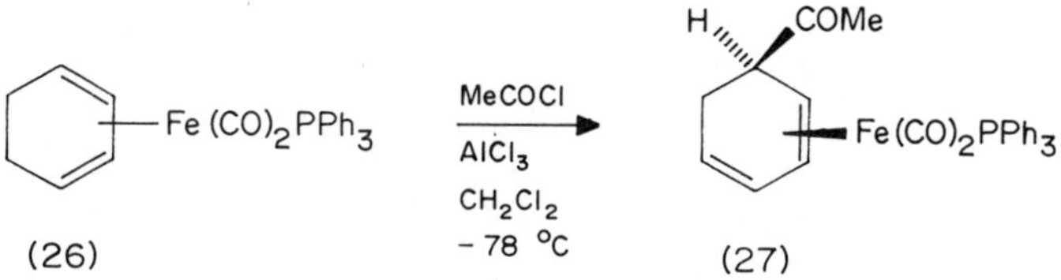

With cycloheptatriene complexes a different situation arises, since there is both an uncoordinated double bond and a diene–Fe(CO)$_3$ unit. Normally the uncomplexed olefin is more reactive and in this case it is more so, since there is an added driving force in the formation of the stable dienyl–Fe(CO)$_3$ cation (see Chapter 8). This accounts for the observed regiochemistry, and also allows the diene complex to be formylated under conditions where other complexes are unreactive.

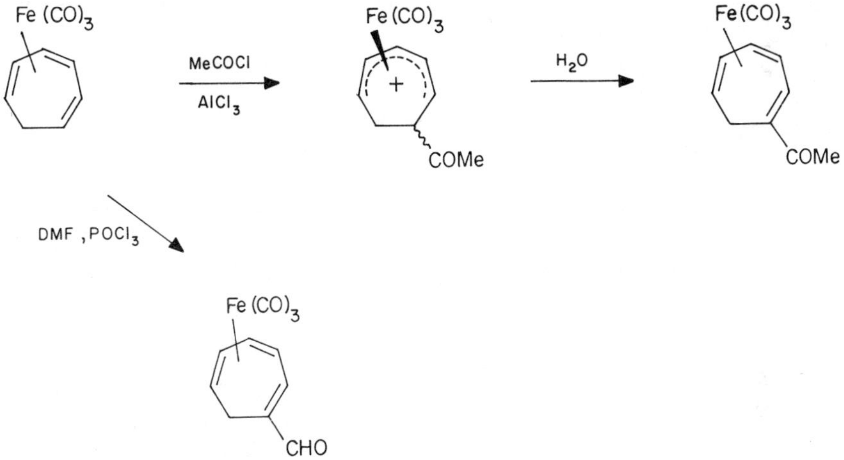

The simplest electrophile, H$^+$, also reacts with diene–Fe(CO)$_3$ complexes. Solutions of butadiene–Fe(CO)$_3$ in acids show broad signals due to dynamic equilibrium between protonated iron species and π-allyl complexes. The Fe–H species is evidenced by high-field proton NMR absorptions (*ca.* $\delta = 6$ p.p.m.).

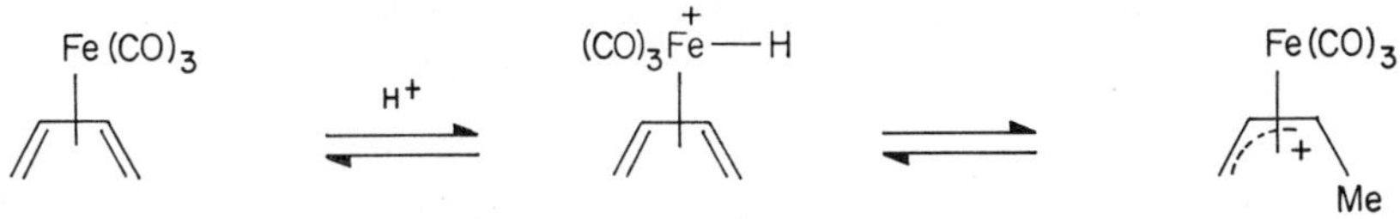

The allyl–Fe(CO)$_3$ species **28** is coordinatively unsaturated. Under appropriate conditions it will disproportionate to give the stable allyl–Fe(CO)$_4$ complex **29** which can be isolated by precipitation with diethyl ether. Better yields of this complex are obtained when the acid treatment is conducted under an atmosphere of carbon monoxide, provided the temperature is not allowed to rise.

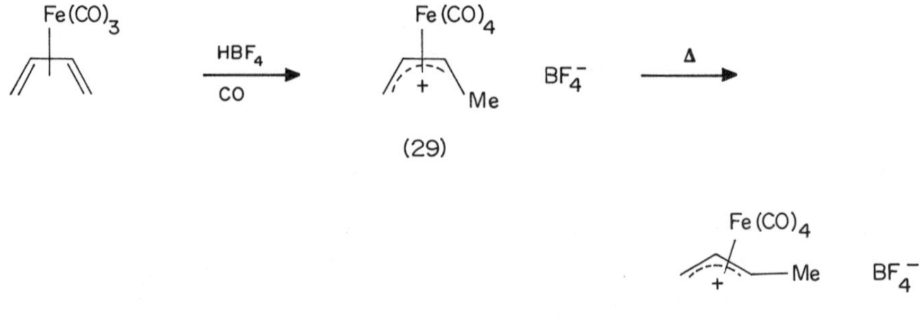

(29)

(30)

If the reaction temperature is allowed to reach *ca.* 50 °C, conversion of **29** to the more stable **30** occurs. π-Allyl–Fe(CO)$_4$ complexes derived from cyclohexadiene–Fe(CO)$_3$ have never been isolated, despite a number of attempts, suggesting inherent instability of these species.

Use of Fe(CO)$_3$ for Diene Protection

Reaction of the terpene myrcene with Fe(CO)$_5$ gives good yields of myrcene–Fe(CO)$_3$ **31**. The isolated double bond of this complex can be made to undergo a number of typical conversions without affecting the diene–Fe(CO)$_3$ grouping. These are summarized below.[27]

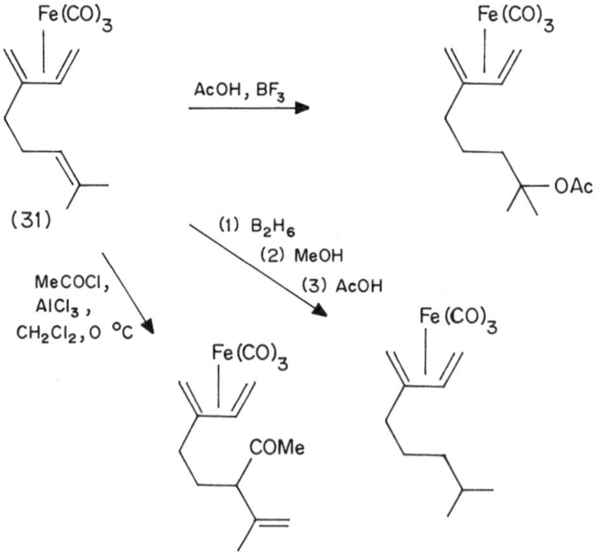

(31)

The enhanced reactivity of the free double bond may be used to form species which will undergo intramolecular reaction with the diene–Fe(CO)$_3$ group, as shown below.[28]

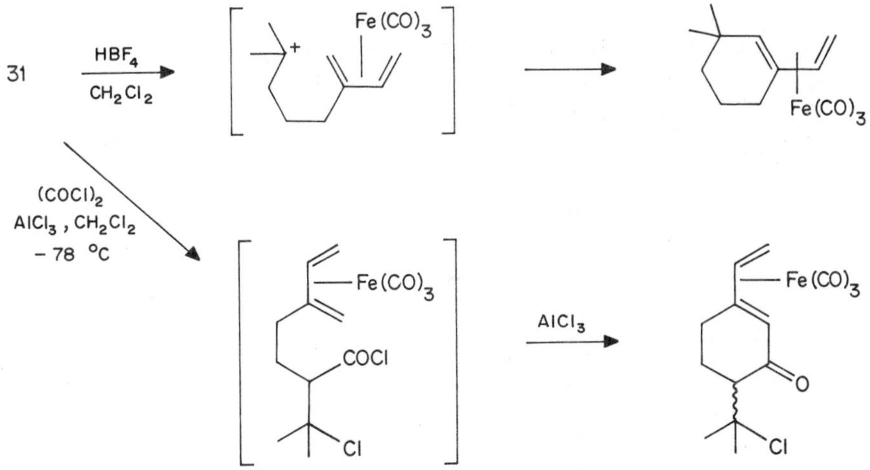

Tricarbonyl(ergosteryl acetate)iron also shows protection of the ring-B diene moiety during a number of useful transformations, shown below. The Fe(CO)$_3$ group can be removed easily with iron(III) chloride to liberate the organic molecules.[23]

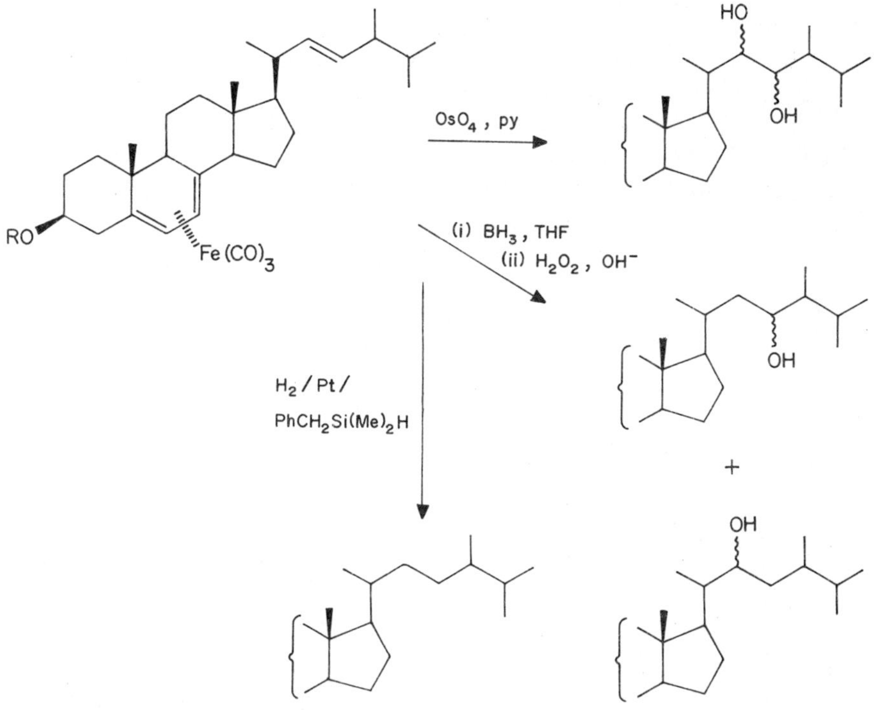

The application of both the steric bulk and coordinating ability of the Fe(CO)$_3$ group is shown in the following conversion of ergosterol to epiergosterol 33. The coordinating ability of the iron prevents the diene of the intermediate **32** from conjugating with the ketone carbonyl, and its steric bulk forces the hydride reducing agent to attack from the usually more hindered β-face.[23]

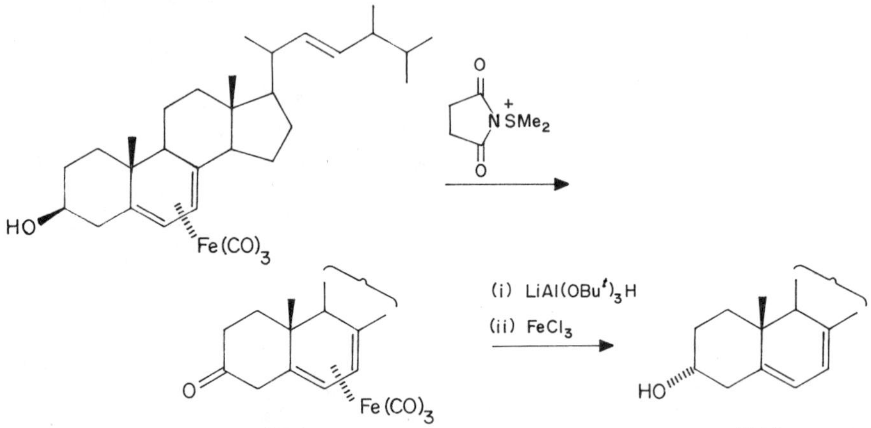

(32)   (33)

An elegant exploitation of the ability of $Fe(CO)_3$ to protect the diene and also activate a triene complex to Friedel–Crafts reaction has been reported by Franck-Neumann et al.[29] Cycloheptatrienone normally does not undergo Friedel–Crafts reactions owing to the deactivating influence of the keto group. However, its $Fe(CO)_3$ derivative **34** readily acetylates to give **35**, owing to the driving force discussed above. The non-complexed double bond in **35** reacts selectively with 2-diazopropane to give, after ring opening of the intermediate, the complex **36**. This was then converted to the natural product β-thujaplicin **37** or β-dolabrin **38**.

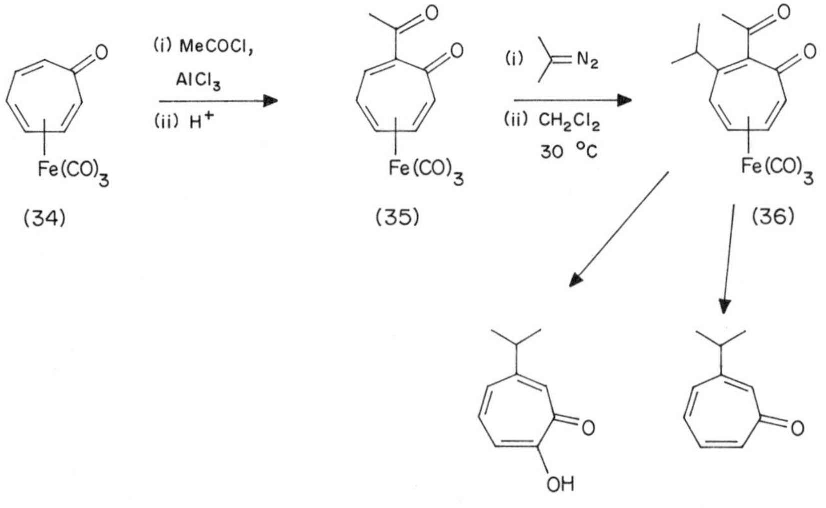

(34)   (35)   (36)

(37)   (38)

The diene–Fe(CO)$_3$ moeity is stable to a large number of organic functional group transformations. This will become apparent in Chapter 8 when the synthetic applications of dienyl–Fe(CO)$_3$ complexes are discussed. The conversion of the diene complexes to the latter by treatment with triphenylmethyl tetrafluoroborate leads to one of the more fruitful areas for synthetic applications:

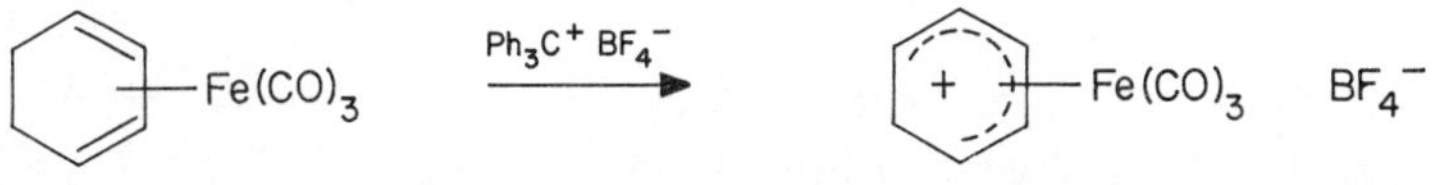

While one normally thinks of diene–Fe(CO)$_3$ complexes as reactive towards electrophiles, some interesting results have been obtained by Semmelhack,[30] who showed that treatment with strong carbon nucleophiles results in addition to the diene ligand to produce an anionic complex which can be protonated, with loss of iron to give a mixture of olefins.

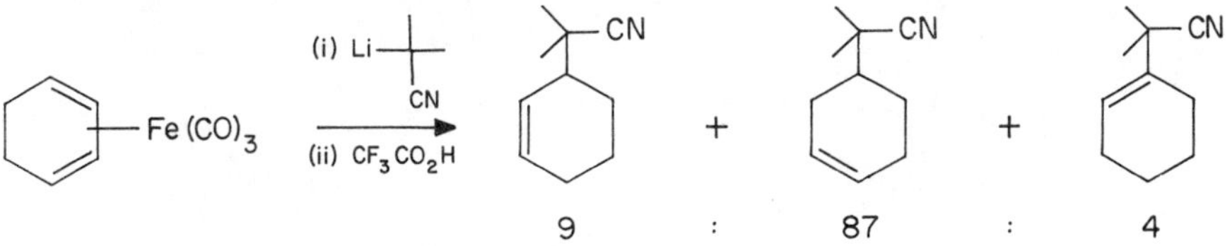

$$9 \quad : \quad 87 \quad : \quad 4$$

Oxidative Cyclization

Treatment of a tricarbonyldieneiron derivative containing an enolizable 1,3-dicarbonyl group in the side-chain with manganese dioxide[31] results in an oxidative cyclization, e.g. **41** → **42** and **43** → **44**.

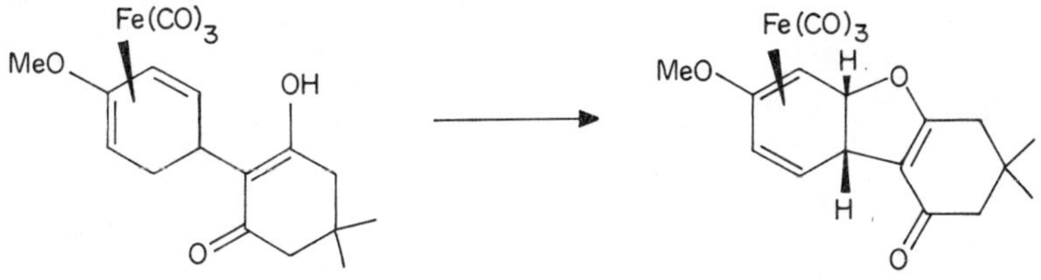

(41)   (42)

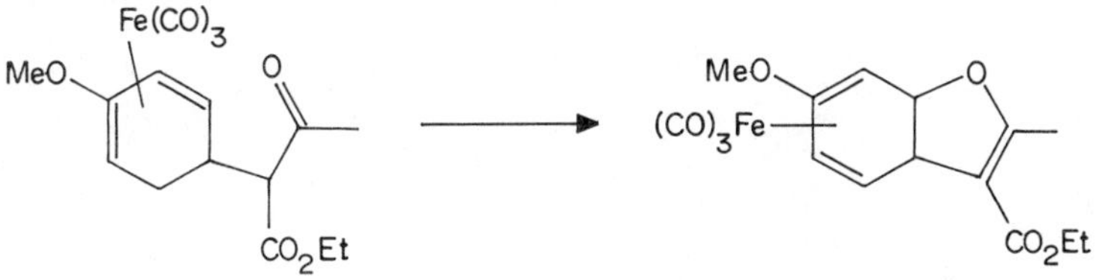

(43)   (44)

Intermolecular oxidative addition of alcohols has also been achieved by treatment of the cyclohexadiene–Fe(CO)$_3$ complex with thallium(III) trifluoro-acetate in the alcohol as solvent.[32]

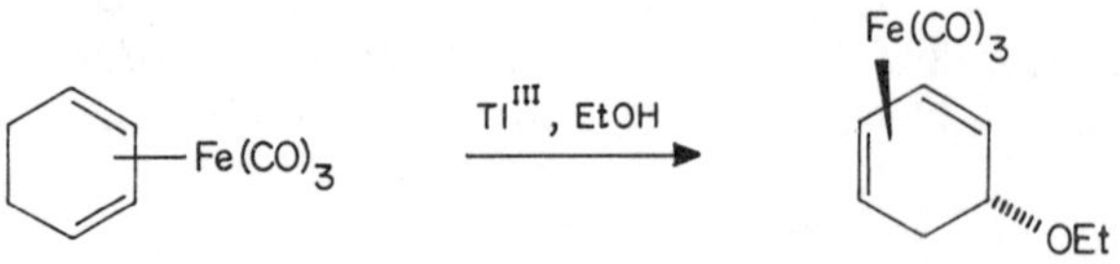

Intramolecular variations of these reactions have been achieved with complexes containing hydroxy-substituted lateral chains.[33] This is useful for synthesizing dienyl cation complexes having structures inaccessible by the usual methods:

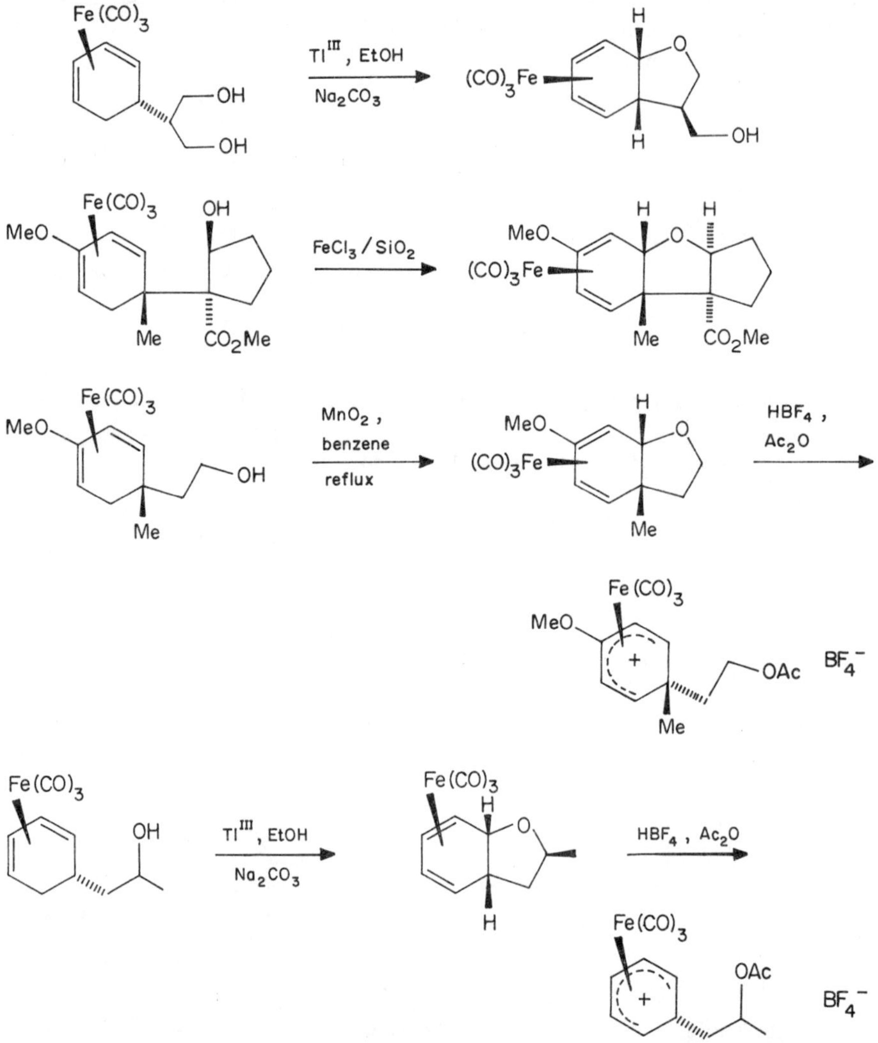

Cyclooctatetraene–Fe(CO)₃

The attachment of an Fe(CO)$_3$ unit to cyclooctatetraene offers a means of moderating the reactivity of this compound and of obtaining a degree of control over reactions which normally cannot be performed. The uncomplexed tetraene gives rise mainly to polymeric material on attempted Friedel–Crafts acylation. Reaction of pentacarbonyliron with cyclooctatetraene under either thermal or photochemical conditions gives good yields of the complex **45** as a red crystalline solid. The complex shows averaging in the NMR spectrum which may be 'frozen out' at low temperature, due to fluxionality, the mechanism of which appears to involve 1,2-shifts of the Fe(CO)$_3$ group around the ring.[34] This may partly account for its resistance to catalytic hydrogenation. The reaction of **45** with electrophiles is not a simple process but does give clean products in good yield. Protonation results in formation of the bicyclo[5.1.0]octadienyl complex **46**, which is smoothly reduced with sodium borohydride to give the neutral complex **47**.

(45)   (46)   (47)

Reaction of **45** with the powerful dienophile tetracyanoethene appears to proceed by a non-concerted mechanism to give, finally, the adduct **48**, the formation of which, by internal nucleophile addition to a cyclooctadienyl cation complex, is consistent with other nucleophile additions to these systems (see Chapter 8). Removal of the Fe(CO)$_3$ results in formation of the dihydrotriquinacene derivative **49**.[35]

(48)

(49)

Interestingly, reaction of **45** with chlorosulfonyl isocyanate (CSI) followed by desulfonylation results in the normal cycloaddition product **50** expected from the uncomplexed diene portion. This difference is probably due to the greater ability of a bis-nitrile moiety to stabilize a carbanoin, resulting in preference for a non-concerted pathway.

$$45 \quad \xrightarrow[\text{(ii) PhSH, py}]{\text{(i) O}=\text{C}=\text{NSO}_2\text{Cl}}$$

(50)

The action of heat on **45** results in what appears to be a thermally disallowed disrotatory cyclization of the uncomplexed diene portion of the molecule to give the bicyclo[4.2.0]octatriene complex **51**. Undoubtedly the metal is affecting the course of electrocyclic reaction by perturbation of the cot MOs, as discussed earlier (Chapter 2), but the exact details have not been worked out.

$$45 \quad \xrightarrow{160\ °C}$$

(51)

Some electrophilic substitution reactions of (cot)Fe(CO)$_3$ have also been examined. Vilsmeier formylation proceeds smoothly to give the 'normal' product **52**, from which the metal can be removed to give formylcyclooctatetraene.

$$45 \quad \xrightarrow[60\%]{\text{POCl}_3, \text{DMF}} \qquad \xrightarrow[80\%]{\text{Ce}^{\text{IV}}}$$

(52)

On the other hand, Freidel–Crafts acetylation results in only a very low yield (5%) of unrearranged acetyl derivative **53**, the major product being the bicyclic derivative **54**, analogous to the product of protonation.[36]

$$45 \quad \xrightarrow{\substack{\text{MeCOCl} \\ \text{AlCl}_3}}$$

(53)   +   (54)

An interesting reaction takes place on treatment of **45** with aluminum chloride in benzene at $10\,°C$, to give the carbonyl insertion product **55**. Treatment of this with carbon monoxide at elevated pressure gives barbaralone **56**.[37]

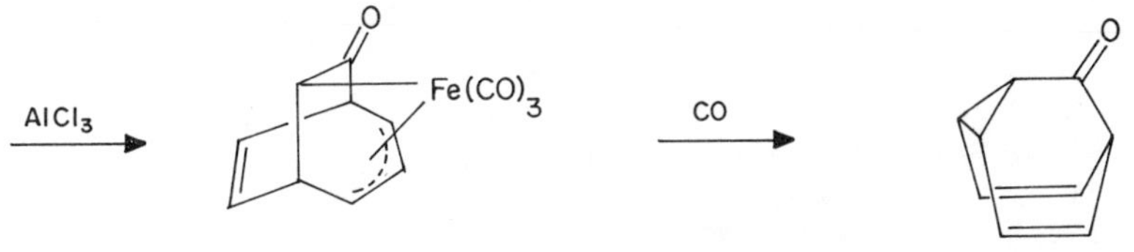

(55)   (56)

Other Metals

Of the other metals which form complexes with 1,3-dienes, the most noteworthy are cobalt, rhodium and iridium, the commonly used counter ligand being η^5-cyclopentadienyl.[38]

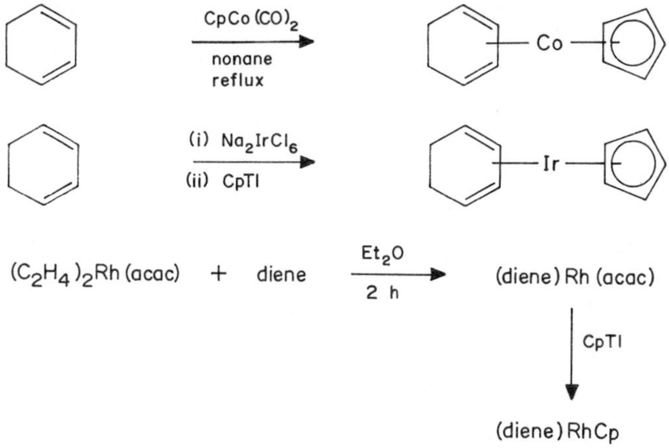

Reaction of these complexes with electrophiles, such as Friedel–Crafts acylating reagents, results in attack on the cyclopentadienyl ligand:

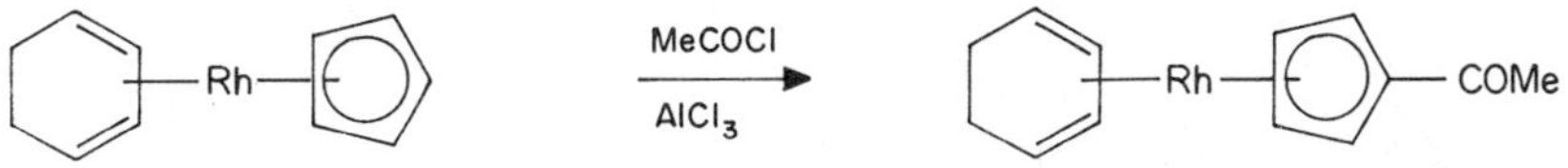

The complexes react with trityl tetrafluoroborate to give cyclohexadienyl complexes by hydride abstraction, and these react with nucleophiles in an analogous fashion to the corresponding iron derivatives (see Chapter 8):

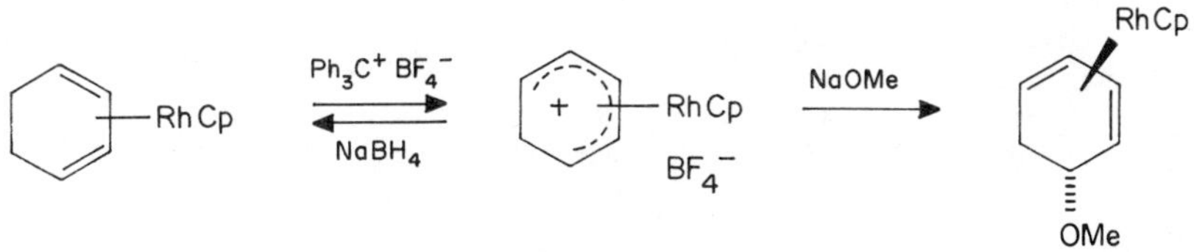

These complexes have not yet found such a range of synthetic applicability as the tricarbonyliron derivatives.

REFERENCES

1. J. K. Stille and D. E. James, *J. Am. Chem. Soc.*, 1975, **97**, 674.
2. R. Palumbo, A. De Renzi, A. Panunji and G. Paiaro, *J. Am. Chem. Soc.*, 1969, **91**, 3874; A. Panunzi, A. De Renzi, R. Palumbo and G. Paiaro, *J. Am. Chem. Soc.*, 1969, **91**, 3879; J. Tsuji and H. Takahashi, *J. Am. Chem. Soc.*, 1965, **87**, 3275; M. Tada, Y. Kuroda and T. Sato, *Tetrahedron Lett.*, 1969, 2871.
3. A. Segnitz, P. M. Bailey and P. M. Maitlis, *J. Chem. Soc., Chem. Commun.*, 1973, 698.
4. M. C. Cornock, R. C. Davis, D. Leaver and T. A. Stephenson, *J. Organomet. Chem.*, 1976, **107**, C43.
5. G. E. Herberich and B. Hesner, *Z. Naturforsch., Teil B*, 1979, **34**, 638.
6. M. Green, J. A. K. Howard, J. L. Spencer and F. G. A. Stone, *J. Chem. Soc., Dalton Trans.*, 1977, 271.
7. M. Green, D. M. Grove, J. L. Spencer and F. G. A. Stone, *J. Chem. Soc., Dalton Trans.*, 1977, 2228.
8. M. Green, J. A. K. Howard, A. Laguna, M. Murray, J. L. Spencer and F. G. A. Stone, *J. Chem. Soc., Chem. Commun.*, 1975, 451; M. Green, J. A. K. Howard, A. Laguna, L. E. Smart, J. L. Spencer and F. G. A. Stone, *J. Chem. Soc., Dalton Trans.*, 1977, 278.
9. J. Chatt and L. M. Venanzi, *J. Chem. Soc.*, 1957, 4735; E. W. Abel, M. A. Bennett and G. Wilkinson, *J. Chem. Soc.*, 1959, 3178.
10. J. C. Trebellas, J. R. Olechowski, H. B. Jonassen and D. W. Moore, *J. Organomet. Chem.*, 1967, **9**, 153; G. Nagendrappa and D. Devaprabhakara, *J. Organomet. Chem.*, 1968, **15**, 225.
11. R. H. Crabtree and G. E. Morris, *J. Organomet. Chem.*, 1977, **135**, 395.
12. R. H. Crabtree, H. Felkin and G. E. Morris, *J. Organomet. Chem.*, 1977, **141**, 205.
13. D. A. Russell and P. A. Tucker, *J. Chem. Soc., Dalton Trans.*, 1976, 841.
14. J. Evans, B. F. G. Johnson and J. Lewis, *J. Chem. Soc., Dalton Trans.*, 1977, 510.
15. E. W. Abel, M. A. Bennett and G. Wilkinson, *J. Chem. Soc.*, 1959, 3178; C. Potvin, J. M. Manoli, G. Pannetier, R. Chevalier and N. Platzer, *J. Organomet Chem.*, 1976, **113**, 273; T. V. Ashworth and E. Singleton, *J. Organomet. Chem.*, 1974, **77**, C31.
16. E. K. von Gustorf and J. C. Hogan, *Tetrahedron Lett.*, 1968, 3191.
17. F. A. Cotton, A. J. Deeming, P. L. Josty, S. S. Ullah, A. J. P. Domingos, B. F. G. Johnson and J. Lewis, *J. Am. Chem. Soc.*, 1971, **93**, 4624.
18. F. A. Cotton, M. D. LaPrade, B. F. G. Johnson and J. Lewis, *J. Am. Chem. Soc.*, 1971, **93**, 4626.
19. P. J. Harris, S. A. R. Knox and F. G. A. Stone, *J. Organomet. Chem.*, 1978, **148**, 327; T. A. Manuel and F. G. A. Stone, *Chem. Ind. (London)*, 1959, 1349; M. A. Bennett and G. Wilkinson, E. O. Fischer and W. Frohlich, *Chem. Ber.*, 1959, **92**, 2995.
20. Review: A. J Pearson, *Transition Met. Chem.*, 1981, **6**, 67.
21. H. Reihlen, A. Gruhl, G. von Hessling and O. Pfrengle, *Justus Liebigs. Ann. Chem.*, 1930, **482**, 161.
22. R. Pettit and G. F. Emerson, *Adv. Organomet. Chem.*, 1964, **1**, 1.
23. D. H. R. Barton, A. A. L. Gunatilaka, T. Nakanishi, H. Patin, D. A. Widdowson and B. R. Worth, *J. Chem. Soc., Perkin Trans. 1*, 1976, 821.
24. Y. Shvo and E. Hazum, *J. Chem. Soc., Chem. Commun.*, 1974, 336.
25. E. O. Greaves, G. R. Knox, P. L. Pauson, S. Toma, G. A. Sim and D. I. Woodhouse, *J. Chem. Soc., Chem. Commun.*, 1974, 257.
26. R. E. Graf and C. P. Lillya, *J. Organomet. Chem.*, 1976, **122**, 377.
27. D. V. Banthorpe, H. Fitton and J. Lewis, *J. Chem. Soc., Perkin Trans. 1*, 1973, 2051.

28. A. J. Birch and A. J. Pearson, *J. Chem. Soc., Chem. Commun.*, 1976, 601; A. J. Pearson, *Aust. J. Chem.*, 1976, **29**, 1841.
29. M. Franck-Neumann, F. Brion and D. Martina, *Tetrahedron Lett.*, 1978, 5033.
30. M. F. Semmelhack, *Pure Appl. Chem.*, 1981, **53**, 2379.
31. A. J. Birch, K. B. Chamberlain and D. J. Thompson, *J. Chem. Soc., Perkin Trans.* 1, 1973, 1900.
32. B. F. G. Johnson, J. Lewis and D. G. Parker, *J. Organomet. Chem.*, 1977. **127**, C37.
33. A. J. Pearson, *J. Chem. Soc., Chem. Commun.*, 1980, 488; C. W. Ong and A. J. Pearson, *Tetrahedron Lett.*, 1980, **21**, 2349; A. J. Pearson and M. Chandler, *Tetrahedron Lett.*, 1980, **21**, 3933; A. J. Pearson and C. W. Ong, *J. Org. Chem.*, 1982, **47**, 3780.
34. F. A. Cotton, *Acc. Chem. Res.*, 1968, **1**, 257.
35. M. Green, S. Heathcock and D. C. Wood, *J. Chem. Soc., Dalton Trans.*, 1973, 1564; L. A. Paquette, S. V. Ley, M. J. Broadhurst, D. Truesdell, J. Fayos and J. Clardy, *Tetrahedron Lett.*, 1973, 2943.
36. B. F. G. Johnson, J. Lewis and G. L. P. Randall, *J. Chem. Soc. A*, 1971, 422.
37. V. Heil, B. F. G. Johnson, J. Lewis and D. J. Thompson, *J. Chem. Soc., Chem. Commun.*, 1976, 270.
38. R. B. King, P. M. Treichel and F. G. A. Stone, *J. Am. Chem. Soc.*, 1961, **83**, 3593; F. A. Cotton and L. T. Reynolds, *J. Am. Chem. Soc.*, 1958, **80**, 269.

CHAPTER 8

η^5-Dienyl Complexes

NON-CYCLICALLY CONJUGATED DIENYL COMPLEXES

Iron, Ruthenium and Osmium

By far the most work has been done on acylic pentadienyl and cyclohexa-, cyclohepta- and cyclooctadienyl complexes of iron, ruthenium and osmium, so most of our attention in this section will be confined to these complexes. We shall also introduce analogous complexes of the cobalt group, and of manganese and chromium.

The preparation of tricarbonylcyclohexadienyliron tetrafluoroborate **1** in 100% yield was first reported in 1960 by Fischer and Fischer.[1] It is formed by hydride abstraction from the cyclohexa-1,3-diene–$Fe(CO)_3$ complex using triphenylmethyl (trityl) tetrafluoroborate.

(1)

Being insoluble in ether, the dienyl complex is readily precipitated and separated from the triphenylmethane by-product.

These cationic derivatives, obtained either as their BF_4^- or PF_6^- salts, are invariably yellow microcrystalline involatile solids. They exhibit two strong absorptions in the IR spectrum at *ca.* 2150 and 2050 cm^{-1}, approximately 100 cm^{-1} to higher frequency than the corresponding absorptions of neutral diene–$Fe(CO)_3$ complexes. This is due to delocalization of the positive charge of the dienyl group on to the metal, which is the reason for the high stability of these compounds, resulting in reduced back-donation of electrons from Fe into the CO π^* antibonding orbitals, and therefore an increased CO π-bond order (see Chapter 1).

The corresponding acyclic pentadienyl–$Fe(CO)_3$ complexes are less easy to prepare. For example, treatment of pentadiene with $Fe(CO)_5$ gives exclusively the complex **2**, which will *not* undergo hydride abstraction to give **3**. The corresponding cisoid pentadiene complex **4** will in fact undergo smooth hydride abstraction with trityl tetrafluoroborate, but since this is obtained usually by borohydride reduction of **3**, it does not constitute a synthetic method.

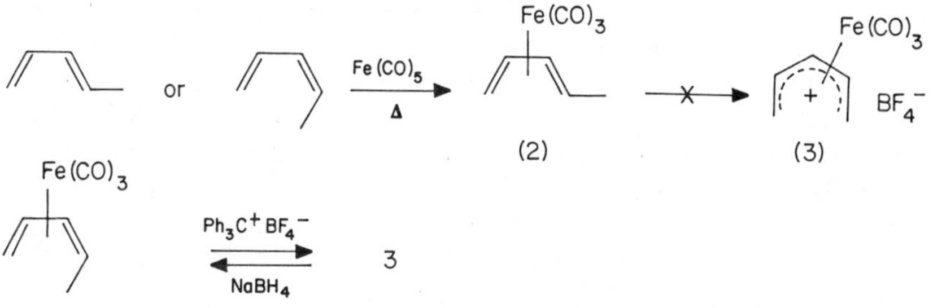

(2)　　　　　　　(3)

(4)

Evidently a cisoid substituted diene structure, automatically produced by cyclohexadiene systems, is essential for H^- abstraction to be successful.

Trans-Pentadienol reacts with $Fe(CO)_5$ thermally to produce reasonable yields of the complex **5**. Treatment of this with anhydrous HBF_4 leads to rearrangement of the complex and loss of a hydroxy group to give **3**. Substituted derivatives, mainly the methylated compounds have been prepared similarly.

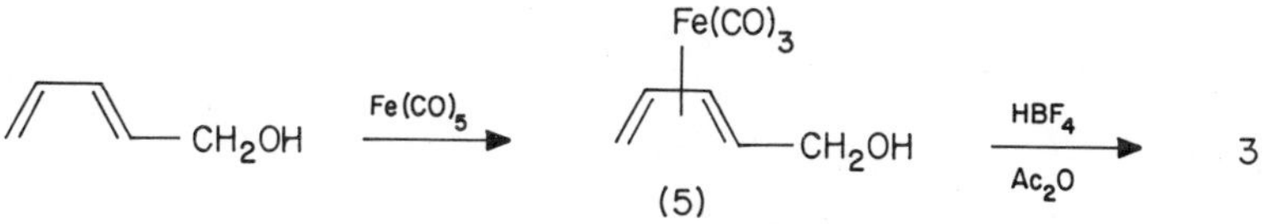

(5)

This appears to be the best method for the preparation of the acyclic complexes, although compounds bearing functionalized substituents or substituents capable of perturbing the electronic system do not appear to have been investigated.

Cycloheptadienyl–$Fe(CO)_3$ complexes may be prepared in two ways. Reaction of cycloheptatriene–$Fe(CO)_3$ derivatives with HBF_4 under anhydrous conditions, followed by precipitation with diethyl ether, leads to good yields of the complex **6**. This may also be prepared by reaction of the cycloheptadiene–$Fe(CO)_3$ complex with trityl tetrafluoroborate in the usual manner. Substituted derivatives do not appear to have been investigated.

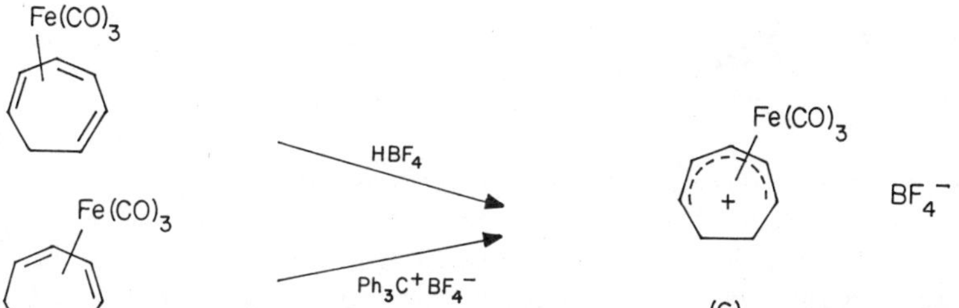

(6)

Cyclooctadienyl–$Fe(CO)_3$ complexes may be prepared in an analogous manner.

Nuclear Magnetic Resonance Spectra of Simple Dienyl–Fe(CO)₃ Complexes

The proton NMR spectra of the simple unsubstituted dienyl complexes show the following trends. The lowest field absorption is found for the 'central' dienyl proton H-3 at *ca.* 7–7.5 p.p.m. (downfield from Me_4Si); the next signal at *ca.* δ 6 p.p.m. is that due to H-2 and H-4, and the highest field signal is that due to the terminal protons. As is the case with neutral butadiene complexes, the 'inner' terminal protons H-1b of acyclic pentadienyl complexes are found at higher field than the 'outer' protons H-1a. The cyclohexadienyl complex shows an interesting pattern for the 6-methylene group. Thus, the *endo*-proton is found at lower field than the *exo*-proton, possibly reflecting a deshielding influence of the $Fe(CO)_3$ group, also found for the uncharged cyclohexadiene–$Fe(CO)_3$ complexes. The *exo*-proton is found as a simple doublet with large geminal coupling to *endo*-H, whilst the *endo*-proton is a doublet of triplets displaying a significant coupling to H-1 and H-5. This infers that the shape of the molecule is that shown in Figure 8.1, since the Karplus relationship for vicinal proton–proton couplings predicts a greater coupling where there is a dihedral angle near 0° than when this angle is near 90°. The ¹H NMR data (δ p.p.m.) for these simple complexes are summarized below.

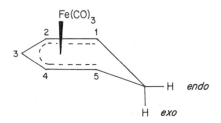

Figure 8.1   Shape of cyclohexadiene–$Fe(CO)_3$ complex

In general, the proton NMR shifts for these cationic complexes are at lower field than the corresponding resonances for diene–$Fe(CO)_3$ derivatives, probably owing to the distribution of charge on to the ligands. Unfortunately, the differences between the shifts, amounting to around 3–4 p.p.m. are approximately within the range which is obtained by magnetic anisotropy effects and consequently tell us nothing about π-electron density distributions in the ligand (charge distribution). For this purpose ^{13}C NMR is much more useful. Carbon-13 NMR shifts (δ p.p.m.) are given below for the simple cyclohexadienyl and pentadienyl complexes. Included for comparison are the data for the uncomplexed cyclohexadienyl cation (obtained by protonation of benzene).[2]

It is evident that all the carbons of the cyclohexadienyl cation are shifted to higher field on complexation to Fe(CO)$_3$, reflecting a delocalization of the charge on to the metal. It is even more interesting that, if we assume an approximate relationship between positive charge and ^{13}C shifts (lower field, greater charge), the differences now being well outside the range due to anisotropy effects, we can see that the charge distribution in complexed and uncomplexed cations is considerably different. Whilst the distribution observed for the free cation is exactly what we would have expected from the valence bond description and Huckel calculation, that found for the complexed system is not. It in fact reflects a donation of electrons from the iron to the dienyl LUMO as discussed in Chapter 1. This has interesting consequences regarding reactivity, to which we shall return later.

Synthesis of Substituted Cyclohexadienyl–Fe(CO)$_3$ Complexes

Organic synthetic methodology based on the reactivity of the parent cyclohexadienyl–Fe(CO)$_3$ complex is likely to be of limited usefulness. Potentially, of greater use would be to exploit the reactivity of these complexes which bear substituents such as methyl, methoxy and CO$_2$Me on the dienyl system. These might also have significant effects on reactivity which could be exploited.

The most obvious way to produce these complexes would be to react an appropriate diene derivative with trityl tetrafluoroborate (or hexafluorophosphate). However, in an unsymmetrically substituted diene complex there are two methylene groups from which H$^-$ might be removed and, in principle, two products for the reaction. The nature of the substituent might well affect the regiochemistry of this reaction, and we now consider a selection of results which throw some light on the factors controlling the selectivity of this process. These are summarized in Tables 8.1 and 8.2.

Basically, hydride abstraction appears to be susceptible to two controlling factors: (1) steric effect of substituents since Ph$_3$C$^+$ is very bulky and (2) elec-

Table 8.1

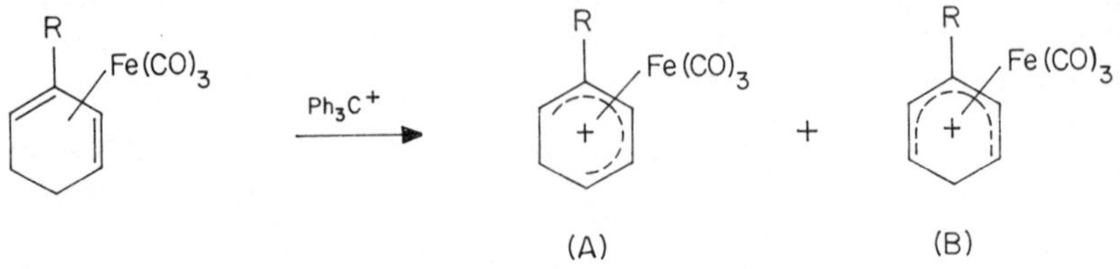

R	Ratio **A:B**
OMe	94:6
Me	60:40
CO$_2$Me	20:80

Table 8.2

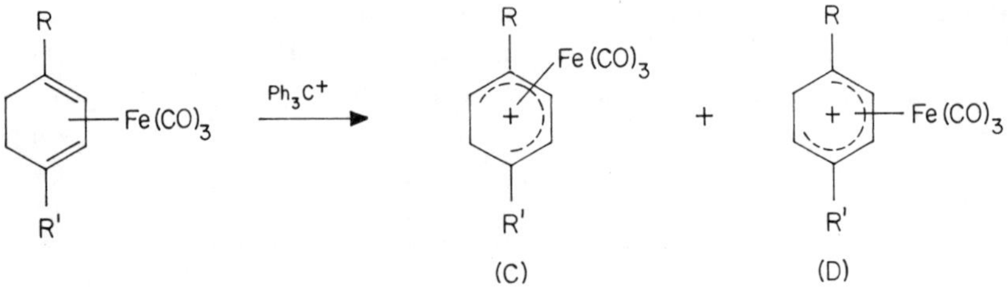

R	R′	Ratio **C:D**
OMe	H	20:80
CO$_2$Me	H	5:95
OMe	Me	95:5

tronic effects. The steric effect is self-explanatory and evidenced in Table 8.2 (first entries). The electronic effects are less obvious and information concerning these can only be usefully obtained from reactions of diene complexes such as those in Table 8.1, and the last entry in Table 8.2, where steric effects at the alternative methylene groups are balanced. The results in Table 8.1 showing the effects of methoxy and CO$_2$Me substituents are at first puzzling. For example, protonation of anisole to give the 3-methoxycyclohexadienyl cation and the *meta*-directing effect of CO$_2$Me groups on benzene nuclei, proceeding via a 2-methoxycarbonyl-cyclohexadienyl cation, suggest that in the uncomplexed series these are the favored, i.e. thermodynamically most stable, cations of the available alternatives. However, hydride abstractions give complexes which contain the less favored dienyl cation. The explanation is relatively straightforward. Since the dienyl cation is bound to Fe by a synergic interaction involving the dienyl HOMO interacting with empty metal d orbitals and the dienyl LUMO interacting with filled metal d orbitals, we might expect that the strongest bond is formed when these pairs of orbitals are relatively close in energy. This is well known from perturbation theory.[3] It is likely that a better bonding interaction will be obtained between the iron d orbitals and a cation having high-energy HOMO and low-energy LUMO, as shown in Figure 8.2.

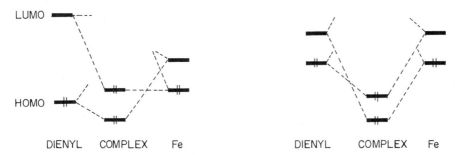

Figure 8.2   Interaction diagrams for Fe bonding to dienyl cation

Table 8.3 gives the calculated HOMO and LUMO energy levels for variously substituted methoxy- and methoxycarbonylpentadienyl cations.[4] It can be seen that the 2-methoxy-substituted cation has the lowest LUMO/highest HOMO of the methoxy series, whereas the 3-methoxycarbonyl system has the lowest LUMO/highest HOMO of the ester series. Therefore, in the absence of competing steric effects, it is likely that a product-controlled reaction would favor the formation of complexes of these cations, as is indeed observed (these are also the least stable free cations, as expected).

There are, of course, limitations on the use of trityl cations to prepare these complexes. The strong dependence of product on electronic factors noted above makes it difficult to prepare specific alkyl-substituted complexes, e.g. 2-methylcyclohexadienyl–Fe(CO)$_3$ (Table 8.1). However, there is an alternative method which makes these compounds accessible, *viz.* acid treatment of the

Table 8.3 Calculated HOMO and LUMO energy levels for substituted methoxy- and methoxycarbonylpentadienyl cations.

Cation	HOMO	LUMO	Stabilization energy
1-MeO	−0.88552	0.09987	−9.80864
2-MeO	−0.855581	0.00000	−9.65792
3-MeO	−1.00000	0.10086	−9.78395
1-CO$_2$Me	−1.11164	−0.27975	−14.91557
2-CO$_2$Me	−1.09302	0.00000	−14.96638
3-CO$_2$Me	−1.00000	−0.29805	−14.80583

appropriately substituted methoxyalkylcyclohexadiene complex. Some examples, together with mechanism of the reaction, are given below.[5]

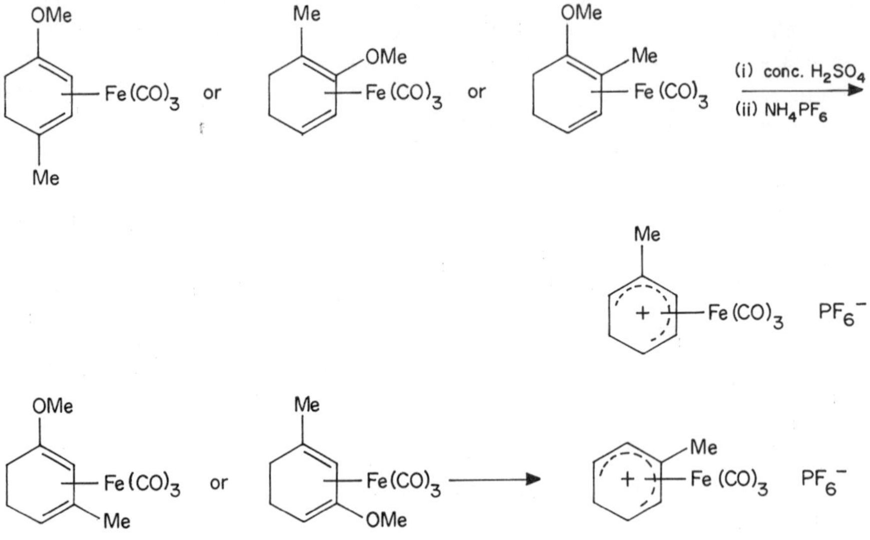

Mechanism :

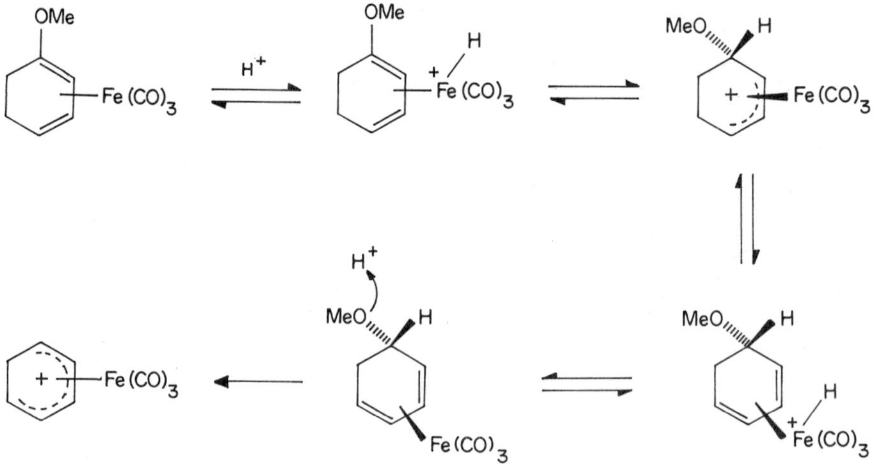

Another limitation is that the trityl cation is very bulky, so that hydride abstraction is completely suppressed from complexes such as **7** and **8**. An indirect route is available, however, in which the complexes are subjected to oxidative cyclization (see Chapter 7) followed by acid treatment and acetylation of the alcohol functionality produced during ring opening, to give the unusually substituted cations **9** and **10**.

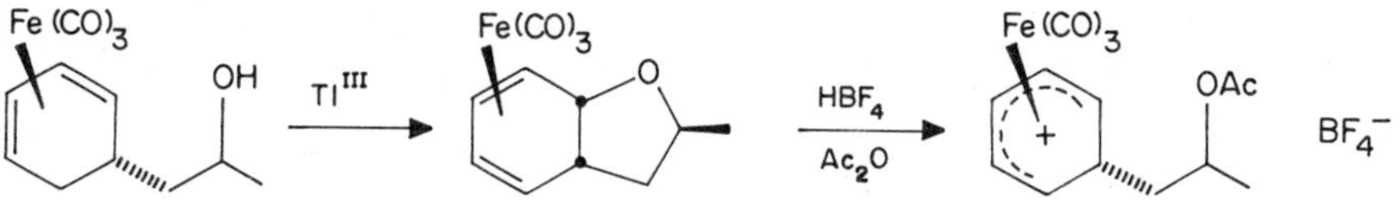

(7)

(9)

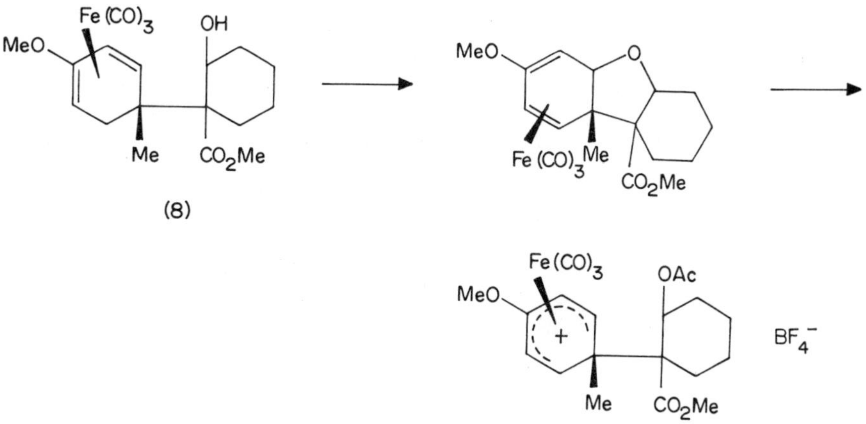

(8)

(10)

The ruthenium and osmium tricarbonyl complexes of the cyclohexadienyl and cycloheptadienyl cation analogous to the iron complexes 1 and 6 have also been prepared by hydride abstraction from the corresponding diene complexes and protonation of triene complexes.

Reactions of Dienyl–M(CO)₃ (M = Fe, Ru, Os)[6]

As might be expected from their cationic nature, dienyl–M(CO)$_3$ complexes (M = Fe, Ru, Os) react with nucleophiles to give neutral complexes. Less obvious is the fact that in most of the cases studied the nucleophile attaches itself directly to the dienyl ligand to give a substituted diene complex, rather than attacking one of the CO ligands (metal carbonyls are known to react with nucleophiles at CO). The reaction is usually very clean. However, on close examination the product obtained is not exactly that which might be expected. Whilst much work has been done on fairly simple acyclic dienyl complexes, most results for substituted complexes have been obtained for cyclohexadienyl systems owing to their ready availability. Simple cycloheptadienyl complexes have been studied, but not in great detail. However, a comparison of the available results is interesting, and here we shall attempt to rationalize the behavior of these three types of complex.

Basically, the observed reactions are summarized in the following equations, including some specific examples.

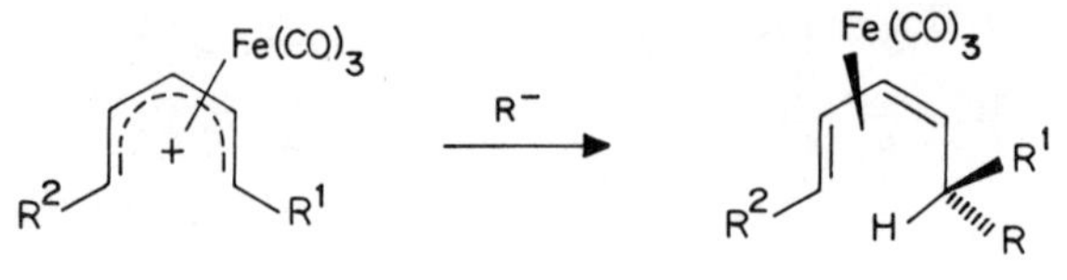

(11)   (12)

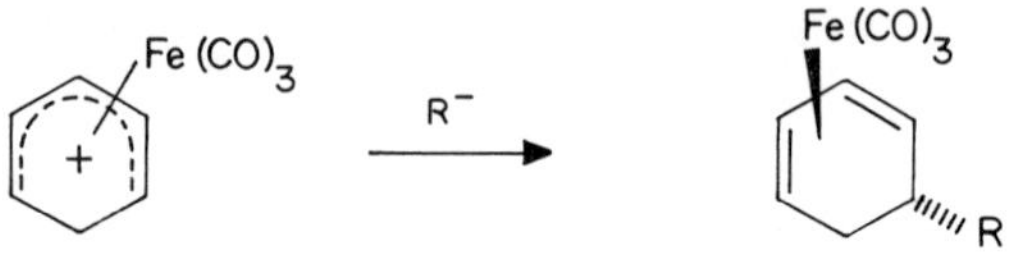

(13)

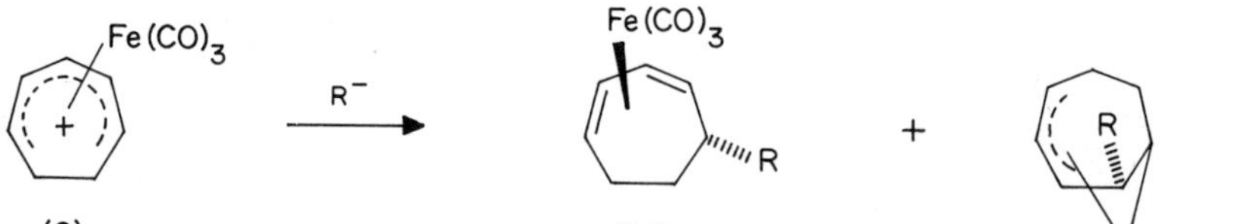

(6)   (14)   +   (15)

Reactions of complex 11, $R^1 = R^2 = H$:

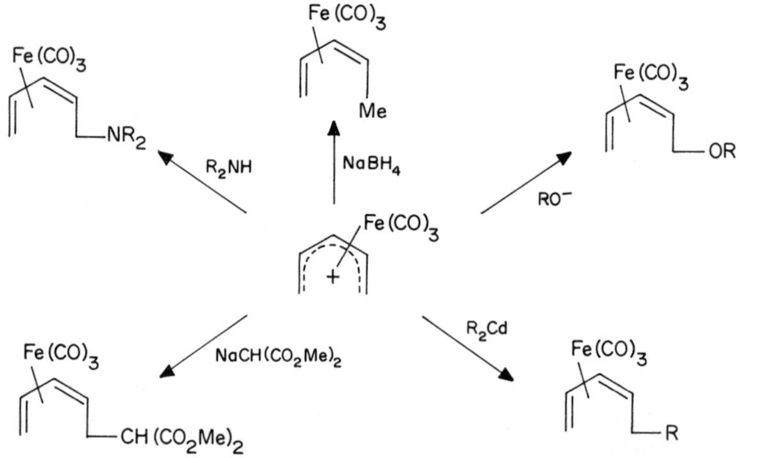

$R = $ i-Pr, Ph, PhCH$_2$, CH$_2$CH$=$CH$_2$

Reactions of complex 1:

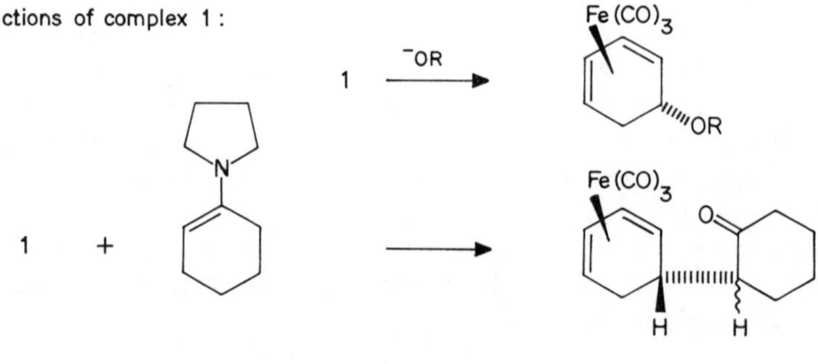

1   +

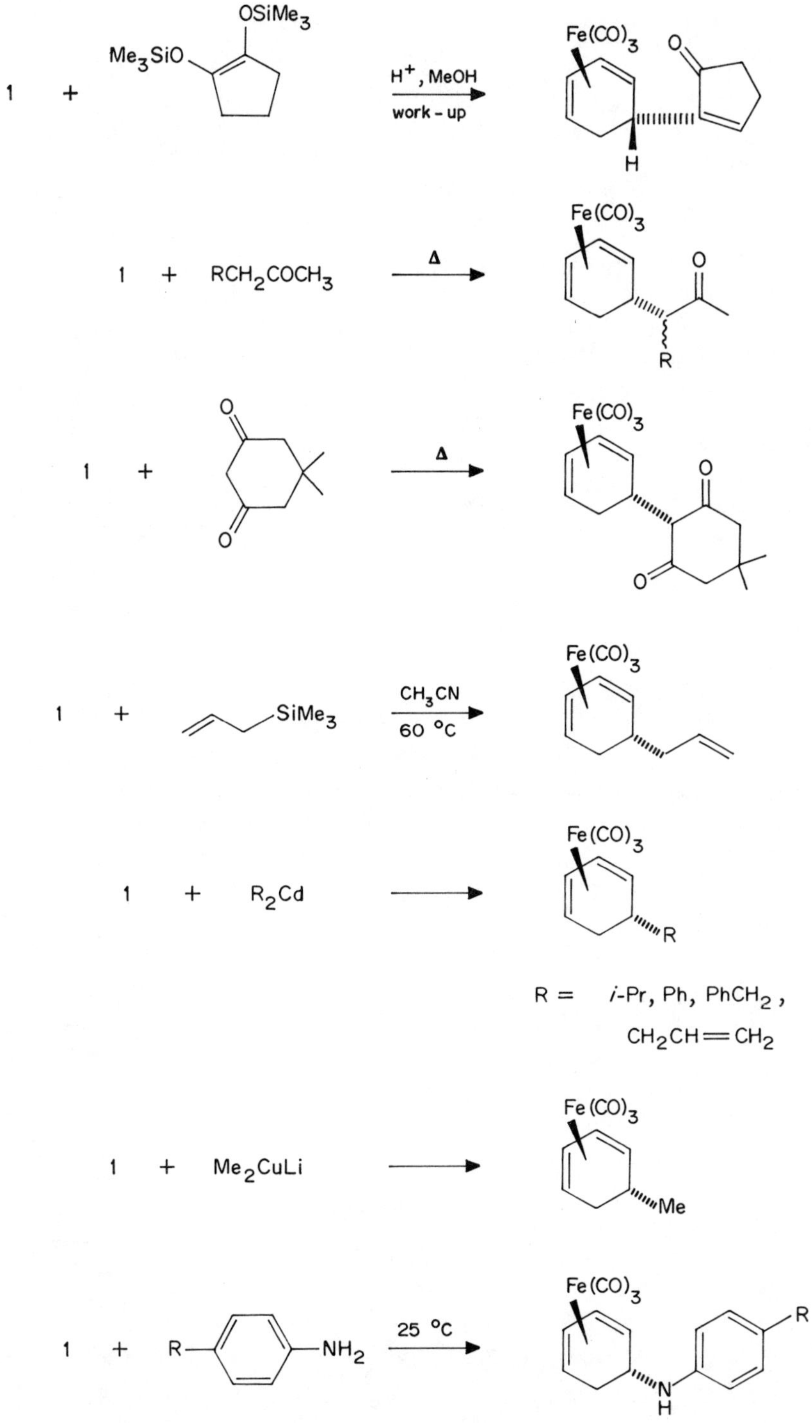

1 +
OSiMe3
Me3SiO
H+, MeOH
work - up
Fe(CO)3
O
H
1 + RCH2COCH3
Δ
Fe(CO)3
O
R
1 +
O
O
Δ
Fe(CO)3
O
O
1 +
SiMe3
CH3CN
60 °C
Fe(CO)3
1 + R2Cd
Fe(CO)3
R
R = i-Pr, Ph, PhCH2, CH2CH=CH2
1 + Me2CuLi
Fe(CO)3
Me
1 + R
NH2
25 °C
Fe(CO)3
R
N
H

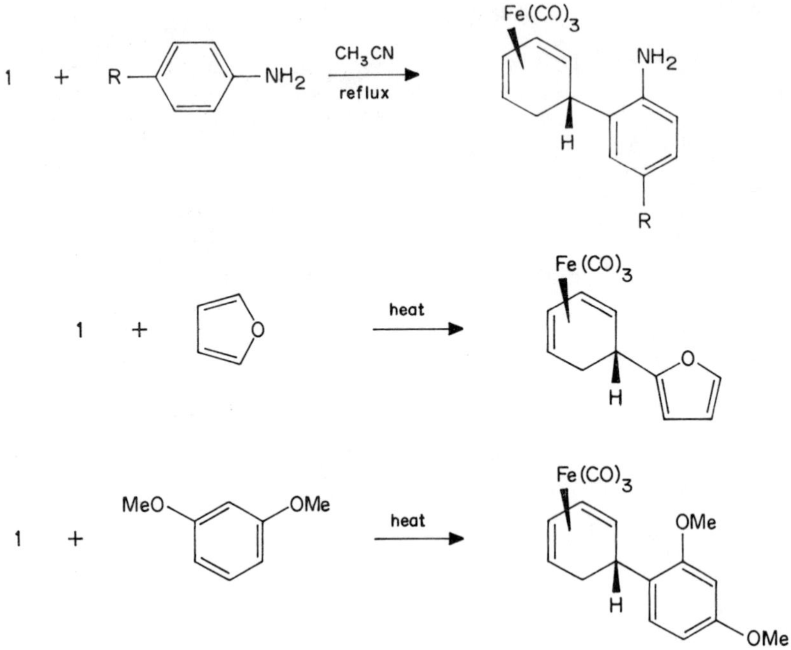

1 + R—⟨benzene⟩—NH₂ →(CH₃CN, reflux) [complex]

1 + [furan] →(heat) [complex]

1 + [MeO—benzene—OMe] →(heat) [complex]

Reactions of complex 6 :

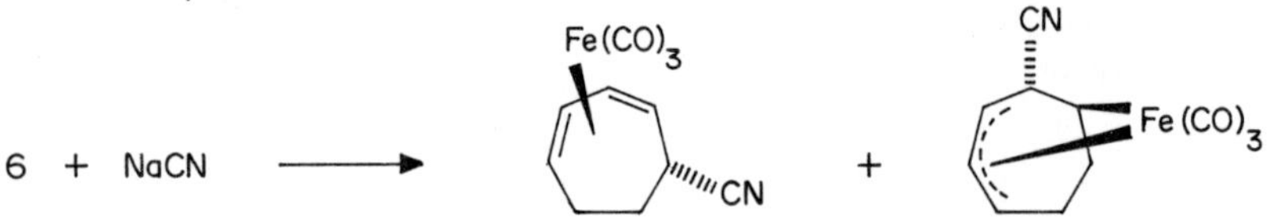

6 + NaCN → [complex] + [complex]

1 : 4

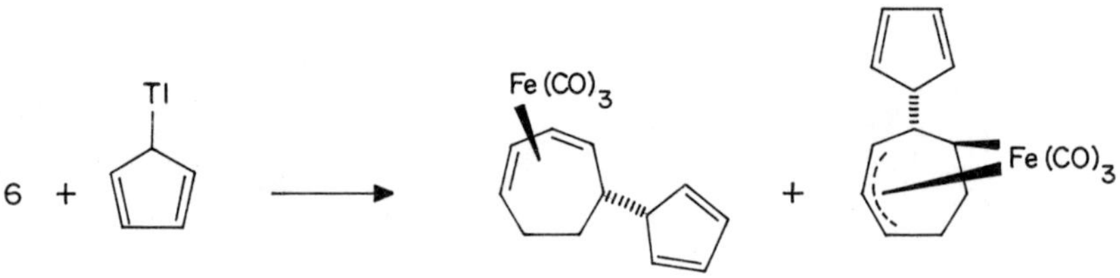

6 + [Tl-cyclopentadienide] → [complex] + [complex]

85 : 15

6 + NaBH₄ → [complex] + [complex]

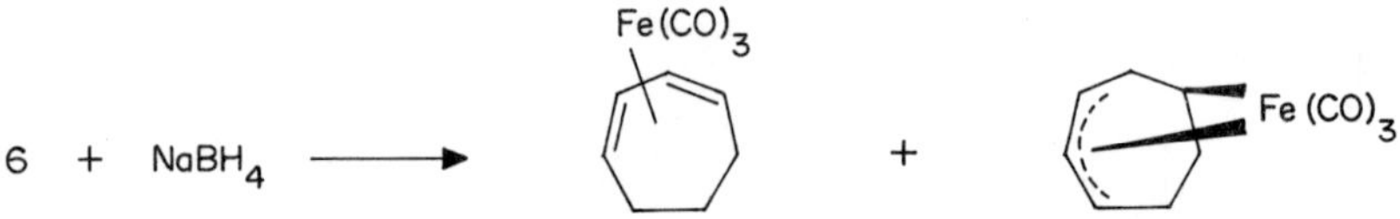

reaction in EtOH : 1 : 2.3

reaction in EtOH /CH₂Cl₂ : 1.2 : 1

The reactions of the pentadienyl and cyclohexadienyl complexes are in fact what we might have expected for the reactivity of a dienyl cation, i.e. nucleophiles add to the terminus of the ligand. However, when we examine the ^{13}C NMR spectrum of **1** (above), we see that the problem is not this simple. These data show that the greatest positive charge is in fact at C-2, so that the we might well have expected a nucleophile to attack at that position. Note that this mode of addition is actually (partially) observed for the cycloheptadienyl complex, so is not a disallowed pathway. The available MO calculations[7] are in agreement with the ^{13}C NMR data, and the charge distributions of **1** together with ^{13}C shifts for comparison are given in Table 8.4.

Table 8.4. Charges and ^{13}C shieldings of complex **1**

	C-1	C-2	C-3
Charge	+ 0.140	+ 0.226	+ 0.159
^{13}C shift (p.p.m.)	63.7	101.4	89.0

It can be seen that ^{13}C shifts correlate very well with charge densities. Why does nucleophile addition to **1** occur solely at the terminal position? Note also that all nucleophiles studied add M–*exo* (i.e. side opposite metal). This might be due to steric hindrance from the metal grouping, but it is also possible that other factors are involved, particularly in view of Hoffman and Eisenstein's recent interpretation of nucleophile additions to olefin–metal complexes (Chapter 5).

A considerable amount of kinetic information is available for these reactions,[8] but so far it has yielded no indications of the factors controlling nucleophile addition. It seems clear, however, that the reaction is not completely under charge control. Probably the nucleophile HOMO interacts with the complex LUMO at the carbon atom with the largest coefficient. Unfortunately, there is no published information on the nature of the LUMO, and it seems unlikely that much progress will be made for sometime. Such calculations as are available[9] in fact show around seven molecular orbitals very close in energy and therefore all candidates as the LUMO. Figure 8.3 shows in diagrammatic form the three lowest energy MOs. The coefficients at each dienyl carbon atom are appro-

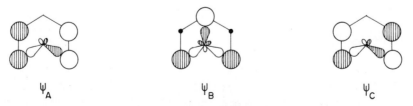

Figure 8.3   Possible dienyl–Fe(CO)$_3$ LUMO combinations

ximately shown by the size of the orbital circle, but these and probably the relative energies of the MOs are undoubtedly sensitive to (a) the method of calculation employed and (b) the nature of the dienyl ligand.

We can see, though, that if the addition is under overall frontier orbital control, then ψ_C would favor addition to the terminal carbons, ψ_B would give addition to mainly C-3 and the terminal carbons, whilst ψ_A would give addition mainly to C-2 and (less) the terminal carbons. Of course, the actual situation may be a combination of the three. More than this cannot be said at present. The results with the cycloheptadienyl complex are interesting in the context of charge *vs.* frontier orbital control. The addition of cyanide gives mainly the product expected from charge-controlled reactions whereas the addition of cyclopentadienyl anion gives mainly that is expected from 'orbital' control. Borohydride gives varying results depending on the solvent, and this too might reflect the balance between charge and orbital control, depending on whether the $NaBH_4$ is more associated in less polar solvents. Unfortunately, no other work has been done to probe the controlling factors (but see later). Note, however, that CN^- is a fairly 'hard' (low-energy) nucleophile and might be expected to be less influenced by orbital interactions,[10] whereas Cp^- is softer and very susceptible to interaction with any LUMO of similar energy.

In can be seen that the picture is not clear. Suffice it to say that in the case of cyclohexadienyl–$Fe(CO)_3$ complexes, which are proving to be synthetically useful (see below), the reaction is *regiospecific* (terminal C atom) and *stereospecific* (opposite Fe).

Substituent Effects on Reactivity

With the ready availability of methoxy-substituted cyclohexadienyl complexes from the corresponding aromatic anisole derivatives it follows that the effect of methoxy substitution on the reactivity of cyclohexadienyl–$Fe(CO)_3$ complexes is fairly easy to study. The 1-substituted complex **16** is readily hydrolyzed to give the cyclohexadienone complex **17** (see also Chapter 2).

(16)   (17)

The 2-methoxy substituted derivative **18** is more interesting from the synthetic point of view, since it reacts regiospecifically with nucleophiles to give products **19**, a few examples being given below.

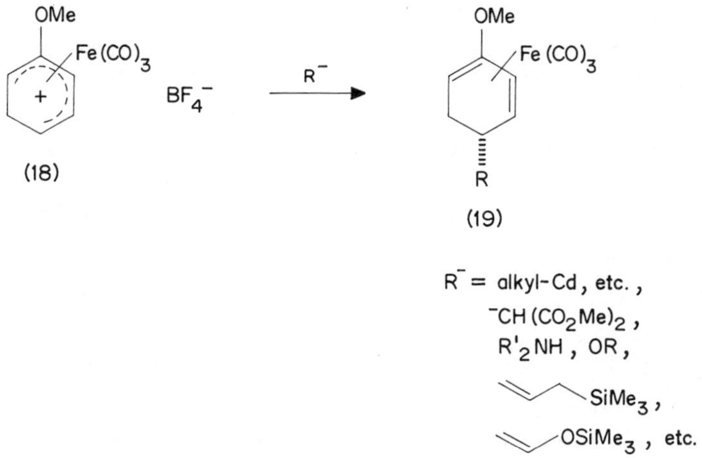
OMe
Fe(CO)3
BF4−
R−
OMe
Fe(CO)3
R
R− = alkyl-Cd, etc.,
−CH(CO2Me)2,
R'2NH, OR,
SiMe3,
OSiMe3, etc.

(18) (19)

The disubstituted complex **20** also reacts regiospecifically at the methylated dienyl terminus to give, e.g., **20**. Thus, the methoxy group is exerting a directing effect which forces nucleophiles to react at the sterically more crowded terminus of the cyclohexadienyl ligand.

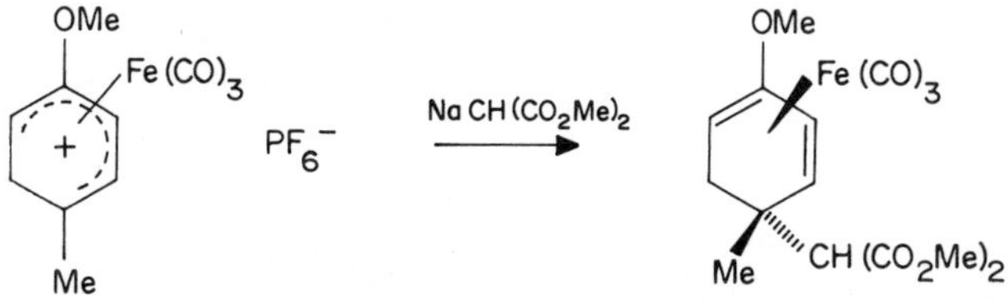
OMe
Fe(CO)3
PF6−
Me
Na CH(CO2Me)2
OMe
Fe(CO)3
Me
CH(CO2Me)2

(20) (21)

Let us examine the mechanism by which this directing effect operates. The ^{13}C NMR shifts for the complex **18** are shown below:

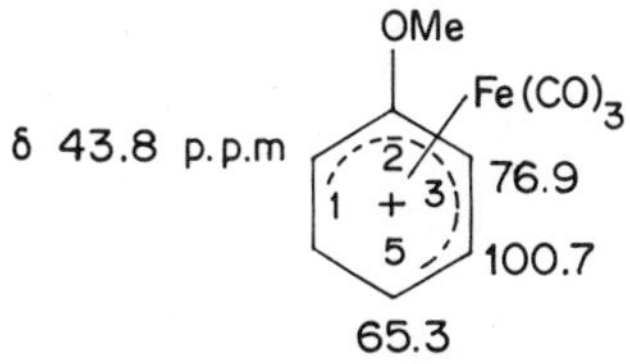
OMe
Fe(CO)3
δ 43.8 p.p.m
1
2
3
5
76.9
100.7
65.3

We can see that of the two termini, that adjacent to the methoxy (C-1) is at higher field, suggesting that there is less positive charge at this position. Consequently, we should expect that of the two termini, C-1 will be less reactive towards nucleophiles, simply owing to the difference in charge. That is, there is an electronic effect which results in a charge control being superimposed on the overall nucleophile addition. This is sufficient to cause nucleophiles to add at the more positive center even when that is substituted, as in complex **20**. We can see how this charge distribution arises by examining the likely effects of 2-OMe on the coefficients of the bonding M.O's of the complex, shown pictorially in Figure 8.4. Thus, the coefficients at C-1 for ψ_1 and ψ_2 are larger than those at C-5. Hence electron density is increased at C-1.

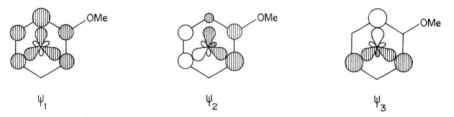

ψ_1 ψ_2 ψ_3

Figure 8.4   Bondings MOs of complex **18** (schematic only)

The orbital ψ_2 seems to be particularly important from this point of view. In fact, no calculations on the complex are available in the literature, the above being adapted from Huckel calculations on uncomplexed dienyl systems. We can particularly note how the orbital involving ϕ_2 of the dienyl system is changed on methoxy substitution:

When a slightly larger substituent is placed *para* to the methoxy group is in complex **22**, a mixture of the regioisomers **23** and **24** is obtained, and their ratio is in fact dependent on the nature of the counter cation associated with the dimethylmalonyl anion. The results for Li, Na and K enolates are shown below.[11]

(22)   (23)   (24)

M^+	ratio **23:24**
Li^+	3.0:1
Na^+	3.8:1
K^+	5.7:1
Na^+ (+ 18-crown-6)	5.7:1

It can be seen that the use of a counter cation which strongly coordinates the enolate gives a lower selectivity for the substituted terminus than does a cation which does not strongly coordinate. This interesting result indicates some kind of secondary controlling effect which is not understood at present, but possibly involves a frontier orbital interaction.

Excellent regiocontrol on these reactions can be obtained by replacing the 2MeO group with the bulkier 2-Pr^{i}O. Thus the diene complex **25** still gives the dienyl complex **26** in 100% yield on treatment with $Ph_3C^+ BF_4^-$, and this now reacts with malonate nucleophiles to give, quantitatively, a single product **27**.[11]

(25)   (26)   (27)

The conversion of complexes of type **21**, **23** and **27** to 4,4-substituted cyclohexenones of general structure **28** suggests a potential usefulness for these complexes, since there are many natural products which can be derived from such intermediates, provided the behaviour is general.

We shall now briefly review such progress as has been made during the period 1977–82 towards target-oriented organic synthesis using complexes of the above type. In many instances functional group transformations are carried out in the presence of the metal which illustrate the stability of the $Fe(CO)_3$ group to a diverse range of reaction conditions and reagents. This is particularly important if this group is to be employed as diene or dienol ether protection.

SYNTHETIC APPLICATIONS OF TRICARBONYL (CYCLOHEXADIENYL) IRON COMPLEXES

Cyclohexenone γ-Cation Equivalents

The realization that the types of complex discussed above are synthetic equivalents of cyclohexenone γ-cations has led to a possible application of retrosynthetic analysis employing that particular synthon. In particular, tricarbonyl(4-methoxy-1-methylcyclohexadienylium)iron hexafluorophosphate **20** is now finding applications as an A-ring synthon for a number of novel approaches to steroids and various terpenoids.

Approaches to Steroid Total Synthesis

The complex **20** is found to react regiospecifically with the tetralone-derived enolate **29** to give a mixture of diastereoisomeric complexes **30** in 95% yield. This mixture is readily converted to the enones **31** and **32**, and these in turn have been converted to the D-homoaromatic steroid derivatives **33, 34, 35**, as shown below.[12]

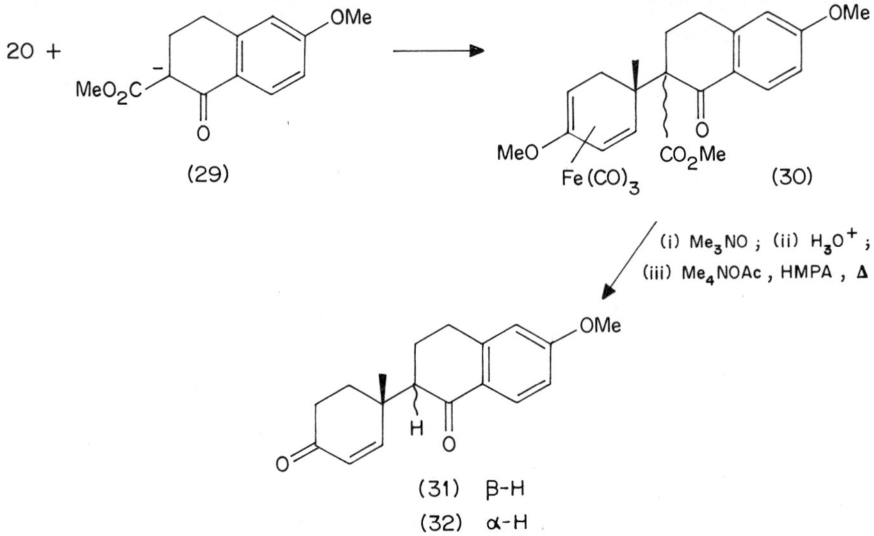

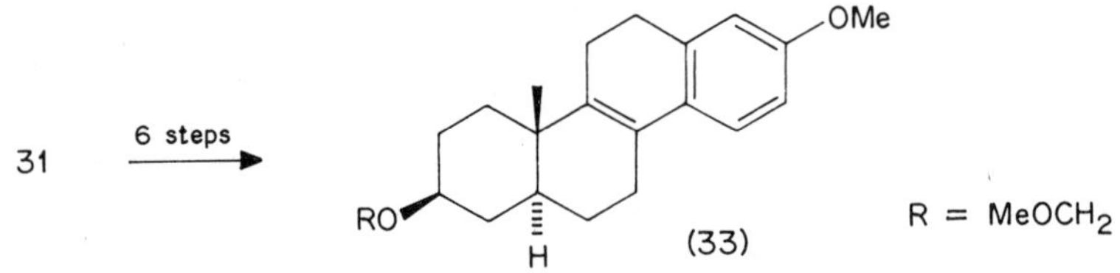

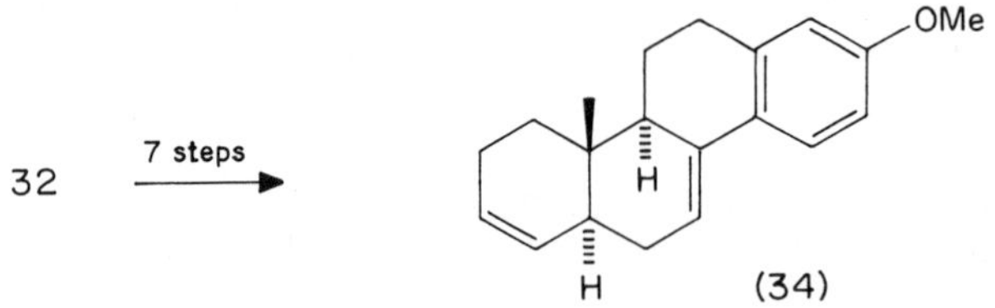

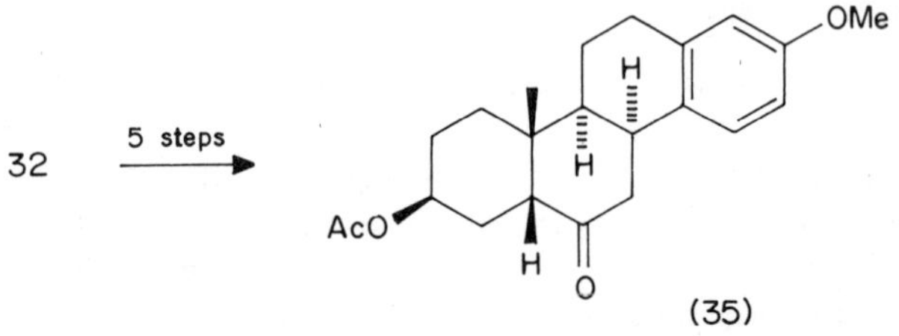

(29)

(30)

(i) Me₃NO ; (ii) H₃O⁺ ;
(iii) Me₄NOAc , HMPA , Δ

(31) β-H
(32) α-H

31 — 6 steps → (33)   R = MeOCH₂

32 — 7 steps → (34)

32 — 5 steps → (35)

Approaches to Trichothecene Synthesis

The trichothecenes are a group of naturally occurring sesquiterpenes exemplified by trichodermol **36**. They show potent cystostatic activity and are potential anti-cancer compounds. The natural products are very toxic, so it is of interest to develop methods for the synthesis of unnatural analogues which might show beneficial alterations in activity. The 12,13-epoxide is known to be essential for activity.

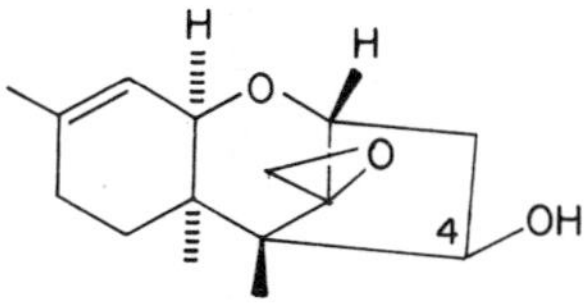

(36)

Complex **20** was shown to react with methyl 2-oxo-1-potassiocyclopentane carboxylate to give a mixture of diastereoisomers **37**, only one of which possesses the relative stereochemistry appropriate to the trichothecene ring system. The separate isomers were reduced quantitatively to the hydroxy esters **38** and **39** and the undesired **39** could be converted to the mixture by treatment with p-toluenesulfonic acid in dichloromethane. This facile interconversion of diastereoisomers [involving inversion of the diene–Fe(CO)$_3$ group] proceeds by the mechanism shown below, and allows the preparation of the desired **38** in good yield.

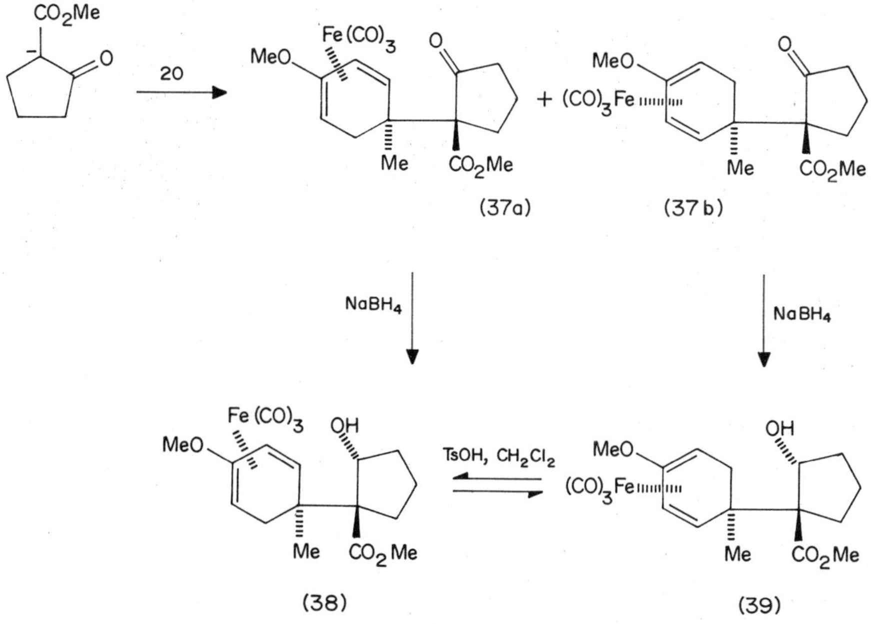

(37a)   (37b)

(38)   (39)

Mechanism :

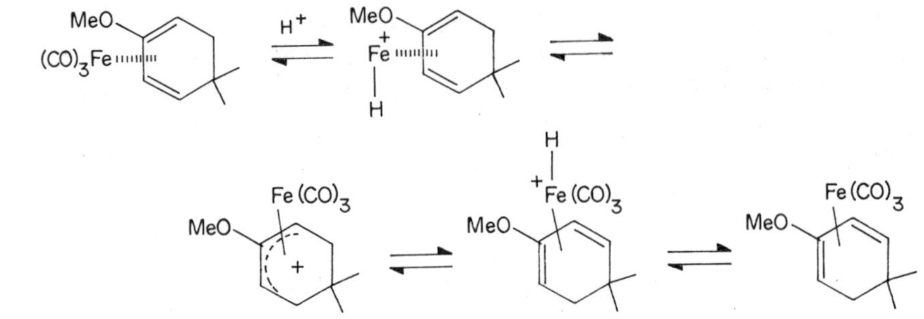

The strategy for conversion of **38** into a trichothecene analogue, which does not possess a hydroxy function at C-4, involved a stereospecific epoxidation and *trans*-diol formation, and since a number of problems were associated with handling the corresponding hydroxy ester in which the $Fe(CO)_3$ group had been removed, it was necessary to perform appropriate transformations on the complex. The sequence illustrates the stability of the diene–$Fe(CO)_3$ group to a number of useful chemical transformations, including Sharpless epoxidation, ester reduction and alcohol dehydration.

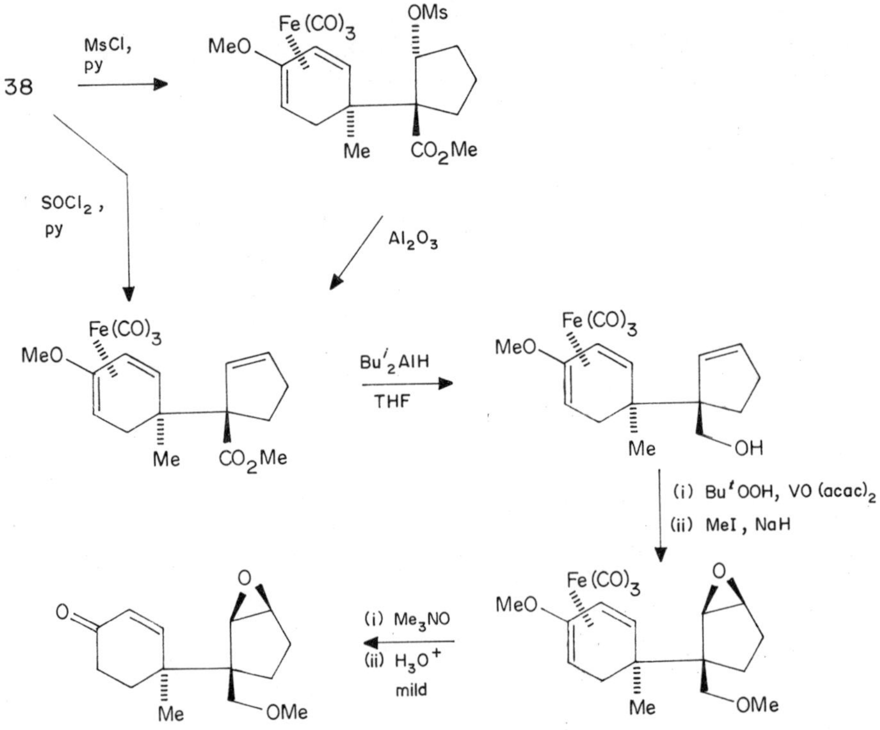

(40)

The intermediate enone **40** underwent regiospecific and stereospecific epoxide opening on treatment with acid under fairly vigorous conditions to give a diol in which only one OH has the stereochemistry which will allow its intramolecular Michael cyclization to give the tricyclic intermediate **41**. This was converted to the 12,13-epoxytrichothecene derivative **42** using standard methodology.

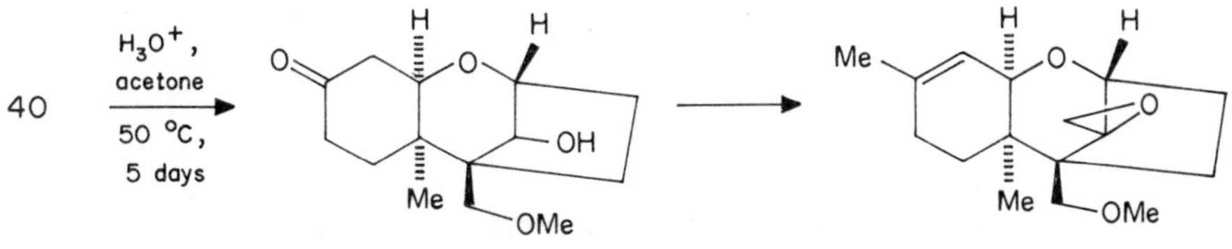

(41)   (42)

Synthesis of Spirocyclic Systems: Inter- and Intramolecular Nucleophile Addition to Dienyl–Fe(CO)₃ Groups[13]

4-Methoxycinnamic acid is readily converted to the diene–Fe(CO)$_3$ complex **43** in *ca.* 55% overall yield. This can then be converted into the dienyl complex **44** in high yield. Reaction of this complex with malonate nucleophiles gives a mixture of regioisomers from attack at C-1 and C-5, the former predominating (*ca.* 85% using dimethyl *potassio*malonate). The desired product is converted to spiro[4.5]decane derivatives **45** using standard techniques. In similar fashion the homologous ester complex **46** is converted into spiro[5.5]undecane derivatives **47**.

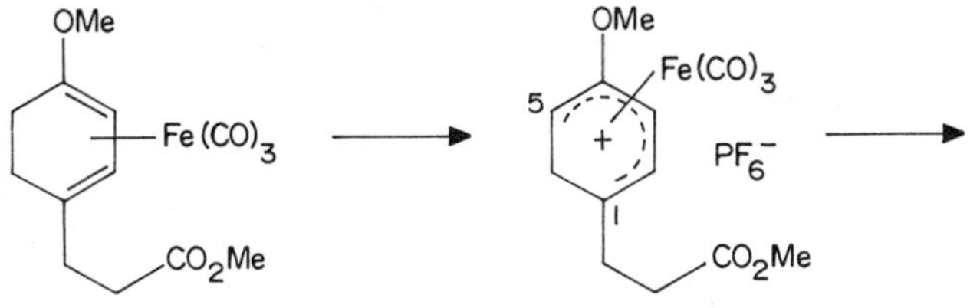

(43)   (44)

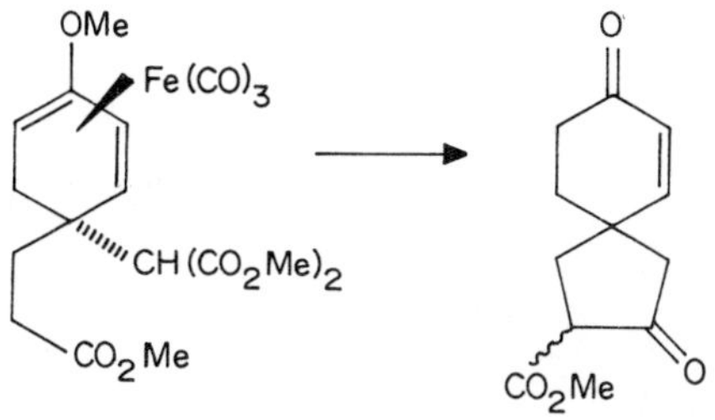

(45)

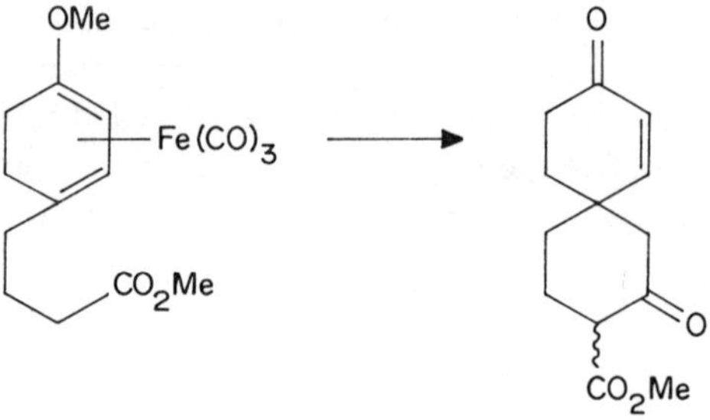

(46)   (47)

One way in which regioselectivity problems can be overcome is to carry out the nucleophile addition *intramolecularly*. To do this it is necessary to obtain dienyl complexes containing an enolizable group in the 1-substituent. A number of such compounds have been prepared as shown below, which again illustrates the variety of transformations which can be carried out in the presence of the diene–$Fe(CO)_3$ group.

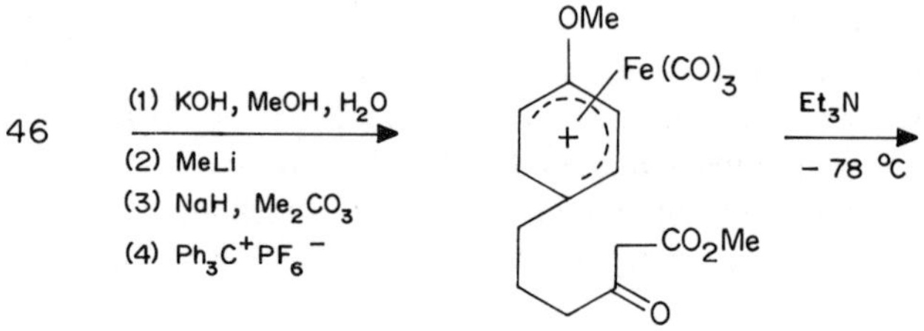

$$46 \quad \xrightarrow[\substack{(1)\ KOH,\ MeOH,\ H_2O \\ (2)\ MeLi \\ (3)\ NaH,\ Me_2CO_3 \\ (4)\ Ph_3C^+PF_6^-}]{}$$

(48)

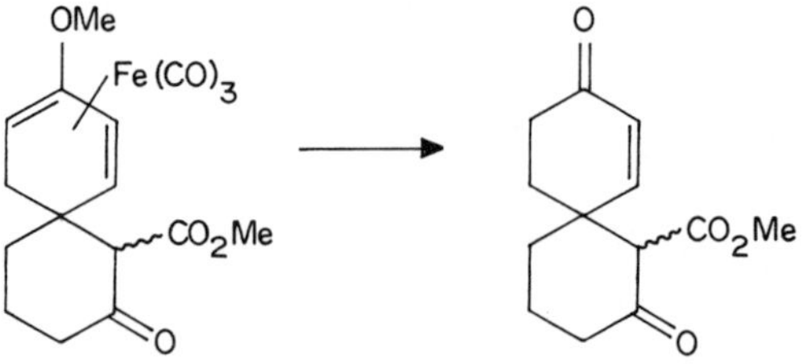

$$\xrightarrow[-78\ ^\circ C]{Et_3N}$$

(49)   (50)

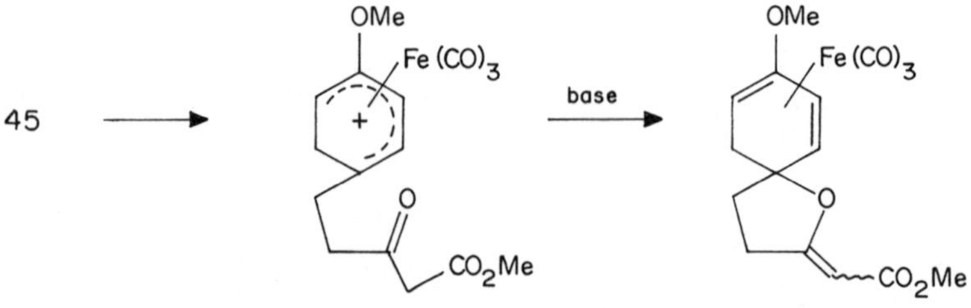

$$45 \quad \longrightarrow$$

(51) $\xrightarrow{base}$ (52)

(53) → Et$_3$N, CH$_2$Cl$_2$ −78 °C → (54)

CHCH$_2$CH$_2$CH(CO$_2$Me)$_2$

X = CN, CO$_2$Me

(55) → base → (56) $n = 3$ or (57) $n = 4$

As can be seen, the keto ester methylene group readily enolizes and deprotonates in **48**, giving rise to internal nucleophile addition to give **49** in 90% yield, which is readily converted through to the enone **50**. This was the first recorded example of an annulation reaction involving these complexes. A problem is encountered with the lower homologue **51** since it prefers to cyclize on oxygen to give **52**, analogous to the corresponding γ-bromo keto esters or ketones in organic chemistry. Attempts to induce cyclization on carbon using the gem diester **53** resulted only in deprotonation of the exocyclic methylene group α-to the dienyl terminus. This indicates that this grouping is relatively acidic (pK_a ca. 13–14). The problem is easily overcome using the malononitrile and cyano ester derivatives **55**, which are less sterically demanding and also have more acidic methine groups. Whilst the only cyclizations of dienyl–Fe(CO)$_3$ complexes so far reported are those giving rise to spirocyclic systems, there is every reason to believe that other modes of annulation can be developed using similar enolizable groups.

Interesting ring-forming reactions can also be developed using tosyloxy-substituted complexes, such as **58**. In these complexes there are two electrophilic centers, one on the dienyl ligand and one at the tosyloxy substituent (a good leaving group). Reaction of these complexes with *excess* of sodiomalononitrile results in addition to the dienyl system, followed by deprotonation of the malononitrile group (by exchange) and intramolecular tosylate displacement to give **56** and **57**. There is still a problem with regioselection and the isolated yields are *ca.* 55%. When a reversible nucleophile, e.g. benzylamine, is used the regioselectivity problem is alleviated by the equilibrium established between 'wrong' regioisomer and dienyl complex, resulting in very high yields of azaspirocyclic complexes **59** and **60** which are readily converted to organic derivatives **61** and **62**.

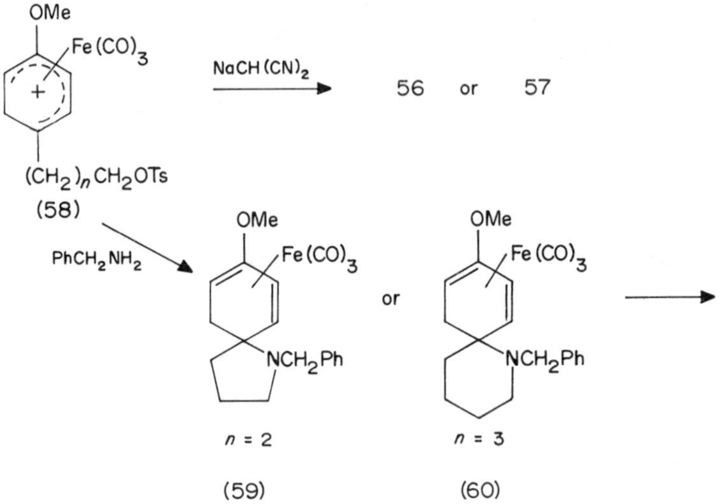

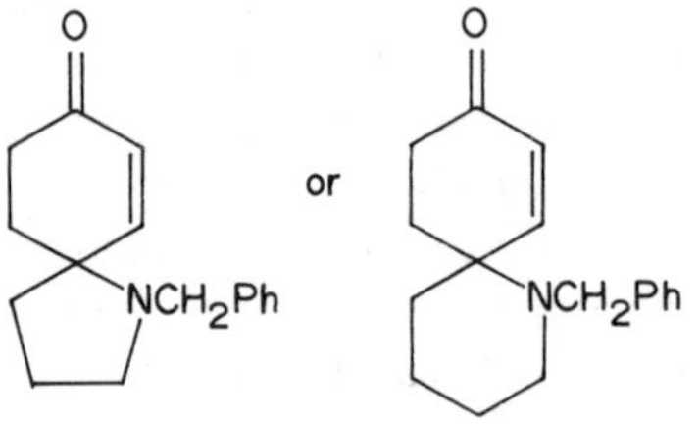

The intermediate **62** has been employed in a formal total synthesis of perhydrohistrionicotoxin **63**,[14] the fully hydrogenated derivative of histrionicotoxin **66**, a neurotoxin isolated from skins of the Colombian frog *Dendrobates histrionicus*. Perhydrohistrionicotoxin is a very useful neurophysiological tool. The intermediate **62** has been converted into **65**, which had previously been converted by Corey's group to **63**.

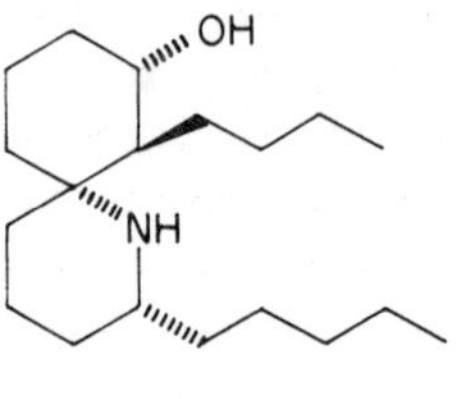

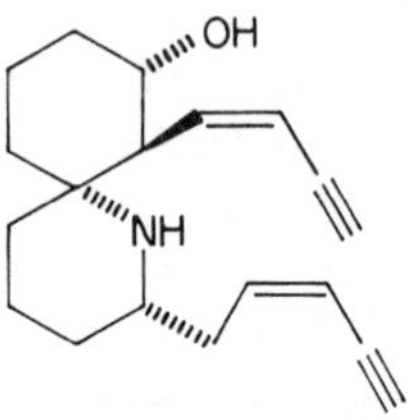

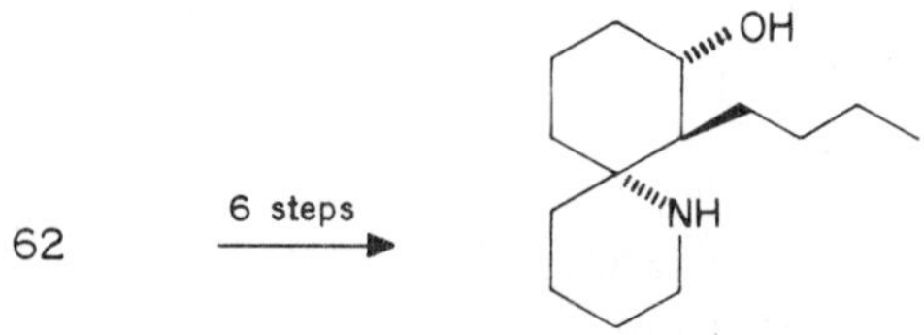

$$62 \xrightarrow{\text{6 steps}} (65)$$

Of course, it should be pointed out here that any regioselectivity problems associated with complexes **44, 58**, etc., can be overcome by replacing the 4-MeO group by 4-OPr^i as described earlier. The work on isopropoxy-substituted complexes was in fact a later development.

Synthesis of Aspidosperma Alkaloids[15]

There are a number of Aspidosperma alkaloids which possess an oxygenated C-18 grouping, e.g. cylindrocarpinol **66** and limaspermine **67**. The usual methods for constructing these molecules utilize intermediates such as **68**, which can in turn be derived from the appropriate 4,4-disubstituted cyclohexenone **69**. Thus, there is potential for synthesis of these alkaloids from the above types of organoiron complex. The total synthesis of (±)-limaspermine has been achieved, starting from the readily available ester complex **43**. Whilst better selectivity may now be obtained using the isopropoxy-substituted complexes discussed above, the results obtained with **43** were acceptable at the time the work commenced.

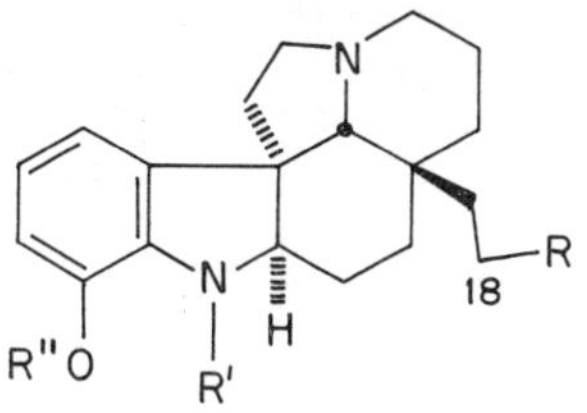

(66) R = OH , R' = H , R" = Me
(67) R = OH , R' = COEt , R" = H
(74) R = H , R' = COMe , R" = Me

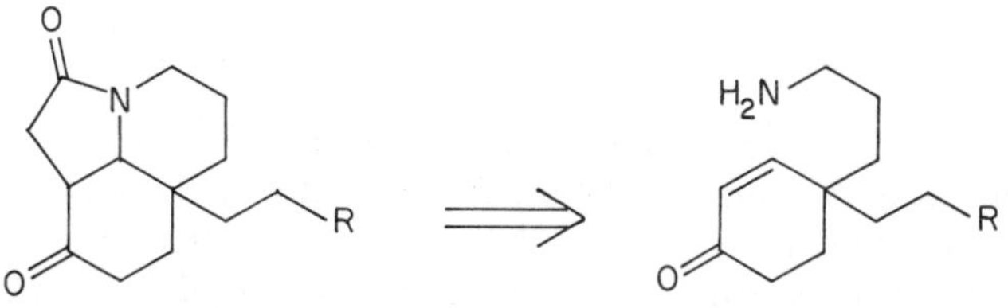

(68)   (69)

The complex **43** is converted to the dienyl salt **70**, containing the protected (phthalimide) primary amine, in *ca.* 75% overall yield. Reaction of this with dimethyl potassiomalonate gives a mixture of regioisomers from which pure **71** can be obtained in 70% yield by crystallization. This intermediate is readily converted to the tricyclic intermediate **72** and thence to (±)-limaspermine **67** using standard procedures.

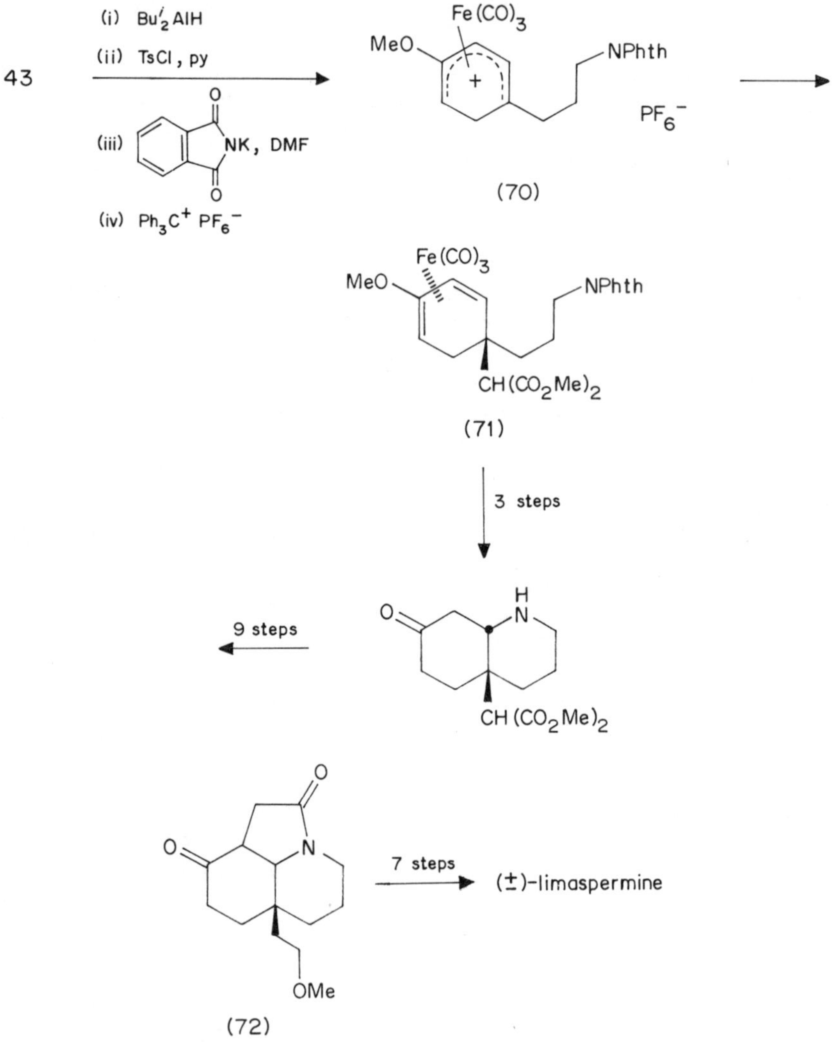

(70)

(71)

3 steps

9 steps

(72)

7 steps → (±)-limaspermine

A more efficient route to intermediate **72** has also been developed. A model study commencing with complex **23** established the methodology for elaborating the decahydroquinoline system of **73**, which is an intermediate previously converted to aspidospermine **74**. The sequence employed again illustrates the diverse number of transformations to which the diene–Fe(CO)$_3$ group is stable.

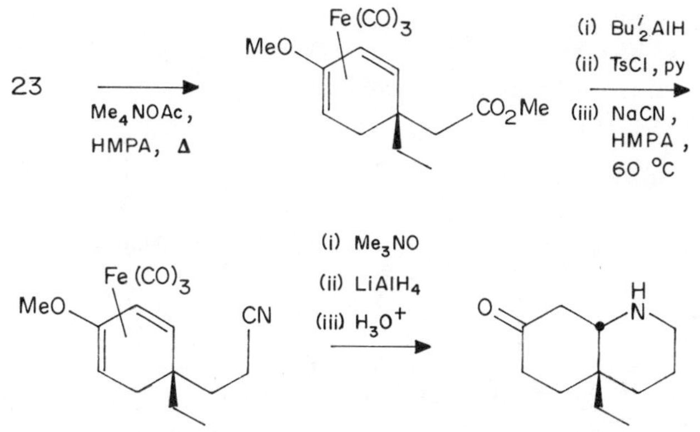
23
Me4NOAc,
HMPA, Δ
Fe(CO)3
MeO
CO2Me
(i) Bu^i_2AlH
(ii) TsCl, py
(iii) NaCN,
HMPA,
60 °C
Fe(CO)3
MeO
CN
(i) Me3NO
(ii) LiAlH4
(iii) H3O+
O
H
N

(73)

Starting with the functionalized dienyl complex **75**, and using identical methodology, the decahydroquinoline intermediate **76** can be constructed, and converted into **72**.

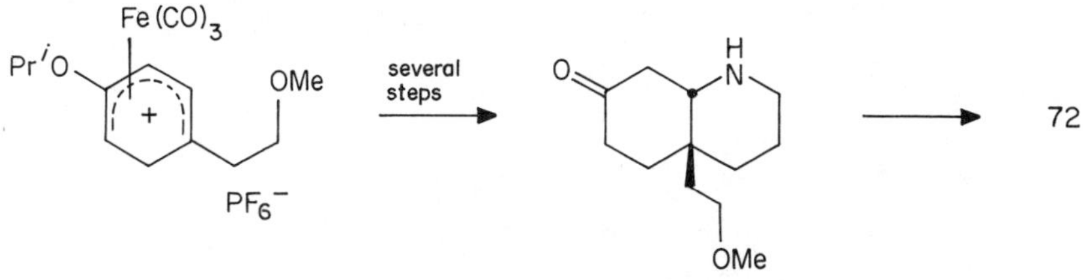
Fe(CO)3
Pr^iO
OMe
+
PF6−
several
steps
O
H
N
OMe
72

(75) (76)

Introduction of Angular Substituents into Bicyclic Systems[16]

The bicyclic dienyl–$Fe(CO)_3$ complexes **77** and **80** are readily prepared from the corresponding tetralin and indane derivatives. These complexes react with high regioselectivity with sodium cyanide and dimethyl sodiomalonate to give **78**, **79**, **81** and **82**. Introduction of methyl substituents into the uncomplexed ring caused steric inhibition of reactions at the angular dienyl terminus. The regiochemistry of the reactions depends on the stereochemistry of the methyl substituent and the results are summarized schematically below.

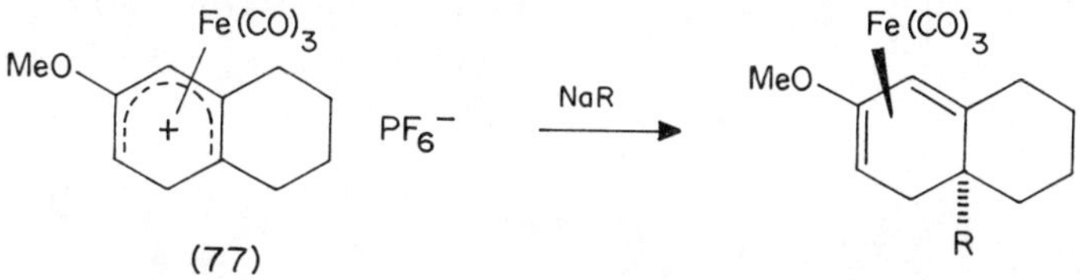
Fe(CO)3
MeO
+
PF6−
NaR
Fe(CO)3
MeO
R

(77)

(78) R = $CH(CO_2Me)_2$

(79) R = CN

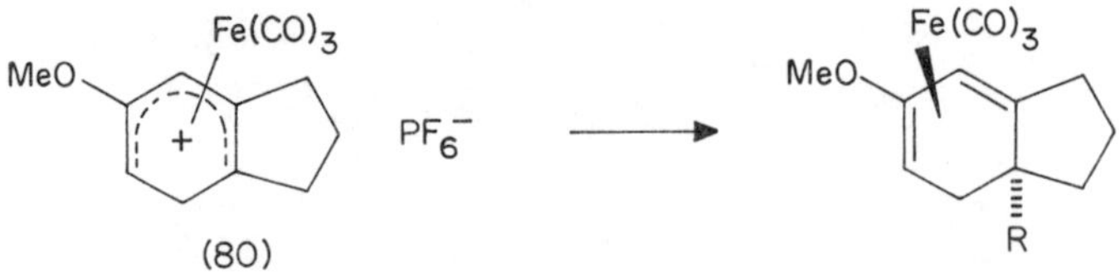

(80)

PF$_6^-$

(81) R = CH(CO$_2$Me)$_2$
(82) R = CN

NaCH(CO$_2$Me)$_2$

NaCH(CO$_2$Me)$_2$

+

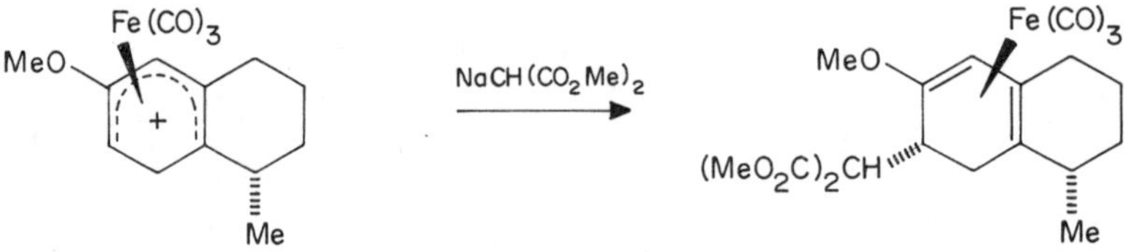

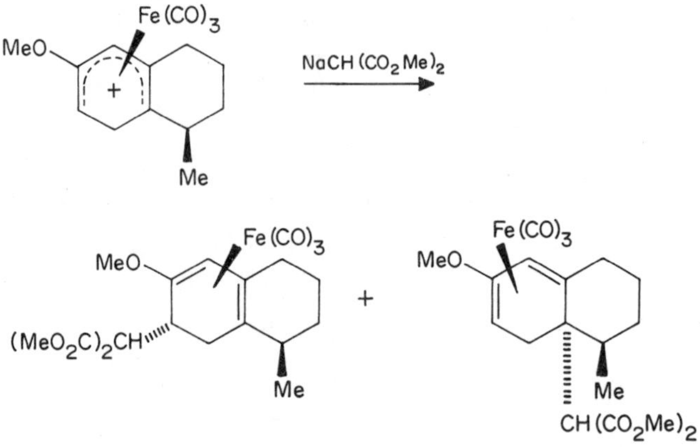

1   1

Tricarbonyl (3 methoxycyclohexadienyl) iron Hexafluorophosphate and Related Compounds

This 3-methoxy-substituted cyclohexadienyl complex **83** is not available using the usual hydride abstraction procedures, for reasons discussed above. It has been prepared by Birch's group, starting from 1,3-dimethoxybenzene.[17]

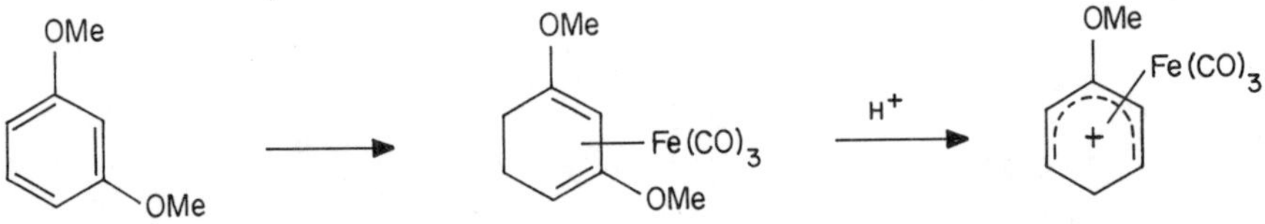

H$^+$

(83)

Reaction of this complex with nucleophiles occurs to give complexes **84**, and removal of iron followed by mild and hydrolysis gives the 5-substituted cyclohexenones **85**.

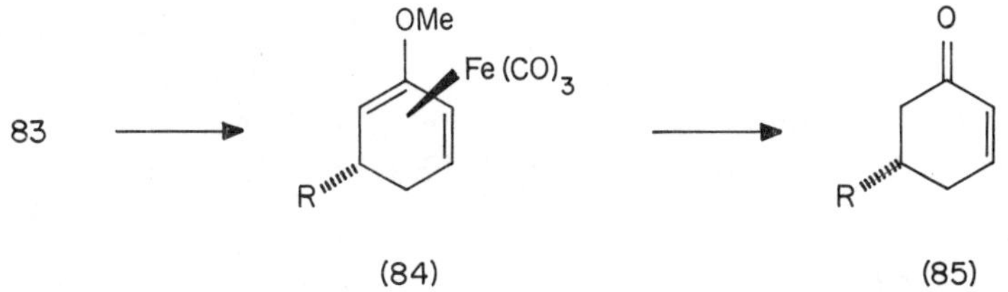

The Complex **83** is therefore the synthetic equivalent of the cation **86**.

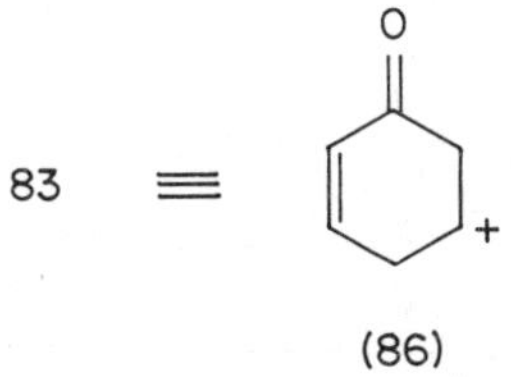

Similar, but more highly substituted complexes have been prepared, e.g. the dienyl complex **10**. The dienyl termini in this complex are extremely hindered and normal products from nucleophile addition are not obtained. Interestingly, no product from nucleophile addition to non-terminal dienyl positions, and no product from nucleophile attack at the M-*endo* face were observed. Reaction of **10** with dimethyl sodiomalonate gave only **87**, from nucleophile addition to a CO ligand, and the cyclohexadienone **88**, the mechanism of formation of which is unknown. The latter compound can also be formed in high yield by direct treatment of **10** with cerium(IV) ammonium nitrate.

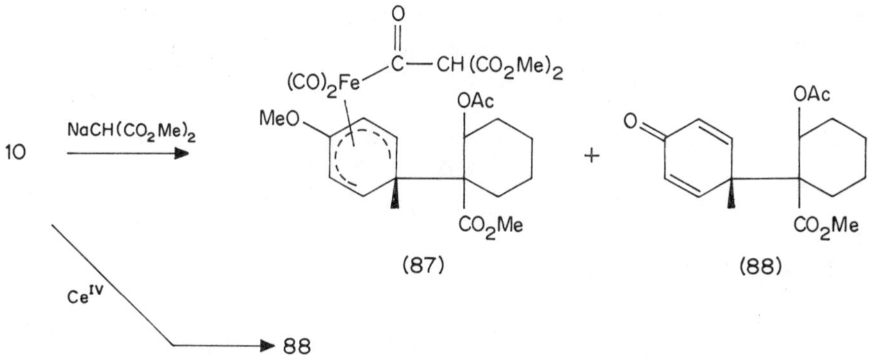

Complexes **89**, similar to **87** have also been observed during the reaction of NaOMe with tricarbonylcyclohexadienylosmium: —

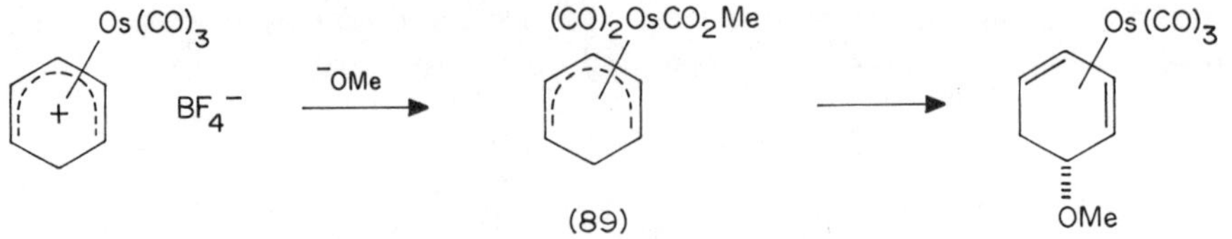

(89)

When the 6-*exo*-substituted dienyl complex is less hindered, some success is obtained during nucleophile addition to the dienyl ligand, e.g. **90→91** and **92→93**.

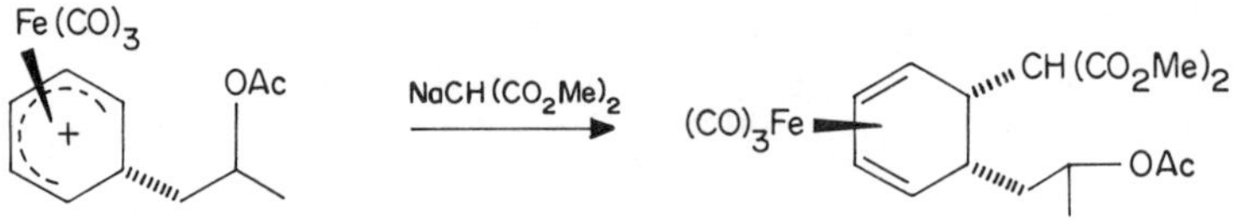

(90)   (91)

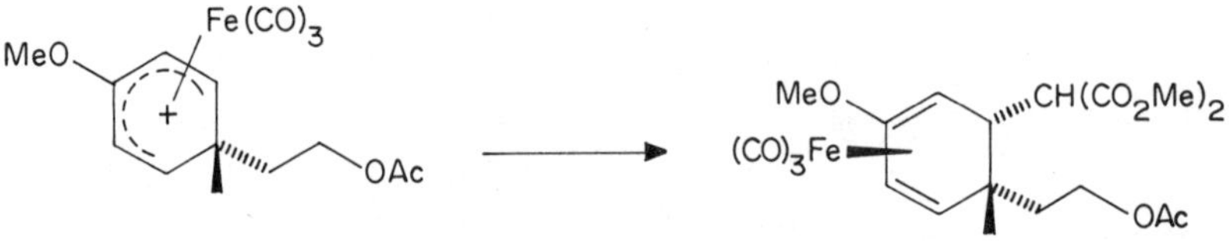

(92)   (93)

Application of Cyclohexadienyl–Fe(CO)₃ Complexes as Aryl Cation Equivalents

Complexes resulting from nucleophile addition to simple cyclohexadienyl–Fe(CO)₃ complexes can usually be subjected to oxidative reaction conditions which will remove the metal and oxidize the resultant cyclohexadiene to a substituted aromatic compound. In some cases the conversion can be achieved in a single step by direct treatment of the complex with palladium on charcoal at elevated temperature. In most cases, however, prior removal of Fe(CO)₃ with trimethylamine *N*-oxide is preferred, followed by oxidation, usually with DDQ. Good yields for this procedure have been reported.[18] Some examples are shown below.

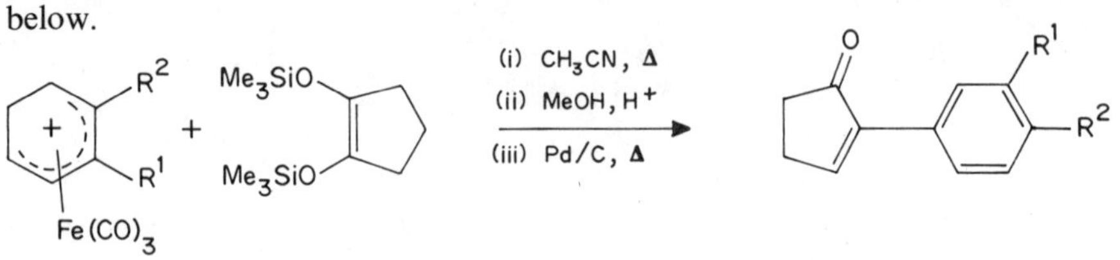

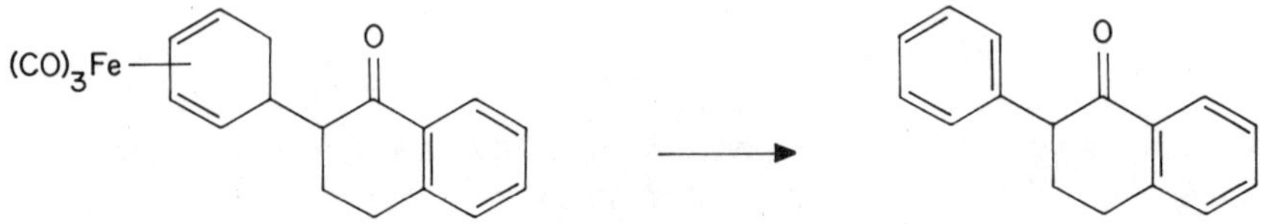

The following synthetic equivalents can therefore be derived and utilized for synthetic planning:

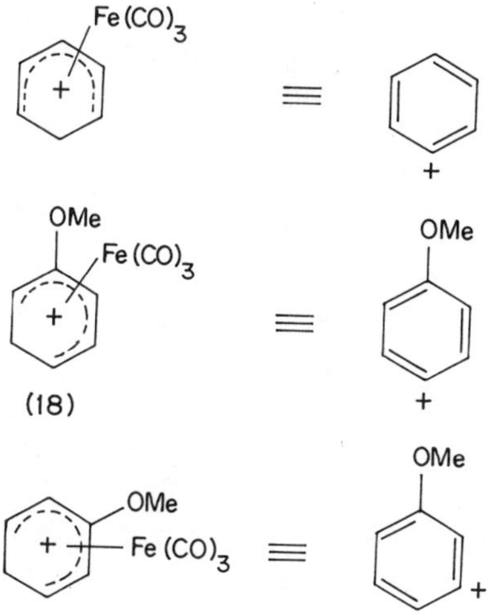

(18)

(83)

This represents considerable versatility of these complexes. The complex **18** has recently been employed[19] for new synthetic approaches to the Sceletium alkaloid *O*-methyljoubertiamine **95**.

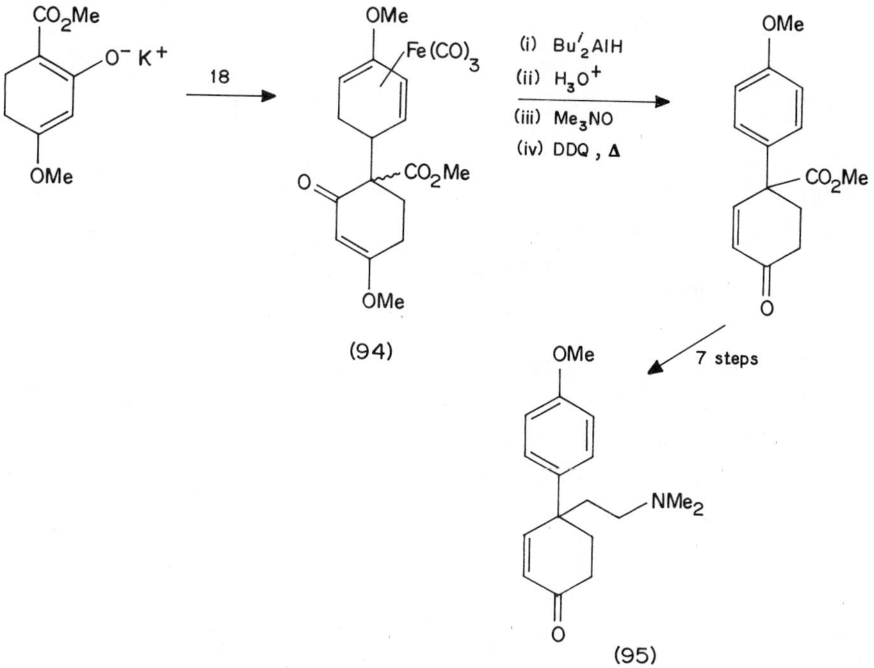

(94)

(i) $Bu_2'AlH$
(ii) H_3O^+
(iii) Me_3NO
(iv) DDQ , Δ

7 steps

(95)

Cobalt, Rhodium, Iridium

Whilst these complexes have not been widely studied, there are a few reports of the preparation and reactivity of simple derivatives, and these will be briefly presented here. There is some scope for further development in this area, particularly with regard to the preparation of substituted derivatives. Some of the problems encountered with the corresponding iron complexes may or may not be overcome using cobalt complexes. The most commonly used ancillary ligand is η^5-cyclopentadienyl, the complexes having the general structure:—

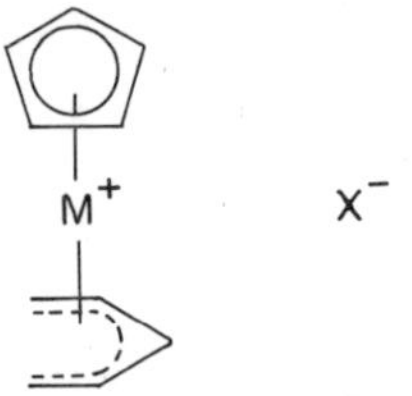

The basic methods of preparation are similar to those for the iron complexes, and these, together with some reactions with nucleophiles, are summarized below.[20]

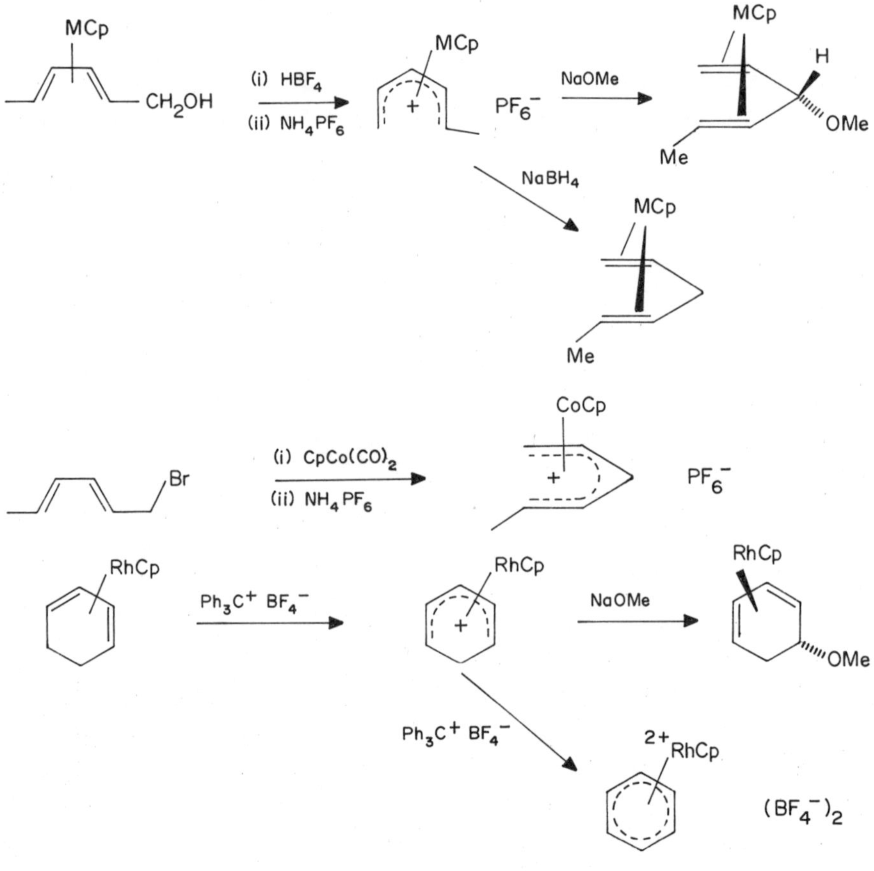

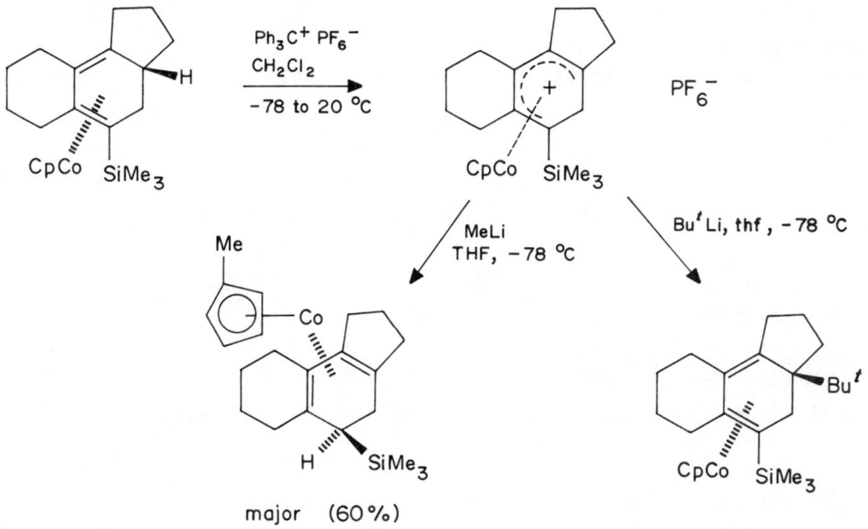

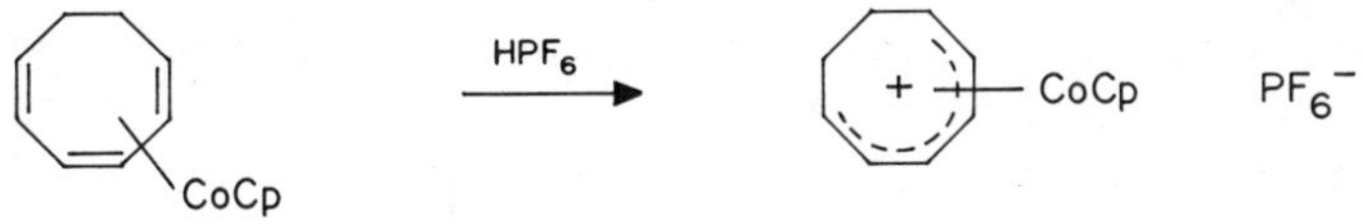

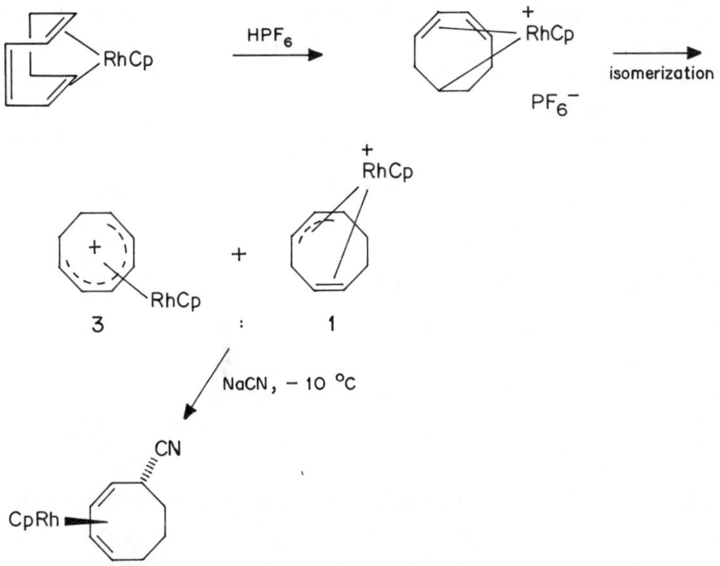

Owing to the limited number of examples available, no general statements concerning reactivity can be made, apart from the fact that, in most cases, although not all, nucleophile addition occurs to the dienyl rather than the cyclopentadienyl ring.

Manganese and chromium also form cyclohexadienyl–$M(CO)_3$ complexes, but these are intermediates formed by nucleophile addition to η^6-arene complexes, and will be discussed in Chapter 9.

η^5-CYCLOPENTADIENYL METAL COMPLEXES

Complexes containing the η^5-cyclopentadienyl ligand can be roughly divided into two types: (a) metallocenes, which contain no other ligands, and (b) complexes containing η^5-Cp and other ligands, such as CO, NO, PR_3, etc. The cyclopentadienyl ligand is useful as an ancillary ligand in complexes whose main function lies in the reactivity of, e.g. an olefin, alkyl or other ligand. Examples of these have been met throughout this book.

The metallocenes are by now an extremely well known class of compounds, exemplified by ferrocene, the discovery and characterization of which initiated the explosive growth of organometallic chemistry witnessed during the past 30 years. Ferrocene and its analogues in fact occupy a favored place among metallocenes, there being few parallels in the chemistry of other Cp_2M complexes with that displayed by the iron triad. This is a direct consequence of the variation of the number of electrons in the valence shell of the metals from the 'magic' number of eighteen. Complexes of metals to the left of iron in the Periodic Table are electron deficient, and these compounds, such as titanocene and molybdenocene, often have dimeric or polymeric structures and undergo reactions in which extra ligands are added to the coordination sphere. Complexes of metals to the right of iron are electron rich, and the chemistry of, e.g., cobaltocene, rhodocene and nickelocene is dominated by reactions which lead to a net loss of electrons in the coordination sphere.

Since a great deal of chemistry has now been done on ferrocene, we shall treat this and the related ruthenocene and osmocene in some detail, and then briefly consider other metallocenes.

Ferrocene, Ruthenocene and Osmocene

An excellent review by Rosenblum[21] covering the basic chemistry of these species appeared in 1965. Since space does not permit such a detailed presentation here, the interested reader is referred to this review.

Methods of Preparation

There are a number of methods for the preparation of these complexes, which in some cases are employed for the synthesis of other metallocenes. The first preparation of ferrocene **96** by Kealy and Pauson[22] utilized the reaction of cyclopentadienylmagnesium bromide with $FeCl_3$ in diethyl ether–benzene solution (not applicable to metals whose halides are insoluble in these organic solvents).[22] Subsequently, alkali metal cyclopentadienides, readily prepared from cyclopentadiene and the metal, metal hydride or metal acetylide in liquid ammonia or thf, have also been used in reactions with the metal halide. Ferrocene itself is an orange crystalline solid, m.p. 174 °C.

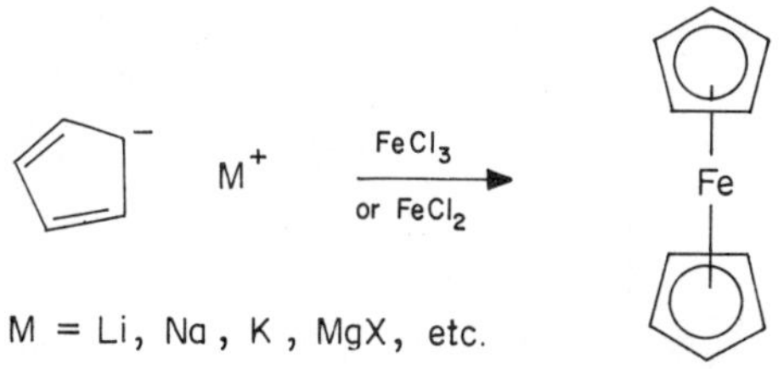

$$M = Li, Na, K, MgX, etc.$$

(96)

By far the most useful method for the preparation of ferrocene is to simply treat a solution of cyclopentadiene in diethylamine with anhydrous iron(III) or iron(II) chloride. This method is also applicable to other metallocenes. The iron chloride can be generated by using iron powder and replacing the amine with its hydrochloride.

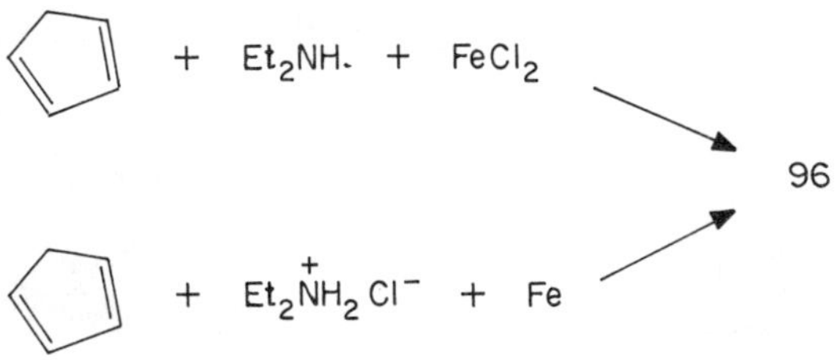

Direct reaction of the metal (or metal oxide, chloride) and diene at elevated temperatures, in the presence of various metal oxides, also furnishes ferrocene. This was the basis of the early preparation by Miller et al.[23]

The free metal or its salts may be replaced by pentacarbonyliron in this reaction. Indeed, reaction of metal carbonyls with cyclopentadiene provides a fairly general method for metallocene preparation. The reaction, with $Fe(CO)_5$, appears to proceed with intermediate formation of the dimeric complex 97, which will be discussed later.

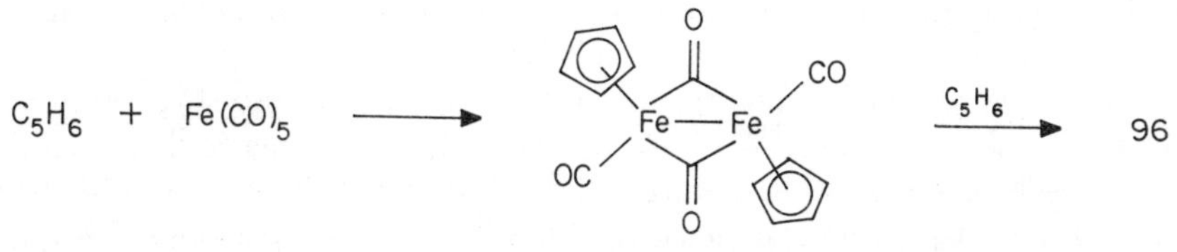

Ruthenocene 98 and osmocene 99 have been prepared using similar methodology to that employed for 96. Thus, reaction of cyclopentadienylmagnesium bromide with ruthenium(III) acetylacetonate gives 98, whilst reaction of sodium cyclopentadienide with $RuCl_3$ or $OsCl_4$ in thf or dme gives **98** and **99**, respectively.

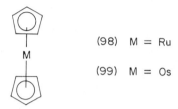

(98) M = Ru

(99) M = Os

Structure and Spectroscopic Properties

X-ray structure analyses of ferrocene and its simple derivatives have revealed that this complex exists in the staggered or antiprismatic conformation shown below, in which nonbonded interactions between heteronuclear carbon atoms are minimized. On the other hand, in ruthenocene and osmocene the significantly greater distance between the rings leads to a reduction of these interactions to such an extent that in the crystal the molecules adopt the eclipsed (prismatic) conformation.

antiprismatic 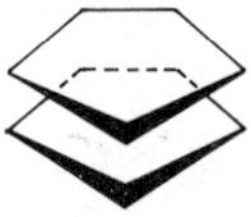   prismatic

The barrier to rotation of the Cp rings appears to be low in ferrocene, as evidenced by the non-existence of isomeric heteroannularly, disubstituted ferrocenes. The potential barrier to ring rotation has been estimated as 2–5 kcal mol^{-1} from NMR line width studies.

The mass spectra of metallocenes are simple, showing $C_{10}H_{10}M^+$, $C_5H_5M^+$ and M^+. Weaker metal—ligand bonds sometimes lead to lower intensity for the parent molecular ions, but this is not reliable. Substituted ferrocene derivatives tend to show mass spectral fragmentation more akin to the corresponding aliphatic systems than to benzylic compounds.

In the infrared and Raman spectra, ferrocene shows the following bands: C—H stretch 3085, antisymmetric C—C stretch 1411, antisymmetric ring breathing 1108, C—H bend parallel to plane of ring (∥) 1002, C—H bend perpendicular to plane of ring (⊥) 811, antisymmetric ring tilt 492, antisymmetric ring—M stretch 478 and ring—M bend 170 cm^{-1}. The two absorptions at *ca.* 1100 and 1000 cm^{-1} are useful for structural assignments of polysubstituted ferrocenes. Derivatives in which one ring is substituted tend to give both these bands, whilst those having both rings substituted lack the absorptions. This may be complicated by the presence of substituents which themselves absorb in this region. The peaks around $1000\,cm^{-1}$ are also useful in distinguishing '*ortho-*' and '*meta*' acetylalkyl-, acetylaryl- or diarylferrocenes: 1,2-disubstituted compounds show a single peak, whereas 1,3-disubstituted compounds show two peaks.

In the electronic spectrum, ferrocene shows absorptions at 440 and 325 nm, due to d–d type transitions. These bands are sensitive to ring substitution. The corresponding bands for ruthenocene and osmocene are found at shorter wavelengths.

In the proton NMR spectrum ferrocene exhibits a singlet at $\delta\,4.04$ p.p.m., and ruthenocene and osmocene show lower field shifts at $\delta\,4.42$ and 4.71 p.p.m., respectively. Substituted derivatives show the kind of unsymmetrical spectra expected and this is extremely useful, of course, for distinguishing homo- and heteroannularly disubstituted complexes, since the fluxionality of the molecules invariably leads to a single absorption for the unsubstituted ring.

Chemical Properties

Ferrocene, ruthenocene and osmocene are susceptible to attack by electrophilic reagents. NMR spectra in acidic media show the presence of metal–protonated species. For example, ferrocene–H^+ shows the high-field absorption ($\delta - 12.09$ p.p.m.) due to a metal-bound proton. This is obtained as a broad, unresolved multiplet, and coupling with the ring protons leads to these being shown as a doublet ($J = 1.3\,Hz$) at $\delta\,4.97$ p.p.m. Rapid proton exchange with solvent leads to loss of this coupling in ruthenocene–H^+, which now shows sharp singlets for both absorptions ($\delta\,5.33$ and -17.2 p.p.m.). Protonation is therefore occurring on a metal 'lone pair' orbital, i.e. one of the nonbonding e_{2g} orbitals.

The availability of the e_{2g} electrons undoubtedly also accounts for the ease of oxidation and the facile electrophilic (Friedel–Crafts-type) substitution reaction of these complexes. For example, oxidations of a large number of metallocenes, through loss of one or more electrons from the metal atom, are readily effected electrolytically, photolytically and also using a variety of organic and inorganic oxidants. Ferrocene, ruthenocene and osmocene give the paramagnetic ferricenium, ruthenicenium and osmicenium cations, respectively. The cations can be reduced using, e.g., tin(II) chloride, sodium hydrogen sulfite and ascorbic acid. Ferrocene and its relatives undergo very facile

$$Cp_2M \quad \xrightarrow[\text{(M'\,BF}_4)]{-\,e^-} \quad Cp_2M^+ \quad BF_4^-$$

Friedel–Crafts acylation reactions. Indeed, ferrocene itself is acetylated 3.3×10^6 more rapidly than benzene. The monoacetyl compound **100** undergoes acetylation to give the heteroannularly disubstituted compound **101** about 200 times more rapidly than benzene.

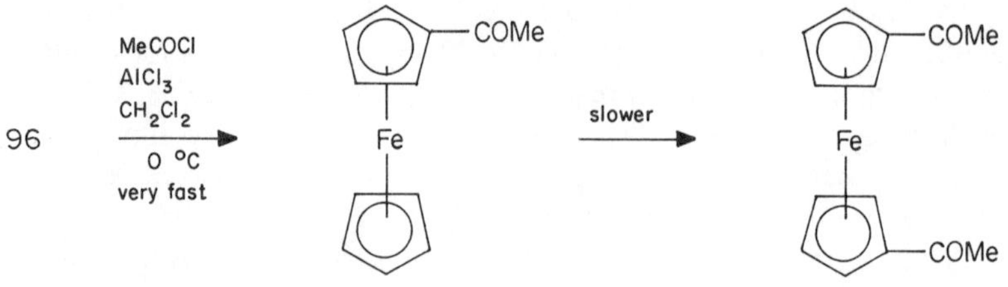

(100)   (101)

The deactivation of the second ring in **100** relative to **96** makes it possible to isolate the monoacetyl compound relatively easily.

There is general agreement that acylation takes place first at the metal (via the e_g 'lone pairs') followed by acyl transfer to the ring and then deprotonation. This is supported by the above observation of protonation and is similar to the reactivity of diene–$Fe(CO)_3$ complexes (Chapter 7).

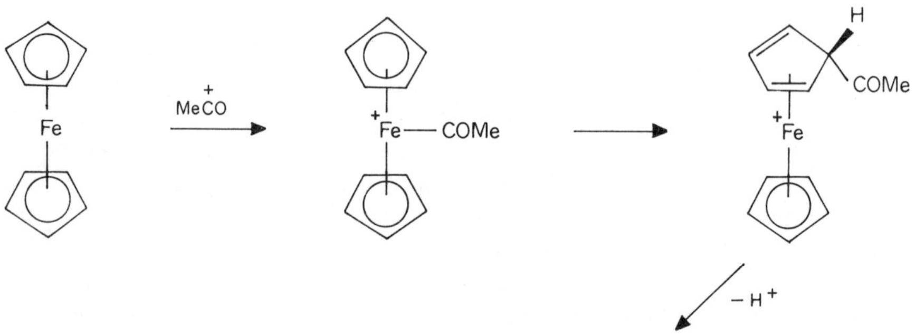

100

The facility of this reaction also allows an intramolecular heteroannular reaction to take place with β-ferrocenylpropionic acid **102**.

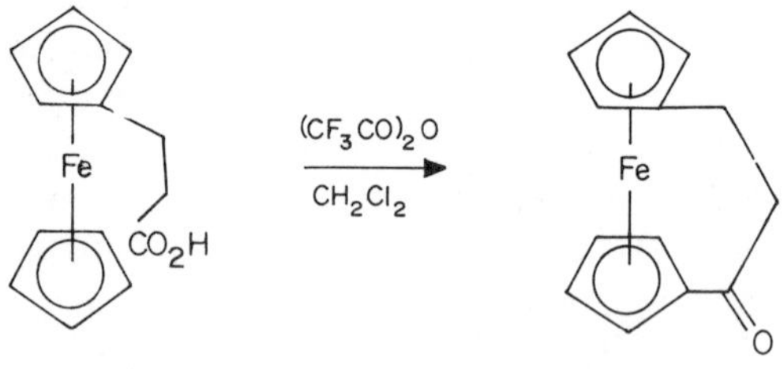

(102)   (103)

The reactivity towards acylation of the iron group metallocenes follows the order Fe > Ru > Os. Formylation of ferrocene to give **104** may be achieved under normal Vilsmeier conditions, again indicative of its high reactivity, since this reaction is normally applicable only to highly activated benzene derivatives. Only a monoformyl derivative is produced, owing to the deactivation effect of CHO.

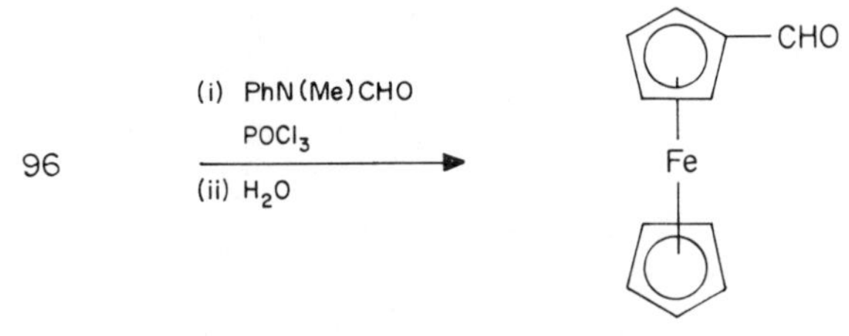

(104)

The aldehyde **104**, a dark red, crystalline substance, may also be prepared by a number of other routes, shown below.

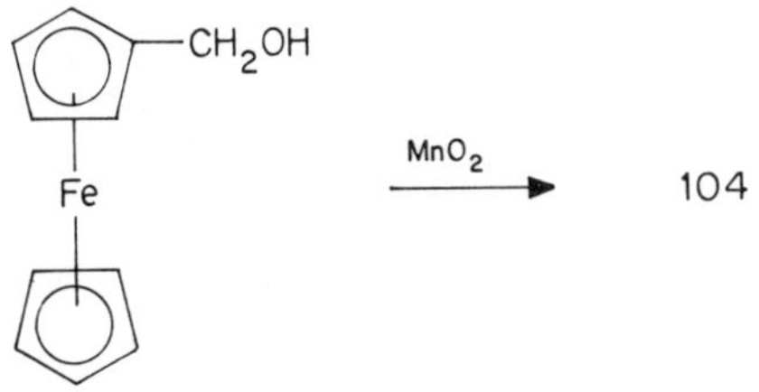

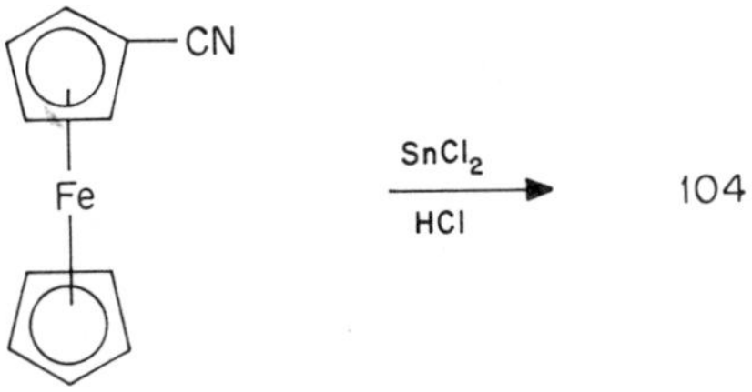

$$96 \xrightarrow[\text{Lewis acid}]{Cl_2CHOEt} 104$$

Alkyl-substituted ferrocenes can be acylated to give mixtures of isomeric ketones, and the predominant product generally tends to reflect the steric requirements of both the alkyl substituent and the acylating reagent. The steric requirements of substituents on the ferrocene ring appear to be considerably smaller than in benzene derivatives. Some examples are given below.

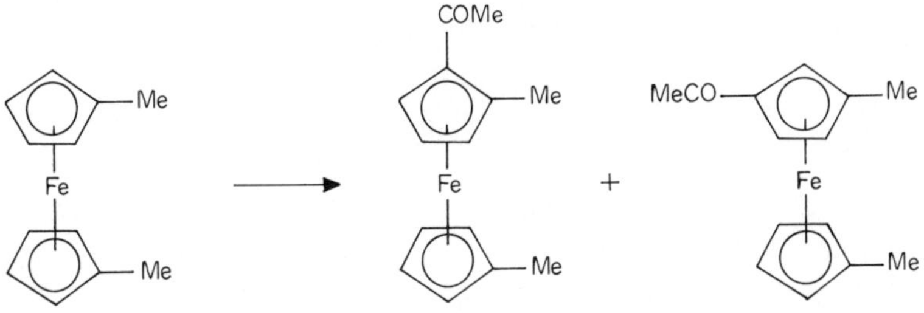

1 2·3

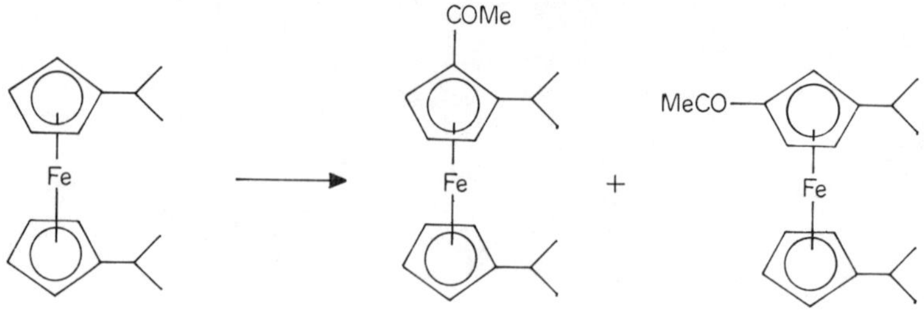

Compare :

sole product

The overall reactivity of ferrocene is little affected by alkyl substitution, and also there is little difference in the reactivity of the rings in monoalkylferrocenes, illustrated below for ethylferrocene.

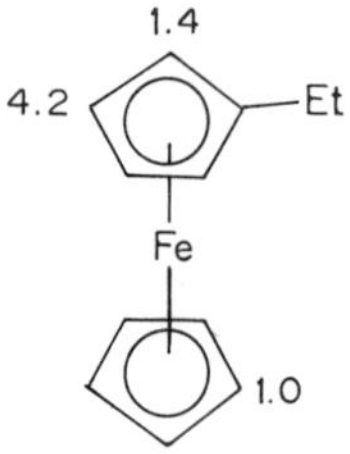

Relative site reactivities in Friedel—Crafts acetylation

Some site reactivities during the acetylation of aryl-substituted ferrocenes are shown below (approximate).

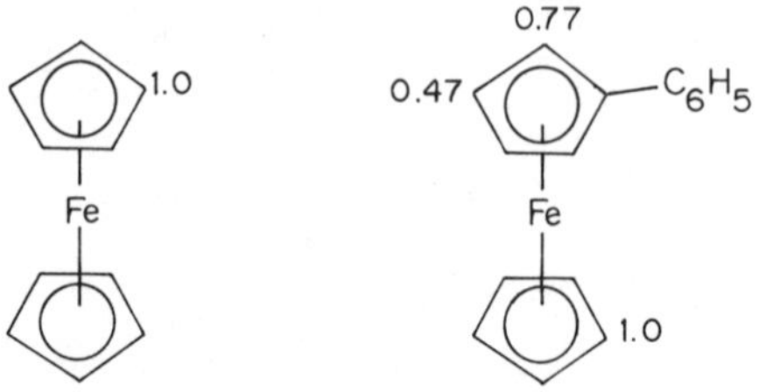

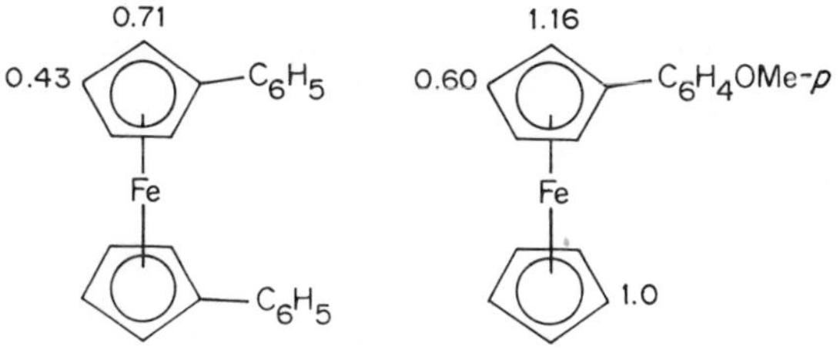

The acyl- and formylferrocene derivatives undergo most of the usual organic transformations expected for these functional groups. Some typical reactions are shown below, although this listing is not complete. The electron-releasing capacity of the ferrocenyl group results in a diminished reactivity of the carbonyl group.

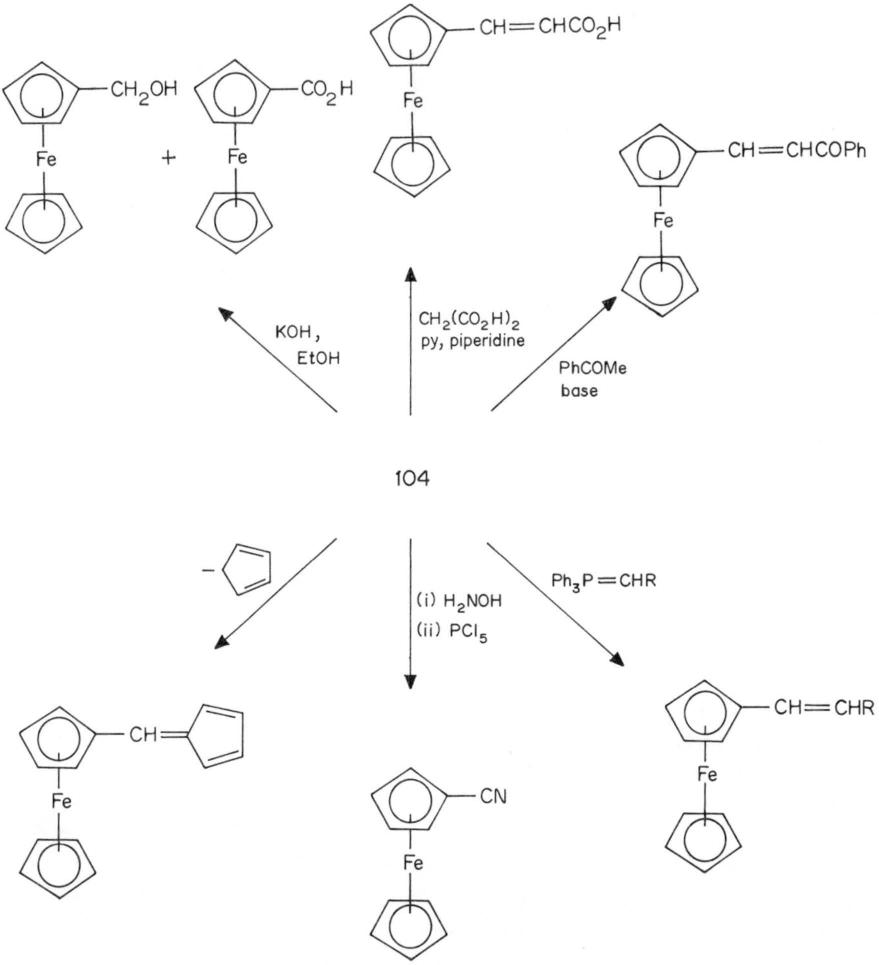

104

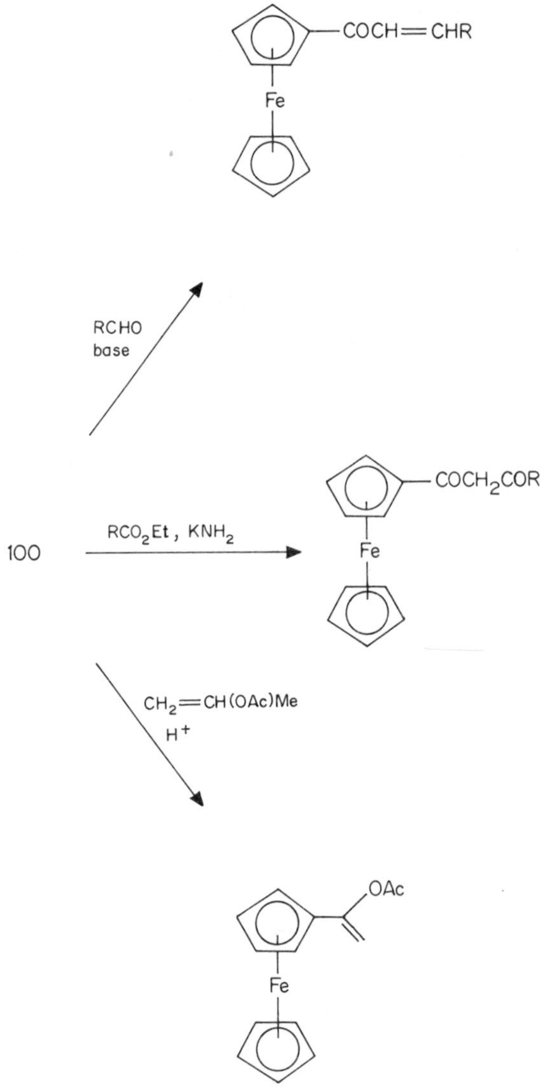

As expected from reactions of aromatic organic compounds, Friedel–Crafts alkylation of ferrocene does not occur cleanly, being accompanied by the usual carbonium ion rearrangements.

Ferrocene and its derivatives can be metallated by treatment with, e.g., butyllithium. The resultant ferrocenyllithium is of considerable importance for the preparation of substituted ferrocenes not accessible by direct methods (e.g. nitro compounds can not be prepared by nitration since this causes oxidation of the ferrocene). Some reactions are shown below.

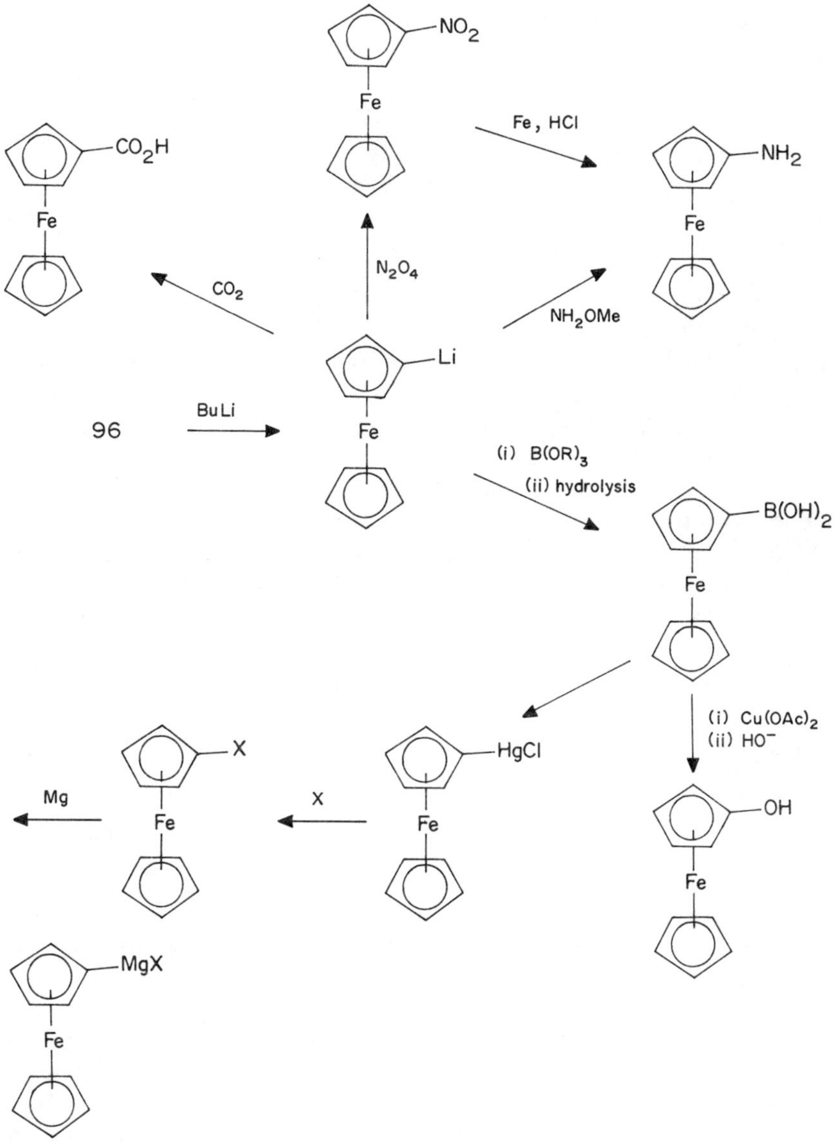

Stabilization of α-Carbonium Ions and Neighbouring Group Participation by the Ferrocenyl Group[24]

The remarkable ability of the ferrocenyl group to stabilize α-carbonium ions is shown by the facile acetoxylation of vinylferrocene **105** in acetic acid to give the acetate **107**, a reaction which undoubtedly proceeds via the intermediate carbonium ion **106**.

(105)   (106)   (107)

The α-carbonium ion appears to be stabilized by overlap of the filled non-bonding iron orbitals (e_{2g}) with the vacant carbon 2p orbital as indicated below, together with direct overlap of the 2p orbital with the π-system of the metallocene rings. As a consequence of this effect, solvolysis of α-acetoxy compounds proceeds very readily by a unimolecular mechanism. Indeed, the solvolysis of the acetate **107** proceeds approximately seven times faster than the solvolysis of triphenylmethyl acetate, so the intermediate α-carbonium ion **106** is much more stable than the trityl cation. Another interesting observation is that the analogous osmocene and ruthenocene complexes react more rapidly, the order (relative rate) being osmocene complex (5.37) > ruthenocene complex (1.36) > ferrocene complex **107** (1.0) for the reaction

This order is the reverse of that found for basicity or Friedel–Crafts acylation. It is probably due to the physically larger d orbitals of osmium (or ruthenium) being more able to overlap with the carbon p orbital.

An equally interesting phenomenon is the neighbouring group participation exhibited by certain α-acetoxy-substituted complexes. As an example, consider the *exo*- and *endo*-acetates **108** and **109**, respectively. Both solvolyze by a unimolecular mechanism to give the *exo*-alcohol **110**, due to addition of water to the cation from the less hindered side, but the *exo*-acetate solvolyzes 2500 times faster than the *endo*-acetate.

(108) (*exo*)   (110)   (109) (*endo*)

This is a classical neighbouring group participation effect in which assistance is only realized when the assisting group (Fe) can effect displacement of the leaving group by 'backside' attack. An elegant explanation can be invoked using frontier orbital considerations, and comparing with the simple case of an S_N2 reaction. The latter reaction proceeds with inversion because the nucleophile HOMO can only interact successfully with the backside lobe of C—X bond (X = leaving group). A net antibonding interaction occurs with the front lobe.[25]

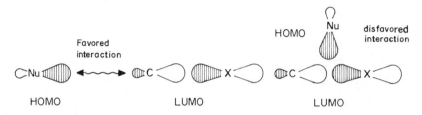

In similar fashion we can see that for assisted solvolysis, stabilization of the *developing* carbonium ion by an iron e_{2g} orbital can only be achieved when the leaving group is arranged *anti* to the metal, as shown below. This interaction will lower the energy of the transition state for solvolysis of the *exo*-acetate but will slightly raise the energy of the transition state from the *endo*-acetate.

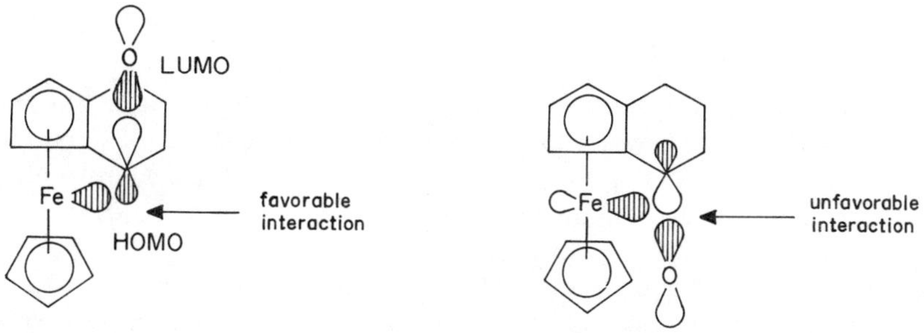

This provides a useful explanation of the observation that the corresponding ruthenium *exo*-acetate solvolyzes faster than **108**, whereas the ruthenium *endo*-acetate solvolyzes slower than **109**. The higher energy e_{2g} orbital of ruthenium will be closer in energy to the C—OAc LUMO, so that perturbation theory would predict that a greater stabilizing interaction would occur for the *exo* leaving group, whilst a greater *destabilizing* interaction should occur for the *endo* isomer.

The above effects leading to stabilization of α-carbonium ions also results in very facile dehydration (E1) of α-hydroxy compounds to the corresponding olefins. The complexes **111** and **112** also undergo solvolysis reactions. In this case different products are formed, and it can readily be seen that **111** can form a stabilized carbonium ion, with neighbouring group participation from the metal, and this then reacts with acetate anion present to give the observed primary acetate. In **112**, however, neighbouring group participation is not possible, with the result that it reacts 2780 times slower than **111**. The product arises by a typical ring expansion of the primary tosylate, well known in organic chemistry.[26]

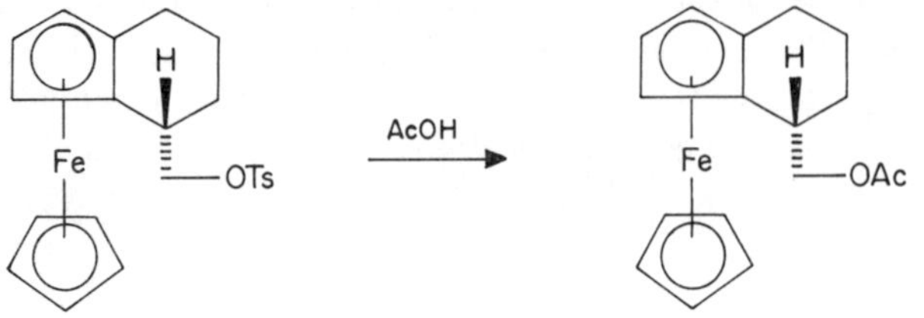

(111)

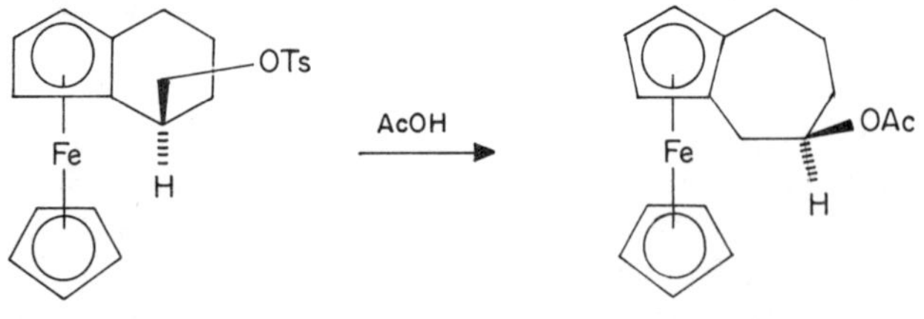

(112)

Cyclopentadienyl Metal Carbonyl Derivatives of Fe, Ru and Os

The deep red crystalline complex $Cp_2Fe_2(CO)_4$ **97** has already been mentioned in connection with the preparation of ferrocene. Owing to its ready preparation and ease of handling owing to its stability to oxidation, the chemistry of this complex has been widely studied. Evidence for the presence of a metal—metal bond is obtained from the X-ray structure, which shows a fairly short Fe—Fe distance [2.49 Å, compared with 2.46 Å for $Fe_2(CO)_9$], and also from the observed diamagnetism of the complex. In solution, the infrared spectrum shows a multiplicity of bands for bridging CO groups. This suggests that the structure is no longer the centrosymmetric arrangement found in the solid state, and variable-temperature IR studies suggest the following equilibria:

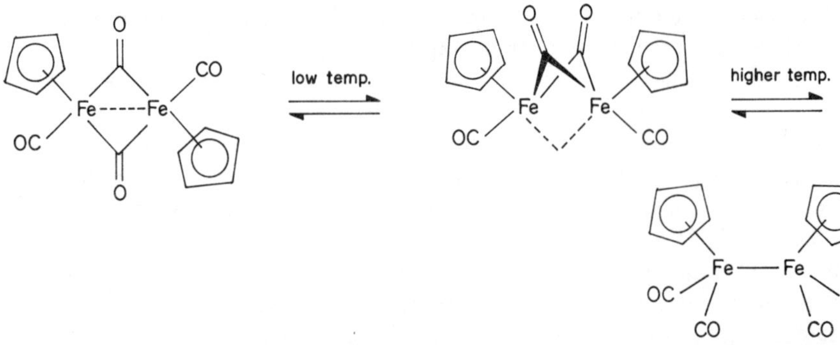

Support for this comes from the observation that this complex, and also the ruthenium and osmium analogs, show appreciable dipole moments in benzene solutions. Also, the osmium analogue appears to have a non-bridged structure shown below, even in the solid state, evidenced by the absence of bridging CO bands in the IR spectrum.

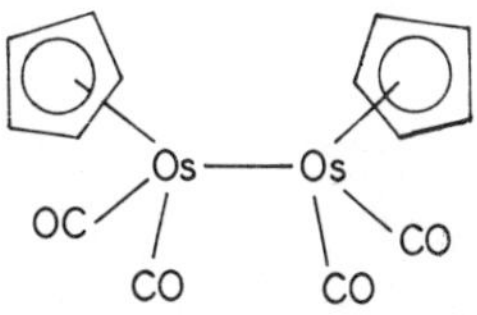

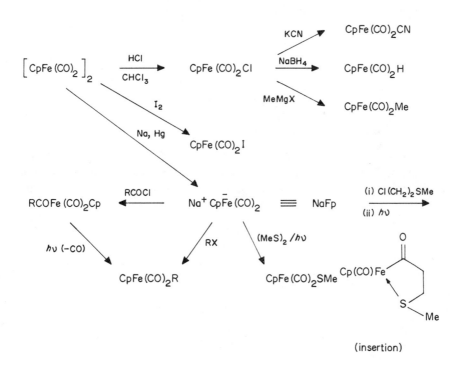

The iron complex **97** has been usefully applied to the preparation of a number of CpFe(CO)$_2$—('Fp'—) derivatives, some of which have already been presented in Chapters 3 and 5. Some of the reactions of the dimer are shown below.

Whilst NaFp reacts with a number of alkyl halides to give CpFe(CO)$_2$R, reaction with t-butyl chloride proceeds with elimination of isobutene to produce the yellow, air-sensitive, unstable hydride HFe(CO)$_2$Cp:

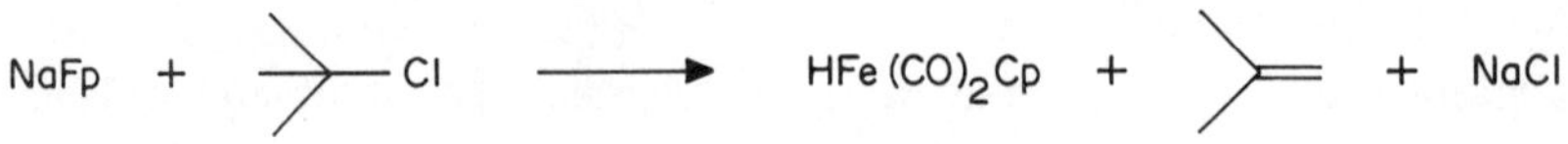

$$\text{NaFp} \; + \; \text{(CH}_3\text{)}_2\text{C—Cl} \; \longrightarrow \; \text{HFe(CO)}_2\text{Cp} \; + \; \text{(CH}_3\text{)}_2\text{C=CH}_2 \; + \; \text{NaCl}$$

The σ-bonded methyl complex $MeFe(CO)_2Cp$ undergoes carbonyl insertion on treatment with, e.g., phosphine derivatives, as discussed in Chapter 2.

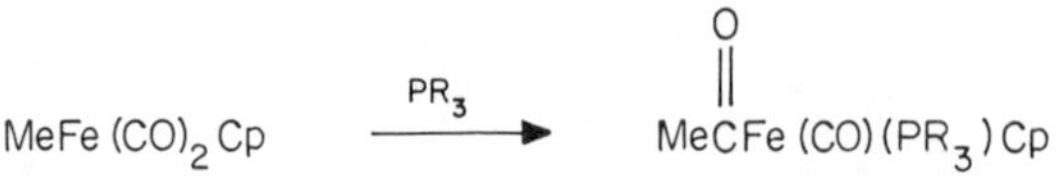

$$\text{MeFe(CO)}_2\text{Cp} \; \xrightarrow{\; PR_3 \;} \; \overset{\displaystyle O}{\overset{\displaystyle \|}{\text{MeC}}}\text{Fe(CO)(PR}_3)\text{Cp}$$

Treatment of the halide $CpFe(CO)_2Cl$ with aluminum chloride and carbon monoxide under pressure results in the formation of the cationic complex $CpFe(CO)_3{}^+$, which can be precipitated as its hexafluorophosphate from aqueous solution. The same complex may also be prepared by the following sequence, thus avoiding the use of carbon monoxide under pressure:

$$\text{NaFe(CO)}_2\text{Cp} \; + \; \text{Me}_2\text{NCOCl} \; \xrightarrow{\; thf \;} \; \text{Me}_2\overset{\displaystyle O}{\overset{\displaystyle \|}{\text{NC}}}\text{Fe(CO)}_2\text{Cp}$$

$$\Big\downarrow \text{MeOH}$$

$$\text{CpFe(CO)}_3{}^+ \; \xleftarrow{\; HCl \;} \; \text{MeOCFe(CO)}_2\text{Cp} \quad (\overset{\displaystyle \|}{}\underset{\displaystyle O}{})$$

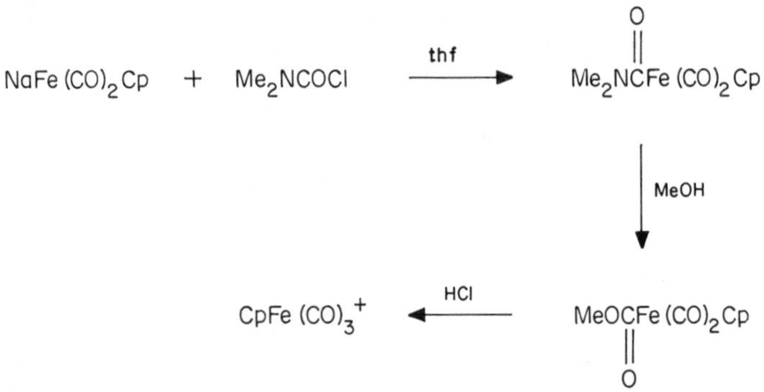

An interesting contrast is shown between $CpFe(CO)_3{}^+$ and the corresponding $CpFe(CO)_2PPh_3{}^+$ in their behavior towards nucleophiles. Thus, the tricarbonyl complex reacts by displacement of CO, whilst the phosphine derivative undergoes addition of nucleophile to the Cp ligand:

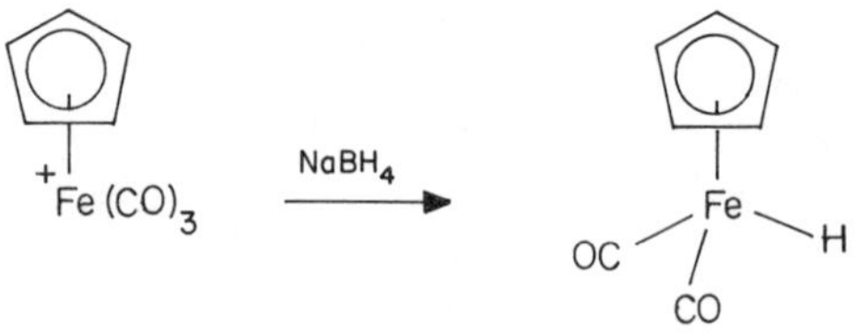

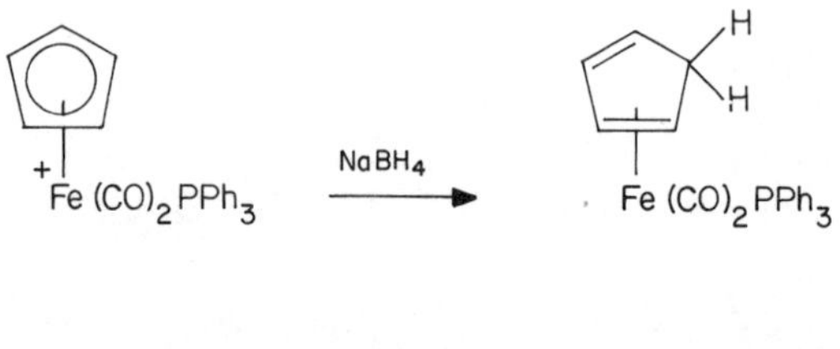

Other Metallocenes

There are, of course, a large number of metallocenes known, and it is impossible to cover their chemistry fully. We shall content ourselves with a brief presentation of the salient features of metallocenes derived from transition metals other than the iron group.

Titanium, zirconium and hafnium[27]

Whereas elements of the first transition series from vanadium to nickel form stable bis(π-cyclopentadienyl) complexes similar to ferrocene, the corresponding Cp_2Ti and Cp_2Zr are unknown as discrete chemical compounds. However, these molecules are implicated as intermediates in a number of reactions with olefins, hydrogen, CO, etc. Many attempts to produce titanocene, for example by reduction of the readily obtained (η-C_5H_5)$_2$TiCl$_2$ with Na, Na/Hg, etc., resulted in the formation of dimeric species, e.g. **113**.

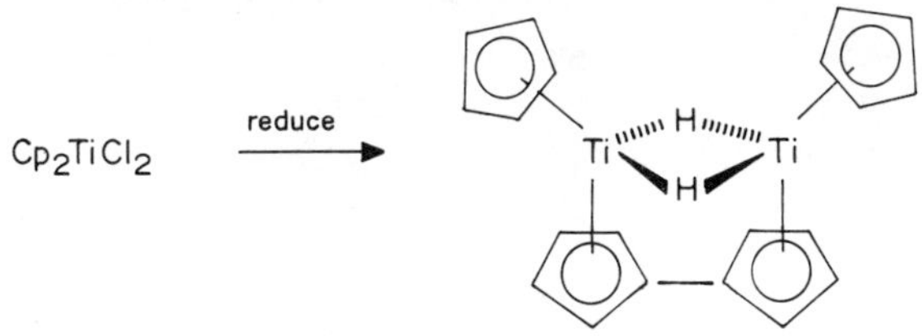

(113)

The formation of **113** can be explained as involving the intermediacy of (η-C_5H_5)$_2$Ti, since MO calculations have indicated that the most stable arrangement for titanocene is the bent configuration **114**, which confers carbene-like properties on the molecule. Rearrangement of such a 'carbene' by a hydrogen abstraction from a Cp ligand could lead to **113**, as shown below.

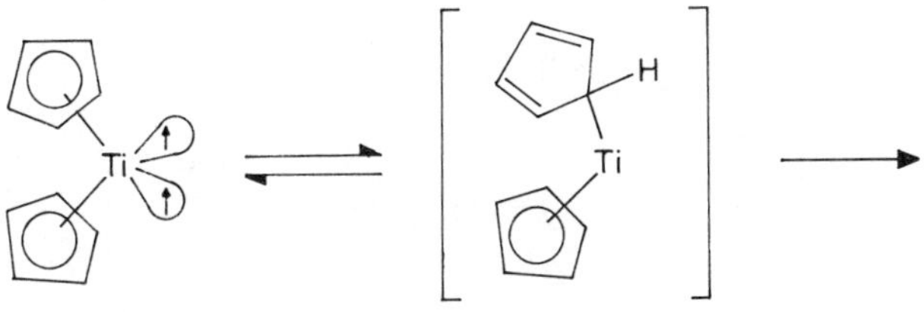

(114)

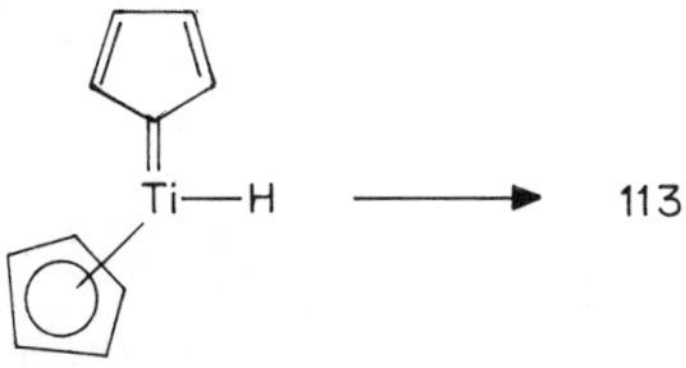

Reduction of $(\eta\text{-}C_5H_5)_2TiCl_2$ with 2 equivalents of potassium naphthalene at $-80\,°C$, warming to room temperature, affords the dimeric species **115** as a grey–black powder which can be converted to the thf-solvated species **116** by slow diffusion of isopentene into a thf solution.

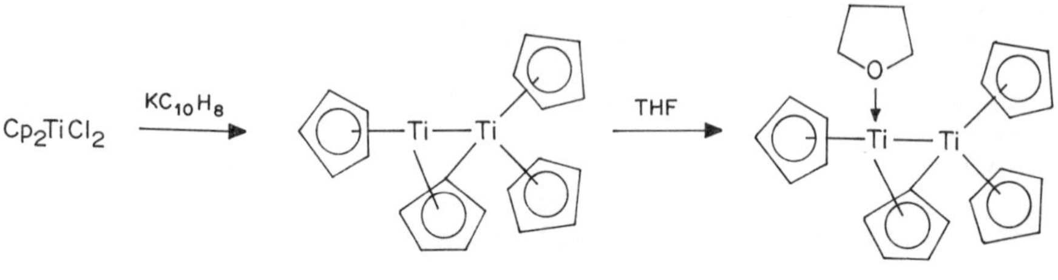

(115)   (116)

The problems associated with attempted preparation of titanocene have not prevented the synthesis of the permethyltitanocene derivative. When a solution of the complex **117** is heated, a complex of structure **118** is produced, and reaction of this with H_2 causes hydrogenolysis to give the dihydride **119**, obtained as orange crystals. This is converted to the corresponding dinitrogen complex **120**, and at room temperature evacuation of solutions of this complex causes liberation of nitrogen gas. Removal of solvent yields permethyltitanocene **121** as a yellow–orange crystalline solid.

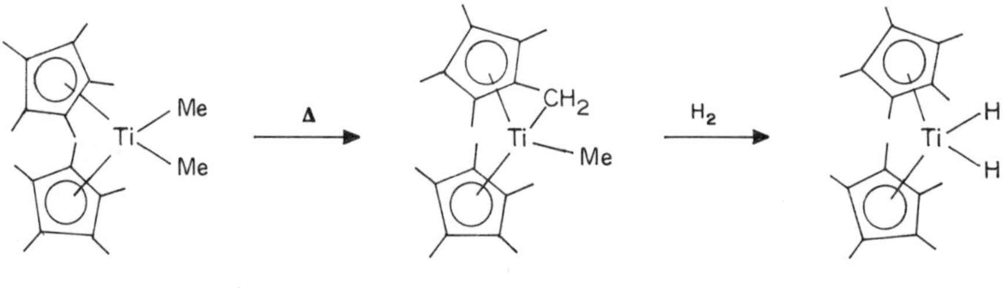

(117)   (118)   (119)

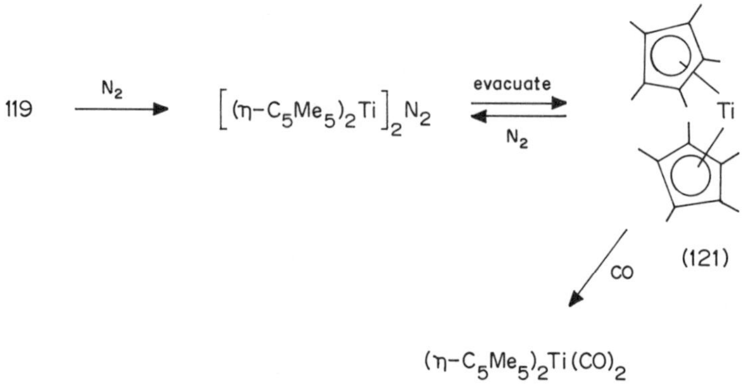

$$119 \xrightarrow{N_2} \left[(\eta\text{-}C_5Me_5)_2Ti \right]_2 N_2 \underset{N_2}{\overset{\text{evacuate}}{\rightleftharpoons}}$$

(121)

$$(\eta\text{-}C_5Me_5)_2Ti(CO)_2$$

This compound is sufficiently stable for characterization and chemical studies.

However, on standing solutions of **121** an equilibrium is established with the tautomer **122**. On prolonged storage, the complex decomposes irreversibly to give the violet paramagnetic complex **123**.

121 ⇌ →

(122) (green) (123)

The corresponding permethylzirconocene has not been isolated, but attempted preparation by a similar route to the above gives solutions containing mainly [η-C_5Me_5][η-$C_5Me_4CH_2$]ZrH, analogous to **122**, which may contain small amounts of the permethylzirconocene derivative.

In view of the above behavior of titanium complexes, the reported preparation of zirconocene and hafnocene must be treated with some skepticism. Indeed, reduction of (η-C_5Me_5)$_2$ZrCl$_2$ with potassium naphthalene under conditions similar to those used for production of the titanium complex **115** affords μ-(2-η^1:1-2-η^2-naphthyl)hydridobis[bis(η-cyclopentadienyl)zirconium](Zr–Zr) **124**.

(124)

Low-valent titanocene and zirconocene derivatives, particularly those related to the dinitrogen complex **120**, have been investigated as potential models for nitrogen fixation, and some successful reductions of molecular nitrogen have been achieved. These aspects are dealt with in Pez and Armor's review[27] and will not be discussed here. Titanium and zirconium metallocenes containing π-acid ligand are dealt with later (see also Chapter 3 on hydrozirconation).

Vanadium, Niobium and Tantalum

Vanadocene **125** is a well known compound first prepared in the 1950s, subsequent to the discovery of ferrocene. Whilst tetrakis(η-cyclopentadienyl)niobium **126** has been known for a number of years, only recently has evidence been presented supporting the existence of Cp_2Nb.[28] The cyclopentadienyl ligand has been used with tantalum complexes (see, e.g., Chapter 4, carbenes), as is indeed the case with many transition metal organometallics. There does not appear to be such a rich chemistry of the cyclopentadienyl ligands for these complexes as for ferrocene, etc., and we shall present here only the preparations of the metallocenes and more common derived complexes, summarized in the following equations.[29]

$$VCl_4 \ + \ NaCp \ \text{or} \ CpMgBr \ \longrightarrow$$

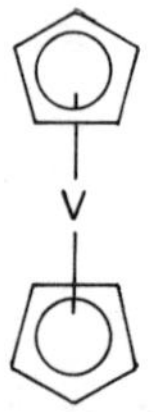

violet crystals,
m.p. 167 – 168 °C

(125)

$$125 \xrightarrow{\text{4 N HNO}_3} Cp_2V^{2+}$$

$$Cp_2V \ + \ CO \ \longrightarrow \ C_5H_5V(CO)_4$$

orange crystals,
m.p. 138 °C

$\diagdown$ 2 Na

$$Na_2\left[C_5H_5V(CO)_3 \right]$$

$$CpMgCl \ + \ VCl_4 \ \longrightarrow \ Cp_2VCl_2$$

pale green crystals

$$CpMgBr \ + \ NbBr_5 \ \longrightarrow \ Cp_2NbBr_3$$

dark reddish brown crystals

$$CpMgBr \ + \ TaBr_5 \ \longrightarrow \ Cp_2TaBr_3$$

rust-colored crystals

$$NbCl_5 \xrightarrow[\text{CpNa}]{\text{excess}}$$

$$TaCl_5 \xrightarrow[\text{CpNa}]{\text{excess}} Cp_4Ta$$

$$CpV(CO)_4 \quad + \quad 1,3-diene \longrightarrow CpV(CO)_2(diene)$$

$$26-27\%$$

$$Cp_2VCl \xrightarrow[\text{BPh}_4^-]{\text{H}_2\text{O}} Cp_2V^+ \ BPh_4^-$$

$$Cp_2VR \quad + \quad 2\ CO \longrightarrow$$

$$R = Me,\ CH_2Ph$$

but

$$Cp_2VPh \quad + \quad 2\ CO \longrightarrow$$

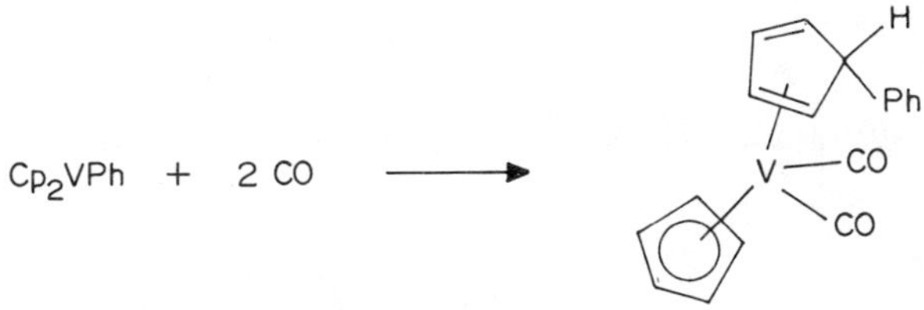

Chromium, Molybdenum and Tungsten[30]

Chromocene(Cp$_2$Cr) is the only known example of the simple biscyclopentadienyls of Group VI. The staggered structure **126** is adopted in the solid state, and it has a magnetic moment of 3.2 BM, showing two unpaired electrons.

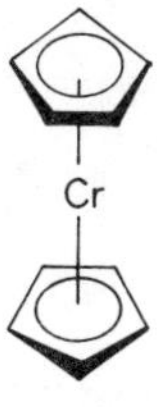

(126)

There is evidence that Cp_2Mo and Cp_2W are produced by photolysis of the dihydrides Cp_2MH_2, although these complexes cannot be isolated. The evidence comes from the following observations:

$$Cp_2MoH_2 \ + \ CO \ \xrightarrow{h\upsilon} \ Cp_2Mo(CO) \ + \ H_2$$

$$Cp_2MoH_2 \ + \ C_2H_2 \ \xrightarrow{h\upsilon} \ Cp_2Mo(C_2H_2) \ + \ H_2$$

$$Cp_2MoH_2 \ + \ PR_3 \ \xrightarrow{h\upsilon} \ Cp_2MoPR_3 \ + \ H_2$$

In the absence of added ligand, photolysis of Cp_2MoH_2 produces $(Cp_2Mo)_x$, formally derived by oligomerization of Cp_2Mo.[31]

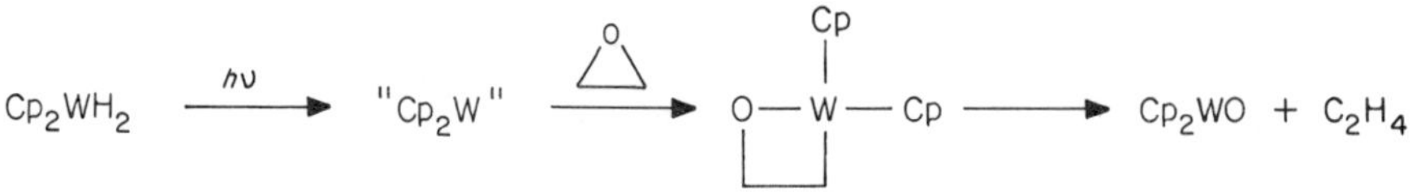

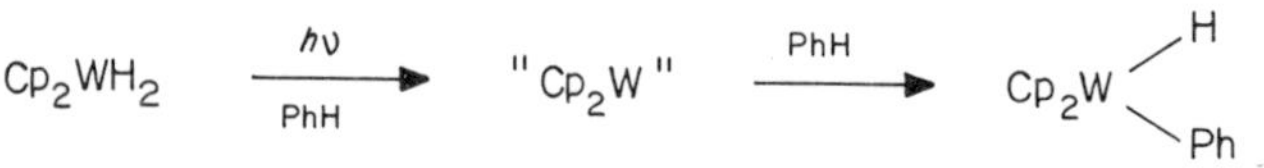

$$Cp_2WH_2 \ \xrightarrow{h\upsilon} \ "Cp_2W" \ \xrightarrow{\triangle} \ O-W-Cp \ \longrightarrow \ Cp_2WO \ + \ C_2H_4$$

$$Cp_2WH_2 \ \xrightarrow[PhH]{h\upsilon} \ "Cp_2W" \ \xrightarrow{PhH} \ Cp_2W\underset{Ph}{\overset{H}{<}}$$

The synthesis of chromocene, using the conventional method:

$$CrCl_3 \ + \ 3 \ NaC_5H_5 \ \xrightarrow[25\ °C]{thf} \ Cp_2Cr$$

needs rigorous exclusion of air.[32] Vacuum sublimation of the product gives approximately 60% yields of red crystals of extremely air-sensitive product, m.p. 173 °C.

Chromocene may also be prepared by reaction of hexacarbonylchromium with cyclopentadiene at elevated temperatures and pressure. The tungsten and molybdenum hydride derivatives are prepared in yields up to 50% by the following reactions:[33]

$$WCl_6 \ + \ NaC_5H_5 \ + \ NaBH_4 \ \longrightarrow \ Cp_2WH_2$$

$$MoCl_5 \ + \ NaC_5H_5 \ + \ NaBH_4 \ \longrightarrow \ Cp_2MoH_2$$

In contrast to ferrocene and related compounds, most of the reactivity studies on these complexes have centered around reactivity of the metal rather than the cyclopentadienyl ligand. The hydrides Cp_2MH_2 (M = Mo, W) undergo insertion reactions with activated acetylenes:

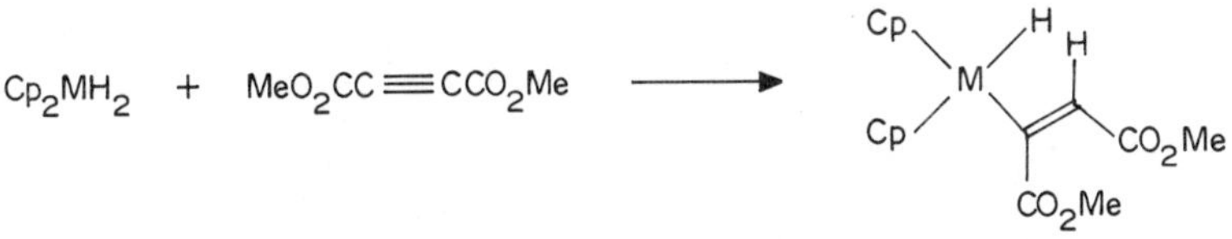
Cp2MH2 + MeO2CC≡CCO2Me
Cp
Cp
M
H
H
CO2Me
CO2Me

solely *cis* adduct

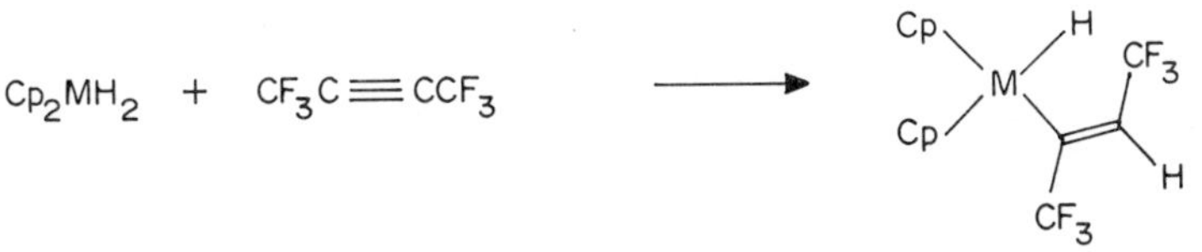
Cp2MH2 + CF3C≡CCF3
Cp
Cp
M
H
CF3
H
CF3

solely *trans* adduct

In contrast, reaction with, e.g., diphenylacetylene leads to displacement of *both* hydrides to give the acetylene π-complex.

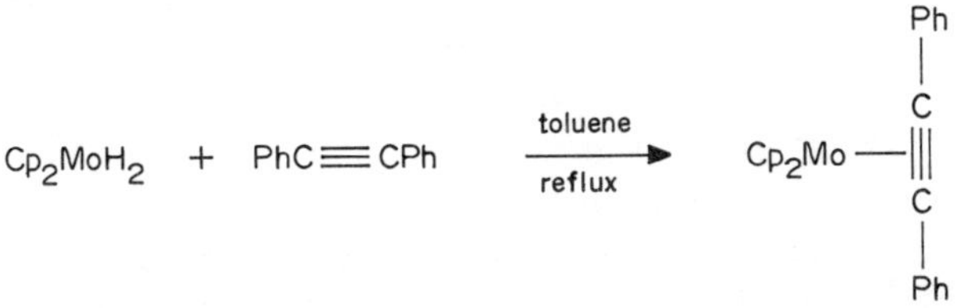
Cp2MoH2 + PhC≡CPh
toluene
reflux
Cp2Mo
Ph
C
C
Ph

Biscyclopentadienylchromium, on reaction with hexafluorobut-2-yne, gives an analogous η^2-acetylene complex:

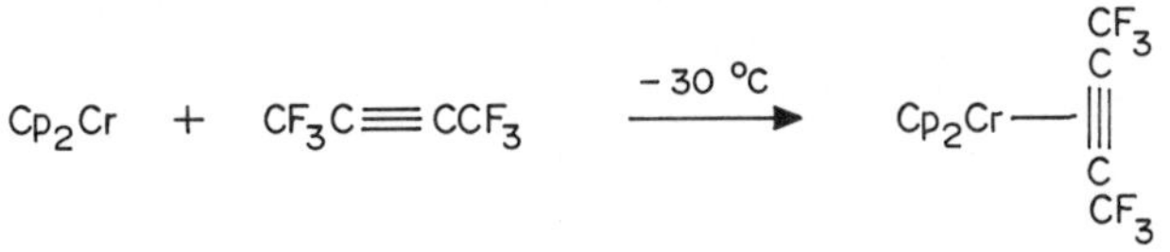
Cp2Cr + CF3C≡CCF3
- 30 °C
Cp2Cr
CF3
C
C
CF3

Further reactions of the hydrides, reflecting more the chemistry of the M—H bonds than the Cp—M system, are shown below.[34]

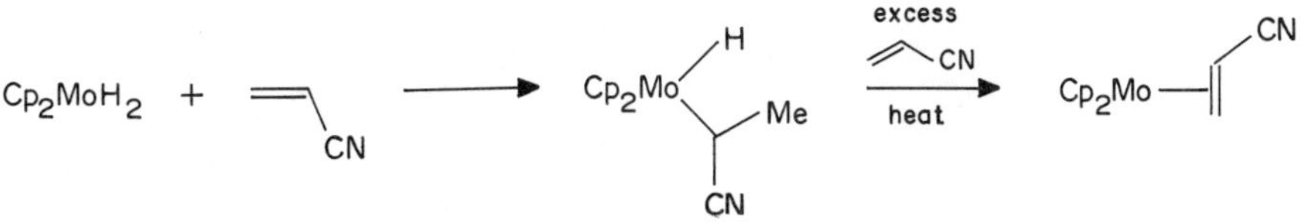
Cp2MoH2 +
CN
Cp2Mo
H
Me
CN
excess
CN
heat
Cp2Mo
CN

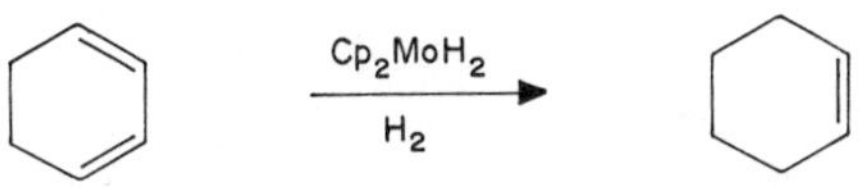
Cp2MoH2
H2

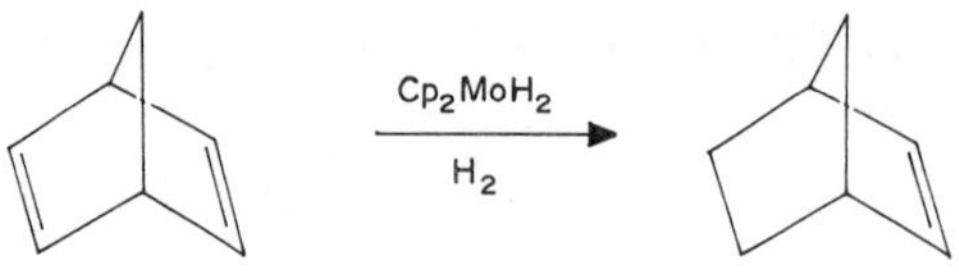

$$PhCHCl_2 \xrightarrow[20\ ^{\circ}C]{Cp_2MoH_2} PhCH_2Cl \xrightarrow[80\ ^{\circ}C]{Cp_2MoH_2} PhCH_3$$

$$CCl_4 \xrightarrow[-10\ ^{\circ}C]{Cp_2MoH_2} CHCl_3 \xrightarrow[30\ ^{\circ}C]{Cp_2MoH_2} CH_2Cl_2$$

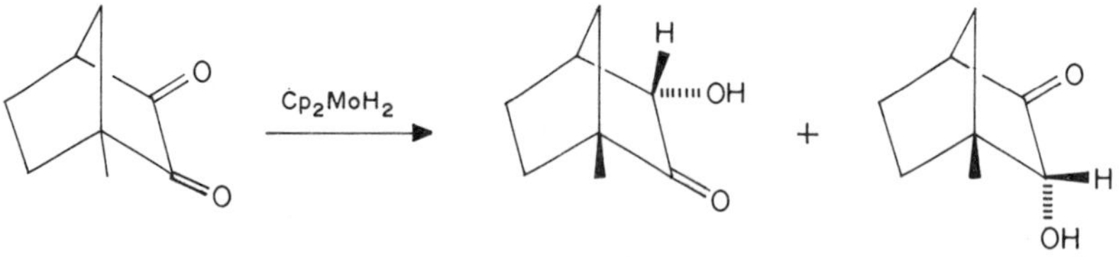

9 : 1

The reactivity of alkyl halides toward Cp_2MoH_2 is in a relative order similar to that observed for trialkyltin hydrides. Methyl and ethyl iodide react even at room temperature, whereas allyl and benzyl chlorides are inert, requiring elevated temperatures, so that polyhalogenated compounds can be reduced in a stepwise manner. The corresponding reactions of Cp_2WH_2 are slower. Reaction with α-diketones is strongly influenced by steric effects, the ratio of products obtained in the above reaction contrasting with a 1:1 ratio produced by zinc–acetic acid reduction.[35]

The dihydrides $Cp_2MH_2(M = Mo, W)$ may be converted into Cp_2MX_2 (e.g. X = Cl), and these complexes undergo nucleophilic substitution at the metal, e.g.

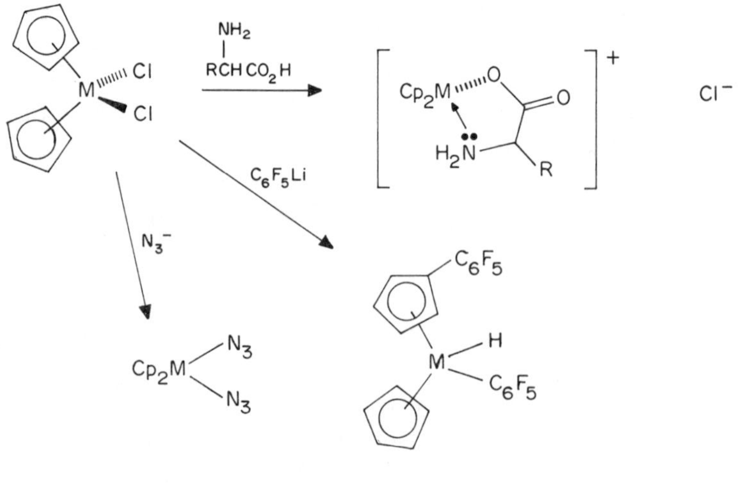

Treatment of the dihydrides, e.g. Cp_2MoH_2, with Grignard reagents, $RMgX$, in diethyl ether leads to the formation of ether-insoluble crystalline yellow compounds. The crystal structure of **127** (R = cyclohexyl) has been determined.

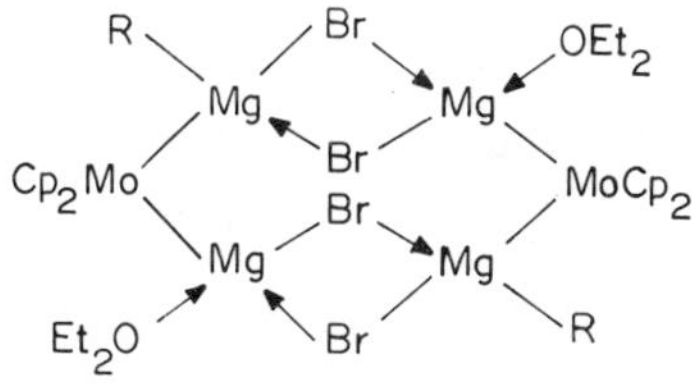

(127)

These odd-looking inorganic Grignard compounds are found to be useful for the preparation of other complexes based on the Cp_2Mo system, but in low yield.[36]

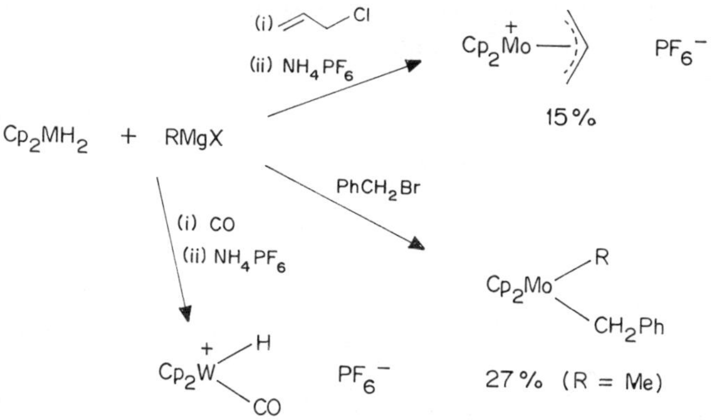

Manganese, Technetium and Rhenium

Treatment of anhydrous $MnCl_2$ with Na^+Cp^- in thf gives rise to bis-cyclopentadienylmanganese (manganocene), obtained as brown crystals, m.p. 172–173 °C. Although X-ray studies show the complex has the 'ferrocene' structure, the chemical properties suggest that the bonding is primarily of an ionic nature. Thus, the complex is readily hydrolyzed to give cyclopentadiene and manganese hydroxide, it reacts with CO_2 giving carboxylic acid salts and it reacts instantaneously with $FeCl_2$ in diethyl ether to give ferrocene. Solid solutions in Cp_2Mg are paramagnetic, with five unpaired electrons per Mn atom, indicating that the metal is present as the Mn(II) ion. Tricarbonylcyclopentadienyl-manganese has been prepared in *ca.* 60% yield by treatment of sodium cyclopentadienide with $MnBr_2$ followed by CO at 800 p.s.i. and 250 °C, and also by the action of CO under pressure on Cp_2Mn.[37]

The complex is obtained as yellow crystals, m.p. 76.8–77.1 °C.

$$Cp_2Mn \quad \xrightarrow{CO} \quad CpMn(CO)_3$$

Whereas Cp_2Mn is inert to Friedel–Crafts acylation conditions, $CpMn(CO)_3$ undergoes acylation analogously to ferrocene.

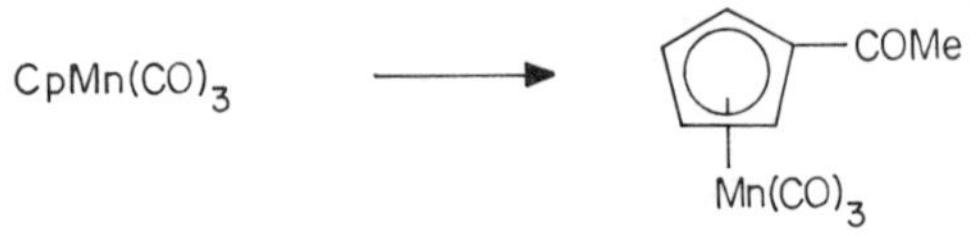

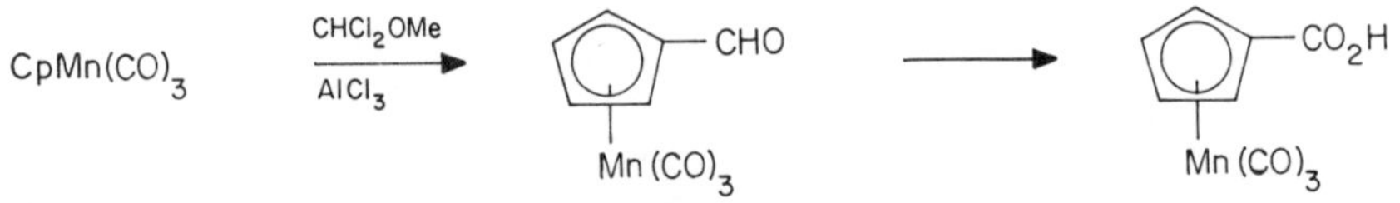

Ligand substitution reactions of $CpMn(CO)_3$ are achieved fairly readily. Thus, irradiation in the presence of ethene or cyclohexa-1,3-diene results in low yields of the appropriate monoolefin complexes.

$$CpMn(CO)_3 + C_2H_4 \xrightarrow{h\nu} CpMn(CO)_2(C_2H_4)$$

m.p. 116 – 118 °C

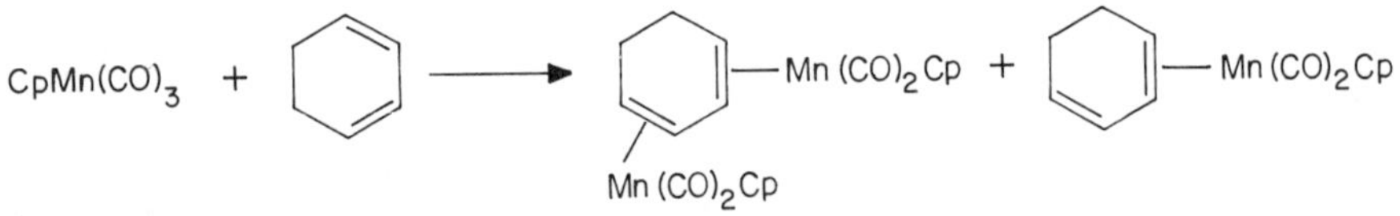

Biscyclopentadienylrhenium hydride is conveniently prepared by reaction of NaCp (excess) on $ReCl_5$, followed by heating the resultant purple product at 120–170 °C. Sublimation gives the hydride as yellow crystals, m.p. 161–162 °C (yield 20%). It is stable to water but reacts moderately quickly with air. Treatment with acid results in the formation of $Cp_2ReH_2{}^+$. The hydrides have the bent metallocene structure.[38]

Treatment of Cp_2ReH with CO at 250 atm and 100 °C gives a 45% yield of a pale yellow crystalline complex, m.p. 111–112 °C, formulated as the dicarbonyl(η^5-cyclopentadienyl)(η^2-cyclopentadiene)rhenium **128**, on the basis of its IR and hydrogenation behavior.[39]

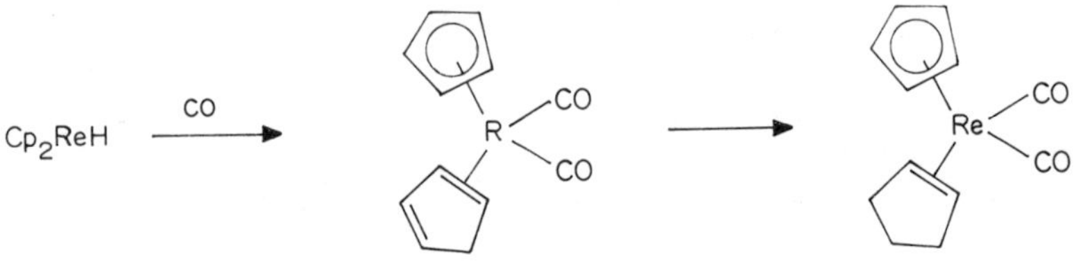

(128)

The tricarbonyl complex $CpRe(CO)_3$ can be prepared as a white crystalline compound, m.p. 104–105 °C, by heating decacarbonylrhenium with dicyclopentadiene:

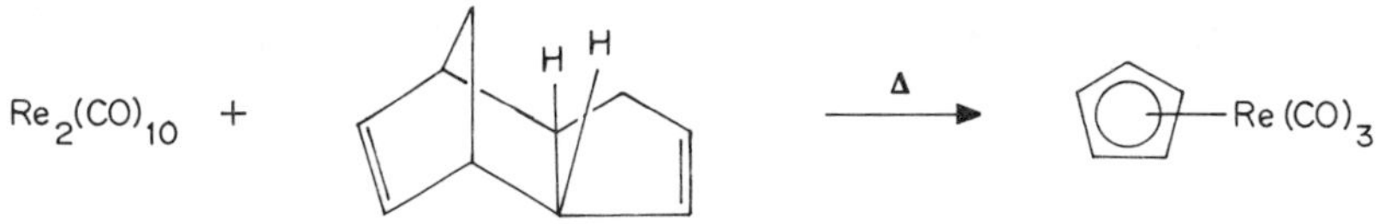

The hydride Cp_2ReH also reacts with halogens (X_2) to give $[Cp_2ReX_2]^+$ $(X = Cl, Br, I)$, and active acetylenes are known to insert into the Re—H bond:[40]

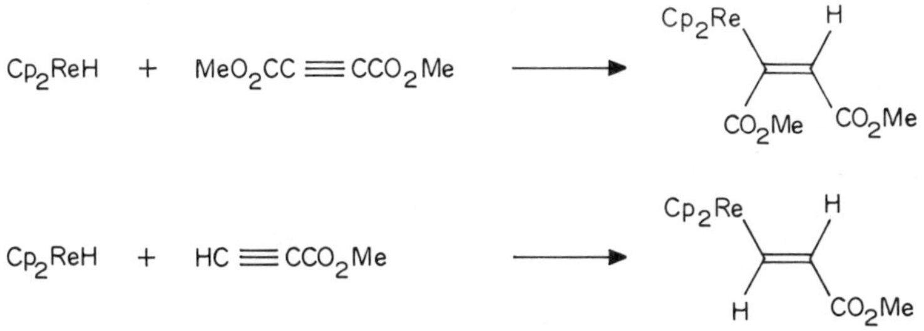

Reactions of Cp_2ReH with butyllithium in the presence of 1,1,4,7,7-penta-methyldiethenetriamine (pmdt) gives the Re lithiated complex $Cp_2ReLi(pmdt)$, which is a useful reagent for the preparation of rhenium alkyls. For example, reaction with methyl chloride gives the orange crystalline Cp_2ReMe (65%), whilst ethyl bromide gives Cp_2ReEt in low yield and allyl bromide gives $Cp_2ReCH_2CH{=}CH_2$ (43%).

$$Cp_2ReEt \xrightarrow{Ph_3C^+ BF_4^-} \left[Cp_2Re(C_2H_4) \right]^+ \quad BF_4^-$$

(129)

Reaction of Cp_2ReEt with trityl tetrafluoroborate gives the π-olefinic cation complex **129** in 41% yield:

$$Cp_2ReLi(pmdt) + RX \longrightarrow Cp_2ReR$$

Further chemical studies of **129** have not been reported.[41]

Cobalt, Rhodium and Iridium[42]

Cobaltocene, Cp_2Co, is readily prepared by reaction of $Co(NH_3)_6Cl_2$ with Na^+Cp^- at 35–40 °C, followed by heating to 65 °C. The metallocene is isolated as dark violet crystals, m.p. 173–174 °C, which are rapidly oxidized by air. The compound is paramagnetic (19e), with one unpaired electron, and is readily oxidized to the 18-electron cobalticinium complex **130**. Its reactivity is as might be expected for a free radical species, whilst the cationic cobalticinium derivative undergoes nucleophile addition to one of the cyclopentadienyl ligands. These aspects do not appear to have been developed as a method of synthesizing substituted (organic) cyclopentane derivatives. Some typical reactions are shown below.

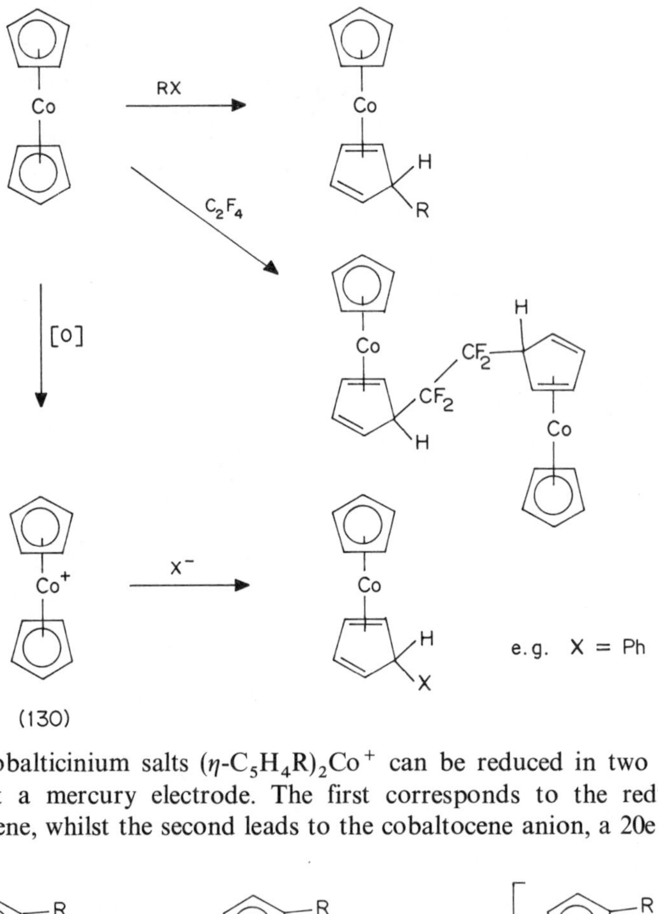

(130)

The cobalticinium salts $(\eta\text{-}C_5H_4R)_2Co^+$ can be reduced in two reversible stages at a mercury electrode. The first corresponds to the reduction to cobaltocene, whilst the second leads to the cobaltocene anion, a 20e species:

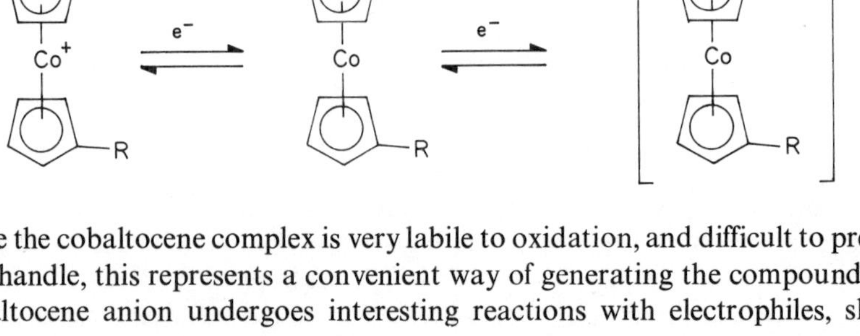

Since the cobaltocene complex is very labile to oxidation, and difficult to prepare and handle, this represents a convenient way of generating the compound. The cobaltocene anion undergoes interesting reactions with electrophiles, shown below, [43] giving a convenient method for the preparation of monosubstituted cobalticinium derivatives.

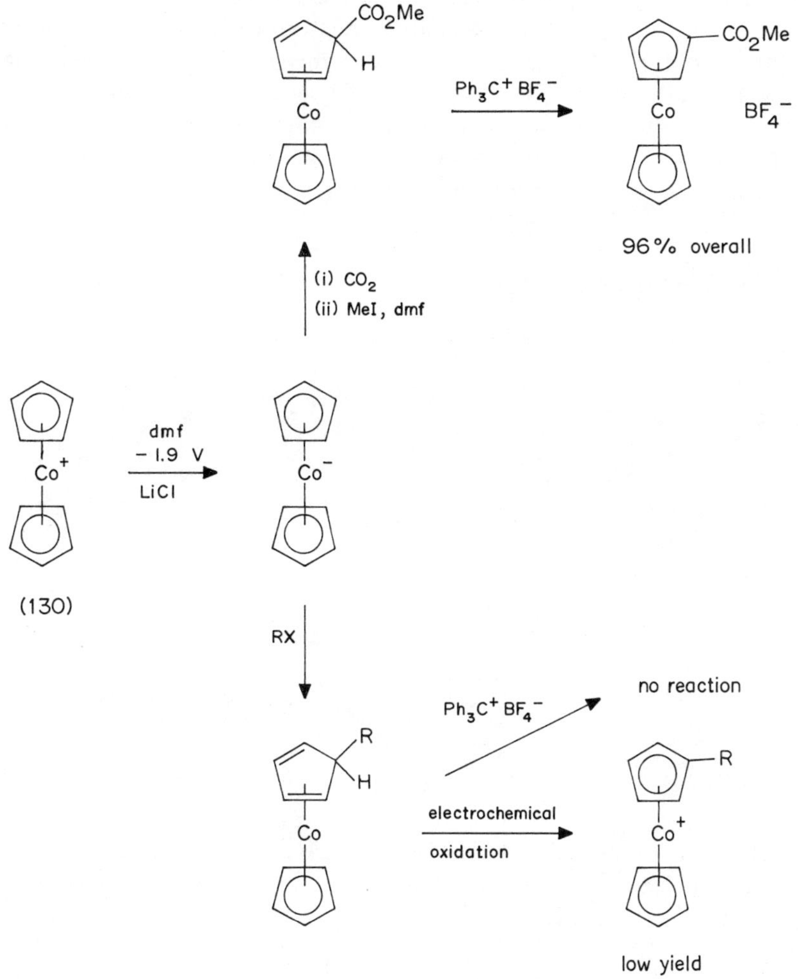

96% overall

The cobalticinium group has been found to stabilize negative charge on an α-carbon atom, owing to the anion ↔ fulvene resonance show below:

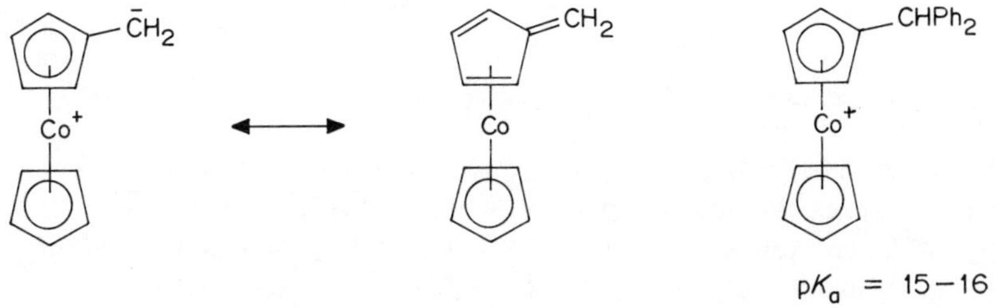

$pK_a = 15-16$

Thus the pK_a of benzhydrylcobalticinium ion is 15–16 at 25 °C, approximately 10^{11} times more acidic than that of triphenylmethane.[44] As might be expected, the cobalticinium group greatly decreases the stability of a carbonium ion in the α-position, and it has been found that the relative stabilities for $R(Ph)_2C^+$ are cobalticinium 1, H 10^2, phenyl 5×10^8 and ferrocenyl 10^{16}. There appears to be little if any delocalization of positive charge over the cobalticinium group.[45]

Some interesting reactions are observed for complexes which carry a leaving group, or a group which can support negative charge β- to one of the Cp rings, e.g.[46]

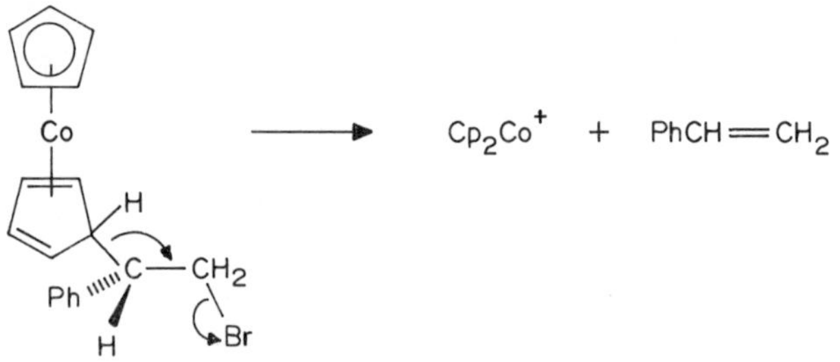

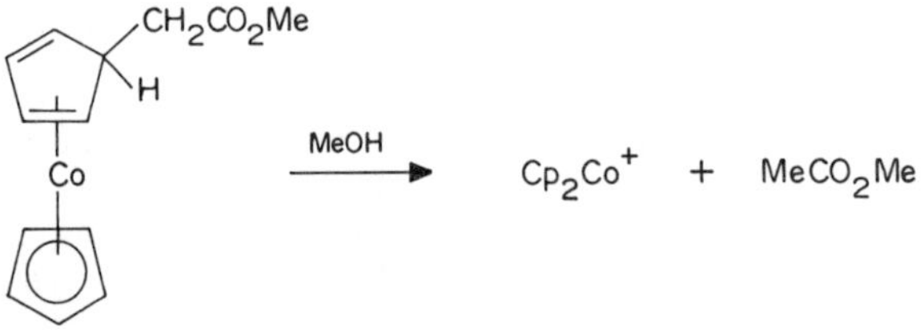

The driving force for these eliminations is undoubtedly the formation of the stable cobalticinium cation. Ring expansion occurs readily for complexes having leaving group α- to the Cp ring:[47]

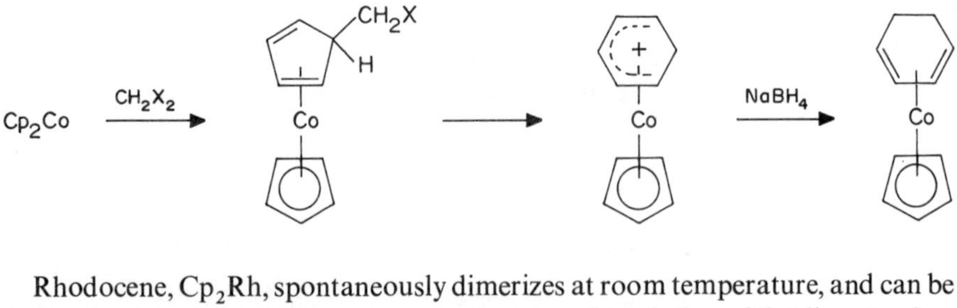

Rhodocene, Cp_2Rh, spontaneously dimerizes at room temperature, and can be isolated only at either high temperature (causing dissociation of the dimer) or low temperature (preventing dimerization). Treatment of rhodocenium cations with the usual reducing agents gives the cyclopentadiene complexes:

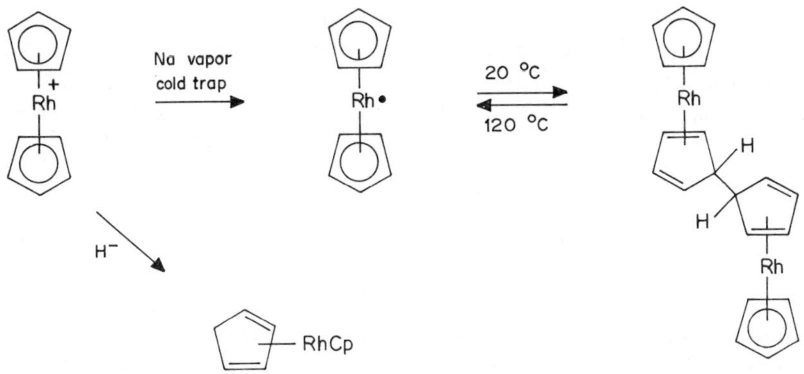

Electrochemical reduction of the rhodocenium ion furnishes rhodocene, which has a half-life for dimerization of about 2s at room temperature. Evidence has been presented for the transient existence of Cp_2Rh^-, analogous to Cp_2Co^-, during these reductions.[48]

Treatment of Cp_2Co with CO at elevated temperature and pressure, or reaction of $Co_2(CO)_8$ with cyclopentadiene at room temperature, results in the formation of $CpCo(CO)_2$, obtained as a distillable liquid. The rhodium and iridium analogues may be prepared as shown below.

$$Cp_2Co \;+\; CO \;\xrightarrow{\;\Delta\;}\; CpCo(CO)_2$$

$$Co_2(CO)_8 \;+\; C_5H_6 \;\xrightarrow{\;\Delta\;}\; CpCo(CO)_2$$

$$\left[Rh(CO)_2Cl\right]_2 \;+\; NaCp \;\longrightarrow\; CpRh(CO)_2$$

$$Ir(CO)_3Cl \;+\; NaCp \;\longrightarrow\; CpIr(CO)_2$$

Replacement of the CO ligands by other π-acid ligands is readily achieved, see e.g. the preparation of diene complexes (Chapter 7). The complexes readily undergo oxidative addition reactions, e.g.

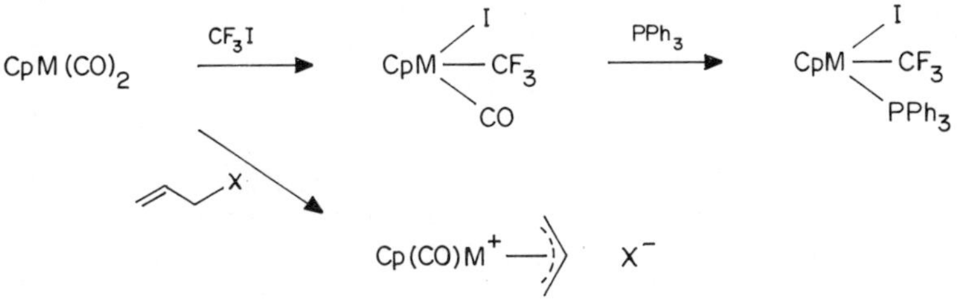

The trifluoromethyl products undergo some interesting reactions with Grignard reagents. For example, the cobalt complex, on reaction with MeMgI, gives the σ-methyl complex and a smaller amount of diiodide complex. The iridium complex gives only the diiodide. All of the reactions produce only low yields.

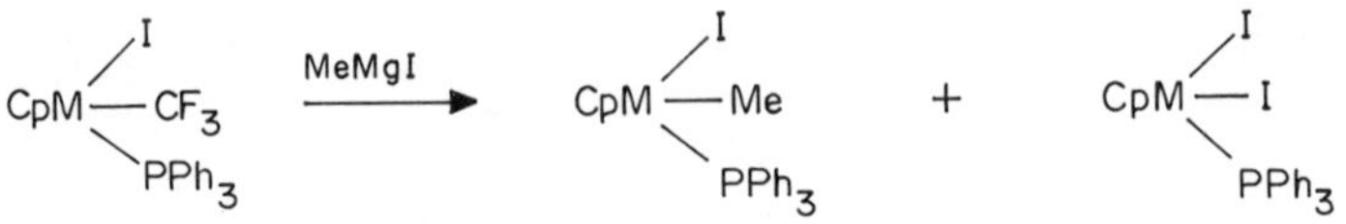

Reduction of CpCo(CO)$_2$ with sodium amalgam results in the formation of the dimeric radical anion complex **131**.

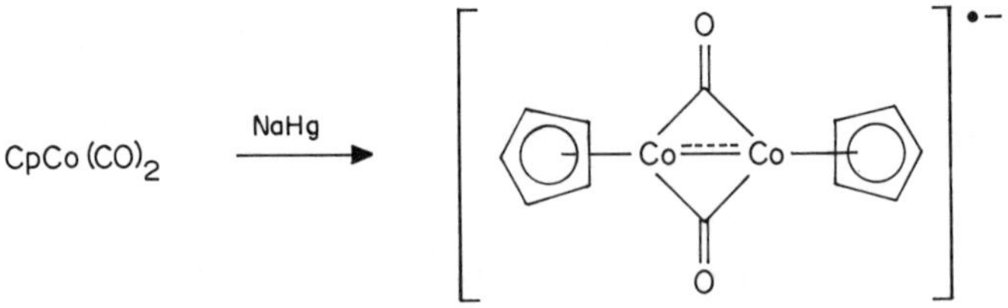

(131)

Much interest has been shown in the reaction of CpCo(CO)$_2$ with alkynes, sometimes in attempts to obtain intermediates which shed light on the mechanism of acetylene oligomerization. Recall Vollhardt's estrone synthesis presented in Chapter 2. Among the usual products obtained are cyclobutadiene, cyclopentadiene and metallacyclopentadiene complexes. Some more recent results, which are meant to give a few representative examples, not an exhaustive survey, are summarized below.[49]

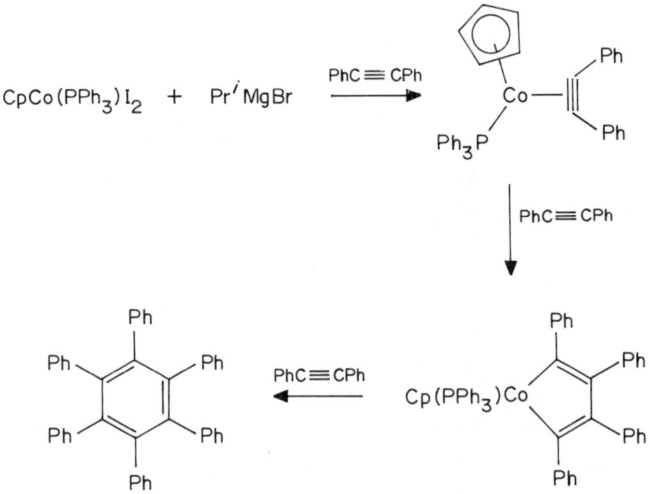

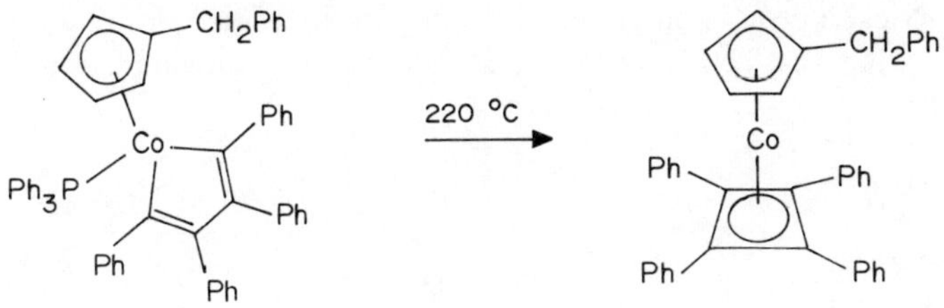

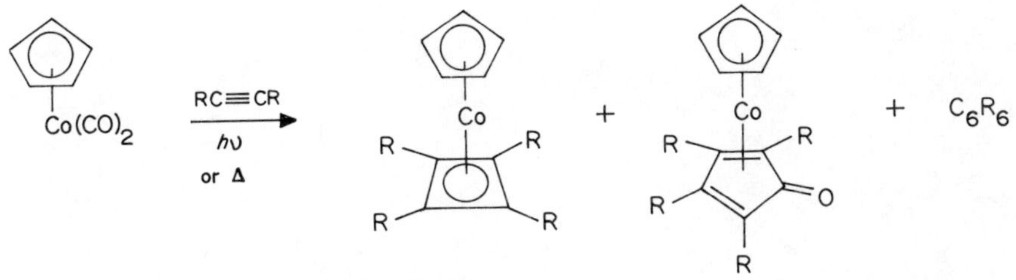

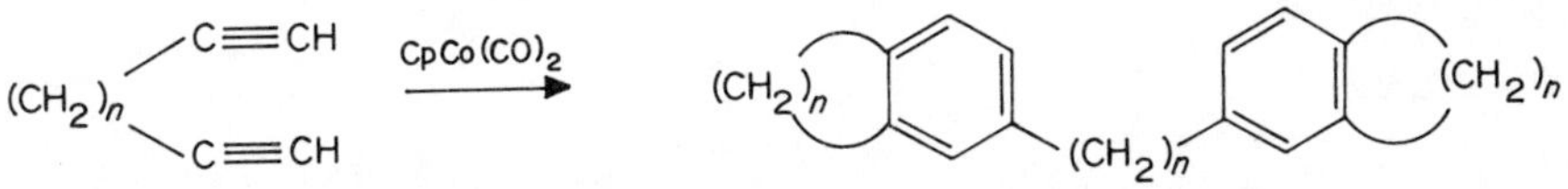

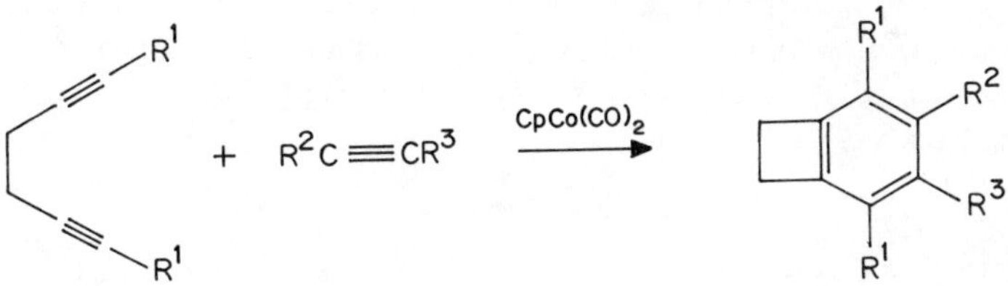

Nickel, Palladium and Platinum[50]

Reaction of hexamminenickel(II) chloride with sodium cyclopentadienide in thf solution, or treatment of nickel(II) bromide with cyclopentadiene in the presence of diethylamine, yields biscyclopentadienylnickel as a dark green, crystalline solid, readily purified by vacuum sublimation. Nickelocene is a 20-electron complex, having the expected staggered structure, and the two electrons extra to the next rare gas configuration are unpaired, as indicated by the magnetic moment (2.86 BM). These occupy a degenerate pair of antibonding molecular orbitals which are predominantly nickel in character. Oxidation removes one of these electrons to give the isolable yellow Cp_2Ni^+, which is unstable and has a 19-electron configuration. Cyclic voltammetry at low temperature shows the existence of the 18-electron complex Cp_2Ni^{2+}, but this is too unstable to allow isolation. Nickelocene is fairly reactive and readily undergoes reactions which convert it to an 18-electron complex, shown by the following examples.

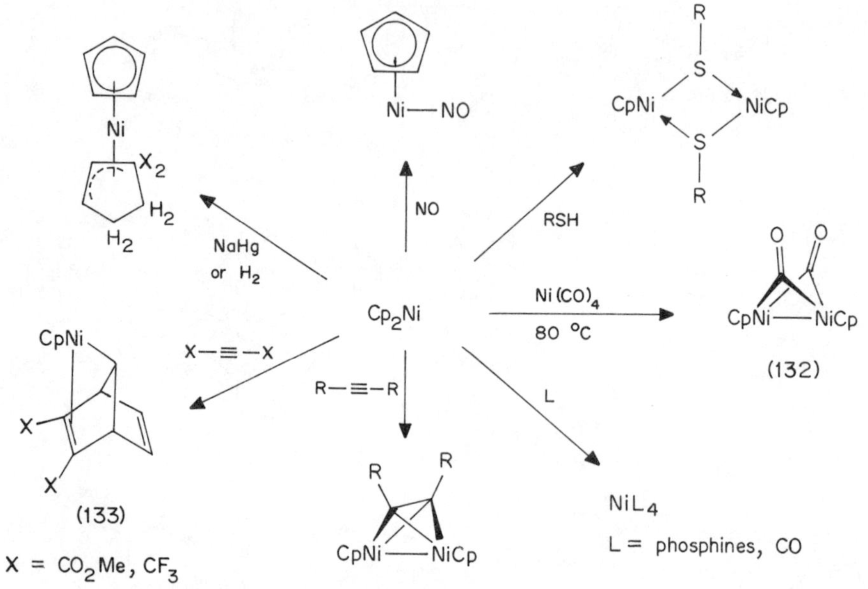

(132)

(133)

X = CO₂Me, CF₃

The cyclopentadienyl nickel carbonyl complex **132** undergoes reaction with perfluoroalkyl iodides, e.g. CF_3I, to give the unstable $CpNi(CO)I$ and the fairly stable perfluoroalkyl complex, e.g. $CpNi(CO)CF_3$. Other alkyl derivatives, which are very unstable, can be obtained by chemistry analogous to that of many other metal–Cp complexes, i.e.

$$\left[CpNi(CO)\right]_2 \xrightarrow[thf]{KHg} K\left[CpNiCO\right] \xrightarrow{MeI} MeNiCp(CO)$$

Reaction of NaC_5H_5 with $Pd(acac)_2$ gives a polymeric red complex, $(Cp_2Pd)_n$, (probably a dimer), whilst reaction of NaC_5H_5 with $PtCl_2$ gives the more stable green complex $(Cp_2Pt)_2$, shown by X-ray analysis to have the structure below:

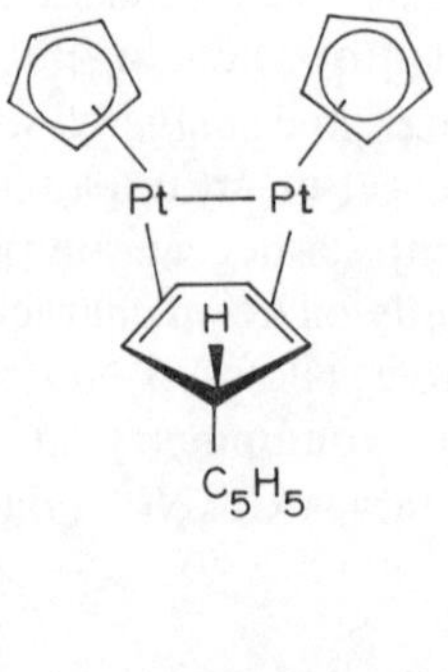

Similar complexes are isolated from the reaction of metal halides (MX_2, M = Ni, Pd, Pt) with NaCp in the presence of excess of cyclopentadiene:

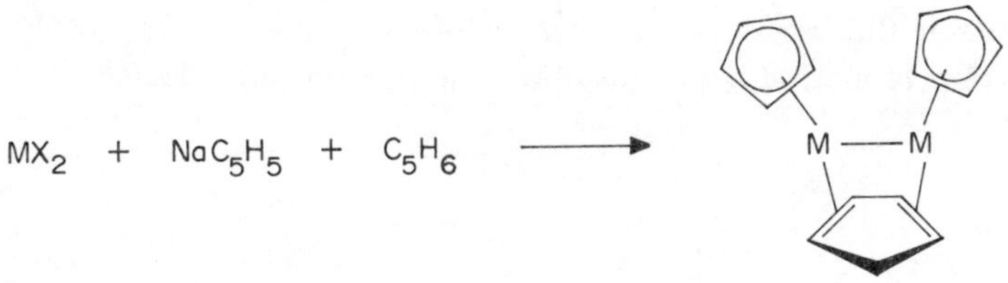

M = Ni, Pd, Pt

The reaction shown above of Cp_2Ni with active acetylenes deserves comment. The product **133** is formally the result of 1,4-addition of the acetylene to one of the Cp ligands. A plausible mechanism, involving coordination of the acetylene to nickel, is shown below.

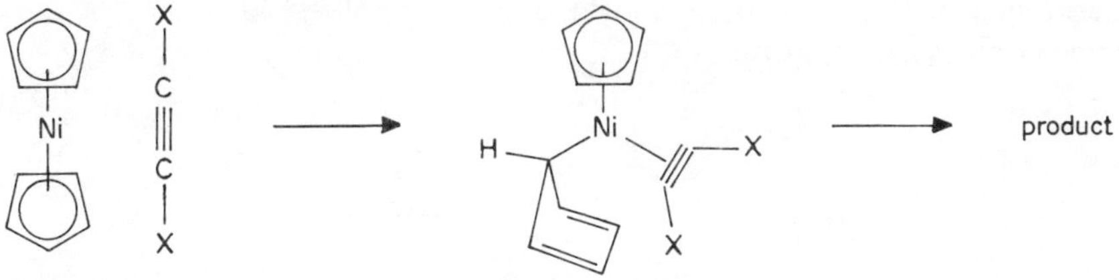

(134)

Note the difference between this reaction and that with acetylenes that do not have electron-withdrawing substituents (above). This type of ring addition reaction appears to require an acetylene which is able to function as a powerful dienophile.

The reaction of nickelocene with dienophilic olefins follows a slightly different course, to give the products of formal 1,2-addition to the Cp ring, e.g.

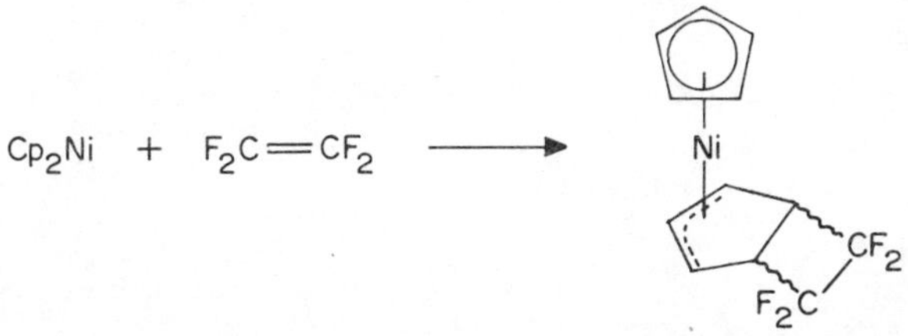

$$Cp_2Ni \ + \ F_2C{=}CF_2 \longrightarrow$$

The stereochemistry of this reaction is not known, so it is not possible to decide whether coordination of the alkene to Ni is involved. Note, however, that acetylenes coordinated to Ni, as in intermediate **134**, have a free double bond, orthogonal to the complexed one, which can enter into the cycloaddition reaction (coordinated olefin would not).

The number of ring substitution reactions known for nickelocene is limited, owing to its high reactivity, making the study of Friedel–Crafts reactions extremely difficult.

Protonation of nickelocene has been reported to occur directly at the Cp ligand, rather than at the metal first (followed by H^+ transfer to Cp).[51] The product of protonation is the unstable cationic diene complex **135**.

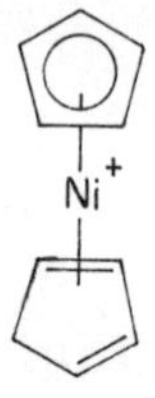

(135)

The reactivity of diene cations, which can be prepared via **135**, towards nucleophiles has recently been studied, and is exemplified below.[52] The diene ligand is the reactive group.

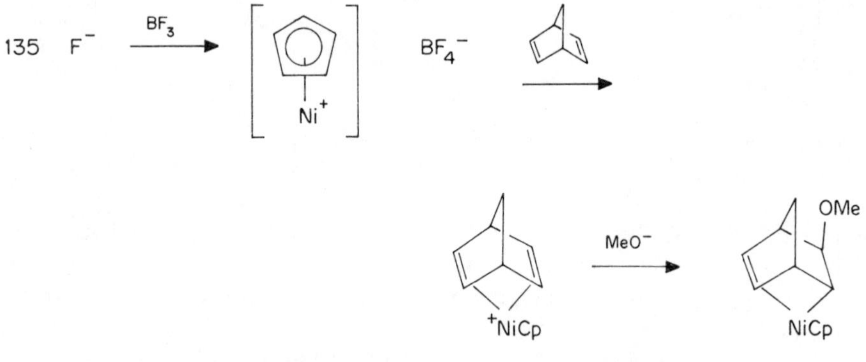

Reaction of nickelocene with azobenzene occurs with insertion into the *ortho* C—H bond of the aromatic compound:

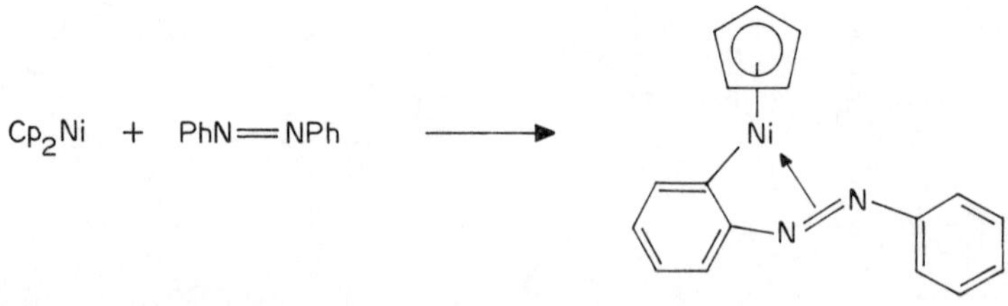

Treatment of Cp_2Ni with allylmagnesium chloride results in the formation of the π-allyl complex:

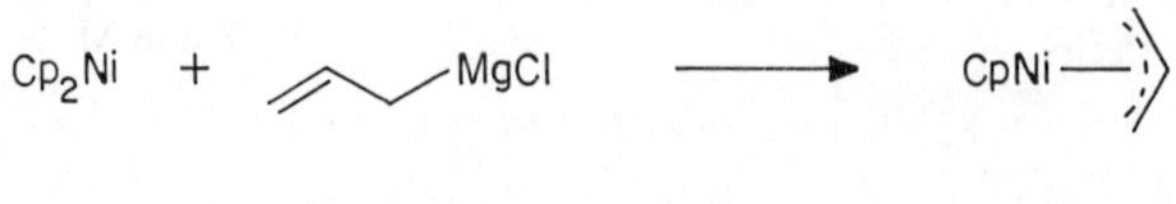

Substituted allyl complexes are found to result from the reaction of Cp_2Ni with butadiene:

mixture of
stereoisomers

This section would hardly be complete without mention of the entertaining 'triple-decker' complexes, resulting from the interaction of nickelocene with carbonium ions under very mild conditions:[53]

100 %

This complex has been fully characterized by X-ray crystallography.[54]

REFERENCES

1. E. O. Fischer and R. D. Fischer, Angew. Chem., 1960, **72**, 919.
2. Complexes: A. J. Birch, P. W. Westerman and A. J. Pearson, *Aust. J. Chem.*, 1976, **29**, 1671; G. A. Olah, S. H. Yu and G. Liang, *J. Org. Chem.*, 1976, **41**, 2383. Cations: G. A. Olah, R. H. Schlosberg, R. D. Porter, Y. K. Mo, D. P. Kelly and G. D. Mateescu, *J. Am. Chem. Soc.*, 1972, **94**, 2034.
3. M. J. S. Dewar and R. C. Dougherty, *The PMO Theory of Organic Chemistry*, Plenum Press, New York, 1975, p. 19.
4. A. J. Pearson and A. McElroy, Cambridge University, unpublished data, 1978–79.
5. A. J. Birch and M. A. Haas, *J. Chem. Soc. C*, 1971, 2465.
6. A. J. Pearson, in Comprehensive Organometallic Chemistry (Ed. G. Wilkinson, E. W. Abel and F. G. A. Stone), Pergamon Press, 1982, Ch. 58.
7. D. W. Clack, M. Monshi and L. A. P. Kane-Maguire, *J. Organomet. Chem.*, 1976, **107**, C40.
8. A. J. Birch, D. Bogsanyi and L. F. Kelly, *J. Organomet. Chem.*, 1981, **214**, C39; L. A. P. Kane-Maguire, *J. Chem. Soc. A*, 1971, 2187; L. A. P. Kane-Maguire and C. A. Mansfield, *J. Chem. Soc., Dalton Trans.*, 1976, 2187; 2192; M. Gower, G. R. John,

L. A. P. Kane-Maguire, T. I. Odiaka and A. Salzer, *J. Chem. Sec., Dalton Trans.*, 1979, 2002; G. R. John and L. A. P. Kane-Maguire, *J. Chem. Soc., Dalton Trans.*, 1979, 873; 1196.

9. R. Hoffman and O. Eisenstein, personal communications.
10. See, for example, G. Klopman, *J. Am. Chem. Soc.*, 1968, **90**, 223.
11. A. J. Pearson, T. R. Perrior and D. C. Rees, *J. Organomet. Chem.*, 1982, **226**, C39; A. J. Pearson, P. Ham, C. W. Ong, T. R. Perrior and D. C. Rees, *J. Chem. Soc., Perkin Trans.* 1, 1982, 1527.
12. A. J. Pearson, E. Mincione, M. Chandler and P. R. Raithby, *J. Chem. Soc., Perkin Trans.* 1, 1980, 2774; A. J. Pearson and G. C. Heywood, *Tetrahedron Lett.*, 1981, **22**, 1645; E. Mincione, A. J. Pearson, P. Bovicelli, M. Chandler and G. C. Heywood, *Tetrahedron Lett.*, 1981, **22**, 2929; A. J. Pearson, G. C. Heywood and M. Chandler, *J. Chem. Soc., Perkin Trans.* 1, 1982, 2631.
13. A. J. Pearson, *J. Chem. Soc., Perkin Trans.* 1, 1979, 1255; 1980, 400; A. J. Pearson, P. Ham and D. C. Rees, *J. Chem. Soc., Perkin Trans.* 1, 1982, 489.
14. P. Ham, *PhD Thesis*, Cambridge University, 1982; A. J. Pearson, P. Ham and D. C. Rees, *Tetrahedron Lett.*, 1980, **21**, 4637.
15. A. J. Pearson and D. C. Rees, *J. Am. Chem. Soc.*, 1982, **104**, 1118; *Tetrahedron Lett.*, 1980, **21**, 3937; A. J. Pearson, *Tetrahedron Lett.*, 1981, **22**, 4033; A. J. Pearson and D. C. Rees, *J. Chem. Soc., Perkin Trans.* 1, 1982, 2467; A. J. Pearson D. C. Rees and C. W. Thornber, *J. Chem. Soc., Perkin Trans.* 1, 1983, 619.
16. A. J. Pearson, *J. Chem. Soc., Perkin Trans.* 1, 1977, 2069; 1978, 495.
17. L. F. Kelly, A. S. Narula and A. J. Birch, *Tetrahedron Lett.*, 1980, **21**, 871; L. F. Kelly, P. Dahler, A. S. Narula and A. J. Birch, *Tetrahedron Lett.*, 1981, **22**, 1433; A. J. Birch, L. F. Kelly and D. J. Thompson, *J. Chem. Soc., Perkin Trans.* 1, 1981, 1006.
18. L. F. Kelly, *PhD Thesis*, Australian National University, 1981.
19. A. J. Pearson, D. V. Gardner and I. C. Richards, *J. Chem. Soc., Chem. Commun.*, 1982, 807.
20. P. Powell and L. J. Russell, *J. Chem. Res. (S)*, 1978, 283; P. Powell, *J. Organomet. Chem.*, 1977, **165**, C43; B. F. G. Johnson, J. Lewis and D. J. Yarrow, *J. Chem. Soc., Dalton Trans.*, 1972, 2084; E. D. Sternberg and K. P. C. Vollhardt, *J. Am. Chem. Soc.*, 1980, **102**, 4841; J. Evans, B. F. G. Johnson and J. Lewis, *J. Chem. Soc., Dalton Trans.*, 1972, 2868.
21. M. Rosenblum, *Chemistry of the Iron Group Metallocenes, Part I*, Wiley–Interscience, New York, 1965.
22. T. J. Kealy and P. L. Pauson, *Nature (London)*, 1951, **168**, 1039.
23. S. A. Miller, J. A. Tebboth and J. F. Tremaine, *J. Chem. Soc.*, 1952, 632.
24. Review: W. E. Watts, *J. Organomet. Chem. Libr.*, 1979, **7**, 399.
25. I. Fleming, *Frontier Orbitals and Organic Chemical Reactions*, Wiley, New York, 1976.
26. M. J. Nugent, R. Kummer and J. H. Richards, *J. Am. Chem. Soc.*, 1969, **91**, 6141; M. J. Nugent, R. E. Carter and J. H. Richards, *J. Am. Chem. Soc.*, 1969, **91**, 6145.
27. Reviews: G. P. Pez and J. N. Armor, *Adv. Organomet. Chem.*, 1981, **19**, 1; J. A. Labinger, *J. Organomet. Chem.*, 1979, **180**, 187; P. C. Wailes, P. S. R. Coutts and H. Weigold, *Organometallic Chemistry of Titanium, Zirconium and Hafnium*, Academic Press, New York, 1974.
28. A. N. Nesmeyanov, D. A. Lemenovskii, V. P. Fedin and E. G. Perevalova, *Dokl. Akad. Nauk. SSSR*, 1979, **245**, 609.
29. G. Wilkinson, F. A. Cotton, J. M. Birmingham, *J. Inorg. Nucl. Chem.*, 1956, **2**, 95; E. O. Fischer and W. Hafner, *Z. Naturforsch., Teil B*, 1954, **9**, 503; E. O. Fischer and S. Vigoureux, *Chem. Ber.*, 1958, **91**, 2205; G. Wilkinson and J. M. Birmingham, *J. Am. Chem. Soc.*, 1954, **76**, 4281; E. O. Fischer, H. P. Kogler and P. Kuzel, *Chem. Ber.*, 1960, **93**, 3006; F. Fachinetti and C. Floriani, *J. Chem. Soc., Chem. Commun.*, 1975, 578; 1974, 516.
30. Review: K. W. Barnett and D. W. Slocum, *J. Organomet. Chem.*, 1972, **44**, 1.

31. G. L. Geoffroy and M. G. Bradley, *Inorg. Chem.*, 1978, **17**, 2410; M. Berry, S. G. Davies and M. L. H. Green, *J. Chem. Soc., Chem. Commun.*, 1978, 99; C. Giannotti and M. L. H. Green, *J. Chem. Soc., Chem. Commun.*, 1972, 1114.
32. E. O. Fischer, W. Hafner and H. O. Stahl, *Z. Anorg. Allg. Chem.*, 1955, **282**, 47; G. Wilkinson, F. A. Cotton and J. M. Birmingham, *J. Inorg. Nucl. Chem.*, 1956, **2**, 95.
33. M. L. H. Green, J. A. McCleverty, L. Pratt and G. Wilkinson, *J. Chem. Soc.*, 1961, 4854.
34. A. Nakamura and S. Otsuka, *J. Am. Chem. Soc.*, 1972, **94**, 1886.
35. A. Nakamura, *J. Organomet. Chem.*, 1979, **164**, 183; M. L. H. Green and P. J. Knowles, *J. Chem. Soc., Perkin Trans.* 1, 1973, 989.
36. M. L. H. Green, G. A. Moser, I. Packer, F. Petit, R. A. Forder and K. Prout, *J. Chem. Soc., Chem. Commun.*, 1974, 839.
37. T. S. Piper, F. A. Cotton and G. Wilkinson, *J. Inorg. Nucl. Chem.*, 1955, **1**, 165; E. O. Fischer, *Z. Naturforsch., Teil B*, 1954, **9**, 618.
38. M. L. H. Green, L. Pratt and G. Wilkinson, *J. Chem. Soc.*, 1958, 3916.
39. M. L. H. Green and G. Wilkinson, *J. Chem. Soc.*, 1958, 4314.
40. M. Dubeck and R. A. Schell, *Inorg. Chem.*, 1964, **3**, 1757.
41. R. I. Mink, J. J. Welter, P. R. Young and G. D. Stucky, *J. Am. Chem. Soc.*, 1979, **101**, 6928.
42. Review: J. E. Sheats, *J. Organomet. Chem. Libr.*, 1979, **7**, 461.
43. N. El Murr and E. Laviron, *Tetrahedron Lett.*, 1975, 875.
44. J. E. Sheats, W. Miller and T. Kirsch, *J. Organomet. Chem.*, 1975, **91**, 97.
45. J. E. Sheats, E. J. Sabol, Jr., D. Z. Denney and N. El Murr, *J. Organomet. Chem.*, 1976, **121**, 73.
46. G. E. Herberich and G. Greiss, *J. Organomet. Chem.*, 1971, **27**, 113.
47. G. E. Herberich, G. Greiss and H. F. Heil, *J. Organomet. Chem.*, 1970, **22**, 723.
48. N. El Murr, J. E. Sheats, W. E. Geiger, Jr., and J. D. L. Holloway, *Inorg. Chem.*, 1979, **18**, 1443.
49. H. Yamazaki and N. Hagihara, *J. Organomet. Chem.*, 1967, **7**, P22; H. Yamazaki and Y. Wakatsuki, *J. Organomet. Chem.*, 1978, **149**, 377; 385; K. P. C. Vollhardt and R. G. Bergman, *J. Am. Chem. Soc.*, 1974, **96**, 4996.
50. Review: K. W. Barnett, *J. Organomet. Chem.*, 1974, **78**, 139.
51. G. K. Turner, W. Klaui, M. Scotti and H. Werner, *J. Organomet. Chem.*, 1975, **102**, C29.
52. G. Parker, A. Salzer and H. Werner, *J. Organomet. Chem.*, 1976, **67**, 131.
53. H. Werner and A. Salzer, *Synth. Inorg. Metal-Org. Chem.*, 1972, **2**, 239.
54. E. Dubier, M. Textor, H. -R. Oswald and A. Salzer, *Angew Chem.*, 1974, **86**, 125; *Angew Chem., Int. Ed. Engl.*, 1976, **13**, 135.

η^6-Arene and η^6-Triene Complexes

ARENE COMPLEXES

A wide range of η^6-arene transition metal complexes, both neutral and charged, are known. Perhaps the most studied, owing to their ease of preparation, handling and usefulness for synthetic organic chemistry, are the arene–chromium tricarbonyl derivatives. Also well known are the bisarene–metal complexes analogous to the metallocenes, although these are perhaps less useful synthetically. A book devoted to arene–metal complexes has already appeared.[1] A number of reviews dealing with the application of arene–metal complexes to organic synthesis have appeared in the last 10 years .[2]

In this chapter we shall first present a fairly detailed account of the chemistry and synthetic utility of arene–Cr(CO)$_3$ complexes, followed by a description of the known chemistry of related systems. In general terms, the attachment of a metal tricarbonyl unit to an aromatic compound can have several effects: (1) activation of the aromatic ring to nucleophilic attack; (2) enhancement of aryl–H acidities; (3) steric inhibition of attack on functional groups from the same side as M(CO)$_3$; (4) stabilization of negative and or positive charge on carbon atoms α and β to the arene–M moiety. The first of these effects is probably the most important and is now at a fairly advanced stage of exploitation using arene–Cr(CO)$_3$ complexes, owing in large part to the efforts of Semmelhack. While complexes of metals other than chromium show similar effects, their preparation often requires conditions not compatible with functional groups and often low yields are obtained. For this reason they have received less attention.

Chromium, Molybdenum and Tungsten

These metals all form arene–M(CO)$_3$ complexes fairly readily. Typically, heating or irradiating the metal carbonyl, M(CO)$_6$, either with the neat aromatic compound or with the arene in a suitable solvent, e.g. dioxane, heptane or acetonitrile, gives the desired complex. The M(CO)$_3$ group can be removed either by exposure to air and sunlight, or by treatment with a mild oxidizing agent such as iodine or cerium(IV) ammonium nitrate. This is an important aspect if the complexes are to be used as stoichiometric reagents for organic synthesis—it must be possible to remove the metal in high yield after performing appropriate operations on the complex.

Under normal circumstances, an aromatic compound may be considered as a nucleophilic species, undergoing electrophilic substitution reactions. The situation can be reversed by placing electron-withdrawing groups on the benzene ring. For example, it is well known that the halide group in 2,4-dinitrophenyl bromide may be displaced by nucleophiles such as methoxide. Unfortunately, it is necessary to use drastic conditions for nitration and also the desired final product may not be a nitro derivative, so that it would be necessary to remove the NO_2 group efficiently at some stage. These are the main reasons why 'classical' nucleophilic substitution on aromatic compounds has not found wide use in synthesis.

Attachment of the $Cr(CO)_3$ group to an aromatic ring leads to activation to attack by nucleophiles, since the metal group acts as a kind of electron sink. As intermediates for organic synthesis these complexes have several features which commend them. They are air-stable solids, readily purified and easily handled, and can be prepared in good to excellent yields under neutral reaction conditions, thereby allowing many functional groups to be appended to the arene. Many complexes are in fact commercially available. The metal moiety is readily removed.

Structure

A large number of X-ray crystal structure determinations have been carried out on arene–$Cr(CO)_3$ complexes, revealing two kinds of structures, *viz.* those in which the Cr—CO bond vectors are eclipsed with ring carbon atoms (**A**), and those having a staggered arrangement (**B**):

(A)   (B)

Staggered conformations are observed for the complexes of benzene, acetophenone, hexamethylbenzene, *o*- and *m*-toluate anions, phenanthrene, anthracene and naphthalene, among others, while eclipsed conformations are found for complexes of anisole, toluidine, methyl benzoate, *o*-methoxyacetophenone, and others. It has been suggested that substitutions on the ring may control the conformation adopted by mesomeric electron repulsion or withdrawal.[3]

Both staggered and eclipsed conformations have been shown to exist in solution by NMR studies, and again the conformation adopted depends on steric and electronic effects in the complex.

Reactivity of Arene–Cr(CO)₃ complexes

Addition of Nucleophiles

Reaction of a simple complex, benzene–Cr(CO)$_3$, with reactive nucleophiles produces first of all an anionic cyclohexadienyl–Cr(CO)$_3$ complex. As with dienyl–Fe(CO)$_3$ complexes, the nucleophile attacks the face of the ligand remote from the metal, to give the *exo*-substituted product. The resultant anion can be manipulated in two ways: (a) treatment with a mild oxidizing agent removes the metal and aromatizes the ring to give a substituted benzene derivative; (b) treatment with an appropriate acid (e.g. CF$_3$CO$_2$H) produces a diene–Cr(CO)$_3$ complex, which is unstable and loses the metal easily on treatment with iodine to give a substituted diene. Usually a mixture is obtained from the latter procedure, owing to facile rearrangement of the intermediate diene–Cr(CO)$_3$ complex, but a particular diene usually predominates, as shown below.

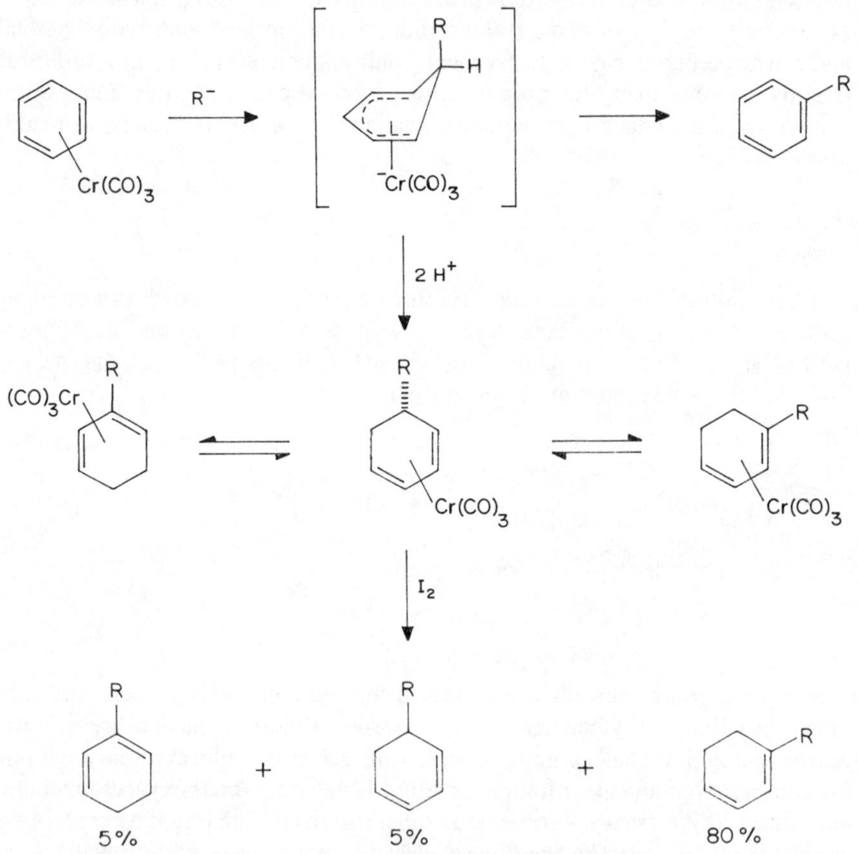

The types of nucleophile which may be used for this particular reaction are limited. Only very reactive carbanions are successful, and a few examples are summarized below.

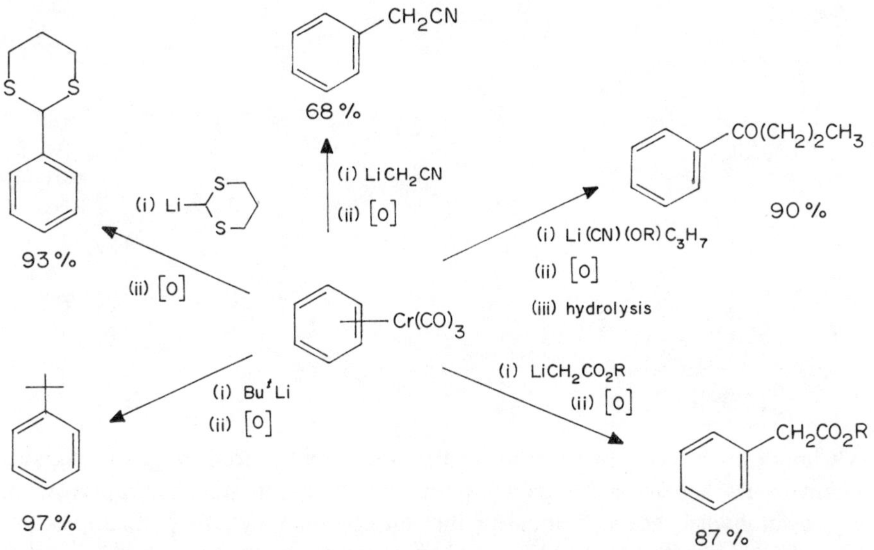

No reaction is observed with stabilized enolate anions from, e.g., dimethyl malonate, methyl acetoacetate or simple ketone enolates (but see later), or *n*-butyllithium or methyllithium. The first three are simply not reactive enough, while the alkyllithium reagents have a pronounced tendency to deprotonate the aryl ring.

The effects of substituents on nucleophile addition are interesting and useful. A methoxy substituent on the aromatic ring, as in anisole–$Cr(CO)_3$, directs nucleophiles mainly into the *meta* position. Minor amounts of *ortho*-substituted product are sometimes observed, but invariably no *para*-substituted product is obtained. A methyl substituent, as in toluene–$Cr(CO)_3$, does not exert such a pronounced directing effect, except with sterically demanding nucleophiles, and interestingly the major products are *ortho*- and *meta*-substituted derivatives, with very little *para*. Some examples are shown below.

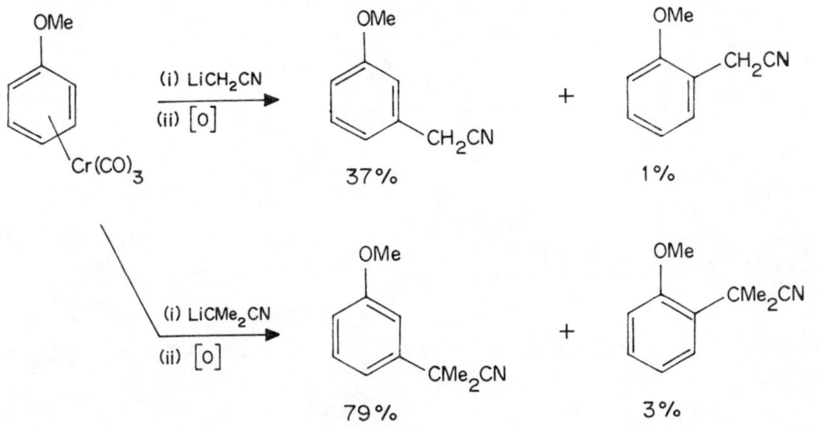

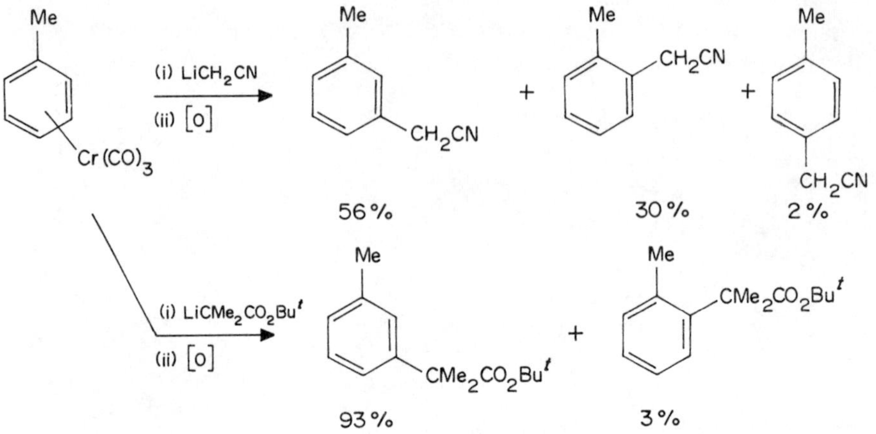

Me → (i) LiCH$_2$CN (ii) [o] → Me ... CH$_2$CN **56%** + Me CH$_2$CN **30%** + Me CH$_2$CN **2%**

Me → (i) LiCMe$_2$CO$_2$Bu^t (ii) [o] → Me ... CMe$_2$CO$_2$Bu^t **93%** + Me CMe$_2$CO$_2$Bu^t **3%**

Complexes carrying two methoxy substituents on the arene ring, although we might expect them to be deactivated by electron donation, still react rapidly with, e.g., cyanohydrin acetal anions (giving nucleophilic acylation). Examples are shown below. Note that the highly unfavorable addition of nucleophile *para* to a methoxy substituent dominates the chemistry of complexes derived from 1,2-dimethoxybenzene.

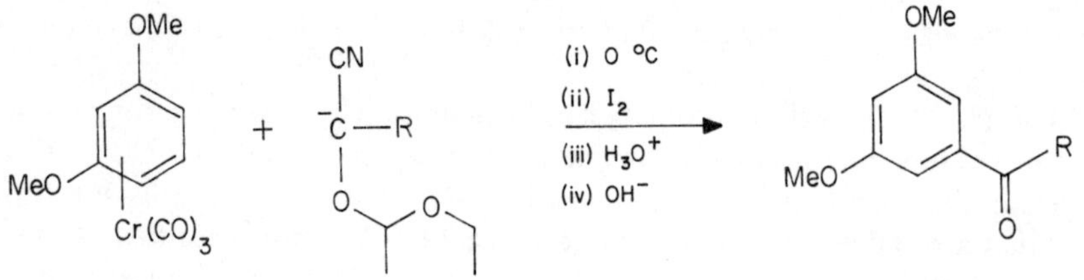

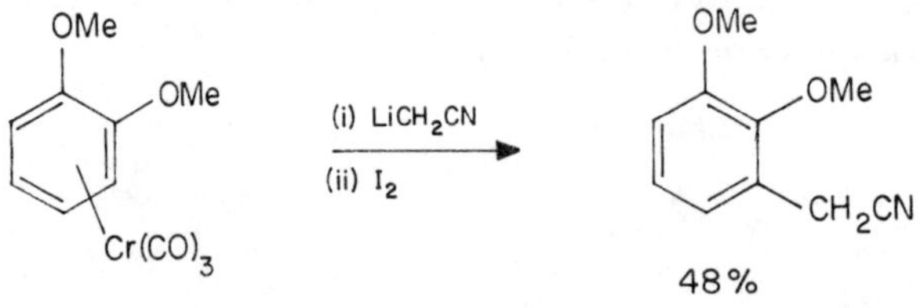

OMe ... MeO ... Cr(CO)$_3$ + CN–C–R (O–CH–O–ethyl) → (i) 0 °C (ii) I$_2$ (iii) H$_3$O$^+$ (iv) OH$^-$ → OMe ... MeO ... C(=O)–R

OMe OMe Cr(CO)$_3$ → (i) LiCH$_2$CN (ii) I$_2$ → OMe OMe CH$_2$CN **48%**

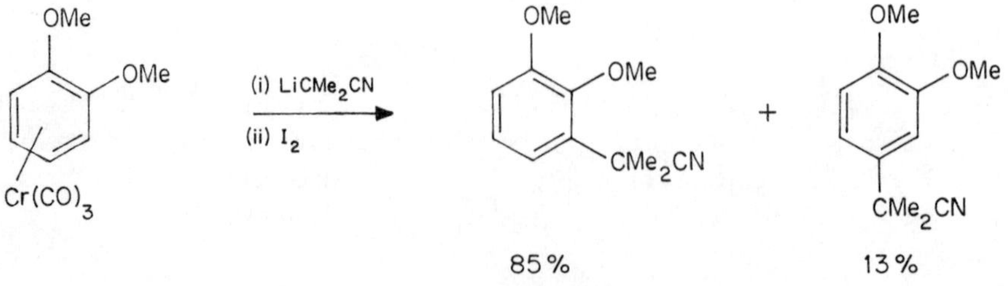

OMe OMe Cr(CO)$_3$ → (i) LiCMe$_2$CN (ii) I$_2$ → OMe OMe CMe$_2$CN **85%** + OMe OMe CMe$_2$CN **13%**

These nucleophile additions may also be performed intramolecularly, as illustrated by the simple examples shown below.

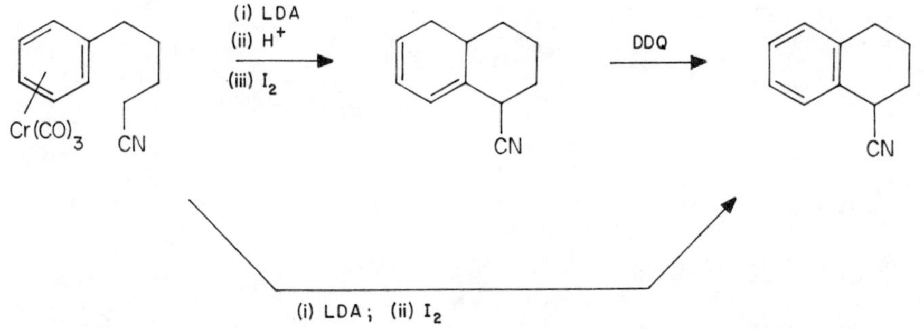

With higher homologues there appears to be an equilibrium set up between fused-ring and spirocyclic dienyl intermediates, since the proportion of spirocyclic product in the mixture increases if the reaction is allowed to proceed for longer times prior to quenching:

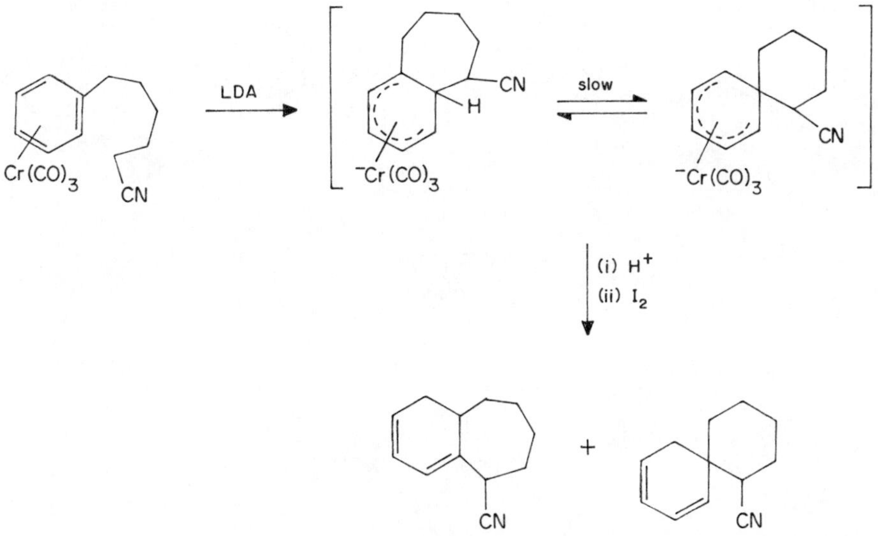

While spiro[4.5]decane derivatives, which are important intermediates for the synthesis of a variety of spirocyclic sesquiterpene natural products, are not obtainable using the monosubstituted complexes above, it is possible to utilize the *meta*-directing capacity of a methoxy substituent in order to drive the reaction (incompletely) in favor of spirocyclization. This is illustrated by the total synthesis of the natural product acorenone by Semmelhack's group, shown below.[4]

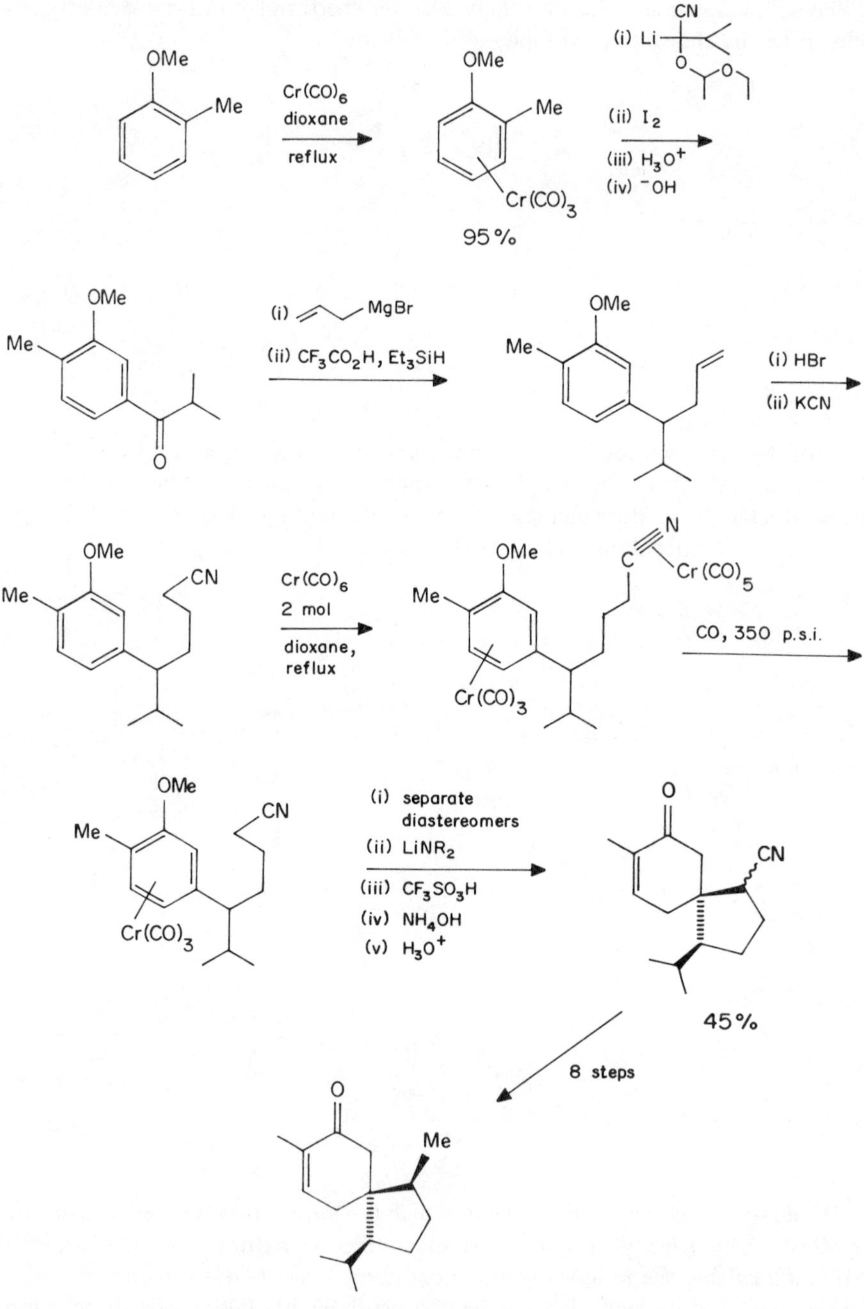

Although it is not possible to arylate less reactive nucleophiles using simple arene–Cr(CO)$_3$ complexes, this problem may be overcome using complexes of aryl halides. Chlorobenzene–Cr(CO)$_3$ and fluorobenzene–Cr(CO)$_3$ undergo reactions with a wide variety of nucleophiles, summarized below.

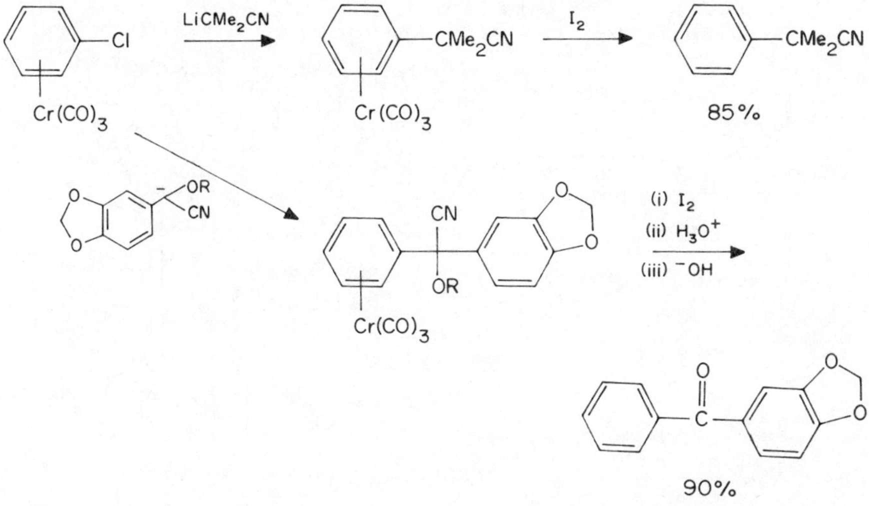

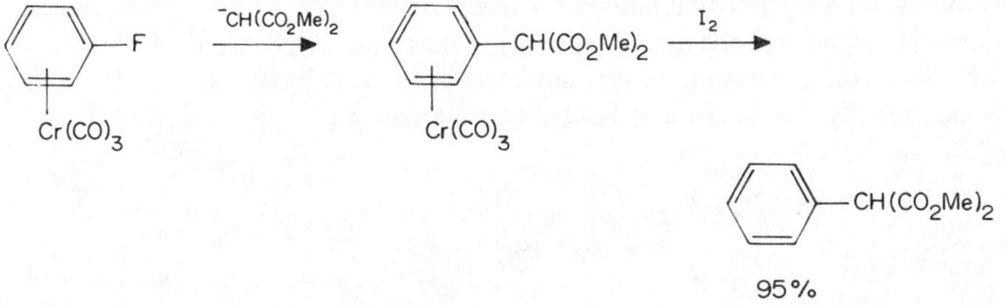

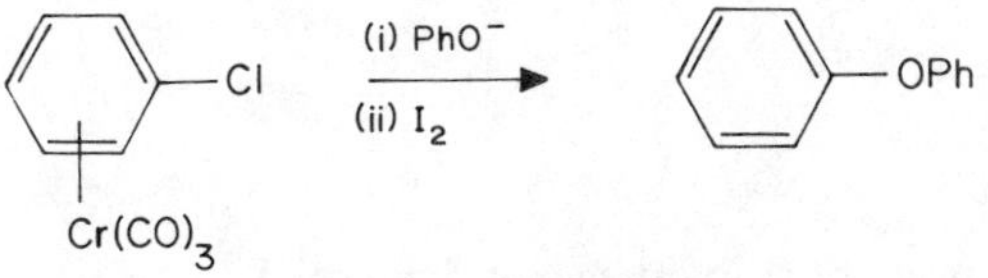

Note that the dimethylmalonate anion displaces chloride from chlorobenzene–Cr(CO)$_3$ but better yields are obtained using the fluorobenzene complex. The order of reactivity of halobenzene complexes is F > Cl > Br > I. Even the comparatively unreactive phenoxide anion will displace halogen. Thus the reaction is one of nucleophilic displacement of halide from an aryl halide, normally difficult to achieve, except with nitroaryl halides. Very reactive carbanions fail in this reaction, since they tend to add irreversibly to the arene ring to give cyclohexadienyl complexes which are stable to rearrangement. Hence there appears to be some reversibility or capacity for rearrangement in the nucleophile addition. When the incoming group ends up on the halogen-bearing carbon atoms the better leaving group, i.e. halide anion, departs in preference giving the substituted arene complex:

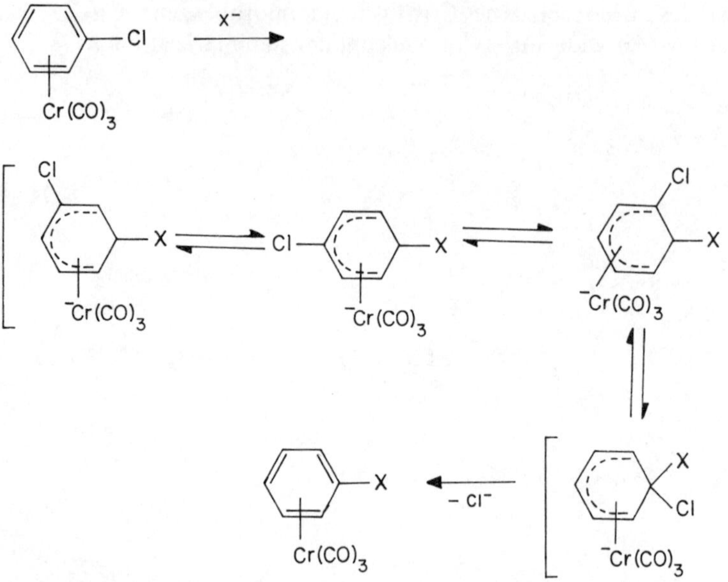

Although very reactive anions do react, rearrangement evidently does not occur. Evidence for the above equilibration process is provided by the results obtained from quenching the reaction at an early stage by oxidation with iodine, which results in a mixture of benzene derivatives, e.g.

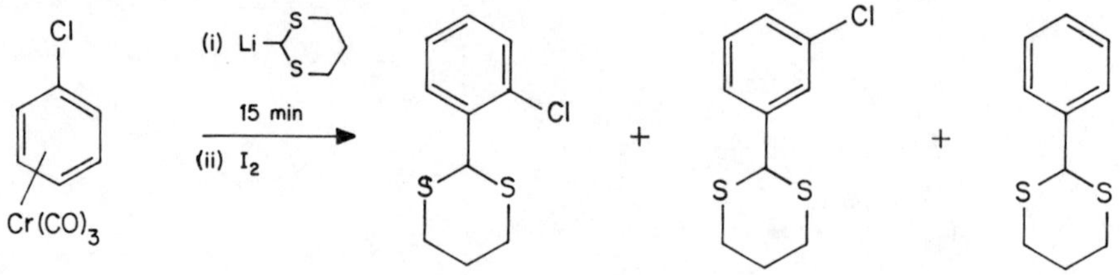

This behavior has been employed in a short synthesis of demerol 1.

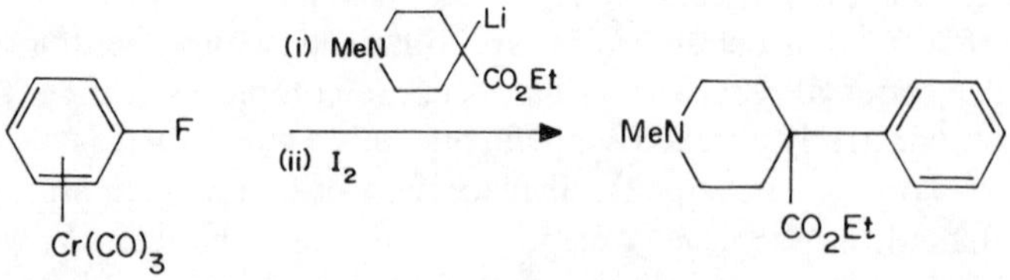

(1)

The above additions of reactive nucleophiles are not limited to simple benzene derivatives. It has been shown that reaction of indole with $Cr(CO)_6$ produces a single complex from coordination of the six-membered ring. A similar complex can be prepared from benzofuran, and these compounds show a pronounced tendency to react with nucleophiles at the 4-position. Attachment of a bulky group on the nitrogen or on the indole 3-position can be used to direct nucleophiles, by steric effect, to either the 4- or 7-position selectively, as shown below.

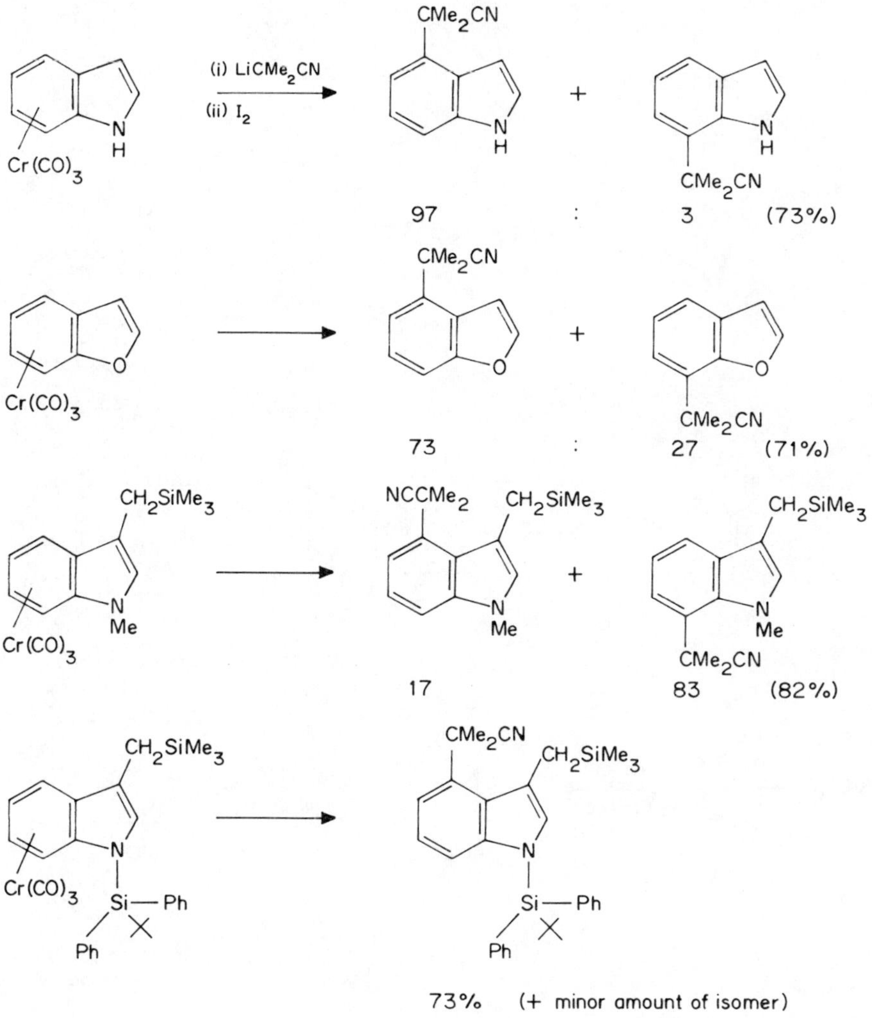

97 : 3 (73%)

73 : 27 (71%)

17 83 (82%)

73% (+ minor amount of isomer)

Ring Deprotonation[5]

Abstraction of ring protons of arene–Cr(CO)$_3$ and (arene)$_2$Cr by lithium bases, etc., to give chromium–aryl–lithium derivatives has been found particularly useful for the preparation of substituted arene–Cr(CO)$_3$ complexes not easily obtained by the usual procedures. The rate of metallation of the complexes exceeds that of uncomplexed arenes. Both direct and indirect lithiation procedures are shown below.

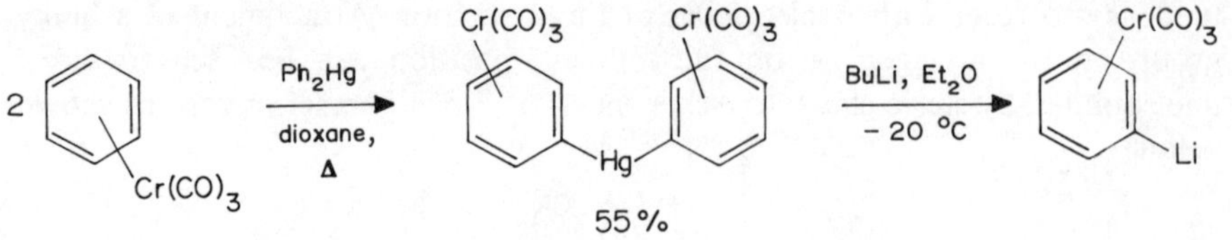

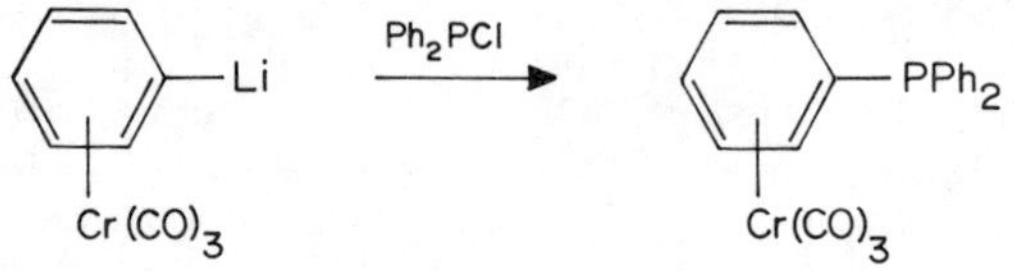

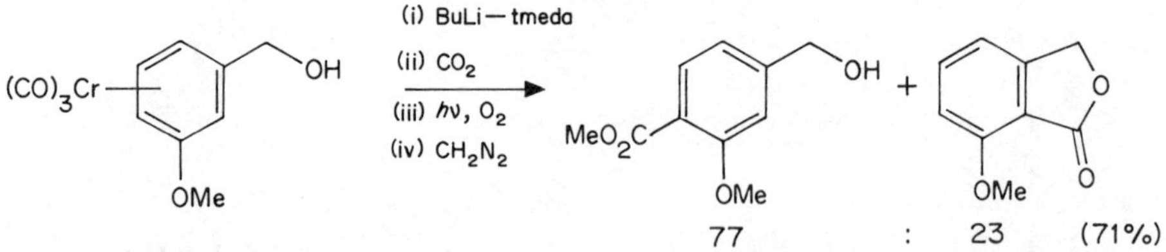

2 [benzene-Cr(CO)₃] $\xrightarrow[\Delta]{Ph_2Hg \; dioxane}$ [bis-arene Hg complex] 55% $\xrightarrow[-20\,°C]{BuLi, \; Et_2O}$ [aryl-Li Cr(CO)₃]

[aryl-Li Cr(CO)₃] $\xrightarrow{Ph_2PCl}$ [aryl-PPh₂ Cr(CO)₃]

[benzyl alcohol Cr(CO)₃ OMe]
(i) BuLi—tmeda
(ii) CO₂
(iii) hν, O₂
(iv) CH₂N₂
$\longrightarrow$ [MeO₂C substituted product OMe] $+$ [isobenzofuranone OMe]

77 : 23 (71%)

Compare:

[benzyl alcohol OMe] $\longrightarrow$ [MeO₂C product OMe] $+$ [isobenzofuranone OMe]

10 : 90

[X substituted (CO)₃Cr arene] $\xrightarrow[-78\,°C]{Bu^nLi \; tmeda}$ [X, Li, Cr(CO)₃ arene] $\xrightarrow{(i)\;CO_2 \;\; (ii)\;CH_2N_2}$ [X, CO₂Me, Cr(CO)₃ arene]

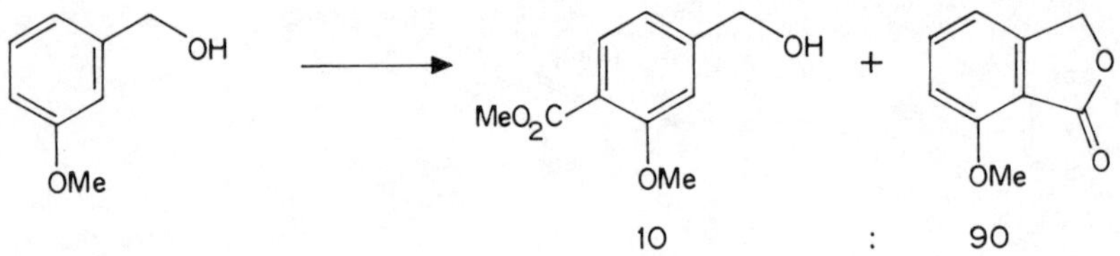

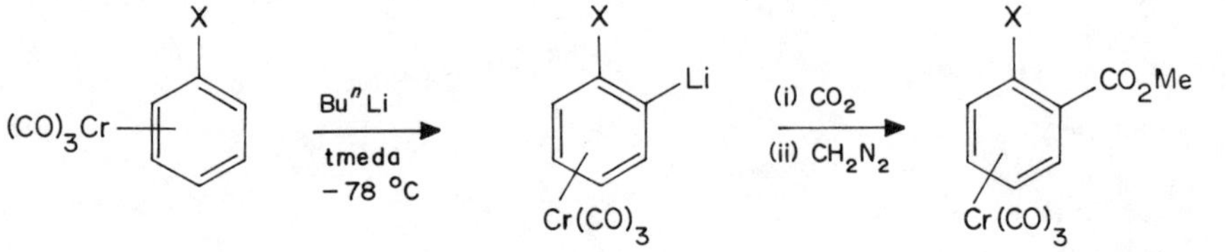

X = H, 72%
X = OMe, 86%
X = F, 99%

When this behavior is coupled with the intramolecular nucleophile addition, some interesting reactions are obtained, e.g.

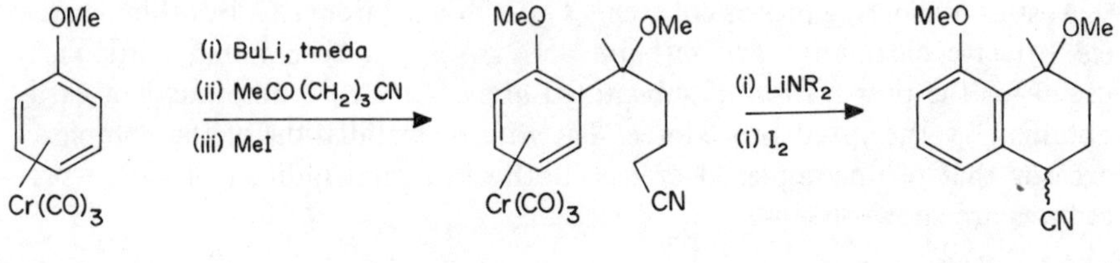

[OMe arene Cr(CO)₃] $\xrightarrow{\substack{(i)\;BuLi,\;tmeda \\ (ii)\;MeCO(CH_2)_3CN \\ (iii)\;MeI}}$ [MeO, OMe, Cr(CO)₃, CN intermediate] $\xrightarrow{\substack{(i)\;LiNR_2 \\ (i)\;I_2}}$ [MeO, OMe, CN bicyclic product]

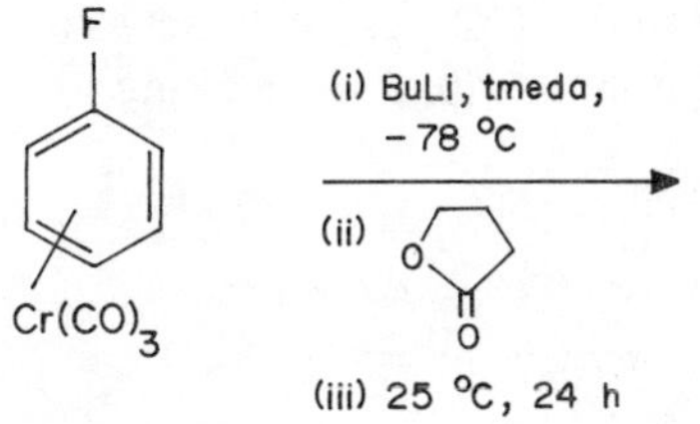

The general tendency is for lithiation to occur *ortho-* to a methoxy or halide substituent. The resultant aryllithium derivative can be reacted with a wide range of electrophiles—only a few have been shown here.

Deprotonation of α-methylene groups

Facile removal of proton from benzylic groups of substituted benzene–$Cr(CO)_3$ complexes has been observed on treatment with, e.g., sodium hydride. The benzylic protons on the complex are found to be around 100 times more acidic than those in uncomplexed systems.[6] This is amply illustrated by the behavior of methyl phenylacetate and its $Cr(CO)_3$ derivatives. While the free ligand is practically inert to NaH/CH_3I (in dmf at 25 °C), it has been found that the $Cr(CO)_3$ complex yields the dimethylated product in 97% yield after 5 min.[2a] Some examples are given below.[7]

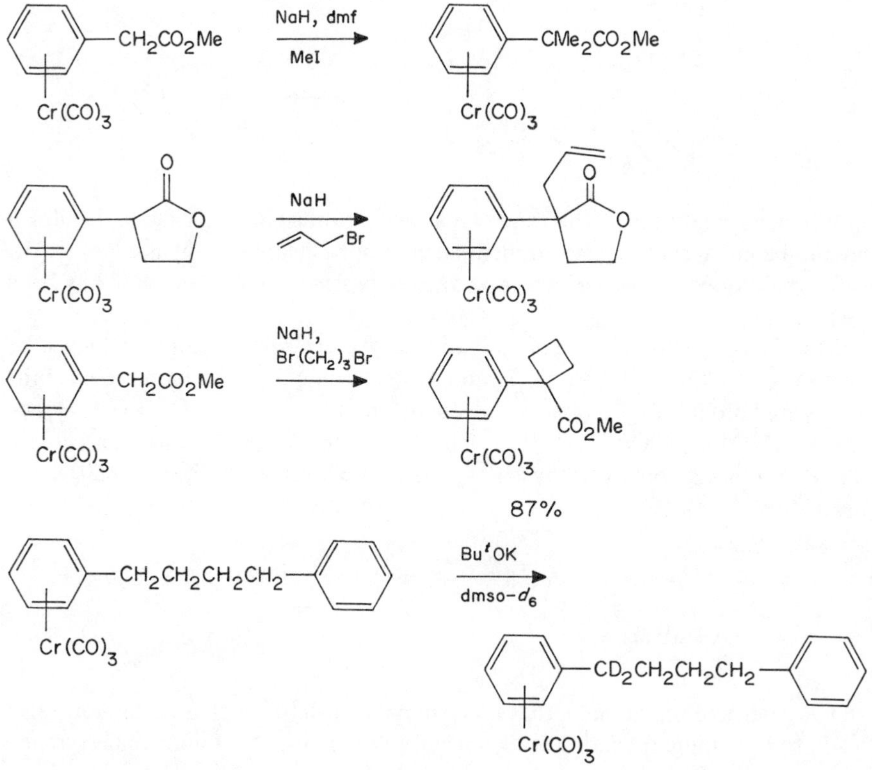

Undoubtedly, this effect arises from the ability of the $Cr(CO)_3$ group to act as an 'electron sink' as in the aforementioned nucleophile additions, making the following canonical forms important:

Less obvious is the apparent stabilization of a carbanion β- to the aromatic group which has been noted. The following examples are indicative of such an effect.

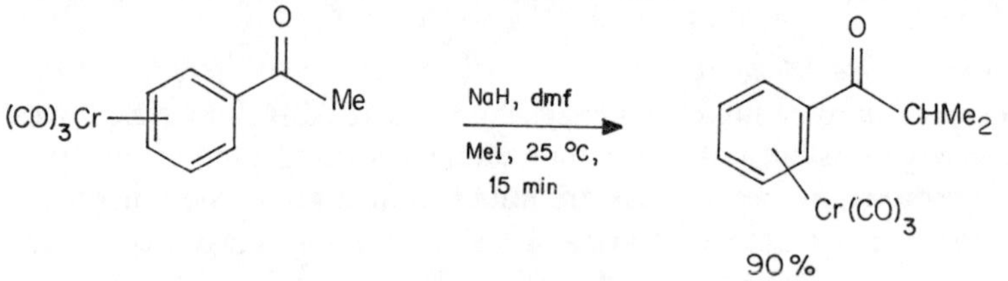

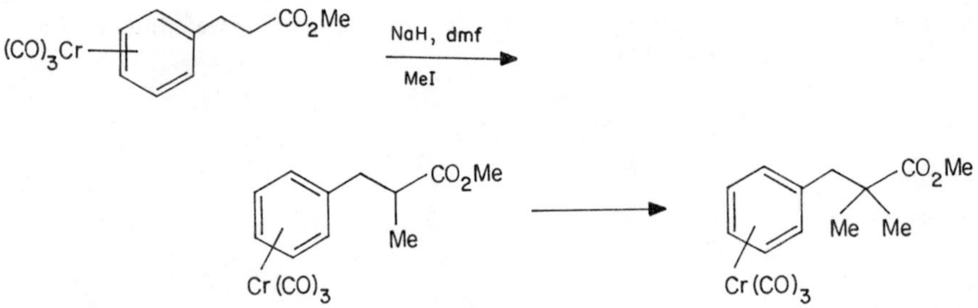

Whilst we might expect this kind of reactivity for the uncomplexed molecules, it should be noted that, e.g., acetophenone cannot be cleanly methylated under the same conditions. However, no good rationalization of this behavior has been forthcoming.

Modification of the degree of activation of the aryl species can be achieved by replacing $Cr(CO)_3$ with $Cr(CO)_2L$, and this grouping has been found to have the following order of electron-withdrawing ability: $Cr(CO)_2CS > Cr(CO)_3 > Cr(CO)_2P(OPh)_3 > Cr(CO)_2PPh_3$. This leads to the ability to control the alkylation of, e.g., phenylacetic esters to give predominantly the monoalkylated derivative:

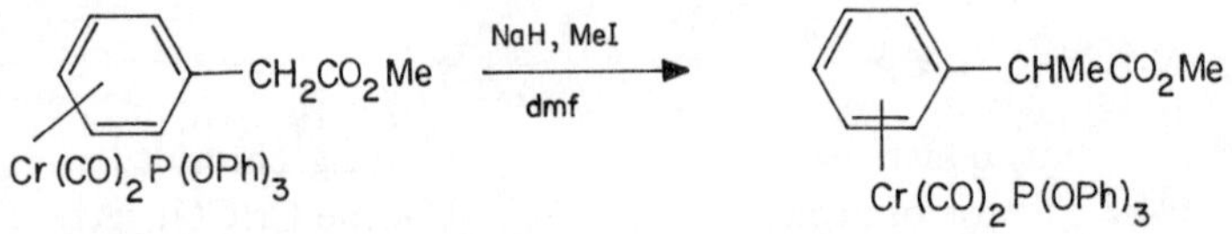

The ability of the arene–$Cr(CO)_3$ systems to stabilize such α-carbanions also leads to a 'conjugate' addition of nucleophiles. Thus, it is found that styrene–

$Cr(CO)_3$ derivatives undergo additions of nucleophiles at the terminal carbons of the double bond. This is illustrated by the following example:[8]

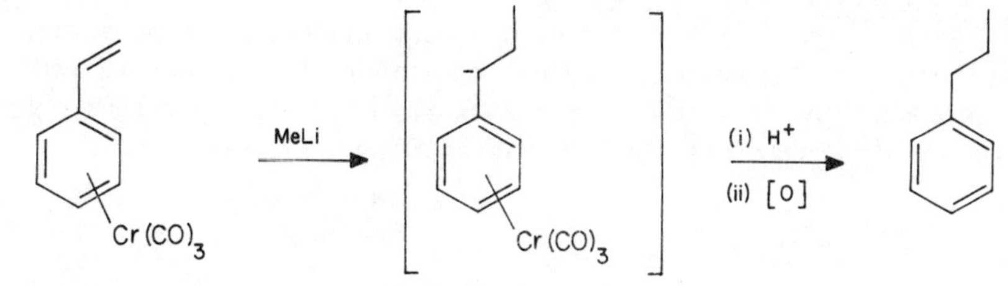

Stabilization of α-Carbonium Ions.

The arene–$Cr(CO)_3$ unit can perform a dual purpose in stabilizing both carbanions and carbonium ions at the benzylic position. Thus, the $Cr(CO)_3$ unit may act as an electron sink through its bonding to the arene unit, or the metal may donate electrons into vacant orbitals (cf. the ferrocenyl systems). The exact mechanism by which carbonium ion stabilization occurs is not known, although it is probable that a direct interaction occurs between a filled metal d orbital and the empty carbon p orbital. This has a profound effect on, e.g., solvolysis of α-hydroxy groups, through neighboring group participation in much the same way as discussed in Chapter 8 for ferrocene derivatives. Some typical reactions are given below.

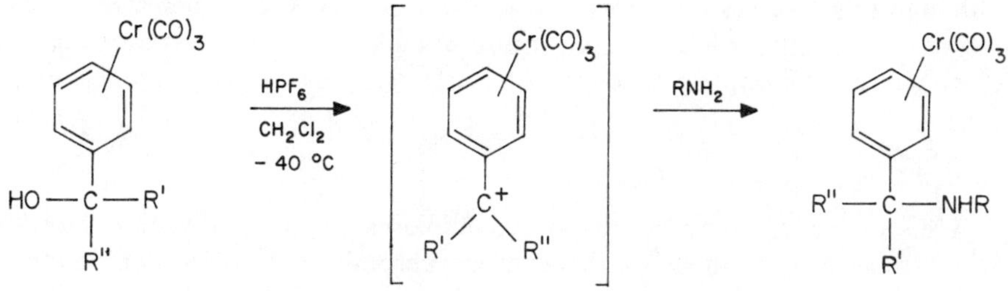

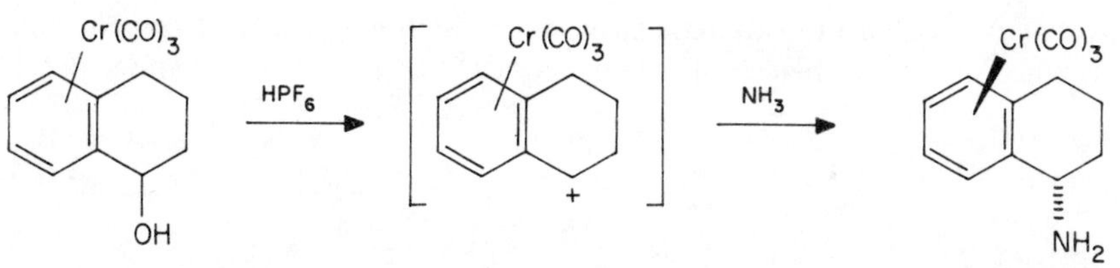

Thus, the carbonium ions are sufficiently stable to allow generation at low temperature and subsequent reactions with simple nucleophiles. The synthetic utility of this behavior has not been explored. The participation by arene–$Cr(CO)_3$ also leads to increase in S_N1 solvolysis rates, e.g. chloride solvolyses are accelerated by a factor of around 10^4. The effect of the $Cr(CO)_3$ group on S_N2 displacements is more complex and may also depend on steric hindrance effects, effects of solvation, etc.

Consequences of the Steric Requirements of Cr(CO)₃

These aspects have mainly been studied by Jaouen, and have been reviewed.[2a,d] The fact that the $Cr(CO)_3$ group is fairly bulky leads to some interesting consequences regarding reactions of proximal functional groups in conformationally rigid or cyclic systems. Thus, reduction of the indanone derivative **2** occurs exclusively by attack from the M-*exo* direction, while electrophilic alkylation of the same indanone also occurs along the *exo* vector:

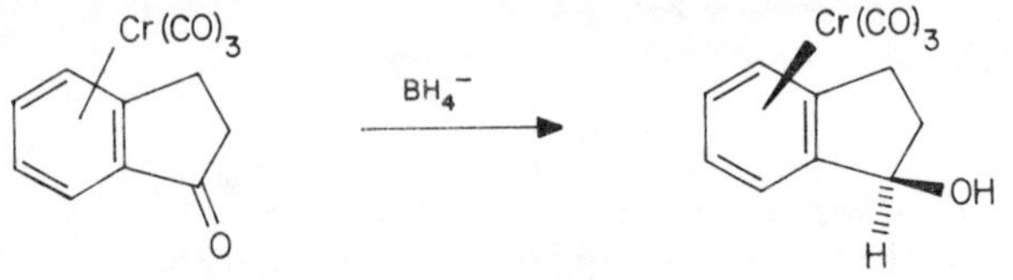

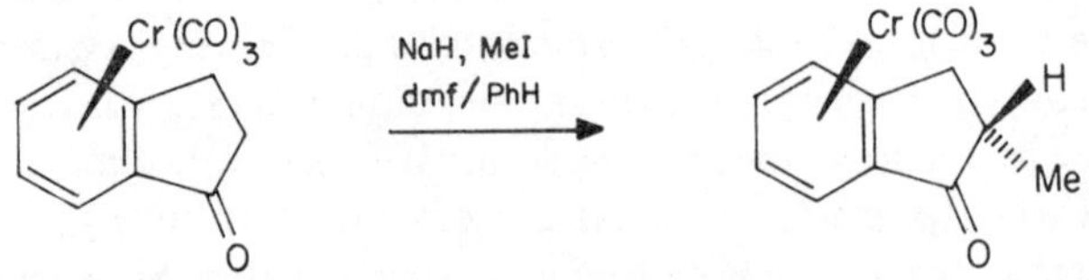

The fact that these complexes are asymmetric, whereas the starting indanones are not, leads to some methods for producing optically active indanone derivatives, although this involves resolution of racemic mixtures, and is therefore not as attractive as methods which might utilize some kind of asymmetric induction during the complexation procedure. This is obviously an area for future research.

Bis-benzene Complexes

A discussion of arene–chromium π-complexes would hardly be complete without some mention of the bis-benzene chromium sandwich compounds. Heating a mixture of $CrCl_3$, $AlCl_3$ and Al powder with benzene, followed by hydrolytic work-up and alkaline dithionite reduction, gives dibenzenechromium, obtained as brown–black solid, m.p. 284–285 °C (the Fischer–Hafner process).

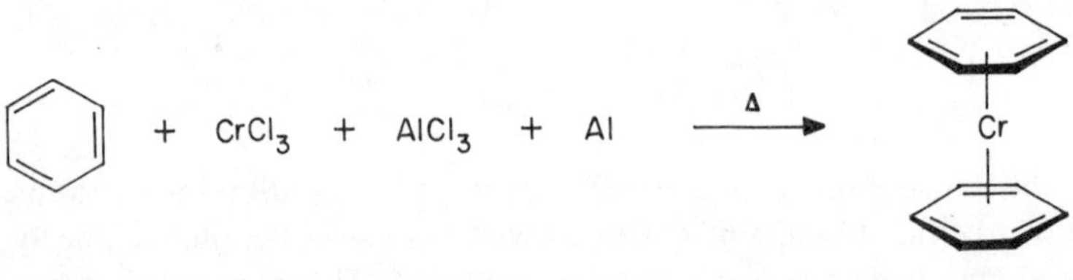

An alternative synthesis involves the reaction of $CrCl_3$ with PhMgBr in an exact ratio of 1:3 in tetrahydrofuran at -25 °C, to give firstly the σ-bonded triphenylchromium(III) tetrahydrofuranate, Ph₃Cr·3thf. Treatment of this with diethyl ether to remove thf leads to the formation of a black pyrophoric substance which is converted to dibenzenechromium by aqueous hydrolysis under a nitrogen atmosphere. Dibenzene-molybdenum and -tungsten derivatives can be prepared similarly but these compounds are less stable to oxidation and thermal decomposition than are the chromium compounds.

A striking feature in the chemistry of dibenzenechromium is its inertness to electrophilic substitution. Attempts to achieve such reactions results only in oxidation to $(PhH)_2Cr^+$. This is also formed by air oxidation of $(PhH)_2Cr$. Metallation can be achieved, however, and the derived sodio derivative will react with aldehydes and ketones, summarized below.

Molybdenum and tungsten form both bis-arene complexes and arene–ML_n complexes. While the bis-arene derivatives can be prepared by the Fischer–Hafner method, the yields are lower than for Cr (yields decrease in the order $Cr > Mo > W$). The arene–metal tricarbonyl complexes may also be formed by direct reaction of $M(CO)_6$ with the arene, although often the yields are low. It should be remembered, however, that in many cases no attempts have been made to optimize the yields of complexes. The use of trisubstituted metal carbonyl derivatives, e.g. (diglyme)$Mo(CO)_3$ or $(CH_3CN)_3W(CO)_3$, allows milder reaction conditions and often results in higher yields. A large number of arene–$M(CO)_3$ (M = Mo, W) complexes are known,[2d] but the reactivity of the arene ligand has not been intensively studied.

Titanium and Zirconium

As reduction does not proceed beyond the Ti(II) stage, application of the Fischer–Hafner aluminum reduction method does not yield bis-arene–titanium complexes. Dibenzenetitanium has been successfully prepared by condensation of benzene and titanium vapor at liquid nitrogen temperature, and is obtained as a red diamagnetic compound.

Reaction of an appropriate aromatic compound with $TiCl_4$, aluminum metal and aluminum chloride yields compounds of structure **3**.

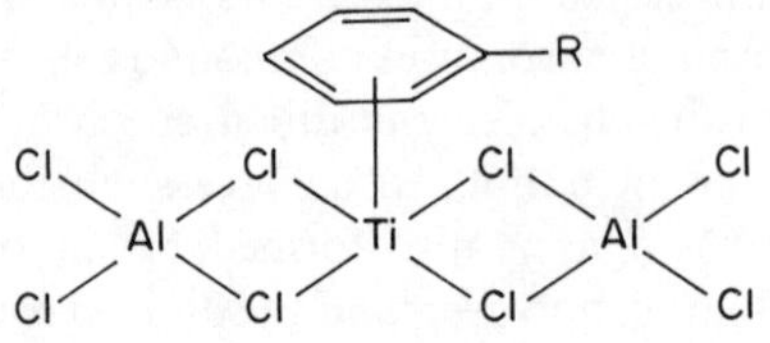

(3)

Under similar conditions, but at 130 °C, followed by careful hydrolysis, hexamethylbenzene affords a dark violet paramagnetic (one unpaired electron) solid, which has been assigned the structure

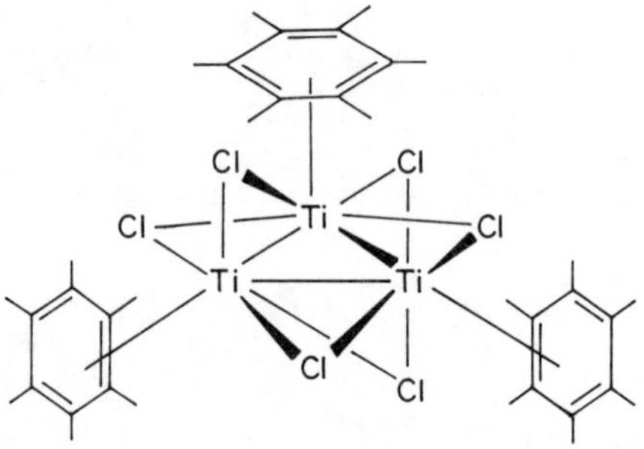

The corresponding zirconium complex was prepared in a similar manner.

Vanadium, Niobium and Tantalum

Bis-arene–vanadium complexes have been prepared by the Fischer–Hafner method, e.g.

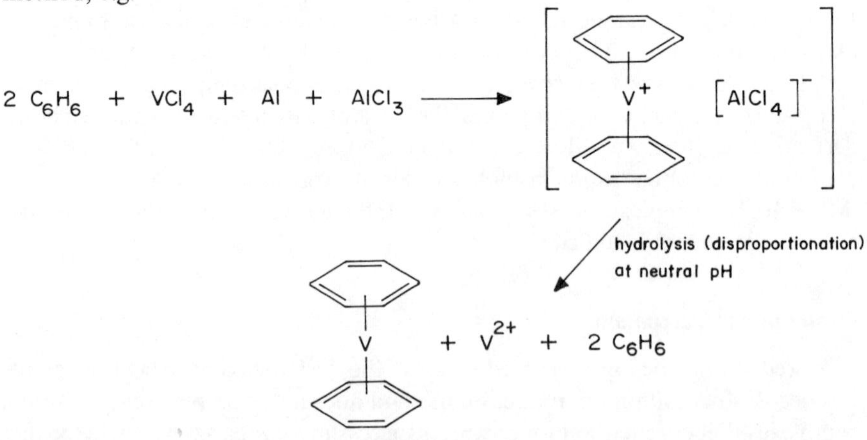

$$2 \ C_6H_6 \ + \ VCl_4 \ + \ Al \ + \ AlCl_3 \longrightarrow$$

hydrolysis (disproportionation) at neutral pH

$$+ \ V^{2+} \ + \ 2 \ C_6H_6$$

That the reaction proceeds with the intermediacy of the cationic complex shown, which disproportionates on hydrolytic work-up, is demonstrated by isolation of the iodide **4** from mesitylene when the reaction mixture is hydrolyzed at 0 °C in the presence of lithium iodide:

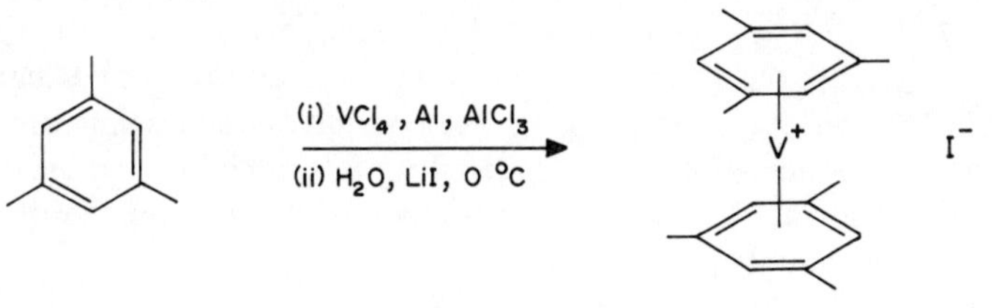

(i) VCl_4, Al, $AlCl_3$
(ii) H_2O, LiI, 0 °C

V^+ I^-

(4)

Dibenzenevanadium may also be prepared by the Grignard method, e.g.

$$PhMgBr + VCl_3(thf)_3 \longrightarrow (C_6H_6)_2V$$

Both $(arene)_2V$ and $(arene)_2V^+$ are paramagnetic, with one and two unpaired electrons, respectively.

Reaction of hexacarbonylvanadium with a range of benzene derivatives, followed by anion exchange, furnishes the cationic arene–vanadium carbonyl complexes, e.g.[9]

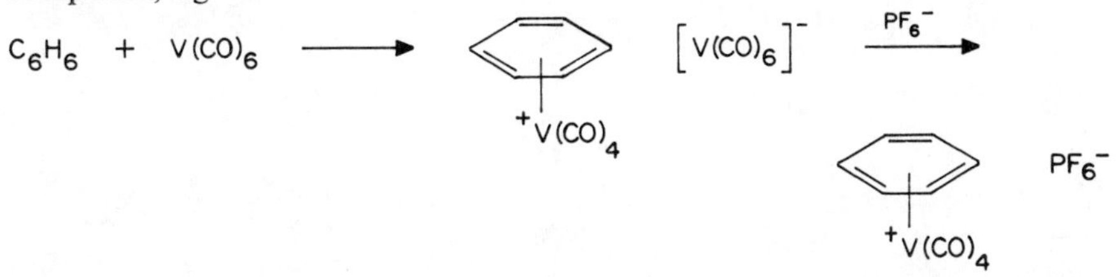

$$C_6H_6 + V(CO)_6 \longrightarrow \overset{+}{V}(CO)_4 \quad [V(CO)_6]^- \overset{PF_6^-}{\longrightarrow} \overset{+}{V}(CO)_4 \quad PF_6^-$$

These complexes have been shown to react with $NaBH_4$ by addition of hydride to the aromatic ring, giving neutral dienyl complexes:

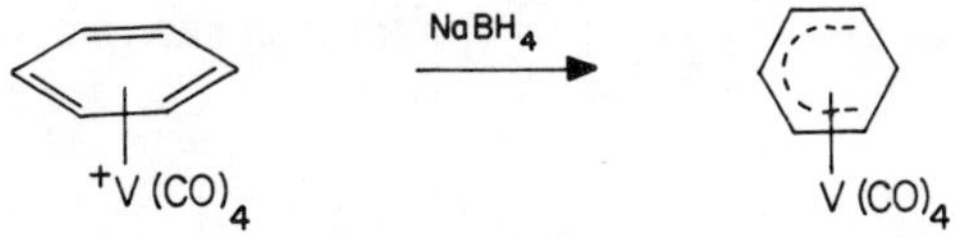

$$\overset{+}{V}(CO)_4 \overset{NaBH_4}{\longrightarrow} V(CO)_4$$

Displacement of a CO ligand occurs on treatment with sodium iodide. Some subsequent reactions of the iodide are listed below.

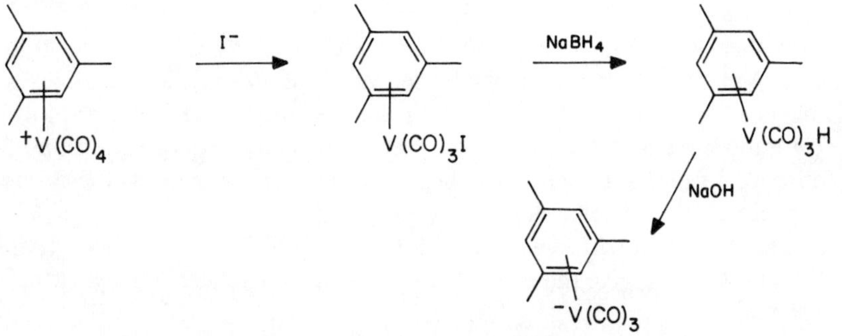

$$\overset{+}{V}(CO)_4 \overset{I^-}{\longrightarrow} V(CO)_3I \overset{NaBH_4}{\longrightarrow} V(CO)_3H$$

NaOH

$$\overset{-}{V}(CO)_3$$

Manganese and Rhenium

Perhaps the most interesting complexes in this group, from the point of view of potential synthetic applications, are the arene–manganese tricarbonyl derivatives, first prepared in 1957, which have received (limited) attention since around 1970. The usual preparative method involves heating the aromatic compound with pentacarbonylmanganese bromide or chloride in the presence of aluminum trichloride, followed by treatment of an aqueous solution of the product with hexafluorophosphoric acid:

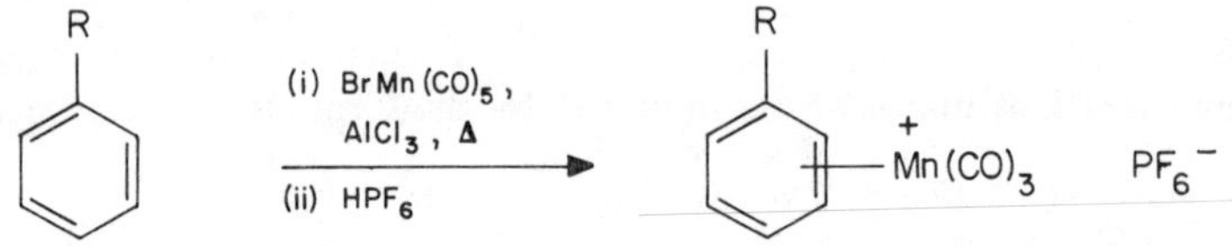

This tends to give variable yields, and it is unlikely that a wide range of functional groups, e.g. in the R substituent, would tolerate the reaction conditions. More recently it has been shown that better yields of the arene complexes can be obtained by ligand exchange with the inexpensive tricarbonyl(methylcyclopentadienyl)manganese:[10]

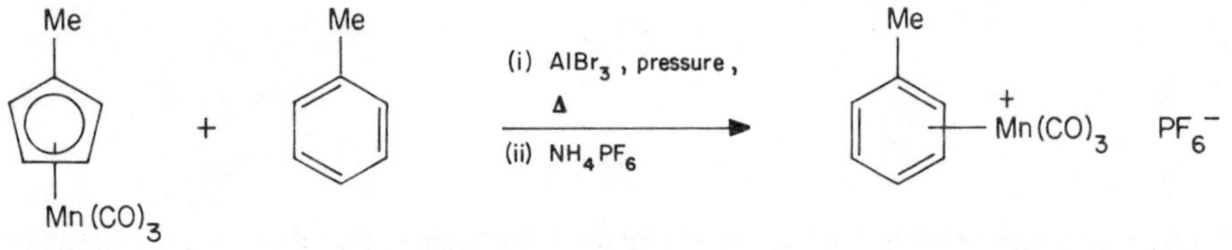

Pauson and co-workers recently described a very mild, although expensive, method of preparation using pentacarbonylmanganese perchlorate:[11]

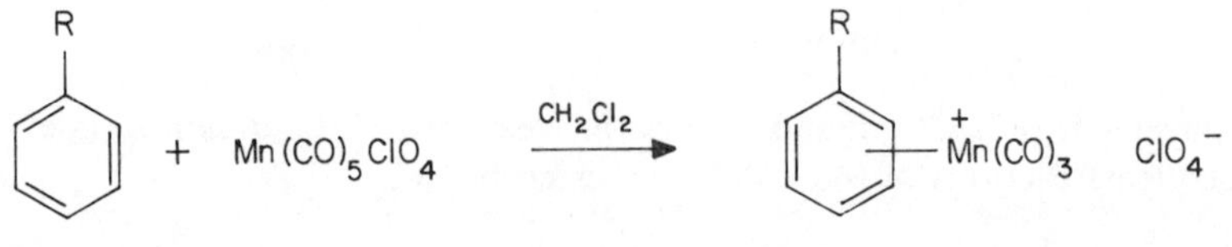

A number of complexes are now known, including those of benzene, toluene, anisole, p-methylanisole, chlorobenzene and naphthalene. The complexes are diamagnetic and are isoelectronic with arene–$Cr(CO)_3$ derivatives. However, they are more reactive than the latter compounds and undergo addition of less reactive nucleophiles. Despite this, their application to organic synthesis has not been explored to the same degree.

The complexes react with, e.g., PhLi, MeLi, $NaCH(CO_2Me)_2$, H^-, N_3^-, MeO^-, NCS^- and CN^- to give the neutral exo-substituted cyclohexadienyl–$Mn(CO)_3$ complexes, e.g.[12]

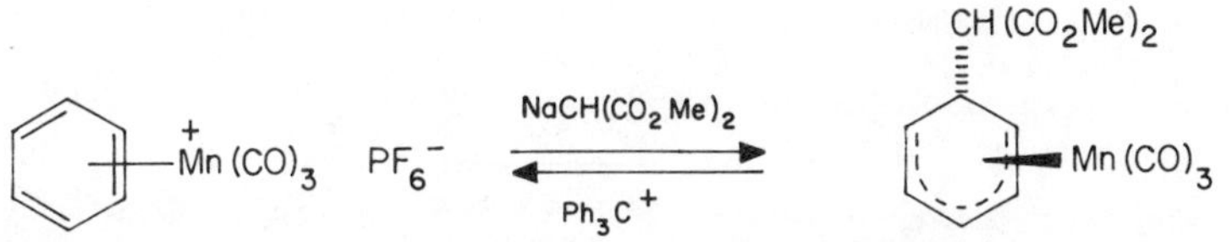

Attempted conversion of the dimethyl malonate adduct to a substituted benzene complex, by treatment with Ph_3C^+, leads only to loss of the malonate group and regeneration of benzene–$Mn(CO)_3$. Some of the products from reversible nucleophile addition (e.g. MeO^-) are unstable, presumably owing to dissociation of the nucleophile.

Interestingly, the cyanide addition product undergoes rearrangement with loss of CO ligand and formation of an Mn—CN bond, e.g.[13]

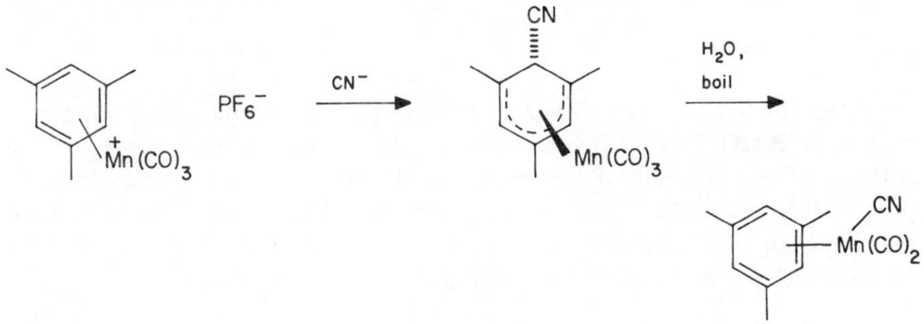

The addition of nucleophiles to the cationic complexes is accompanied by the expected changes in the IR spectrum, the arene complex having absorptions usually at *ca.* 2070 and 2015 cm^{-1}, whilst the dienyl complex absorbs at *ca.* 2020 and 1948 cm^{-1} [cf. the changes for dienyl–$Fe(CO)_3 \rightarrow$ diene–$Fe(CO)_3$, Chapter 8].

Reaction of arene–$Mn(CO)_3$ complexes with trialkylphosphines occurs with loss of CO and formation of an M—P bond:

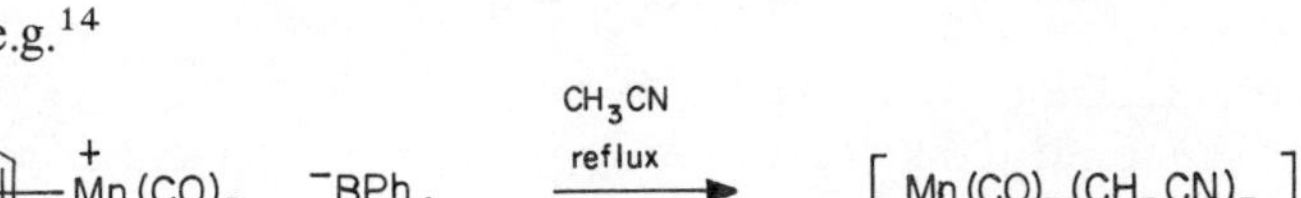

Under some conditions the arene group can be displaced with a suitable ligand, e.g. acetonitrile. This might in fact provide a convenient method of obtaining the substituted benzene derivative from a complex after appropriate synthetic reaction, e.g.[14]

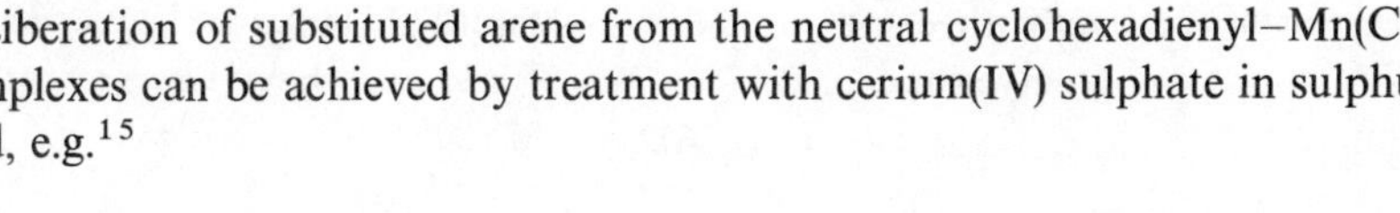

Liberation of substituted arene from the neutral cyclohexadienyl–$Mn(CO)_3$ complexes can be achieved by treatment with cerium(IV) sulphate in sulphuric acid, e.g.[15]

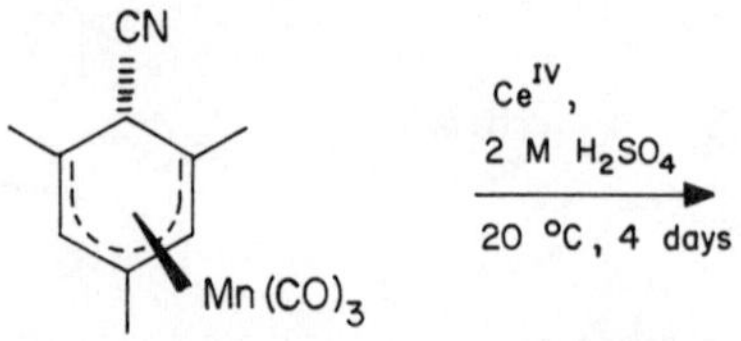

Reactions of arene–Mn(CO)$_3$ with primary amines occurs by addition of RNH$_2$ to a carbonyl ligand, to give carboxamide complexes, although these are too unstable to allow isolation, undergoing facile loss of RNH$_2$ to regenerate the arene–Mn(CO)$_3$ complexes.[16]

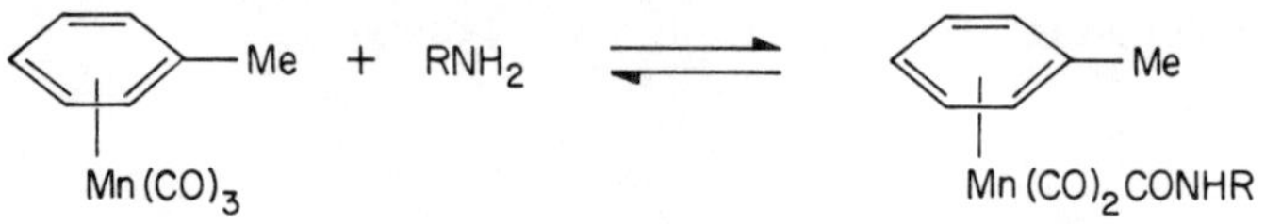

Only a limited amount of work has been carried out on arene complexes bearing functional substituents. Most of the results parallel those for the corresponding chromium derivatives, except that the high reactivity leads to poorer regioselectivity. Thus, methoxy and dialkylamino substituents tend to direct nucleophiles to the *meta* position, whilst halogen substituents are displaced by reversible nucleophiles.[17] Some results are summarized below.

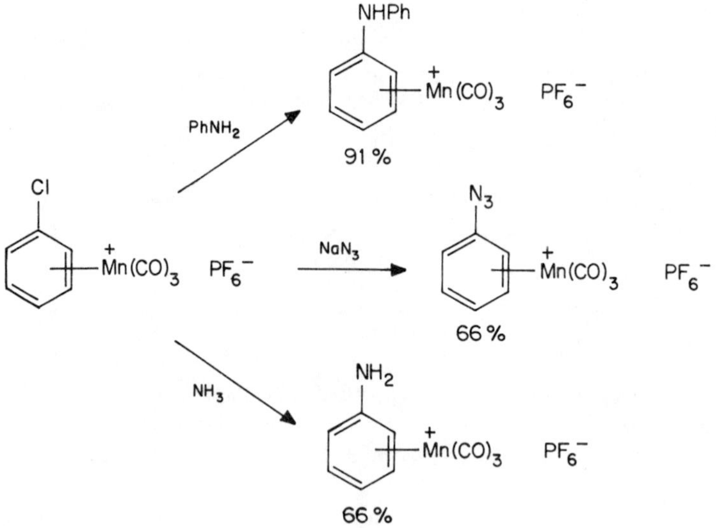

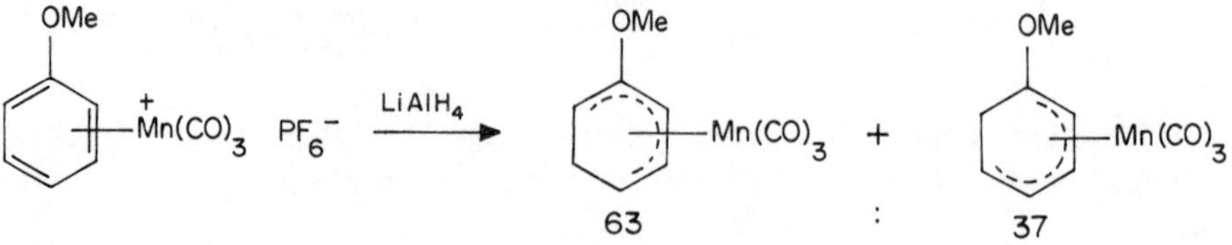

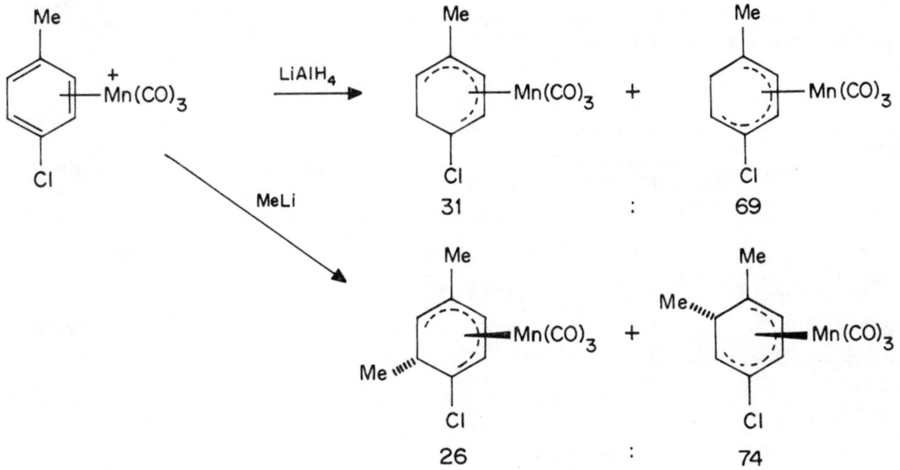

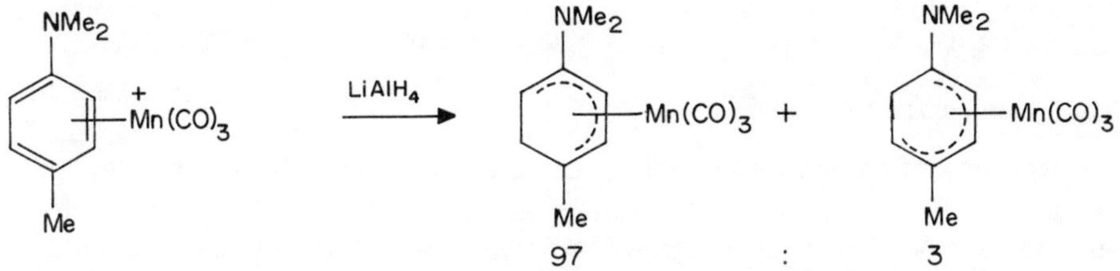

It may be noted that 'irreversible' nucleophiles (H^-, Me^-) do not displace the halogen, but rather add to non-substituted positions, predominantly *meta* to the halide. The dimethylamino group serves as a powerful director of nucleophiles, and whilst this is probably due to electronic rather than steric effects, this remains to be proved.

With highly substituted arene complexes, reduction with lithium aluminum hydride occurs by addition of H^- to both ring positions and carbonyl ligands. Similarly, alkyllithiums may add to both ring and CO ligands:[18]

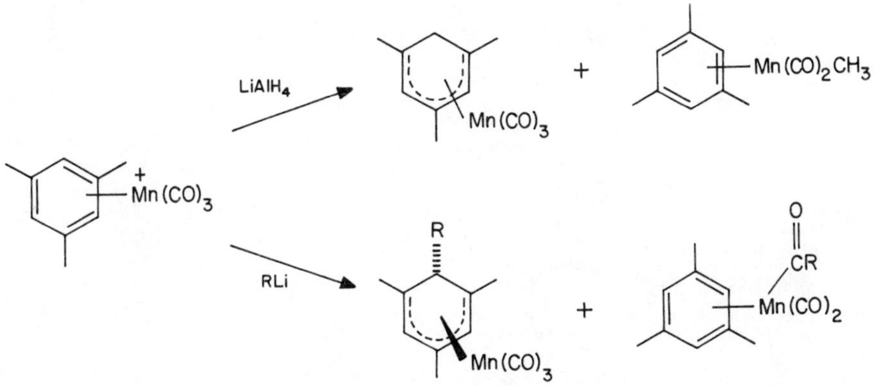

The azidotoluene complex undergoes an interesting ring contraction reaction, presumably via a nitrene intermediate:[19]

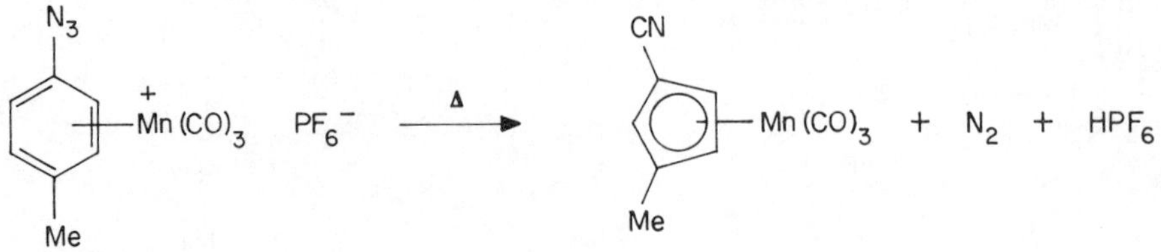

Direct reaction of arenes with $Re(CO)_5Br$ or $Re_2(CO)_8Br_2$ also yields cationic arene–$Re(CO)_3$ complexes, the counter ion being produced as a bromorhenium carbonyl dimer:[20]

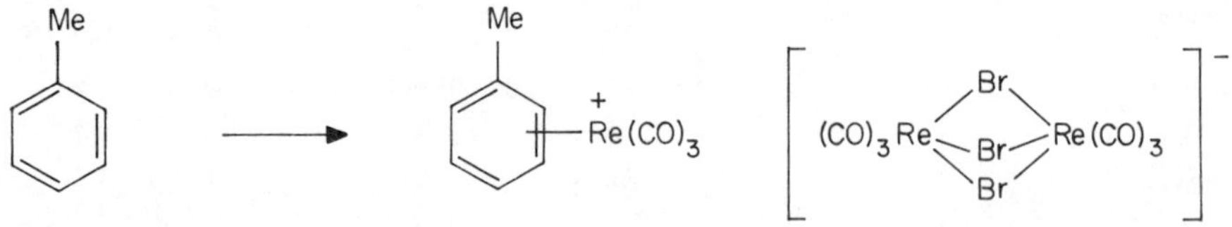

Iron, Ruthenium and Osmium

Two types of arene complex are of interest in the iron group of metals. The metal is formally in the $+2$ oxidation state for both of these, the cyclopentadienyl–M–arene and bis(arene)M complexes, which will be discussed in that order.

The arene–Fe–Cp cationic derivatives are prepared by displacement of one of the cyclopentadienyl ligands of ferrocene by, e.g. benzene, in the presence of $AlCl_3$:

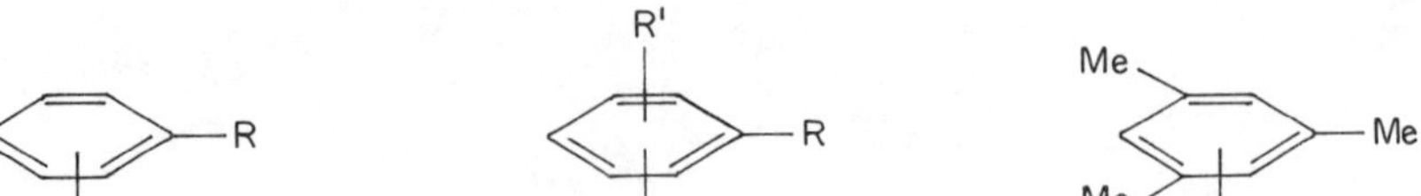

A wide variety of substituted arenes can be employed, giving the appropriate complexes, some examples of which are shown below:

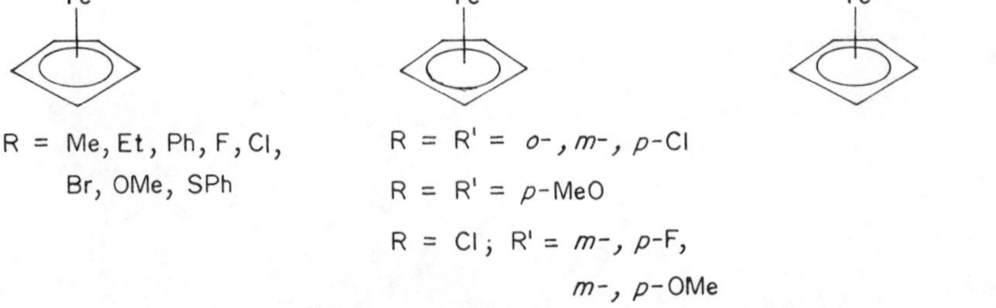

R = Me, Et, Ph, F, Cl,
Br, OMe, SPh

R = R' = o-, m-, p-Cl

R = R' = p-MeO

R = Cl; R' = m-, p-F,
m-, p-OMe

As might be expected these complexes undergo reaction with nucleophiles. Perhaps less obviously, all nucleophiles studied add to the arene ligand, rather than the cyclopentadienyl ligand, to give cyclopentadienyl(η^5-cyclohexadienyl)Fe complexes. Reduction with sodium borohydride gives the expected dienyl complex, treatment of which with trityl cation (or *N*-bromosuccinimide) leads to hydride abstraction (or oxidation) to regenerate the arene complex:

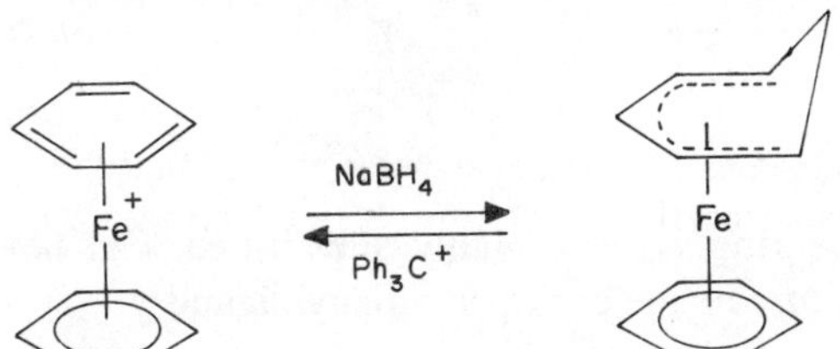

With regard to substituted arene ligands, the results of nucleophile addition, such as they are, follow a similar trend to those of the corresponding arene–$Mn(CO)_3$ cations. Thus, the halogen of chlorobenzene complexes is readily displaced by reversible nucleophiles (MeO^-, PhO^-, RS^-, NH_2^-, etc.) but not by irreversible nucleophiles (H^-), which lead to attack predominantly *ortho* to the halogen giving cyclohexadienyl complexes. The methoxy group in anisole–FeCp derivatives direct nucleophiles mainly into the *meta* position, and the directing effect appears to be stronger than for the Mn complexes. The CO_2Me group directs mainly *ortho*.

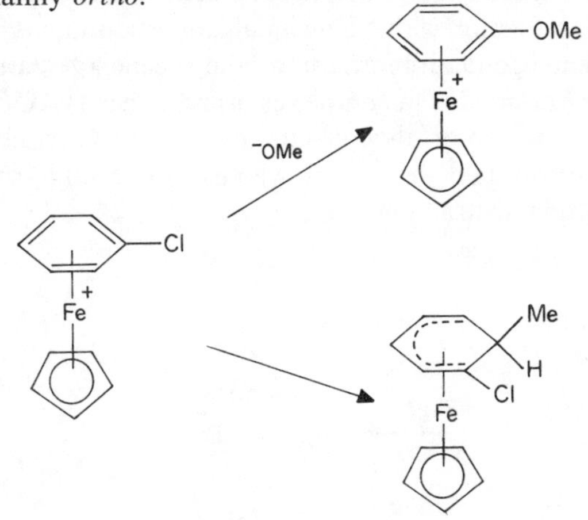

major (4 : 1 , *ortho / meta*)

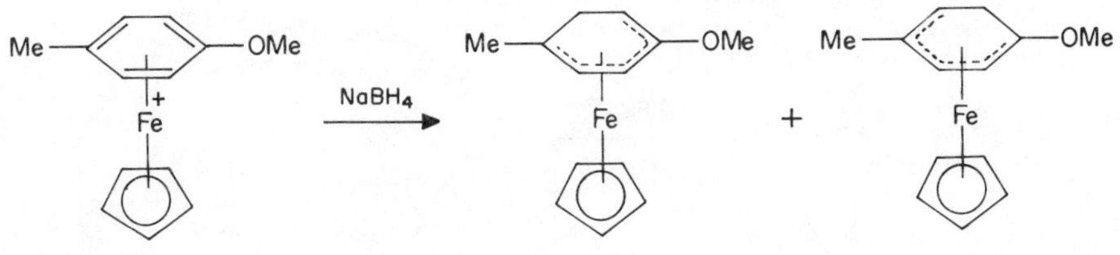

ca. 4 : 1

Direct alkylation has been shown to occur *exo* to the metal, as expected from other similar 18e complexes.[21]

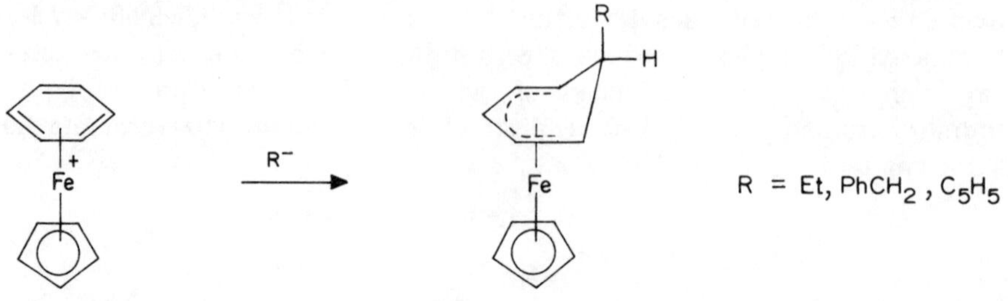

$$R = Et, PhCH_2, C_5H_5$$

When the arene ring is very highly substituted, e.g. hexamethylbenzene, addition of R^- occurs at the cyclopentadienyl ligand:

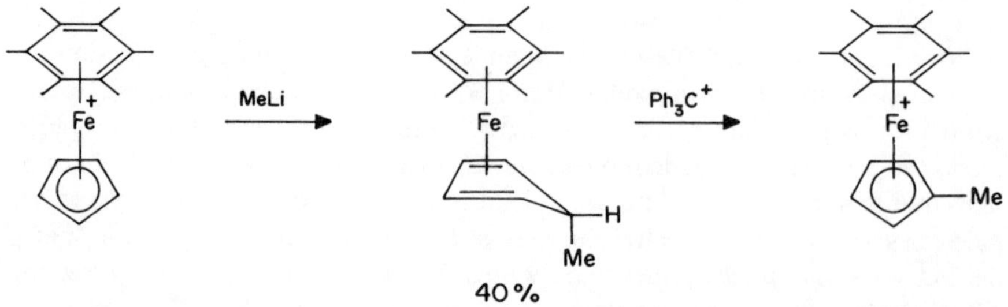

40%

Reaction of non-halogenated arene derivatives with nucleophiles such as OH^-, OR^- or R_2N^-, or reduction with NaHg amalgam or sodium naphthalide, results in loss of the arene ligand and formation of ferrocene. Oxidation of the products of alkylation of arene–FeCp complexes, using either Ph_3C^+ or NBS, may lead to hydride abstraction or alkyl group abstraction. Alkyl abstraction occurs for the benzyl-substituted derivative, whereas the methyl-substituted complex undergoes hydride abstraction:[22]

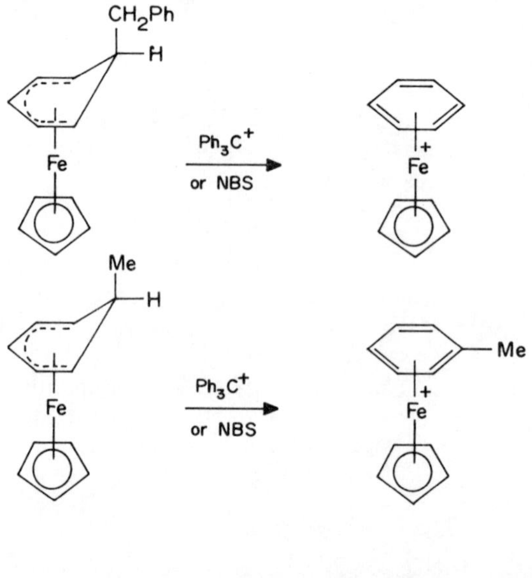

Interesting behavior is shown by cyclohexadienyl–FeCp complexes in which the cyclopentadienyl carries an α-hydroxyalkyl group. Treatment with acid causes loss of OH (as expected from the stability of α-ferrocenyl cations), but the final product is one in which H has migrated from the cyclohexadienyl ring:[23]

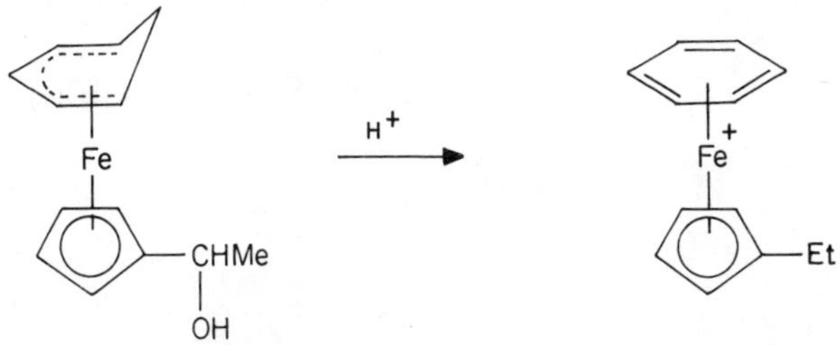

Reaction of $FeCl_2$, $FeBr_2$ or $FeCl_3$ (which is reduced to $FeCl_2$ by arene solvent) with arenes in the presence of aluminum chloride results in the production of diamagnetic bisarene complexes of the type $[(arene)_2Fe]^{2+}$. The analogous ruthenium complexes, $[(arene)_2Ru]^{2+}$, and the osmium complex $[(mesitylene)_2Os]^{2+}$ are also known.

Reduction of $[(Me_6C_6)_2Fe]^{2+}$ with sodium dithionite gives the paramagnetic monocation $[(Me_6C_6)_2Fe]^+$, which can be further reduced to the very sensitive neutral complex $[(Me_6C_6)_2Fe]$ using alkaline dithionite. The ruthenium complex $[(Me_6C_6)_2Ru]^{2+}$ is reduced more readily to the neutral $(Me_6C_6)_2Ru$, which might at first be thought to be a 20-electron complex. X-ray analysis in fact reveals that this compound has a structure in which one of the arene ligands is η^4-bound, so the complex is an 18-electron species:

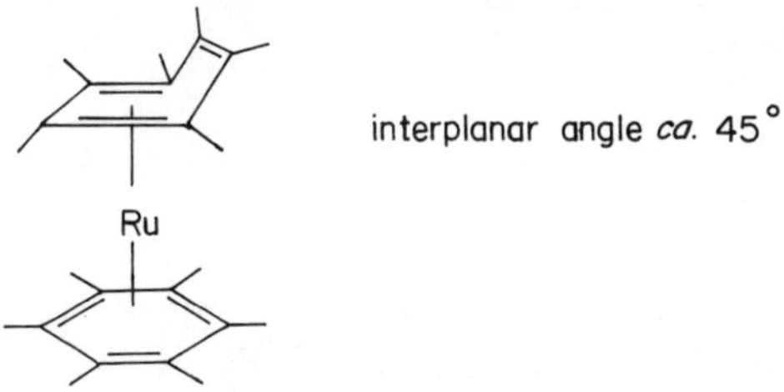

interplanar angle *ca.* 45°

The dications have been found to react with an (as yet limited) range of nucleophiles, as shown below.[24]

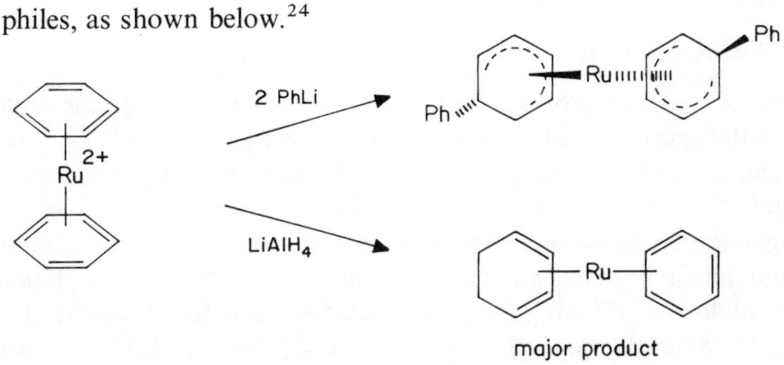

major product

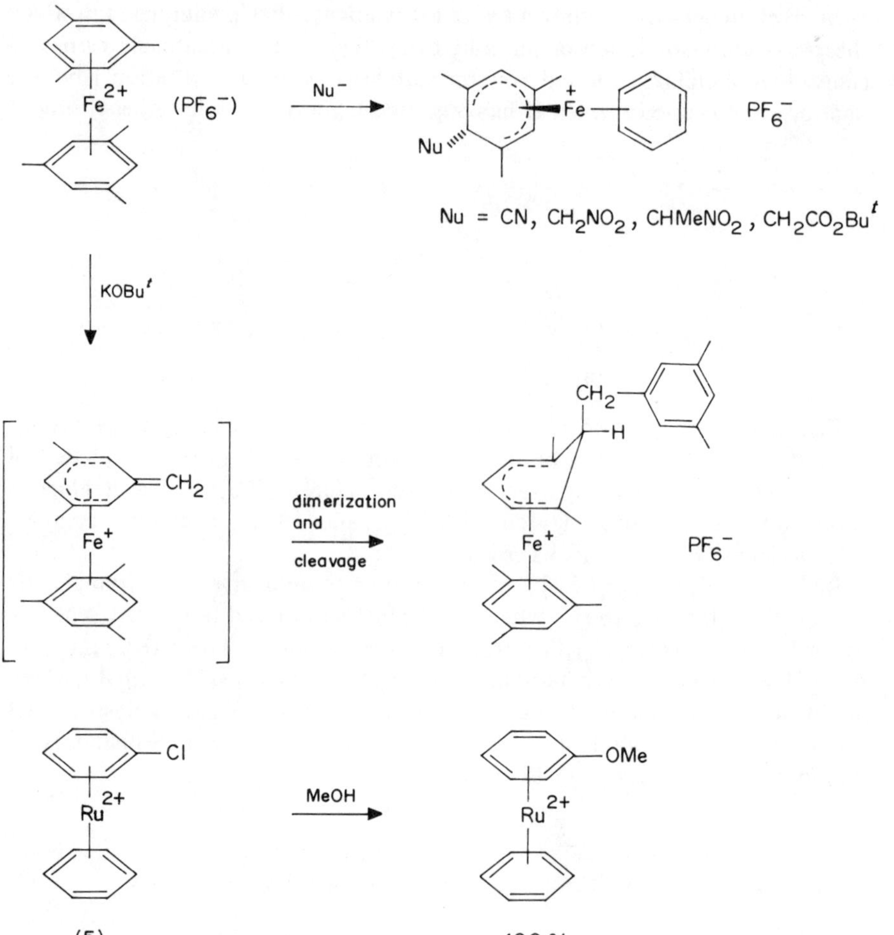

Nu = CN, CH_2NO_2, $CHMeNO_2$, $CH_2CO_2Bu^t$

(5)   100 %

It may be noted that displacement of chloride in the complex **5** is much more rapid than with $(ClC_6H_5)Cr(CO)_3$, undoubtedly owing to its high positive charge.

Cobalt, Rhodium and Iridium

A range of arene complexes of these metals is known, although their reactivity and potential synthetic utility have not been investigated much.

Bisarene complexes $[(Me_6C_6)_2M]^{2+}$ (M = Co, Rh] can be prepared by reacting $CoCl_2$ or $RhCl_3$ with hexamethylbenzene in the presence of $AlCl_3$. A paramagnetic (two unpaired electrons) monocation, $[(Me_6C_6)_2Co]^+$, is formed if aluminum metal is added to the reaction mixture, whereas a diamagnetic rhodium analogue, $[(Me_6C_6)_2Rh]^+$, is formed on reduction of the dication with Zn/HCl. A similar diamagnetic cobalt complex, $[(Me_6C_6)_2Co]^+$, is formed on reaction of dimesitylcobalt(II) with dimethylacetylene. Further reduction of $[(Me_6C_6)_2Co]^{2+}$ to the neutral paramagnetic $(Me_6C_6)_2Co$ (unpaired electron) is achieved with sodium in liquid ammonia.

Mixed sandwich complexes are also known, e.g.

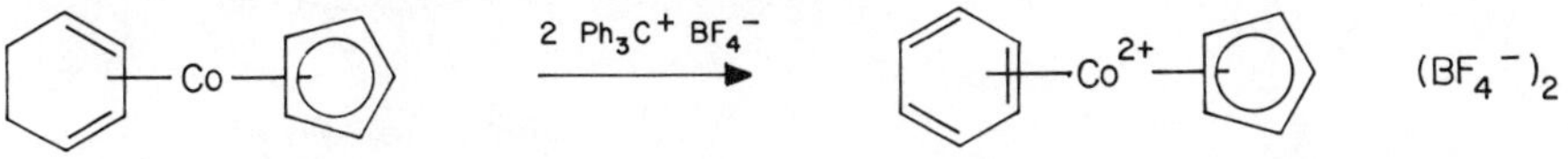

$$2\ Ph_3C^+\ BF_4^-$$

$$(BF_4^-)_2$$

(see also Chapter 7)

$(\pi\text{-Cp})M(OCOCF_3)_2(H_2O)$ $\xrightarrow{\text{arene}}$

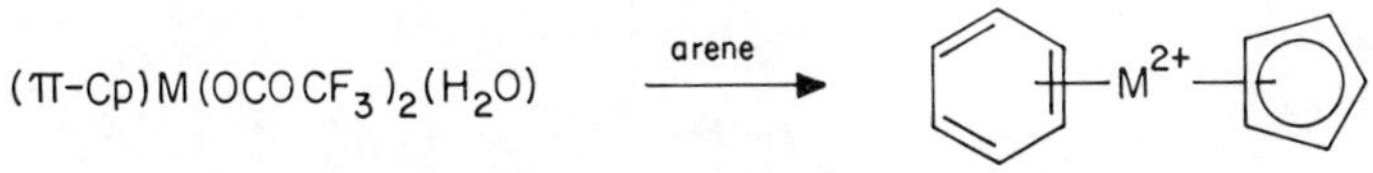

$M = Rh,\ Ir$

The rhodium and iridium complexes undergo hydride addition to the arene ring, e.g.

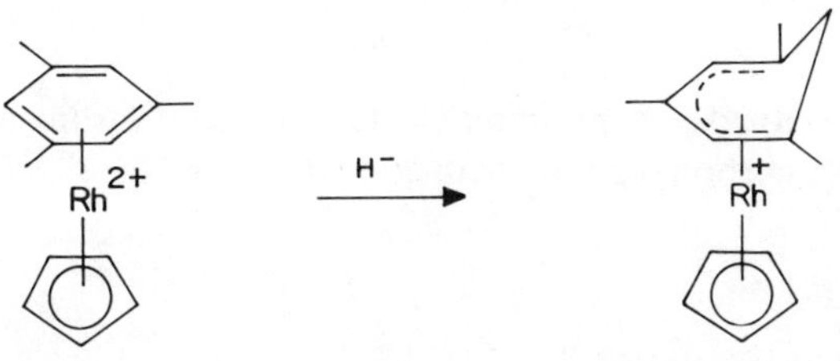

$$H^-$$

Arene–M–Cp complexes in this series are also formed by trimerization of an acetylene. In this way the neutral Rh(I) complex **6** may be prepared.

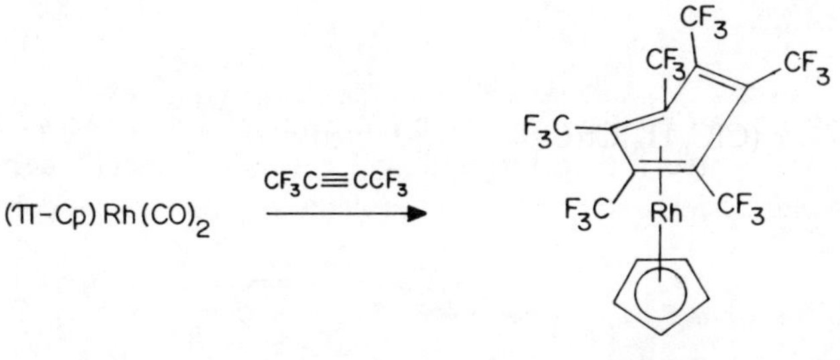

$(\pi\text{-Cp})Rh(CO)_2$ $\xrightarrow{CF_3C\equiv CCF_3}$

(6)

However, an X-ray structure determination reveals that the product is not an η^6-arene derivative, but rather an 18-electron η^4-complex in which the aromaticity of the $(CF_3)_6C_6$ ring is lost.

Substituted cyclobutadienes have also been found to be useful counter ligands in this series, illustrated by the preparation of diamagnetic $[(\text{arene})(\pi\text{-}C_4Ph_4)Co]^+$ cations:

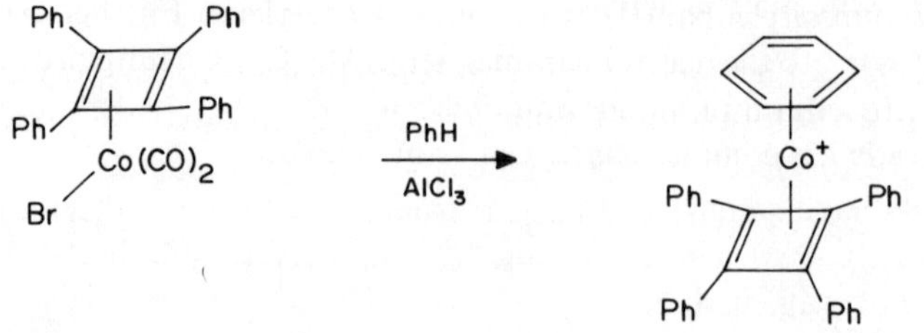

This complex undergoes addition of simple nucleophiles at the arene ligand, and the resulting cyclohexadienyl complexes can be oxidized to (substituted) arene complexes with NBS:

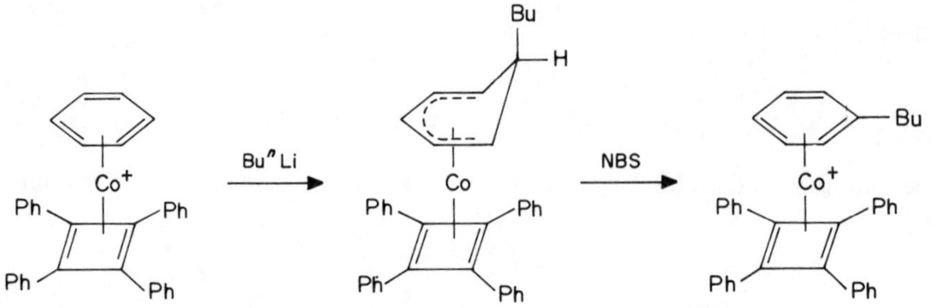

More exotic complexes containing clusters of metal atoms are obtained by reaction of cobalt carbonyls with aromatic compounds, e.g.

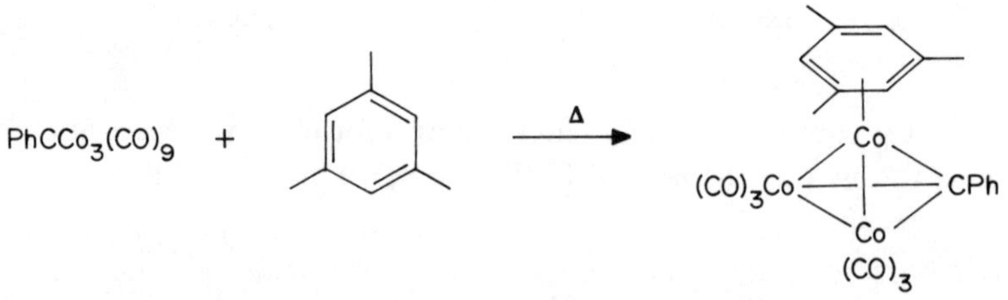

A diene ligand can be displaced from [M(diene)$_2$]$^+$ complexes (M = Rh, Ir) with an arene to give the corresponding [(arene)M(diene)]$^+$ complex. This behavior has also been extended to monoolefin and phosphine complexes:

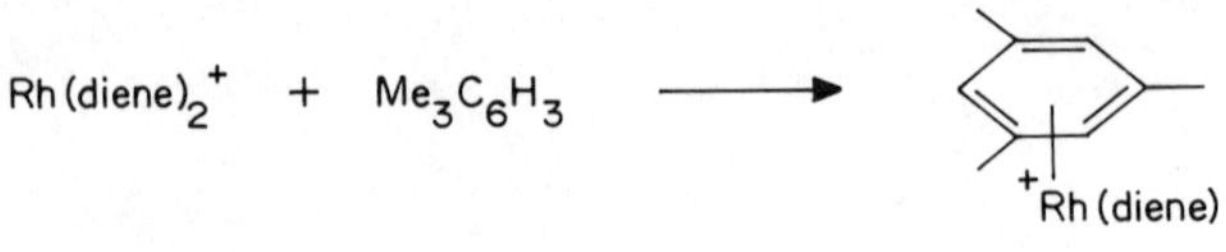

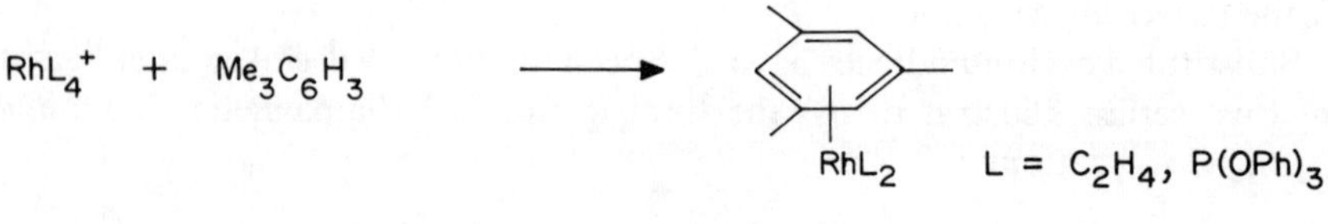

Nickel, Platinum and Palladium

Reaction of hexafluorobut-2-yne with $Ni(cod)_2$ leads to cyclotrimerization of the alkyne with the formation of neutral $[(CF_3)_6C_6]Ni(cod)$ complexes. The cod ligand may be displaced by tertiary phosphines and arsines.

Complexes having unusual binuclear structures are produced in the reaction of $PdCl_2$, $AlCl_3$ and Al with an arene. The complex produced depends on the proportions of reagents used:

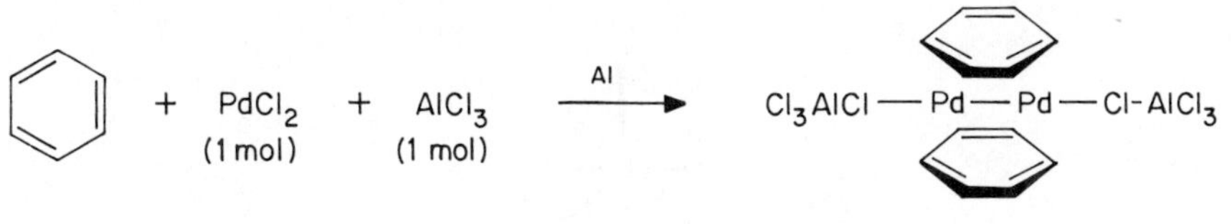

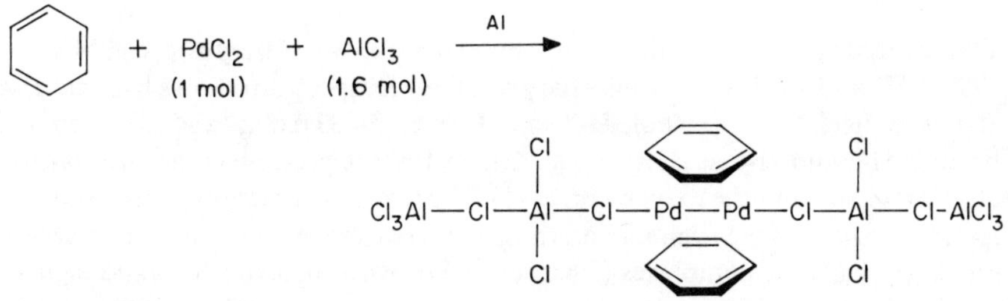

Reaction of $NiBr_2$ with hexamethylbenzene/$AlBr_3$ gives the paramagnetic (two unpaired electron) cationic $[(Me_6C_6)Ni]^{2+}$, isoelectronic with the Fe(0) and Co(I) complexes.

η^6-TRIENE AND DERIVED COMPLEXES

The book by Deganello[25] provides a useful survey which includes η^6-complexes of seven-, eight- and twelve-membered rings. This section concentrates on the chemistry of η^6-complexes of cycloheptatriene, cyclooctatriene, cyclooctatetraene, cyclononatriene and cyclododecatriene, and the derived η^7-trienyl complexes where these are known. Most of the discussion is limited to complexes of chromium, molybdenum and tungsten, with a brief treatment of other metals. We shall further restrict ourselves to (triene)$M(CO)_x$ and (triene)M(Cp) derivatives.

Seven-Membered Rings

Chromium, Molybdenum and Tungsten

Refluxing the metal hexacarbonyl in pure cycloheptatriene or in a high-boiling solvent affords the (η^6-triene)$M(CO)_3$ complex, an 18-electron species in which the non-planar nature of the ligand is maintained:

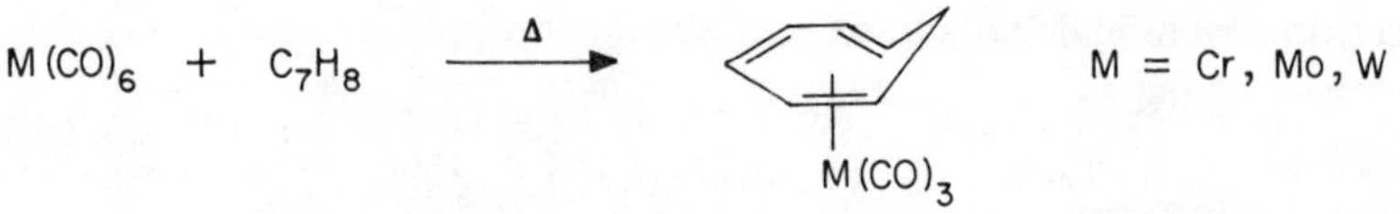

$$\text{M (CO)}_6 \quad + \quad C_7H_8 \quad \xrightarrow{\Delta} \quad \underset{\text{M (CO)}_3}{\includegraphics} \qquad \text{M = Cr, Mo, W}$$

Low yields of the tungsten derivative are obtained by this procedure, but this may be improved by using $W(CO)_3(CH_3CN)_3$ instead of the hexacarbonyl. The complexes adopt a quasi-octahedral arrangement shown below:

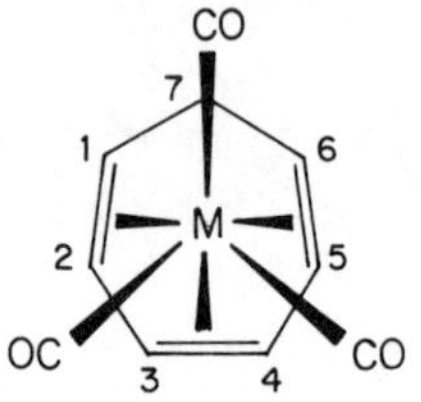

The infrared spectra show three absorptions for the $M(CO)_3$ group, e.g. M = Cr, 1991, 1921 and 1893 cm^{-1}, whilst the proton NMR spectrum shows H-3 and H-4 at lowest field (δ 6.01 p.p.m.), H-2 and H-5 at δ 4.83 p.p.m. and the terminal protons H-1 and H-6 at δ 3.40 p.p.m. The methylene group has the *endo* proton at δ 2.90 p.p.m. and the *exo* proton at δ 1.77 p.p.m., due to its projection into the shielding region of the triene. Thus, the general pattern is similar to that observed for diene–Fe(CO)$_3$ complexes (Chapter 7), but with the expected extra signals.

The triene is readily displaced from the molybdenum complex by heating with acetonitrile:

$$\text{Mo(CO)}_3(C_7H_8) \quad + \quad 3\ CH_3CN \quad \longrightarrow \quad \text{Mo (CO)}_3(CH_3CN)_3 \quad + \quad C_7H_8$$

Substituted cycloheptatriene complexes are accessible by reaction of the appropriate ligand with the metal carbonyl. For example, 7-substituted cycloheptatrienes, on treatment with $Cr(CO)_6$ or $Cr(CO)_3(py)_3/BF_3$, usually produce *endo*-substituted complexes, e.g.

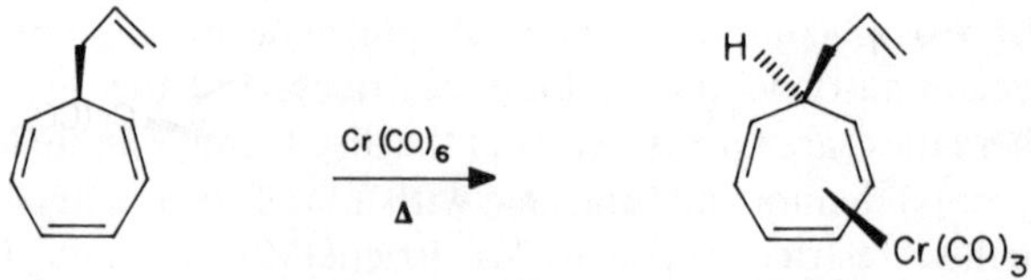

The 7-*endo*-and 7-*exo*-substituted isomers can usually be distinguished by their proton NMR spectra, since protons on an *exo* substituent occur at higher field than *endo* substituent protons. Thus, if both isomers are available, assignments are readily made.

Owing to the relative lability of the M—C$_7$H$_8$ bond, thermal substitution of CO with other ligands is usually not possible. However, this may be achieved, at least for the chromium complex, by photochemical means:

$$\text{Cr (CO)}_3(C_7H_8) \quad + \quad PPh_3 \quad \longrightarrow \quad (CO)_2Cr (PPh_3)(C_7H_8)$$

Treatment of $C_7H_8Mo(CO)_3$ with tetrafluoroboric acid results in protonation of the triene ligand to form a 16-electron cationic dienyl complex. Reaction of this complex with triphenylphosphine occurs at C-3 of the dienyl system (cf. the corresponding iron complexes), followed by disengagement of the metal, as shown below.

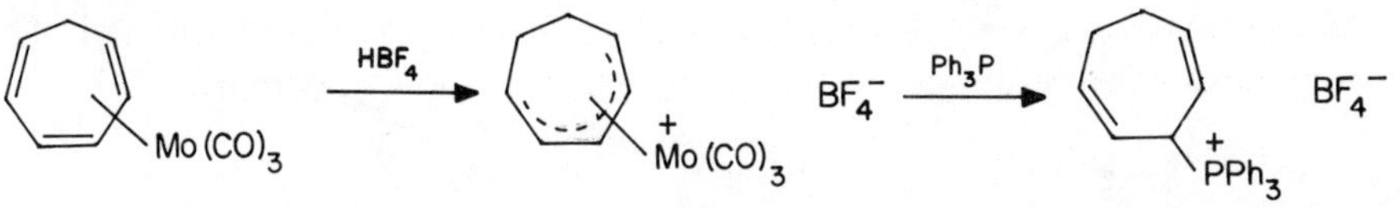

No work appears to have been reported on reactions of the triene complexes with other electrophiles. Hydride abstraction from the cycloheptatriene complexes occurs readily on treatment with trityl tetrafluoroborate to give the tropylium carbonium ion complexes:

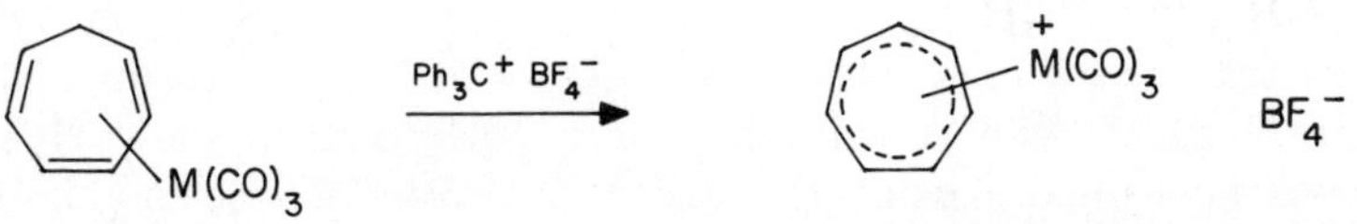

These compounds show the expected single resonance in the proton NMR, and only two strong bands due to $M(CO)_3$ in the IR spectrum (2007 and 1959 cm^{-1} for M = Mo in a KBr disc). The tropylium ring is shown by X-ray methods to be planar. One of the carbonyl ligands may be replaced by, e.g., a phosphine for the Mo and W complexes, but not for M = Cr, which undergoes addition of R_3P at the ligand.

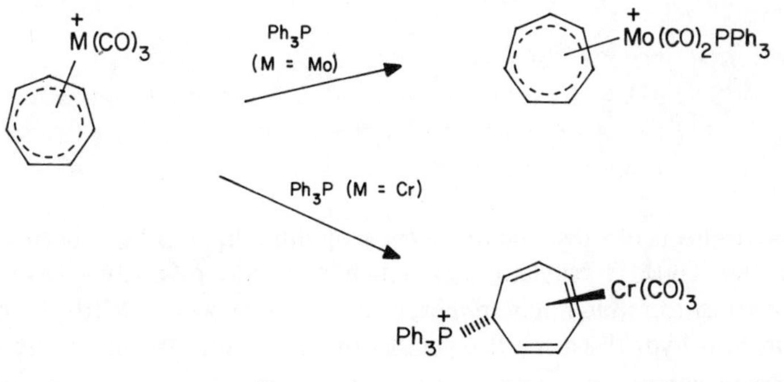

As might be expected, the η^7-complexes react with nucleophiles, to give 7-exo-substituted cycloheptatriene complexes, although no attempts have been made to exploit this behavior for synthetic purposes. Some examples are shown below.

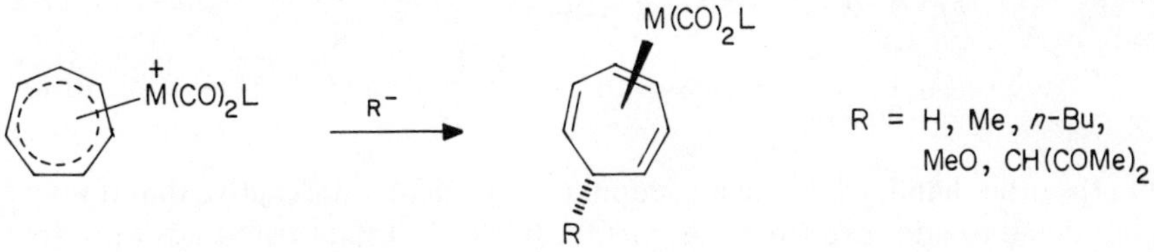

R = H, Me, n-Bu, MeO, $CH(COMe)_2$

The rate of attack of these nucleophiles is independent of the metal. Addition of nucleophiles is not completely general, however, there being three anomalous pathways:

(1) Ring contraction occurs with certain nucleophiles in the presence of excess base (or nucleophile):

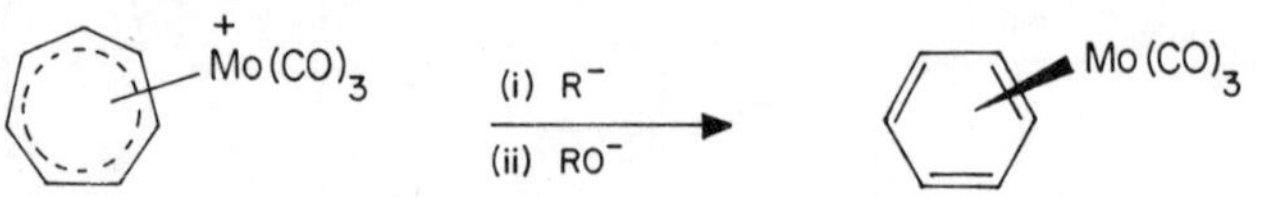

e.g. $R = C_5H_5$, MeC_5H_4, $^-CH(CO_2Me)_2$; $^-OR = ^-OMe$, ^-OEt

(2) Reductive coupling may occur with certain nucleophiles, e.g. PhLi, CN^-, $\bar{O}Ac$, $^-NH_2$, $PhCONH_2$:

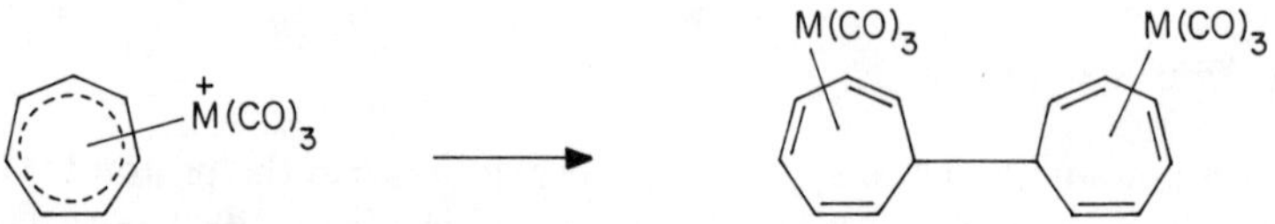

Similar reductive coupling occurs with dienyl–$Fe(CO)_3$ cations with, e.g., MeMgI and PhMgBr.

(3) Displacement of carbonyl may occur with NaX (X = Cl, Br, I).

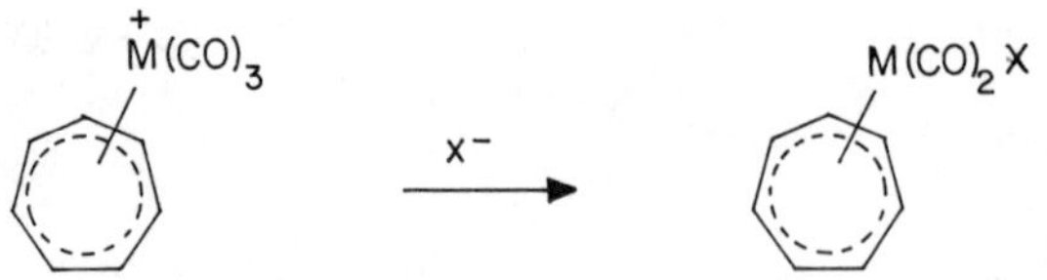

Directive effects of substituents on the tropylium ligands have been studied by Pauson and Todd.[26] Nucleophiles which may add reversibly show different behavior to irreversible nucleophiles, as might be expected. With the latter, e.g. H^- from borohydride, a methoxy substituent is found to direct *meta*, as with arene complexes:

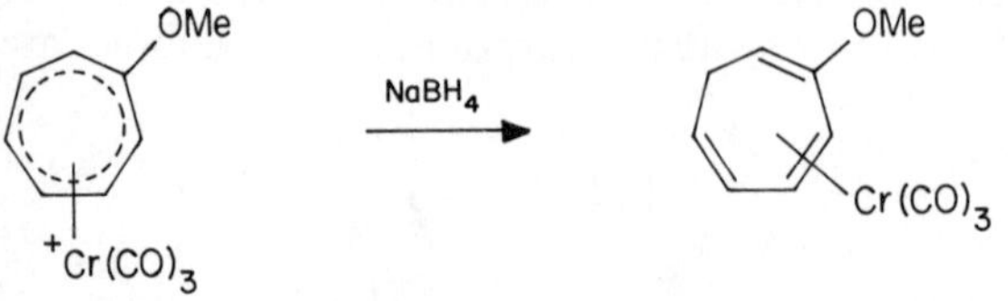

On the other hand, when the nucleophile is sufficiently unreactive that it may be considered to add reversibly, e.g. MeO^-, $^-CH(CO_2Et)_2$, then attack appears to occur at the substituted position, although sometimes only low yields are obtained.

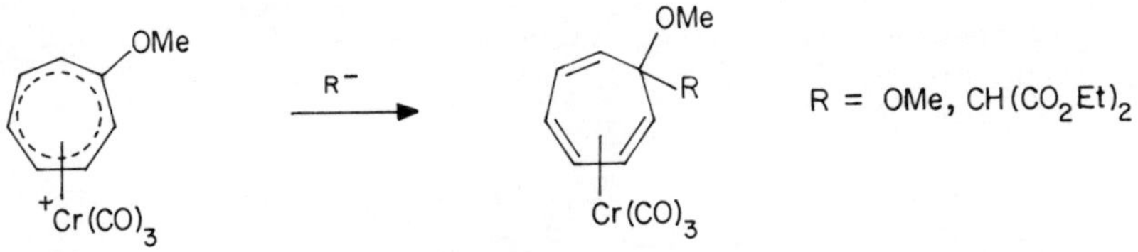

$$R = OMe, CH(CO_2Et)_2$$

With an electron-withdrawing substituent, e.g. CO_2Me, all nucleophiles add predominantly at the 2-position, although the yields are rather variable. Hydride gives small amounts of product from attack at C-4.

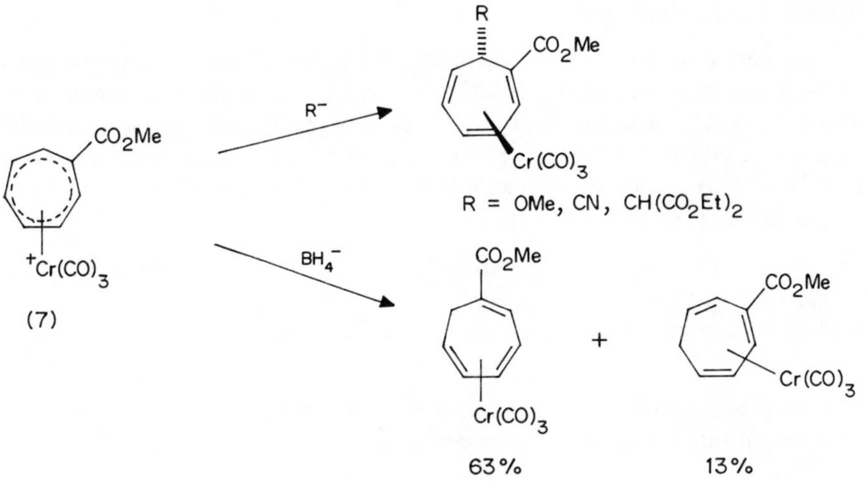

$$R = OMe, CN, CH(CO_2Et)_2$$

63%   13%

The product from methoxide addition to the ester complex **7** undergoes an interesting ring contraction on contact with alumina, giving the methylbenzoate–$Cr(CO_3)$ complex. The same product is also obtained by treatment of the salt with either alumina or hydroxide, although mechanistic details are not available:

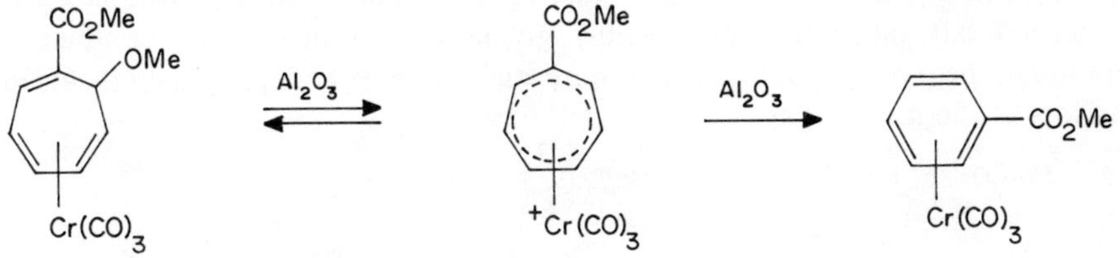

Cyclopentadienyl–M(tropylium) (M = Cr, Mo, W) complexes are also known with the metal in both the +2 and +1 oxidation states. The cationic complex undergoes addition to the tropylium ligand with PhLi, e.g.

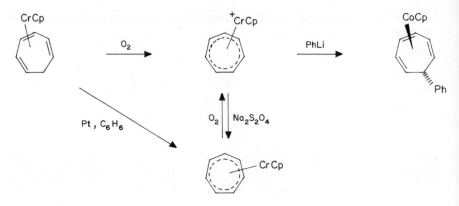

Manganese and Rhenium[27]

Treatment of cycloheptadiene with $Mn_2(CO)_{10}$ in refluxing mesitylene gives the neutral cycloheptadienyl–$Mn(CO)_3$ complex as a crystalline compound, m.p. 64–66 °C, in 70% yield. The $Mn(CO)_3$ carbonyl stretching frequencies are in the range expected (2010, 1950 cm^{-1}) for such a complex. Treatment of this product with Ph_3C^+ BF_4^- affords the cationic η^6-cycloheptatriene–$Mn(CO)_3$ complex (v_{CO} 2028, 1968 cm^{-1}) in 95% yield.

The cation reacts with a fairly wide range of nucleophiles at the triene terminus to afford substituted cycloheptadienyl complexes:

R = H, OMe, OEt, OBu^t, NMe$_2$, Ph, Me, CN,

Interestingly, the dienyl complex undergoes reaction with electrophiles, e.g. Friedel–Crafts acylation, presumably giving the cationic diene complexes, although these could not be isolated pure, and attempts to deprotonate them led to loss of the metal:

The chemistry of these complexes, therefore, parallels to some extent that of diene–$Fe(CO)_3$, derivatives.

The cyclopentadienyl–Mn(cycloheptatriene) complex has been prepared by photochemical displacement of CO ligands from $CpMn(CO)_3$ with cycloheptatriene:

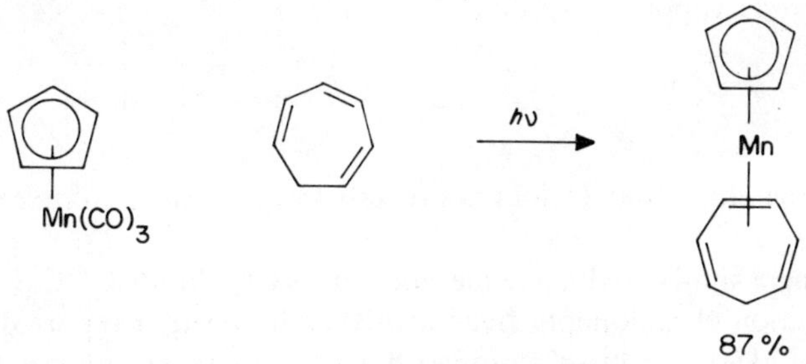

87%

Other Metals

Complexes formed between cycloheptatriene and $Fe(CO)_3$, etc., which are basically η^4 derivatives, have already been discussed.

Reduction of $CpTiCl_3$ with isopropylmagnesium bromide or magnesium metal in the presence of C_7H_8 yields the $CpTi(\eta^7$-$C_7H_7)$ complex. Metallation using butyllithium occurs predominantly, but not exclusively, on the seven-membered ring, and the lithium derivatives react with electrophiles to give substituted complexes. The analogous zirconium complex has also been synthesized.

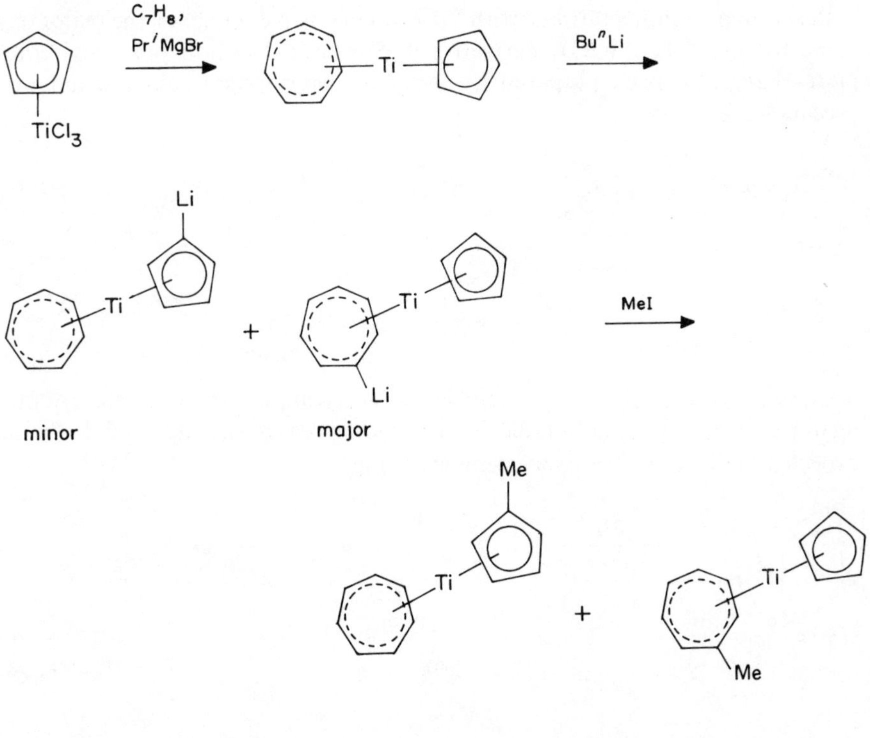

minor   major

Direct reaction of $TiCl_3$ with cycloheptatriene in the presence of a reductant yields not a bis(η^6-C_7H_8) derivative, but instead the unsymmetrical (η^5-C_7H_9)Ti(η^7-C_7H_7) complex. The same complex is obtained by co-condensation of C_7H_8 and Ti vapor at $-196\,°C$.

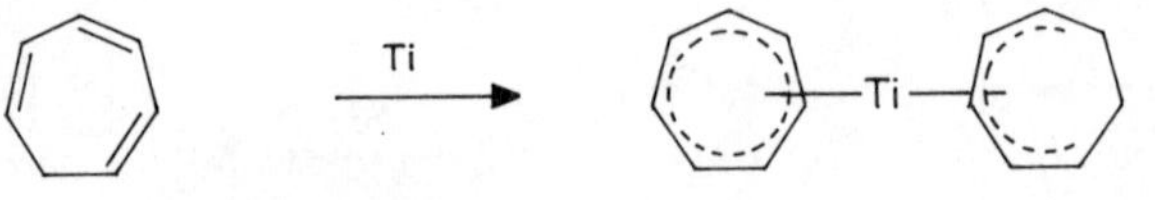

It appears, therefore, that Ti does not readily form η^6-triene complexes (cf. arene complexes).

Employing a similar reductive method with VCl_4 affords (η^6-C_7H_8)$_2$V, whilst co-condensation of cycloheptatriene with vanadium vapor reportedly gives (η^5-C_7H_9)V(η^7-C_7H_7). The bis-η^6-complex **8** reacts with trityl cation in two steps to give mono- and bis-cationic complexes.

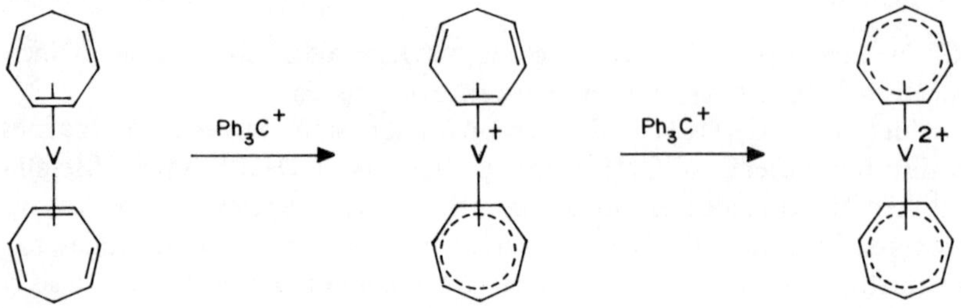

(8)

Reaction of cycloheptatriene with $V(CO)_6$ gives two complexes, the major one being the (η^7-C_7H_7)V(CO)$_3$ derivative **9**. Substituted cycloheptatrienyl complexes have also been prepared by using the appropriate substituted triene precursor.

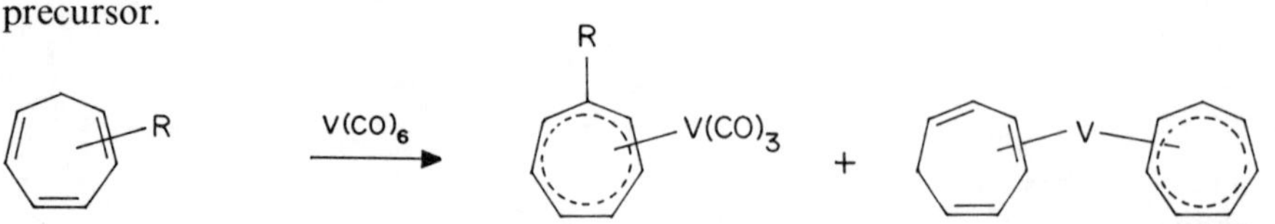

(9)

major ; R = H, Me, Ph,
OR', CO_2R'

minor ; R = H

Cobalt forms a cationic η^6-cycloheptatriene complex **10** when the counter ligand is tetraphenylcyclobutadiene. The product reacts with nucleophiles in the expected manner, at the seven-membered ring.

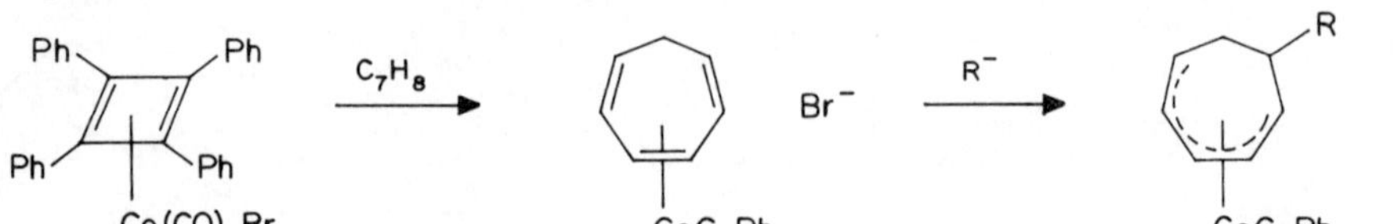

(10)

R = H, OMe, OAc

Reaction of cycloheptatriene with $Co_4(CO)_{10}$ in refluxing hexane also produces η^6-C_7H_8 complexes, the Co cluster being retained.

Eight-Membered Rings

A large number of cyclooctatetraene complexes of varying coordination type are known, and this area is reviewed thoroughly in Deganello's book.[25] We shall concentrate here on the hexahapto complexes obtained for the chromium group. Reaction of the metal hexacarbonyl (M = Cr, Mo, W) with cyclooctatetraene is not a satisfactory method for the preparation of these complexes, but reasonable yields can be obtained using, e.g., $M(CO)_3(NH_3)_3$ or $M(CO)_3(CH_3CN)_3$. In this way all three $(\eta^6$-$C_8H_8)M(CO)_3$ complexes (M = Cr, Mo, W) have been prepared. The IR spectra show three bands due to the CO frequencies (M = Cr; 1996, 1940 and $1912\,cm^{-1}$) and a band at $ca.$ $1670\,cm^{-1}$ attributable to the uncomplexed double bond. At low temperatures ($-40\,°C$) the ^1H NMR spectrum is complex, consistent with the hexahapto structure, but as the temperature is raised the resonances coalesce to give a singlet due to fluxionality of the cot ligand. ($ca.$ δ 5.20 p.p.m. at $80\,°C$). Assignments of peaks show a similar pattern to the complexes met previously, e.g.

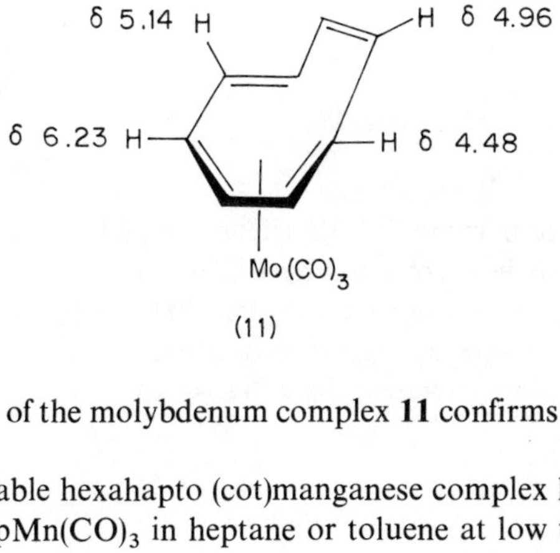

(11)

The X-ray structure of the molybdenum complex **11** confirms the bent structure shown.

A thermally unstable hexahapto (cot)manganese complex has been obtained by irradiation of $CpMn(CO)_3$ in heptane or toluene at low temperature:

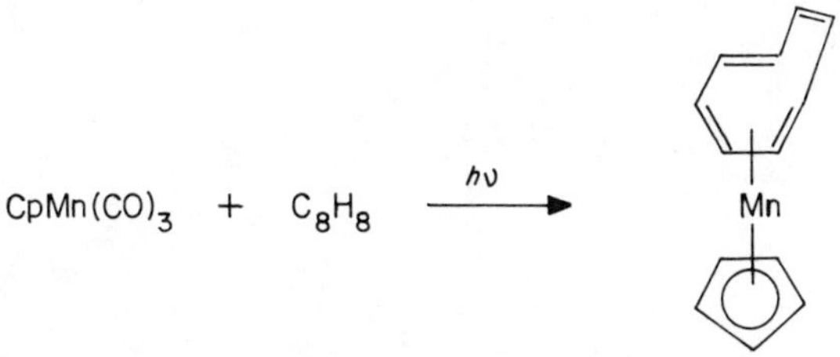

Cyclooctatriene also forms a range of metal complexes, but again many of these have only two of the double bonds coordinated.

Reaction of $CpVCl_3$ with cycloocta-1,3,5-triene and isopropylmagnesium bromide as reducing agent gives a mixture of $\eta^6\text{-}C_8H_{10}$ and $\eta^7\text{-}C_8H_9$ complexes. Dehydrogenation of the reaction mixture with Pt catalyst affords the pure

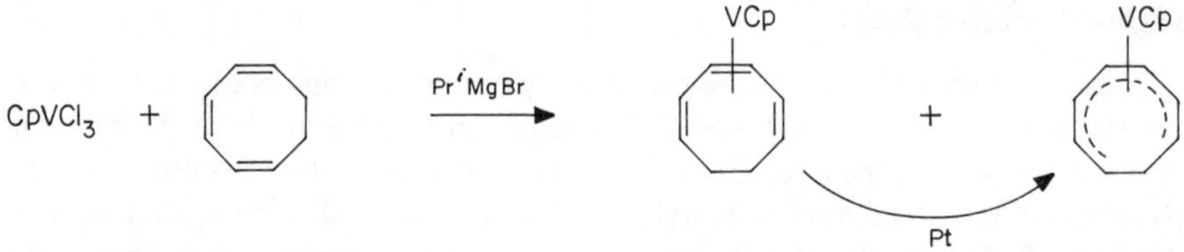

$\eta^7\text{-}C_8H_9$–VCp complex. Again, the most readily formed $\eta^6\text{-}C_8H_{10}$ complexes are with the Group VIA metals (Cr, Mo, W), obtained by direct reaction of the triene with the metal hexacarbonyl, although higher yields (70–90%) can be obtained using $M(CO)_3L_3$ (L = CH_3CN, NH_3, diglyme). Complexes of both cycloocta-1,3,6-triene and cycloocta-1,3,5-triene can be prepared, but the latter are more stable:

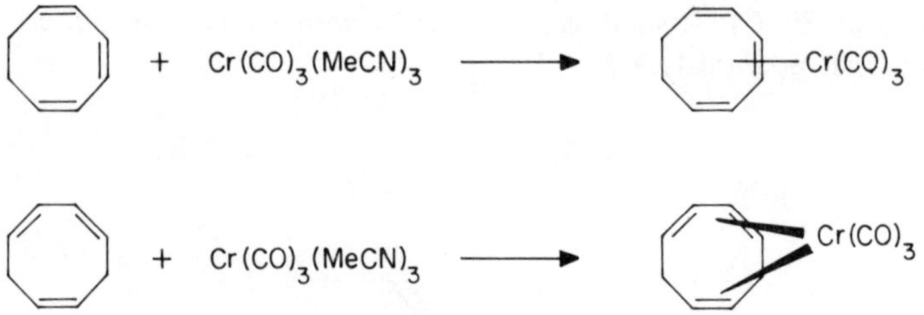

Hydride abstraction from the 1,3,5-triene complexes is readily achieved using $Ph_3C^+ BF_4^-$ to give the corresponding η^7-homotropylium complexes. The same cations of Mo and W, but not Cr, may be obtained by protonation of the η^6-cyclooctatetraene complexes. Addition of a limited range of nucleophiles has been investigated and exclusive reaction at the trienyl terminus has been reported.[28]

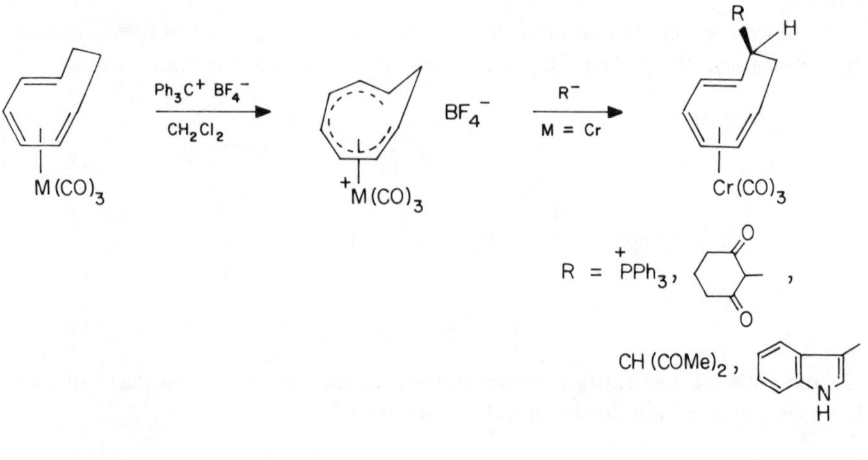

Hexahapto cyclooctatriene complexes of chromium have been prepared using η^5-Cp as counter ligand by direct reaction with C_8H_{10}, and also via the corresponding η^6-cot complexes, as shown below:

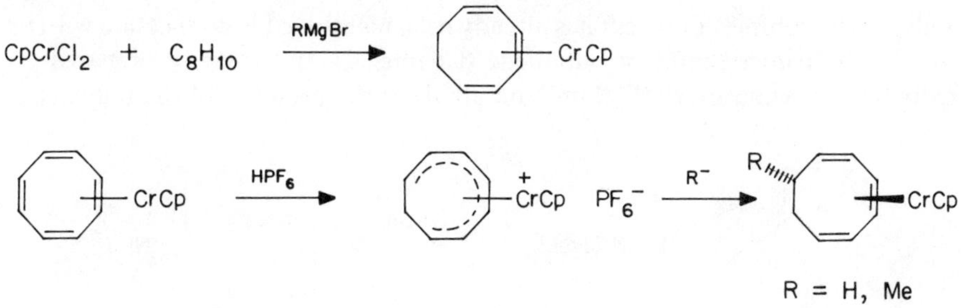

$R = H, Me$

Cyclononatriene Complexes

Treatment of *cis,cis,cis*-cyclonona-1,4,7-triene with hexacarbonylmolybdenum in isooctane at 110 °C gave a 47% yield of crystalline η^6-cyclonona-1,4,7-triene complex, in which the crown configuration of the triene is probably retained. No chemical studies of this compound have been reported.

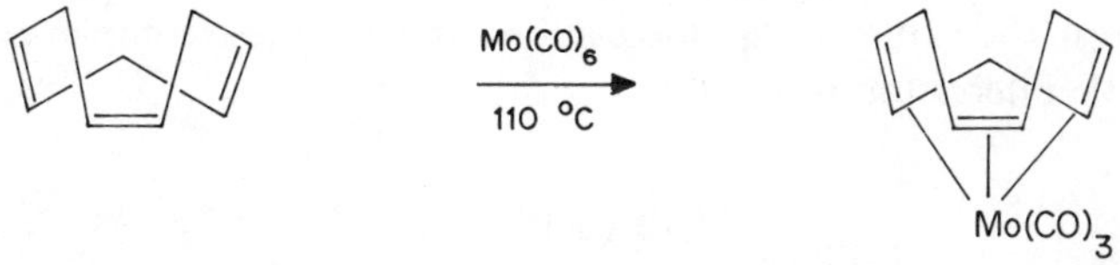

Cyclododecatriene Complexes

The four isomeric forms of cyclododecatriene are illustrated below:

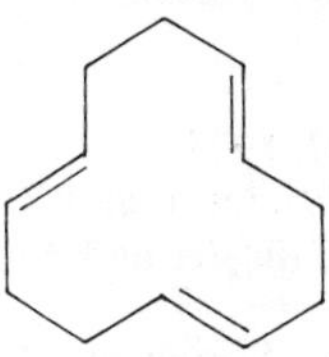

ttt 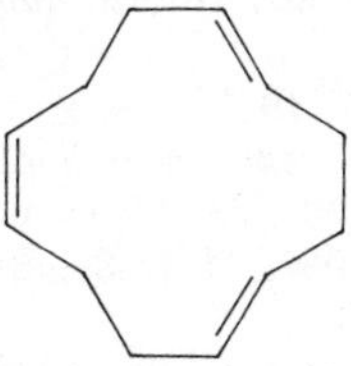   *ttc*

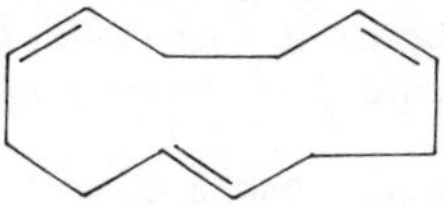

tcc    *ccc*

Perhaps surprisingly, no η^6-cyclododecatriene complexes are obtained by reaction of cdt with $M(CO)_6 (M = Cr, Mo, W)$. The best known complexes are in fact those of nickel, 16-electron complexes of which are known for all four isomers of the triene. These substances are reviewed in depth in the book by Wilke and Jolly.[29] The complex of *ttt*-cdt has already been mentioned in connection with its role in the trimerization of butadiene (Chapter 2). It is readily prepared by reduction of $Ni(acac)_2$ with aluminum alkyls in the presence of the polyolefin:

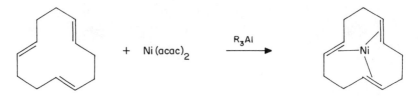

Ligand exchange may be used for the preparation of *ccc*-cdt–Ni:

$$Ni(\textit{ttt}\text{-cdt}) \quad + \quad \textit{ccc}\text{-cdt} \quad \rightleftharpoons \quad (\textit{ccc}\text{-cdt})Ni \quad + \quad \textit{ttt}\text{-cdt}$$

The remaining two complexes, which are unstable at normal temperatures and so not well characterized, have also been obtained using the appropriate cdt isomer in reaction with $NI(acac)_2/R_3Al$.

Apart from their role as oligomerization catalysts, these complexes appear to display little interesting reactivity.

REFERENCES

1. H. Zeiss, P. J. Wheatley and H. J. S. Winkler, Benzenoid–Metal Complexes, Ronald Press, New York, 1966.
2. (a) G. Jaouen, in *Transition Metal Organometallics in Organic Synthesis* (Ed. H. Alper), Academic Press, New York, 1978 Vol. 2; (b) M F. Semmelhack, *Tetrahedron*, 1981, **37**, 3956; (c) M. F. Semmelhack, *Ann. N.Y. Acad. Sci.*, 1977, **295**, 36; (d) G. Jaouen, *Ann. N.Y. Acad. Sci.*, 1977, **295**, 59; W. E. Silverthorn, *Adv. Organomet. Chem.*, 1975, **13**, 47.
3. G. A. Sim, *Annu. Rev. Phys. Chem.*, 1967, **18**, 57.
4. M. F. Semmelhack and A. Yamashita, *J. Am. Chem. Soc.*, 1980, **102**, 5924.
5. M. D. Rausch, *Pure Appl. Chem.*, 1972, **30**, 523; M. Uemura, N. Nishikawa and Y. Hayashi, *Tetrahedron Lett.*, 1980, **21**, 2069; M. F. Semmelhack, J. Bisaha and M. Czarny, *J. Am. Chem. Soc.*, 1979, **101**, 768.
6. A. Ceccon and G. Catelani, *J. Organomet. Chem.*, 1974, **72**, 179; A. Ceccon, *J. Organomet. Chem.*, 1974, **72**, 187.
7. W. S. Trahanovsky and R. J. Card, *J. Am. Chem. Soc.*, 1972, **94**, 2897; G. Jaouen, A. Meyer and G. Simmonneaux, *J. Chem. Soc., Chem. Commun.*, 1975, 813.
8. G. R. Knox, D. G. Leppard, P. L. Pauson and W. E. Watts, *J. Organomet. Chem.*, 1972, **34**, 347.
9. F. Calderazzo, *Inorg. Chem.*, 1966, **4**, 429.
10. P. Bachmann, K.-R. Repp and H. Singer, *Z. Naturforsch. Teil B*, 1977, **32**, 471.
11. K. K. Bhasin, W. G. Balkeen and P. L. Pauson, *J. Organomet. Chem.*, 1981, **204**, C25.

12. A. Mawby, P. J. C. Walker and R. J. Mawby, *J. Organomet. Chem.*, 1973, **55**, C39; G. Winkhaus, L. Pratt and G. Wilkinson, *J. Chem. Soc.*, 1961, 3807.
13. P. J. C. Walker and R. J. Mawby, *Inorg. Chem.*, 1971, **10**, 404.
14. P. J. C. Walker and R. J. Mawby, *Inorg. Chim. Acta*, 1973, **7**, 621.
15. P. J. C. Walker and R. J. Mawby, *J. Chem. Soc., Dalton Trans.*, 1973, 622.
16. R. J. Angelici and L. J. Blacik, *Inorg. Chem.*, 1972, **11**, 1754.
17. P. L. Pauson and J. L. Segal, *J. Chem. Soc., Dalton Trans.*, 1975, 1677 *and* 1683.
18. G. A. M. Munro and P. L. Pauson, *Israel J. Chem.*, 1976–77, **15**, 258.
19. G. A. M. Munro and P. L. Pauson, *J. Organomet. Chem.*, 1978, **160**, 177.
20. R. L. Davis and N. C. Baenziger, *Inorg. Nucl. Chem. Lett.*, 1977, **13**, 475.
21. A. N. Nesmeyanov, N. A. Vol'kenau, L. S. Shilovtseva and V. A. Petrakova, *Izv. Akad. Nauk SSSR, Ser. Khim.*, 1975, 1151.
22. A. N. Nesmeyanov, N. A. Vol'kenau, L. S. Shilovtseva and V. A. Petrakova, *J. Organomet. Chem.*, 1975, **85**, 365.
23. M. M. Khan and W. E. Watts, *J. Organomet. Chem.*, 1976, **108**, C11.
24. J. F. Helling and G. G. Cash *J. Organomet. Chem.*, 1974, **73**, C10; M. A. Bennett and T. W. Matheson, *J. Organomet. Chem.*, 1979, **175**, 87.
25. G. Deganello, *Transition Metal Complexes of Cyclic Polyolefins*, Academic Press, London, 1979.
26. P. L. Pauson and K. H. Todd, *J. Chem. Soc. C*, 1970, 2638.
27. F. Haque, J. Miller, P. L. Pauson, and J. B. P. Tripathi, *J. Chem. Soc. C*, 1971, 743; P. L. Pauson and J. A. Segal, *J. Organomet. Chem.*, 1973, **63**, C13.
28. A. Salzer, *Inorg. Chim. Acta*, 1976, **18**, L31.
29. G. Wilke and P. Jolly, *The Organic Chemistry of Nickel*, Academic Press, 1975. Vols. 1 and 2.

Index

π-Acceptor 2, 4, 33, 164, 180, 232
  ligand in β-elimination 38
Acetic acid synthesis 58
Acetylation 266
  of o-quinodimethane-Fe(CO)$_3$ 67
Acetylene 335, 343
  π-complex 331
  reaction with Fe(CO)$_5$ 68
  trimerization 84, 375
π-Acid 9
  ligand 2
Acidity, of carbene complex 153
Acorenone 353
Acrylonitrile nickel complex 182
Activation energy for reductive elimination 35
Acylation 265
  nucleophilic 352
1,4-Addition, of cuprates 122
Alkene complexation, selective 169
η^2-Alkene complexes 37, 163
  fluxionality 23
Alkene metathesis, mechanism 41
η^2-Alkene–rhodium complexes 177
Alkyl, anionic complexes 103
σ-Alkyl, metal halide complexes 106
σ-Alkyl complexes 100
  reactions of 117
  spectral properties 116
σ-Alkyl–Fe complex 166
Alkyl migration 51
Alkylidene complexes 140
  reactions 154
  titanium 155
Alkylidyne cluster 157
Alkylidyne complex 159
Alkyllithiums 369
  reagents 351

η^2-Alkyne complexes 169, 192
  reaction with electrophiles 197
Alkyne cyclotrimerization 377
Alkyne oligomerization 84
Allyl cation 10
π-Allyl complexes 177, 182, 184, 186, 201, 344
  asymmetric 27
  fluxionality 24
  of cobalt 94
π-Allyl–rhodium complex 82
π-Allyl nickel complexes 79
π-Allyl palladium complexes 112
σ-Allyl 223, 224, 226, 233, 237, 239
σ-Allyl complex 168, 174, 206
  of cobalt 94
Allyl–Fp complexes 124
  reaction with electrophiles 125
Allyl MOs 10
Allylic halides, coupling 207
π-Allyliron hydride 259
Ambident nucleophile 147
Amination 209
Amino acid 158
α-Amino acids 95
Anchimeric assistance 120
Angular substituents 303
Anisole–Cr(CO)$_3$ 351
Anisole–FeCp 371
Annulation reaction 299
Antarafacial, SO$_2$ insertion 56
Antibonding orbitals 9, 20
Arene chromium tricarbonyl, X-ray structure 349
Arene–Cr(CO)$_3$ 348, 366
Arene–FeCp complexes 370
Arene–manganese tricarbonyl 366
Arene–molybdenum complexes 363

390

Arene sandwich complex 17
Arene–vanadium 364
Arsine ligand 3
Aryl cation equivalents 306
Aspidosperma alkaloids 301
Aspidospermine 302
Asymmetric induction 214
    in cyclopropanation 146
Azaspirocyclic 300

Back donation 3, 16, 278
Barbaralone 275
Benzene–chromium tricarbonyl 350
Benzocyclobutadiene–Fe(CO)₃ 61
Benzyl σ-complexes 102
Benzylideneacetone, Fe(CO)₃ complex
    259
Bicyclobutane, opening 148
Bis (alkylidene) complexes 142
Bis-arene, cobalt and rhodium 374
Bis-arene metal complexes 348
Bis-benzene chromium 362
Bond order 3, 11, 13
Bond strength, alkylidene complex 144
Bonding orbitals 9
Butadiene 13, 149, 345
    cyclic trimer 83
    hydrogenation of 93
    trimerization of 388
Butadiene–Fe(CO)₃ 11, 19, 66, 163, 257
    fluxionality 21
    MOs 12
t-Butyl, σ-complexes 103

Canonical forms 360
Carbene 325
Carbene complexes 136
    conjugation in 162
    in α-elimination 39
    in polymerization 76
    reaction with MX₃ 158
    stability 137
    X-ray structure 163
Carbene ligand 5
    rearrangement 151
    transfer 139
Carbene–metal complex 44
Carbocation 194
Carbon monoxide, bonding 2
    insertion 47
    insertion mechanism 48, 49
Carbonium ion 320, 321, 338, 361
Carbonyl insertion 167, 248
Carbonylation 48

Carboxylation 204
    of alkynes 60
Carboxylic acid, from alkene 59
Carbyne complexes 8, 157
    bonding 6
    reactions 159
Charge control 231, 289, 291
Charge densities 289
Chelating 247
Chloride bridge, splitting 178, 186
Chlorobenzene–Cr(CO)₃ 355
Chromium, π-allyl 225
    anionic phenyl complexes 105
    carbene complex 150, 159
    CpCr(NO)₂Me 108
    diene complexes 256
    hexamethyl 103
    triphenyl 102
Chromocene 329, 330
Chugaev salts 136
Claisen rearrangement 220
Cobalt, alkyne complexes 194-7
    π-allyl 237
    σ-alkyl 115
    σ-benzyl complex 116
    cod complex 253
    CpCO(CO)₂ 339, 340
    diene complex 275
    dienyl complexes 308
    NaCo(CO)₄ 237
    tris(π-allyl) 238
Cobalt (III), pentacyano-, as
        hydrogenation catalyst 93
Cobalt carbonyl, in hydroformylation 56
Cobalt catalyst, in diene polymerization
        78
Cobalt group, η²-alkene complexes 176
Cobalt hydridocarbonyl 237
Cobalticinium 336, 338
Cobaltocene 251, 310, 335
    anion 336
Co-dimerization, of olefins 81
Collman's reagent 167, 173
Configuration, during insertion 52
Conformation 312, 349
Conjugate addition 207, 360
Conjugated diene 11
Contraction of d orbitals 190
Coordination number 29
Coordinatively saturated 20, 29, 33
    d⁸ complexes 31
Coordinatively unsaturated 20, 32, 267
Copper, alkyls 121
Coulombic interaction 232

Coupling constants 264
  ^{13}C–H 14
Crystal field 2
Cubane synthesis 62
Cuprate 103
Cyclic voltammetry 341
Cyclization 299
Cycloaddition 254, 274, 343
Cycloalkene complexes 178
Cycloalkyne 192
Cyclobutadiene 62, 340, 375
Cyclobutadiene–Fe(CO)$_3$ 61
  structure 63
Cyclobutadiene–metal complexes 41, 87
Cyclododecatriene 83, 387
Cycloheptadienyl–Fe(CO)$_3$ 279
Cycloheptadienyl–Mn(CO)$_3$ 382
Cycloheptatriene 262, 267
  cobalt complexes 384
  vanadium complexes 384
Cycloheptatriene complexes 378, 379
Cycloheptatriene–Fe(CO)$_3$ 125
Cycloheptatriene–Mn(CO)$_3$ 382
Cycloheptatriene–MnCp 383
Cycloheptatrienone 270
Cyclohexadienone complex 290
Cyclohexadienone–Fe(CO)$_3$ 72
Cyclohexadienyl cation 281
Cyclohexadienyl cobalt complexes 376
Cyclohexadienyl–Cr(CO)$_3$ 350
Cyclohexadienyl–FeCp 371, 373
Cyclohexadienyl–Mn(CO)$_3$ 366
Cyclohexenone γ-cation 293
Cyclohexenones 305
Cyclometallation 111, 131
Cyclononatriene complexes 387
Cycloocta-1,5-diene 247
  from butadiene 82
Cyclooctadienyl cation 273
Cyclooctadienyl–Fe(CO)$_3$ 279
Cyclooctatetraene 25, 84, 240, 273, 385
Cyclooctatetraene–Fe(CO)$_3$ 96
Cyclooctatetraene M(CO)$_3$, NMR 28
Cyclooctatriene 385
trans-Cyclooctene, resolution of 190
Cyclopentadienone–Fe(CO)$_3$ 68
Cyclopentadienides, alkali metal 311
η^5-Cyclopentadienyl 105
  metal alkyls 107
Cyclopentadienyl, bonding 17
Cyclopentadienyl ligand 310
Cyclopentadienylmagnesium bromide 310, 311
Cyclopentane, synthesis 220

Cyclopentene, metathesis 44
Cyclopropanation, Simmons-Smith 145
Cyclopropane opening 148, 186
Cyclopropanes 153
  from carbene complexes 144
Cyclotrimerization, butadiene 203

d^7 Complexes, oxidative addition 33
d^8 Complexes, oxidative addition 30
d^{10} Complexes, oxidative addition 30
d Orbitals 320
Decahydroquinoline 302, 303
Decarbonylation 48, 49, 51, 52, 177
Dehydration 321
De-insertion 113
Demerol 356
Deprotonation, benzylic position 359
  of arene–Cr(CO)$_3$ 357
Desulphurization 54
Dewar benzene 62
Dewar–Chatt–Duncanson 1
Diacetylenes 261
Dibenzene chromium 363
Dibenzenetitanium 363
Dibenzenevanadium 365
Diels–Alder reaction 265
Diene complex 11, 344
  optically active 260
Diene–CR(CO)$_3$ 350
Diene dimerization, mechanism of 82
Diene–Fe(CO)$_3$ 164
  complexes 235
Diene LUMO 13
Diene protection 268
Dienyl 302
Dienyl cation 15, 272
Dienyl complexes 365
  bonding 15
  16-electron 379
Dienyl–Fe(CO)$_3$ 125, 267
Dienyl LUMO 16
Dihydrides, molybdenum and tungsten 332
α-Diketone 332
(–)-Diop 96
Diphenylmercury 250
Diphosphine ligand 96
Dipole moment, carbene complexes 162
  of fulvene complexes 70
Disodium tetracarbonylferrate 66
σ-Donor 2, 4, 33
Double cross experiment 43
Dynamic equilibrium 21

Electrochemical reduction 339

Electrocyclic reaction 274
14-Electron complexes 36
16-Electron complexes 179, 201, 243, 388
  in reductive elimination 35
18-Electron complexes 179, 201, 244, 372
  fluxionality 27
18-Electron rule 17, 49
20-Electron complex 20, 373
Electron density 280
Electron-withdrawing ability 360
Electrophiles 179, 359
  reactions with alkyl complexes 117
Electrophilic attack 266
Electrophilic substitution 274, 363
α-Elimination 37, 44, 100, 140
  in alkene metathesis 42
  mechanism 39
β-Elimination 37, 60, 75, 80, 82, 100, 101,
     103, 128, 133, 141, 155, 156
  deuterium labelling studies 37
  in hydroformylation 56
  reverse 90
Enamines 213
Enolates 213
Entropy of activation 31
Epoxidation 296
  asymmetric 89
  Sharpless 88
Epoxides 165, 169, 207
Equivalence mechanism, NMR 23
Ergosteryl acetate, iron complex 259
Erythronolide 219
Ethene, bonding with iron 1, 8
  hydration 189
  oligomerization 80
  polymerization 74
Ethene complexes, nickel 183
Ethene–Fe(CO)$_4$ 19, 163
  hydration 189
Ethylidene tantalum complex 141
Exchange, olefin ligand 186

Ferrelactone complex 165
Ferricenium cation 313
Ferrocene 310, 330, 361, 370, 372
  acylation 313
  alkyl substitution 315, 316
  bonding in 17
  carbocation stabilization by 320
  electrophilic substitution in 313
  formylation of 314
  metallation of 318
  nitration of 318
  X-ray structure 312

Ferrocenyl, cation 338
  electron release by 317
Fischer–Hafner method 362–4
Fluorobenzene–Cr(CO)$_3$ 355
Fluxionality 21, 22, 264, 385
Fp complexes 323
Fp group, removal of 173
Fp–olefin complexes 167
  nucleophile addition to 170
Free radical 30
  in hydrogenation 94
  in oxidative addition 32
Friedel–Crafts acetylation 274
Friedel–Crafts acylation 237, 273, 334, 382
Friedel–Crafts reaction 366
Frontier molecular orbital 1, 245, 321
Frontier orbitals, and SO$_2$ insertion 55
  control 266
  in oxidative addition 31
  interaction 293
Fulvene 332
  metal complexes 69

Gabaculine 218
Grignard reagents 207, 223 233, 340

Hafnium, π-allyl 223
Hafnocene 327
Heck reaction 127
Heptafuluene 40
Heterodiene 259, 260
Hexacarbonylvanadium 365
Hexa-1, 4-diene 81
Histrionicotoxin 300
HOMO 1, 9, 10, 14, 32, 55, 96, 118, 120,
     173, 174, 231, 236, 263, 266, 283,
     289, 321
  LUMO interaction 232
Homogeneous catalyst 20
Homoleptic, σ-alkyl complexes 100
Homolytic addition, of H$_2$ 93
Homotropylium complexes 386
Humulene 207
Hybridization 8
Hydride 168
Hydride abstraction 168, 174, 180, 255,
     275, 278, 279, 371, 372, 379, 386
  regiocontrol in 281
Hydride–metal complex 40
1,2-Hydride shifts 39
Hydrides, metal-, reaction with alkenes 113
Hydroformylation 56, 176
Hydrogenation 89, 241, 253, 273, 334
  asymmetric 94

Hydrogenation—*contd.*
 mechanism 90
 of activated alkenes 91
 with Co(III) catalyst 93
Hydrogenolysis 326
Hydrozirconation 109, 324
 of acetylenes 111
 in methynolide synthesis 111

Iboga alkaloid 217
Indole 356
Infrared, diene–Fe(CO)₃ 262
Infrared spectra 6
 alkene complexes 9
 alkene–Fe(CO)₄ 164
 alkene–Mn complex 174
 alkyl complexes 117
 arene–Mn(CO)₃ 367
 carbene complexes 143
 COT M(CO)₃ 385
 Cp₂Fe₂(CO)₄ 322
 dienyl complexes 16
 dienyl–Fe(CO)₃ 278
 ethene–Mn(CO)₂Cp complex 175
 ferrocene 312
 Re(CO)₅C₂H₄PF₆ 175
 triene–M(CO)₃ 378
Infrared spectroscopy 3, 4, 51
Insertion 47, 57, 58, 152, 242, 244
 alkene 75
 CO 59, 67, 111, 202, 261, 262, 275,
  324
 olefin 77
 reaction 109, 331
 stereochemistry of SO₂ 55
Iridium, alkene complexes 180
 π-allyl 242
 σ-allyl 116
 σ-aryl 114
 cod complex 253
 complexes, oxidative addition 31
 diene complex 275
 ethene complex 181
Iron, σ-alkyl 114, 115
 π-allyl 235
  Fe(CO)₃I 235
 carbonyl, fulvene complexes 69
 cod complexes 255
 CpFe(CO)₂Br 168
 (η⁵-Cp)Fe(CO)₂(C₂H₄)PF₆ 167
Iron group, alkene complexes 163
Isoelectronic 232
Isomerization, of alkene 253
Isoprenyl 205

Isotactic polymer 74

Ketene 161
Ketones, α,β-unsaturated 171, 175

Labelling, in insertion 51
β-Lactam 165
 synthesis 157
β-Lactone 165
Lactones, via CO insertion 53
Ligand exchange 179
Ligand interaction with metal 2
Ligand substitution 151, 192, 334
Limaspermine 301, 302
Lithium aluminum hydride 369
LUMO 1, 9, 10, 14, 32, 55, 68, 118, 120,
  164, 173, 174, 230, 233, 236, 263,
  266, 281, 283, 289, 321
 coefficients 231

Macrolide 47, 218
Maleic anhydride, cobalt complex 176
Manganese, σ-alkyl 113, 115
 π-allyl 233
 carbene complex 159
 carbonyl hydride 234
 cod complex 256
 dimethyl 101
 EtMn(CO)₅ 174
Manganese group, η²-alkene complexes 174
Manganocene 333
Mass spectral analysis 151
Mass spectrum, diene–Fe(CO)₃ 263
 metallocenes 312
Mechanism, reactions of σ-alkyl complex
  with electrophiles 118
Metal anions 114
Metal–carbon σ bond stability 34
Metallacycles 130
Metallacyclobutane 43, 76, 130, 155, 156
 conformation 45
 from π-allyl complexes 131
 in alkene metathesis 42
Metallacyclobutenes 156
Metallacycloheptane 133
Metallacyclopentadiene 86, 130, 340
Metallacyclopentanes 131
Metallacyclopropane 9
Metallocenes 17, 310
 oxidation 313
Metathesis 77
 alkene 40
 functionalized alkenes 46
 intramolecular 47
 olefin 142, 149

2-Methoxybutadiene 13
Methoxycyclohexadiene complexes 258
3-Methoxycyclohexadienyl, iron complex 304
Methoxy substitution 292
Methylation, of tertiary halides 107
Methylenation 154
Methylmanganese pentacarbonyl 47
Michael addition 166, 183
Michael cyclization 297
Michael reaction 160
Migratory aptitude 50
Migratory insertion 250
MO calculations 325
    dienyl complexes 289
MO theory, and electrophile addition 121
Molecular orbital 6, 173
Molybdenocene 310
Molybdenum, σ-alkyl 114
    alkyne complex 1977
    π-allyl 24, 226
    $Cp_2MoEt(Cl)$ 108
    Cp_2MoH_2 330
    diene complexes 256
    σ-phenyl complexes 105
Myrcene 210, 268

Neighbouring group participation 319, 320, 361
(+)-neomenthyldiphenylphosphine 95
Neopentyl, σ-complexes 102, 103
Neopentyl complexes 105
Nickel, π-allyl complexes 201
    $Ni(acrylonitrile)_2$ 182
Nickel group, η^2-alkene complexes 182
Nickel tetracarbonyl 182, 202
Nickelocene 248, 310, 342, 345, 363
    protonation of 344
Niobium, alkyl complexes 109
    alkylidene complexes 140
    π-allyl 224
Niobium methyl complexes, reaction with ketones 118
Niobium phenyl complexes 105
Nitrene intermediate 370
Nitrogen fixation 327
Nitrosyl complex 229
NMR spectra 6
    alkyl complexes 116
    π-allyl complex 25
    $CpRh(C_2H_4)_2$ 23
    diene–$Fe(CO)_3$ 263
    dienyl–$Fe(CO)_3$ 280
    ethene–$Mn(CO)_2Cp$ 175

ferrocene 313
heptafulvene–$Fe(CO)_3$ complex 71, 72
NMR, ^{13}C, allyl complexes 10
    allyl-$Mn(CO)_4$ 234
    diene complexes 13
    dienyl complexes 16
NMR, allyl–zirconium 25
    carbene complexes 163
    β-effect 14
    iron hydride 267
    nickel, diallyl 26
    studies in insertion 52
Nonacarbonyldiiron 259
Nonbonded interactions 312
Norbornadiene 247
Norbornadiene–$Fe(CO)_3$ 254
Norbornyl, σ-complexes 103
Nucleophile 229, 255, 299, 350, 357, 361, 367, 369, 371, 376, 379, 382, 386
    on π-allyl–Fe 235
    with π-allyl cobalt 242
    with $CpFe(CO)_2L$ cation 324
    with dienyl–CoCp 308
    with dienyl–$M(CO)_3$ 285
Nucleophile addition 173, 212, 248, 254, 273, 306, 336
    and ligand environment 232
    intramolecular 297, 358
    olefin–Pd 186
    rules for 232
    to carbene complex 150
    to CO ligand 305
Nucleophilic substitution 332
Nucleophilicity, of transition metal anions 115

Oestrone, cobalt catalyzed synthesis 85
Olefin, insertion of 113
    Pd complex 210
Olefin dimerizations, Mechanism 80, 81
Olefin–$Fe(CO)_4$ complex, nucleophile addition to 166
η^2-Olefin–Fp complexes 124
Olefin isomerization of 39
Olefin metathesis 131
Oligomerization 80, 131, 215, 216, 340, 388
    diene 202
Optical activity, in insertion 52
Orbital 1
Orbital coefficients 16
Orbital control 290
Organocuprate 121
Osmicenium cation 313

Osmocene 310
Oxidation, of nickelocene 341
Oxidation number and reductive
    elimination 36
Oxidative addition 29, 49, 57, 59, 89, 91,
    120, 128, 206, 211, 216, 339
  cobalt complexes 34
  in preparation of σ-complexes 115
  of H₂ 92
  mechanism 32
Oxidative cyclization 271, 284
Oxo group, as counter ligand 108
Oxygen, reaction with alkyl complexes
    117

Palladium, π-allyl 27, 129, 209
  aryl complex 127
  chloride 211
  cod complexes 248
  (olefin)PdCl₂ dimer 184
  Pd(PPh₃)₄ 127, 216
  vinyl complex 127
Pentacarbonyliron 1, 33, 311
Pentacarbonylmanganese bromide 366
Pentacarbonylmanganese perchlorate 366
Pentadienyl 278
Peptide 150
Pericyclic reaction, on metal complex 96
Perturbation theory 321
Phenyl, σ-complexes 104, 105
Phosphine ligand 3
  optically active 95
Photoelectron spectroscopy 121
Platinacyclobutane complex 45
Platinate, pentakis(trichlorostannyl), as
    hydrogenation catalyst 92
Platinum, η²-alkene complexes 187
  (0), alkene complexes 191
  σ-alkyl 116
  alkyne complex 197, 198, 199
  σ-aryl 118
  (C₂H₄PtCl₂)₂ 187
  cod complexes 248
  complexes, β-elimination 38
  oxidative addition to 30
  (PPh₃)₂Pt(olefin) 191
Poisoning, metathesis catalyst 46
Polybutadienes 79
Polyenes 262
Polymer, via metathesis 41
Polymeric catalyst 222
Polymerization 39, 238, 242, 245
  of dienes 78
Propene, metathesis of 41

Prostaglandin 221
  synthesis 111
Protecting group, Fe(CO)₃ 265
  Fp as 171
  organometallic 151
Protection, alkyne 193
Protonation 198
  of metallocenes 313
Pseudo octahedral 11
Pseudorotation 22

ortho-Quinodimethane–Fe(CO)₃ 66

Reduction, selective alkene 91
Reductive coupling 239, 380
Reductive elimination 34, 38, 57, 58, 59,
    83, 92, 120, 147, 155, 167, 203, 204
  and catalysis 35, 36
  labelling experiment 36
Reformatsky reaction 73
Regioselection 299
Regioselectivity 301
  in arene–Mn(CO)₃ reactions 368
Rehybridization 192
Resolution 362
  of alkenes 190
Resonance, in cyclobutadiene–Fe(CO)₃ 64
Rhenium, alkyls 335
  π-allyl 234
  arene complex 370
  cod complexes 256
  CpRe(CO)₃ 335
  cyclopentadienyl complexes 336
  NaRe(CO)₅ 161
Rhodium, σ-alkyl 116
  alkyne complex 199
  π-allyl 242
  catalyst, hydroformylation 57
  (C₂H₄)₂Rh(acac) 178
  [(C₂H₄)₂RhCl]₂ 177
  (C₂H₄)₂RhCp 179
  cod complexes 252
  diene complex 275
  dienyl complex 308
  ethene complexes 23
  reactions of ethene complex 178
  trisallyl 26
Rhodocene 310, 338
Rhodocenium cation 338
Ring contraction 370, 380, 381
Ring expansion 322
Rotation, of ethene 24
Rubber 78
Ruthenicinium cation 313

Ruthenium, (arene)$_2$ Ru dications 373
  (III) chloride, hydrogenation catalyst 91
  cod complexes 255
Ruthenocene 310

Sandwich complexes 375
α-Santalene 205
Sceletium alkaloid 307
Sharpless epoxidation 296
Shift isomerization 262
Silyl enol ethers 171
Single-cross experiment 42
Slipping, metal 173
Sodium tetracarbonylcobaltate 177
Solvent effect, in CO insertion 50
Solvolysis 361
  ferrocenyl acetates 320
Spirocyclic 353
Spirocyclization 353
Spiro|4.5|decane 297, 353
Spiro|5.5|undecane 297
Stereochemistry, alkene metathesis 45
  of nucleophile addition 170
  reaction of σ-alkyl complex with
     electrophile 119
Stereospecificity, of nucleophile addition
  173
Steric effects, in alkene metathesis 43
  in SO$_2$ insertion 54
Steric factors, in bicyclobutane opening
  149
Steroid 112
Steroid synthesis 67, 83, 88, 294
Substituent effects 290
Sulphur dioxide, insertion 53
  reaction with metal complexes 54
Sulphur nucleophiles 251
Syndiotactic polymer 74
Synergic 5
Synergic interaction 10, 11, 15, 67–9, 192,
  283
Synthetic equivalents 307

Tantalum, alkyl complexes 109
  alkylidene complexes 140
  alkyne complex 197
  π-allyl 224
  bis alkylidene 142
  carbene complex 5
  pentamethyl 101
Tautomer 327
Tetracarbonyletheneiron 1
Titanium, alkyl complexes 108, 109
  π-allyl 223

Cp$_2$TiR(X) complexes 107
cycloheptatrienyl complexes 383
  in methylenation 154
  methyl trichloro complex 106
  phenyl complexes 101
  tetrakis(trimethylsilylmethyl) 100
  tetramethyl 100
  tetraphenyl 102
Titanium group, metallocenes 325
Titanocene 310, 325
  permethyl 326
Toluene–Cr(CO)$_3$ 351
Transannular 239
Transmetallation, of vinyl–Zr complexes
  111
Tricarbonylcyclohexadieneiron 263
Tricarbonylcyclohexadienyliron 15, 278
Tricarbonylcyclohexadienylosmium 305
Tricarbonylcyclopentadienylmanganese
  333
Tetracarbonyletheneiron 8
Tricarbonyl(5-hydroxycyclohexa-1,
  3-diene)iron 73
Tricarbonyl(1-methoxycyclohexa-
  dienyl)iron 73
Tricarbonyl(4-methoxy-1-methylcyclo-
  hexadienylium)iron hexafluorophos-
  phate 293
Tricarbonyl(methylcyclopentadienyl)-
  manganese 366
Tricarbonyl(methyltropylium)iron
  tetrafluoroborate 72
Trichodermol 295
Trichothecene synthesis 295
Triene, M(CO)$_3$ complexes 377
Trimethylamine-N-oxide 261, 306
Trimethylenemethane 64, 222
Trimethylenemethane–Fe(CO)$_3$ 64
Trimethylenemethane–palladium complex
  66
Trimethylsilylmethyl complexes 103, 105
Triple decker complex 345
Tropylium complexes 379
  M–Cp complexes 381
Tungsten, π-allyl 226
  carbene complex 136
  carbyne complex 158, 161
  Cp$_2$WH$_2$ 330
  hexacarbonyl 2, 136, 159
  hexamethyl 100, 101
Tunneling, in alkene complexes 24

Valence bond 6
Valence isomerization 147

Vanadium, π-allyl 224
    Cp₂VRx complexes 108
    NaV(CO)₆ 225
    phenyl complexes 105
    tetrabenzyl 100
Vanadocene 327
Vaska's complexes 31, 181
Vaska's iridium complex 180
Vilsmeier formylation 274
Vinyl carbyne complex 160
Vinyl epoxide 165
Vinyl metal σ-complex 198
Vinylcyclohexene 259
Vinylcyclopropane 164, 186, 209, 261
Vinylferrocene 319
Vinylzirconium complexes 111

Wacker oxidation 87
Wilkinson's catalyst 49, 52, 89, 91
Wittig olefination 154, 156
Wolff rearrangement 50
Woodward–Hoffmann rules 96

X-ray structure, carbene complex 5
    LiCr₂Me₈ THF 104
    Li₃(CrMe₆) 3 diox 104

Zearalenone, synthesis 60
Zeise's salt 163, 187
    trans effect in 189
    X-ray structure 188
Ziegler–Natta polymerization 74
Zirconate, hexamethyl 103
Zirconium, alkyl complexes 109
    alkyl-, transmetallation 110
    Cp₂ZrMe₂ 107
    Cp₂ZrPh₂ 107
    Cp₂ZrR(Cl) complexes 107
    tetramethyl 101
    tetraphenyl 102
Zirconium alkyls, reaction with
    electrophiles 110
Zirconium–carbon bond, cleavage of
    110
Zirconocene 327